Frames and Harmonic Analysis

CONTEMPORARY
MATHEMATICS

706

Frames and Harmonic Analysis

AMS Special Session on
Frames, Wavelets and Gabor Systems

AMS Special Session on
Frames, Harmonic Analysis, and Operator Theory

April 16–17, 2016
North Dakota State University, Fargo, ND

Yeonhyang Kim
Sivaram K. Narayan
Gabriel Picioroaga
Eric S. Weber
Editors

AMERICAN
MATHEMATICAL
SOCIETY

2010 *Mathematics Subject Classification.* Primary 15Axx, 41Axx, 42Axx, 42Cxx, 43Axx, 46Cxx, 47Axx, 94Axx.

Library of Congress Cataloging-in-Publication Data

Names: Kim, Yeonhyang, 1972– editor. | Narayan, Sivaram K., 1954– editor. | Picioroaga, Gabriel, 1973– editor. | Weber, Eric S., 1972– editor.

Title: Frames and harmonic analysis: AMS special sessions on frames, wavelets, and Gabor systems and frames, harmonic analysis, and operator theory, April 16–17, 2016, North Dakota State University, Fargo, North Dakota / Yeonhyang Kim, Sivaram K. Narayan, Gabriel Picioroaga, Eric S. Weber, editors.

Description: Providence, Rhode Island: American Mathematical Society, [2018] | Series: Contemporary mathematics; volume 706

Identifiers: LCCN 2017044766 | ISBN 9781470436193 (alk. paper)

Subjects: LCSH: Frames (Vector analysis) | Harmonic analysis. | Wavelets (Mathematics) | Gabor transforms. | AMS: Linear and multilinear algebra; matrix theory – Basic linear algebra – Basic linear algebra. msc | Approximations and expansions – Approximations and expansions – Approximations and expansions. msc | Harmonic analysis on Euclidean spaces – Harmonic analysis in one variable – Harmonic analysis in one variable. msc | Harmonic analysis on Euclidean spaces – Nontrigonometric harmonic analysis – Nontrigonometric harmonic analysis. msc | Abstract harmonic analysis – Abstract harmonic analysis – Abstract harmonic analysis. msc | Functional analysis – Inner product spaces and their generalizations, Hilbert spaces – Inner product spaces and their generalizations, Hilbert spaces. msc | Operator theory – General theory of linear operators – General theory of linear operators. msc | Information and communication, circuits – Communication, information – Communication, information. msc

Classification: LCC QA433 .F727 2018 | DDC 515/.63–dc23

LC record available at https://lccn.loc.gov/2017044766

DOI: `http://dx.doi.org/10.1090/conm/706`

♾ The paper used in this book is acid-free and falls within the guidelines
established to ensure permanence and durability.
Visit the AMS home page at `http://www.ams.org/`

10 9 8 7 6 5 4 3 2 1   23 22 21 20 19 18

Contents

Preface

Frames were first introduced by Duffin and Schaeffer in 1952 in the context of nonharmonic Fourier series but have enjoyed widespread interest in recent years, particularly as a unifying concept. Indeed, mathematicians with backgrounds as diverse as classical and modern harmonic analysis, Banach space theory, operator algebras, and complex analysis have recently worked in frame theory. The present volume contains papers expositing frame theory and applications in three specific contexts: frame constructions and applications, Fourier and harmonic analysis, and wavelet theory.

In recent years, frame theory has found applications to problems in computer science, data science, engineering, and physics. Many of these applications involve frames in finite-dimensional spaces; one focus of finite frame theory is the construction of tight frames with desired properties such as equiangular tight frames. Other types of frames discussed in these papers include scalable frames, full-spark frames, and fusion frames.

(1) Constructions of Biangular Tight Frames and Their Relationships with Equiangular Tight Frames
(2) Phase Retrieval by Hyperplanes
(3) Tight and Full Spark Chebyshev Frames with Real Entries and Worst-Case Coherence Analysis
(4) Fusion Frames and Distributed Sparsity

Historically there exists a strong connection between operator theory and frame theory. The recent solution of the Kadison-Singer problem is a further illustration of this connection. Modern connections are being formed between frame theory and machine learning.

(5) The Kadison-Singer Problem
(6) Spectral Properties of an Operator Polynomial with Coefficients in a Banach Algebra
(7) Kaczmarz Algorithm, Row Action Methods, and Statistical Learning Algorithms
(8) Lipschitz Properties for Deep Convolutional Networks

There also exists a strong connection between frame theory and harmonic analysis. This is seen in the context of classical Fourier analysis and shift invariant spaces, including in new settings such as on graphs.

(9) Invertibility of Graph Translation and Support of Laplacian Fiedler Vectors
(10) Weighted Convolution Inequalities and Beurling Density
(11) p-Riesz Bases in Quasi Shift Invariant Spaces

This connection between frame theory and harmonic analysis also occurs in the context of spectral measures–those measures which possess an orthogonal basis of exponentials, or, more generally, those measures which possess a harmonic analysis in terms of boundary functions for elements in the Hardy space of the unit disc.

(12) On Spectral Sets of Integers
(13) Spectral Fractal Measures Associated to IFS's Consisting of Three Contraction Mappings
(14) A Matrix Characterization of Boundary Representations of Positive Matrices in the Hardy Space

The (modern) developments of wavelet theory and frame theory are intertwined, particularly in the construction of frames for function spaces. Both have a wide range of practical applications in numerical analysis, signal processing, and image processing. Papers in this volume study the Gibbs phenomenon for wavelet frames, subdivision schemes, and the connection between Markov chains and wavelets.

(15) Gibbs Effects Using Daubechies and Coiflet Tight Framelet Systems
(16) Conditions on Shape Preserving of Stationary Polynomial Reproducing Subdivision Schemes
(17) W-Markov Measures, Transfer Operators, Wavelets, and Multiresolutions

As outlined above, this collection of papers covers a wide variety of topics. As such, this volume will be of interest to researchers in frame theory, as well as approximation theory, data science, representation theory, and functional and harmonic analysis.

Yeonhyang Kim
Sivaram K. Narayan
Gabriel Picioroaga
Eric S. Weber

Participants of the AMS Special Session "Frames, Wavelets and Gabor Systems"

Speakers and titles from the AMS Special Session "Frames, Wavelets and Gabor Systems" from the AMS Central Sectional Meeting, Fargo, ND, April 16–17, 2016.

Roza Aceska
Local sparsity and fusion frames

Radu Balan
The iterative and regularized least squares (IRLS) algorithm for phase retrieval

Laura De Carli
Stability theorems for systems of rect and sinc

Peter G. Casazza
Infinite dimensional phase retrieval

Xuemei Chen
The gap between NSP and RIP

Cheng Cheng
Spatially distributed sampling and reconstruction

Matthew Fickus
Equiangular tight frames from hyperovals

John Isaac Haas
Tight orthoplectic Grassmannian frames

Bin Han
Tight framelets and refinable structure

Christopher Heil
HRT versus the zero divisor conjecture

John Jasper
Tremain equiangular tight frames and strongly regular graphs

Alex Iosevich
On the Fuglede conjecture

Azita Mayeli
Sampling and interpolation on certain nilpotent lie groups

Dustin G. Mixon
The Voronoi means conjecture

Kasso Okoudjou
A new proof of the three-point HRT conjecture

Vignon Oussa
A classification of irreducible admissible groups in dimension three

Judith A. Packer
Wavelets and frames associated to representations of higher-rank graph algebras

Armenak Petrosyan
Frames and Bessel systems generated by the iterative actions of normal operators

Darrin Speegle
Sampling from a continuous frame to obtain a frame

Participants of the AMS Special Session "Frames, Harmonic Analysis, and Operator Theory"

Speakers and titles from the AMS Special Session "Frames, Harmonic Analysis, and Operator Theory" from the AMS Central Sectional Meeting, Fargo, ND, April 16–17, 2016.

Cyrus P. Aryana
Self-adjoint Toeplitz operators associated with representing measures on doubly connected planar regions and their eigenvalues

Marcin Bownik
Improved bounds in Weaver and Feichtinger conjectures

Laura De Carli
Exponential bases on rectangles in $\mathbb{R}^d$

Peter G. Casazza
Frame scalings a condition number approach

Dorin Ervin Dutkay
Fourier series on self-affine measures

Jean-Pierre Gabardo
Beurling density in weighted Fourier spaces

John Herr
Fourier series for singular measures

Calvin F. Hotchkiss
A fast Fourier transform for fractal approximations

Joseph W. Iverson
Group frames with several generators

Palle E. T. Jorgensen
Probability theory of infinite iterated function systems

Ilya A. Krishtal
On the spectral theory of operator polynomials with coefficients in a Banach algebra

Chun-Kit Lai
Non-spectral fractal measures with Fourier frames

Shidon Li
Sparsity-inducing dual frames and sparse signal recovery with coherent frames

Richard G. Lynch
Weaving Schauder frames

Azita Mayeli
Tiling and spectral sets in $\mathbb{Z}_p \times \mathbb{Z}_p$

Kasso Okoudjou
On the boundedness of the bilinear Hilbert transform on modulation spaces

Gabriel Picioroaga
Fourier frames for the Cantor-4 set

Sam L. Scholze
Bridging frame erasures

Keith F. Taylor
Admissible groups in three dimensions

Dongmian Zou
Lipschitz properties for deep networks

Contemporary Mathematics
Volume **706**, 2018
http://dx.doi.org/10.1090/conm/706/14221

Constructions of biangular tight frames and their relationships with equiangular tight frames

Jameson Cahill, Peter G. Casazza, John I. Haas, and Janet Tremain

ABSTRACT. We study several interesting examples of **Biangular Tight Frames (BTFs)** - basis-like sets of unit vectors admitting exactly two distinct **frame angles** (ie, pairwise absolute inner products) - and examine their relationships with **Equiangular Tight Frames (ETFs)** - basis-like systems which admit exactly one frame angle (of minimal coherence).

We develop a general framework of so-called Steiner BTFs - which includes the well-known Steiner ETFs as special cases; surprisingly, the development of this framework leads to a connection with famously open problems regarding the existence of Mersenne and Fermat primes.

In addition, we demonstrate an example of a smooth parametrization of 6-vector BTFs in $\mathbb{R}^3$, where the curve "passes through" an ETF; moreover, the corresponding frame angles "deform" smoothly with the parametrization, thereby answering two questions about the rigidity of BTFs.

Finally, we generalize from BTFs to **(chordally) biangular tight fusion frames (BTFFs)** - basis-like sets of orthogonal projections admitting exactly two distinct trace inner products - and we explain how one may think of them as generalizations of BTFs. In particular, we construct an interesting example of a BTFF corresponding to 16 2-dimensional subspaces of $\mathbb{R}^4$ that "Plücker embeds" into a Steiner ETF consisting of 16 vectors in $\mathbb{R}^6$, which we refer to as a **Plücker ETF.**

1. Introduction

Research into *finite (unit-norm) frames* - basis-like sets of unit vectors - has developed rapidly over the last two decades [**20**, **40**]. While the discipline has found various applications [**44**], possibly the most common use involves the analysis of signals [**24**, **30**, **32**, **48**]. Accordingly, a common criterion for frame design is low *coherence*, as this may lead to improved error reduction in various signal processing applications [**30**, **41**].

The *coherence* of a frame is the maximal value among its set of pairwise absolute inner products. Frames that achieve minimal coherence are called **Grassmannian frames**, a class including the well-studied *equiangular tight frames (ETFs)* [**22**] - which admit exactly one pairwise absolute inner product. Recent advances have shown that *biangular tight frames(BTFs)* - unit-norm frames which admit exactly two pairwise absolute inner products - sometimes form Grassmannian frames [**29**,

2010 *Mathematics Subject Classification.* Primary 42C15.
The authors were supported by NSF DMS 1307685; and NSF ATD 1321779; ARO W911NF-16-1-0008.

46] or can be used to construct Grassmannian frames [7, 13]. Accordingly, this work is a study of the relationship between ETFs and BTFs. In order to illuminate their relationship, we consider interesting examples; moreover, we consider general frameworks for **harmonic BTFs/ETFs** and so-called **Steiner BTFs/ETFs**

In Section 2, we recall basic facts from frame theory and fix notation. In Section 3, we demonstrate a smooth parametrization of 6 vector BTFs in $\mathbb{R}^3$, where the corresponding frame angles transform smoothly with the parameter. This smooth curve of BTFs "passes through" an ETF, and its existence answers two questions regarding the rigidity of BTFs. In Section 4 and Section 5, we develop a framework based on Steiner systems and divisible difference sets, which produces numerous infinite families of so-called **Steiner BTFs**, which admits the **Steiner ETFs** [23] as a special case. Surprisingly, we observe a connection between one of these families and the long-standing open problem(s) [10] regarding the existence of an infinitude of Mersenne primes and Fermat primes. Finally, in Section 6, we generalize from BTFs to **(chordally) biangular tight fusion frames (BTFFs)** - basis-like sets of projections admitting exactly two distinct trace inner products. We define a new type of ETF - a **Plücker ETF** - based on the **Plücker embedding**; as an example, we construct 16 2-dimensional subspaces of $\mathbb{R}^4$ that **Plücker embed** into a Steiner ETF of 16 vectors in $\mathbb{R}^6$.

2. Preliminaries

Given a finite dimensional Hilbert space $\mathbb{F}^m$, with the usual inner product and where $\mathbb{F} = \mathbb{R}$ or $\mathbb{C}$, we fix $\{e_j\}_{j=1}^m$ as the canonical orthonormal basis, and we fix I_m as the $m \times m$ identity matrix. A set of vectors $\mathcal{F} = \{f_j\}_{j=1}^n \subset \mathbb{F}^m$ is a **(finite) frame** if $\text{span}\{f_j\}_{j=1}^n = \mathbb{F}^m$. It is convenient to think of the frame $\mathcal{F}$ in terms of its **synthesis matrix**, $F = [f_1\, f_2 \ldots f_n]$, the $m \times n$ matrix with columns given by the frame vectors, so we identify the set $\mathcal{F}$ with the matrix F as representatives of the same object. We say that $\mathcal{F}$ is **a-tight** if $FF^* = aI_m$ for some $a > 0$, called the **tightness parameter**, where, in general, A^* denotes the adjoint of a matrix A. We say that $\mathcal{F}$ is **unit-norm** if each frame vector has norm $\|f_j\| = 1$.

To abbreviate, a unit-norm, a-tight frame consisiting of n vectors for $\mathbb{F}^m$ is called an **(n, m)-frame**. The tightness parameter for such a frame is determined as $a = \frac{n}{m}$, because

$$n = \sum_{j'=1}^{m} \sum_{j=1}^{n} |\langle e_{j'}, f_j \rangle|^2 = \sum_{j'=1}^{m} \sum_{j=1}^{n} \text{tr}(f_j f_j^* e_{j'} e_{j'}^*) = a \sum_{j'=1}^{m} \|e_{j'}\|^2 = am,$$

which also implies that every such frame satisfies the identity

$$\sum_{j'=1}^{n} |\langle f_j, f_{j'} \rangle|^2 = \frac{n}{m} \text{ for every } j \in \{1, \ldots, n\}. \tag{2.1}$$

Given any unit-norm frame $\mathcal{F} = \{f_j\}_{j=1}^n$, its **frame angles** are the elements of the frame's **angle set**, which we denote and define by $\Theta_{\mathcal{F}} := \left\{ |\langle f_j, f_{j'} \rangle| : j \neq j' \right\}$. We say that $\mathcal{F}$ is **d-angular** if $|\Theta_{\mathcal{F}}| = d$ for some $d \in \mathbb{N}$. In the special case that $\mathcal{F}$ is 1-angular or 2-angular, then we say that it is **equiangular** or **biangular**, respectively. If $\mathcal{F}$ is d-angular frame with frame angles $\alpha_1, \alpha_2, \ldots, \alpha_d$, then $\mathcal{F}$ is **equidistributed** if there exist positive integers $\tau_1, \tau_2, \ldots, \tau_d \in \mathbb{N}$ such that

$$\left| \left\{ j' \in \{1, \ldots, n\} : j' \neq j, |\langle f_j, f_{j'} \rangle| = \alpha_l \right\} \right| = \tau_l$$

for every $j \in \{1, 2, ..., n\}$ and every $l \in \{1, 2, ..., d\}$. In this case, we call the positive integers $\tau_1, \tau_2, ..., \tau_d$ the **frame angle multiplicities** of $\mathcal{F}$ and note that $\sum_{j=1}^{d} \tau_j = n - 1$.

By the lower bound of Welch [**45**], if $\mathcal{F} = \{f_j\}_{j=1}^n$ is a unit-norm frame for $\mathbb{F}^m$, then the frame's **coherence** - the maximal element of its angle set - obeys the inequality,

$$\max_{j \neq j'} |\langle f_j, f_{j'} \rangle| \geq \sqrt{\frac{n-m}{m(n-1)}},$$

and it is well-known [**22**] that $\mathcal{F}$ achieves this bound if and only if $\mathcal{F}$ is an equiangular, tight frame (**ETF**). For convenience, we call this lower bound the **Welch constant** and write $W_{n,m} = \sqrt{\frac{n-m}{m(n-1)}}$.

In general, d-angular (n, m)-frames are not equidistributed, as is demonstrated by the 3-angular *orthoplectic* constructions in [**7**], for example. However, the equidistributed property always holds in the special case that $\mathcal{F}$ is a d-angular (n, m)-frame with $d \leq 2$. This is trivial if $d = 1$, since the frame's angle set is then a singleton. If $\mathcal{F}$ is a biangular tight frame (**BTF**), then straightforward substitution implies the invariance (with respect to the choice $j \in \{1, ..., n\}$) of the distribution of the two squared frame angles occurring as summands in the left-hand side of Equation (2.1); in other words, $\mathcal{F}$ is equidistributed. Furthermore, solving Equation (2.1) in conjunction with the identity $\tau_1 + \tau_2 = n - 1$ yields the frame angle multiplicities.

PROPOSITION 2.1. *If $\mathcal{F}$ is a biangular (n, m)-frame with distinct frame angles α_1 and α_2, then the correponding frame angle multiplicites are*

$$\tau_1 = \frac{n-1}{\alpha_2^2 - \alpha_1^2}\left(\alpha_2^2 - \frac{n-m}{m(n-1)}\right) \text{ and } \tau_2 = \frac{n-1}{\alpha_1^2 - \alpha_2^2}\left(\alpha_1^2 - \frac{n-m}{m(n-1)}\right).$$

We refer to [**15**] for more information about frame theory and its applications.

REMARK 2.2. In order to provide a concise definition of the *equidistributed* property above, it was convenient to define ETFs and BTFs as distinctly different objects, as determined by the cardinalities of their frame angle sets. However, we prefer to think of ETFs as special instances of BTFs; that is, after allowing a slight abuse of the definition, we think of an ETF as a BTF where the two frame angles agree. For the purpose of highlighting the relationship between these two types of frames, we will employ this "abuse of terminology" throughout the remainder of this work, but never without clarification.

Furthermore, one can take copies of any equiangular tight frame (eg, an orthonormal basis) to see that BTFs exist in every dimension; however, we dismiss BTFs obtained in this way as trivial, particularly because the design of frames with low coherence is the primary motivation of this work.

3. A continuum of BTFs in $\mathbb{R}^3$

In this section, we construct a smooth curve of BTFs passing through the space of real $(6, 3)$-frames. Afterwards, we remark on some of the curve's surprising properties and the implications of its existence.

Consider the curve $F : [1, \infty) \to \mathbb{R}^{3\times 6}$ defined by

$$F(t) = \begin{bmatrix} \frac{1}{t^2} & \frac{1}{t^2} & 0 & 0 & \sqrt{\frac{t^4-1}{t^4}} & -\sqrt{\frac{t^4-1}{t^4}} \\ 0 & 0 & \sqrt{\frac{t^4-1}{t^4}} & -\sqrt{\frac{t^4-1}{t^4}} & \frac{1}{t^2} & \frac{1}{t^2} \\ \sqrt{\frac{t^4-1}{t^4}} & -\sqrt{\frac{t^4-1}{t^4}} & \frac{1}{t^2} & \frac{1}{t^2} & 0 & 0 \end{bmatrix}.$$

The following properties of $F(t)$ follow by elementary calculations:

(i) for each $t \in [1, \infty)$, the rows are pairwise orthogonal,
(ii) for each $t \in [1, \infty)$, every row has squared norm equal to 2,
(iii) for each $t \in [1, \infty)$, every column has norm 1 and
(iv) for each $t \in [1, \infty)$, the angle set is

$$\Theta(t) = \left\{ \left| \frac{2 - t^4}{t^4} \right|, \frac{\sqrt{t^4 - 1}}{t^4} \right\}.$$

By our identification of frames with their synthesis operators, $F(t)$ corresponds to a smooth curve, $\mathcal{F}(t)$, of frames passing through the space of real $(6,3)$-frames, where every point on the curve is a BTF. A remarkable property of this curve is that it passes through three frames of noteworthy structural symmetry, and it converges to a fourth noteworthy frame at infinity.

The first ocurs at $t = 1$, where we obtain

$$F(1) = \begin{bmatrix} 1 & 1 & 0 & 0 & 0 & 0 \\ 0 & 0 & 0 & 0 & 1 & 1 \\ 0 & 0 & 1 & 1 & 0 & 0 \end{bmatrix},$$

or two copies of the canonical basis, $\{e_1, e_2, e_3\}$. Although $F(1)$ is an "uninteresting" frame, it is nevertheless highly symmetric and it provides a convenient visualization for our curve's initial point. We imagine that as we "trace out" the parameter from here, this pair of orthonormal bases smoothly deforms to become other interesting frames, while retaining the rigid BTF structure along the way.

At $t = \sqrt[4]{2}$, we obtain the frame

$$F(\sqrt[4]{2}) = \begin{bmatrix} \frac{1}{\sqrt{2}} & \frac{1}{\sqrt{2}} & 0 & 0 & \frac{1}{\sqrt{2}} & \frac{-1}{\sqrt{2}} \\ 0 & 0 & \frac{1}{\sqrt{2}} & \frac{-1}{\sqrt{2}} & \frac{1}{\sqrt{2}} & \frac{1}{\sqrt{2}} \\ \frac{1}{\sqrt{2}} & \frac{-1}{\sqrt{2}} & \frac{1}{\sqrt{2}} & \frac{1}{\sqrt{2}} & 0 & 0 \end{bmatrix},$$

an example of Steiner BTF, as discussed in Section 5. Furthermore, by partitioning its columns, $\{f_j\}_{j=1}^6$, into three consecutive pairs, $\{f_1, f_2\}$, $\{f_3, f_4\}$ and $\{f_5, f_6\}$, we obtain all of the orthonormal bases for the 2-dimensional coordinate subspaces of $\mathbb{R}^3$.

Arguably, the most noteworthy point [**5, 22**] on the curve occurs at $t = \sqrt[4]{\frac{5+\sqrt{5}}{2}}$, because the angle set reduces to the singleton,

$$\Theta\left(\sqrt[4]{\tfrac{5+\sqrt{5}}{2}}\right) = \left\{\tfrac{1}{\sqrt{5}}\right\},$$

meaning that $F\left(\sqrt[4]{\frac{5+\sqrt{5}}{2}}\right)$ is an ETF. This particular ETF is well known [**5, 15**], as it corresponds to a selection of six non-antipodal vertices from a regular icosahedron. In accordance with the characterization of Welch, this frame is a minimizer of coherence among all real $(6,3)$-frames; that is, it is a Grassmannian frame.

Finally, we consider the asymptotics. Evaluating the limits of the coordinates as t approaches infinity yields

$$\lim_{t\to\infty} F(t) = \begin{bmatrix} 0 & 0 & 0 & 0 & 1 & -1 \\ 0 & 0 & 1 & -1 & 0 & 0 \\ 1 & -1 & 0 & 0 & 0 & 0 \end{bmatrix},$$

the canonical basis, $\{e_1, e_2, e_3\}$, unioned with its antipodes, $\{-e_1, -e_2, -e_3\}$. As with the curve's initital point, the curve's limiting frame is not "interesting" but highly symmetric, thereby supplementing our visualization of how the BTFs deform as they transit along the curve.

Thus, we have constructured a curve of highly structured frames (BTFs) whose initial point is a frame with worst possible coherence - two copies of the canonical basis - which then passes through a frame with optimal coherence - an ETF - and then asymptotically approaches another frame with worst possible coherence - the canonical basis along with its negatives. The existence of such a curve answers two questions concerning the existence and properties of BTFs.

3.1. Two questions answered. As with ETFs, biangular tight frames necessitate the equidistributed property, a fairly rigid constraint. Moreover, just as ETFs are known to be difficult to construct [**9, 19, 22, 23, 28, 30, 31, 41–43**], attempts to construct biangular tight frames are accompanied by a similar level of difficulty [**2, 3, 21, 29, 33, 37, 39, 46**]. Because of these similarities, we had, informally speaking, begun to wonder:

"How rigid are BTFs compared to ETFs?"

As it turns out, the existence of our curve, $F(t)$, answers two meaningful formulations of this question. Due to the Welch bound, the square of an ETF's single frame angle is always rational. Based on known examples [**13, 29, 37, 46**], we had begun to suspect a similar statement for BTFs.

QUESTION 3.1. Must the squared frame angles of a BTF be rational - or, at least, nontranscendental?

ANSWER 3.2. Our curve's existence answers this in the negative. The continuum of frame angles, $\{\Theta(t)\}_{t\in[1,\infty)}$, along with elementary properties of the real line indicate the existence of uncountably many inquivalent BTFs occurring along the curve with transcendental squared frame angles.

Incidentally, this observation leads to and answers the next question. We have also pondered the potential cardinality of **inequivalent** BTFs that might occur within a given space of (n, m)-frames, where two biangular (n, m)-frames are **inequivalent** if they have different frame angle sets. There are several instances where two or more inequivalent BTFs coexist within a given space [**13**]. For example, there are at least three inequivalent BTFs for $\mathbb{C}^3$ consisting of 8 vectors [**13**]. However, our limited knowledge of known examples [**13, 29, 37, 46**] of BTFs led us to ponder the following.

QUESTION 3.3. Within a given space of (n, m) frames, must the number of inequivalent BTFs be finite - or, at least, countable?

ANSWER 3.4. Again, our curve's existence answers this in the negative. The continuum of frame angles, $\{\Theta_t\}_{t\in[1,\infty)}$, implies the existence of uncountably many inequivalent BTFs coexisting within the space of $(6, 3)$-frames.

Note that, since the real $(6,3)$-frames are a subset of the complex $(6,3)$-frames, we may view our curve $F(t)$ as a complex enitity. In particular, Question 3.1 and Question 3.3 are answered in the negative for both the real and complex cases.

4. Harmonic BTFs

Given that our thesis is to outline the peculiar connections between BTFs and ETFs, a brief discussion of their relationship when manifesting as so-called *harmonic frames* seems appropriate. Besides their relevance to this paper's overall theme, we will require basic facts about harmonic ETFs and BTFs to construct the *Steiner BTFs* of the next section. Unfortunately, a proper treatment of harmonic frames requires a level of notational complexity that we prefer to avoid in this work, so this section is only intended as a summary of core definitions and results [**13**, **41**, **47**].

In order to define *harmonic frames*, we assume basic familiarity with character theory (see [**6**] for details) and recall our identification of frames with their synthesis matrices. Let $\mathbb{G}$ be an abelian group of order n and let $\mathcal{S}$ be any subselection of m distinct rows from the $n \times n$ character table of $\mathbb{G}$. The resulting $m \times n$ submatrix, H, with columns rescaled to norm 1, is called an **(n, m)-harmonic frame (for $\mathbb{G}$ generated by $\mathcal{S}$)**. By basic properties of character tables [**6**], it is easy to verify [**13**, **15**] that such an object is the synthesis matrix of an (n, m)-frame, $\mathcal{H}$, for $\mathbb{C}^m$. Moreover, every (n, m)-harmonic frame is **flat**, meaning that its entries have constant magnitude $1/\sqrt{m}$; the importance of this property becomes apparent in the next section. Also, we remark that our choice of $\mathcal{S}$ - a combinatorial issue - completely determines $\mathcal{H}$'s frame angle set. Assimilating these facts, assured that existence is not an issue, we offer the following simplified definition, with notational details purposefully suppressed.

Definition 4.1. Given an abelian group $\mathbb{G}$ of order n and a subset $\mathcal{S}$ of $\mathbb{G}$ of m distinct elements, then an **(n, m)-harmonic frame (for $\mathbb{G}$ generated by $\mathcal{S}$)** is a flat (n, m)-frame for $\mathbb{C}^m$ with a frame angle set, $\Theta(\mathbb{G}, \mathcal{S})$, determined by $\mathbb{G}$ and $\mathcal{S}$.

We remark that, under certain conditions on $\mathbb{G}$ or the subselection $\mathcal{S}$, a harmonic frame can manifest as a strictly real frame [**13**]. For more details about harmonic frames, we refer to Waldron's chapter in [**15**] and the references therein.

In order to discuss the relationship between equiangular and biangular harmonic frames, and, ultimately, present the results needed for the next section, we must discuss their frame angle sets. As mentioned, the frame angle set of a harmonic frame depends on the combinatorial relationship between the subselection $\mathcal{S}$ and its ambient group $\mathbb{G}$. In particular, well-known [**17**] combinatorial objects known as **difference sets** play an important role in the this discussion; however, it is convenient if we begin with a generalization, the **bidifference sets**.

Definition 4.2. Let $\mathcal{S} \subset \mathbb{G}$, where $\mathbb{G}$ is an additively written abelian group of order n with identity e and where $\mathcal{S} = \{g_1, ..., g_m\}$ is a subset of m elements. We say that $\mathcal{S}$ is an **(n, m, l, λ, μ)-bidifference set for $\mathbb{G}$ relative to $\mathcal{A}$** if $\mathcal{A} \subset \mathbb{G}$ is a subset of order l with $e \in \mathcal{A}$, every non-identity element of $a \in A$ can be expressed as $a = g_s - g_t$ in exactly λ ways and every element $b \in \mathcal{B} = \mathbb{G} \backslash \mathcal{A}$ can be expressed as $b = g_s - g_t$ in exactly μ ways.

In [**13**], the authors considered a hierarchy of bidifference sets, atop which the *honest* difference sets are a special case of all of the others. We define some of the more well-known [**17**, **35**, **38**] classes from this hierarchy.

DEFINITION 4.3. Let $\mathcal{S} \subset \mathbb{G}$, where $\mathbb{G}$ is an additively written abelian group of order n with identity e and where $\mathcal{S} = \{g_1, ..., g_m\}$ is a subset of m elements.

(1) We say that $\mathcal{S}$ is an (n, m, λ)**-difference set** if it is an (n, m, l, λ, μ) for $\mathbb{G}$ relative to some subset $\mathcal{A}$ with $\lambda = \mu$.
(2) Suppose $\mathbb{H}$ is a subgroup of $\mathbb{G}$ of order l. We say that $\mathcal{S}$ is an (n, m, l, λ, μ)-**divisible difference set for** $\mathbb{G}$ **relative to** $\mathbb{H}$ if it is an (n, m, l, λ, μ)-bidifference set for $\mathbb{G}$ relative to $\mathbb{H}$. We say that $\mathcal{S}$ is an (n, m, l, μ)-**relative difference set for** $\mathbb{G}$ **relative to** $\mathbb{H}$ if it is an $(n, m, l, 0, \mu)$-divisible difference set for $\mathbb{G}$ relative to $\mathbb{H}$.
(3) We say that $\mathcal{S}$ is an (n, m, λ, μ)-**partial difference set for** $\mathbb{G}$ if $\mathcal{S}$ is an (n, m, l, λ, μ)-bidifference set for $\mathbb{G}$ relative to $\mathcal{S} \cup \{0_G\}$, where $l = |\mathcal{S} \cup \{0_G\}|$.

Note that, trivially (or vacuously), every difference set may be viewed as both a divisible difference set and as a partial difference set. Over the last decade, a characterization of equiangular harmonic frames in terms of difference sets has become well-known [**22**, **41**, **47**]. Motivated by this and the hierarchy outlined in the previous definition, the authors of [**13**] studied the frame angle sets of harmonic frames generated by these other types of bidifference sets. We summarize these results [**13**, **41**, **47**] in the following theorem.

THEOREM 4.4. *[*[**13**, **41**, **47**]*] Let $\mathbb{G}$ be an abelian group of order n and let $\mathcal{H}$ be an (n, m)-harmonic frame generated by $\mathcal{S}$.*

(1) *If $\mathcal{S}$ is an (n, m, l, λ, μ)-divisible difference set for $\mathbb{G}$ relative to $\mathbb{H}$, then $\mathcal{H}$ is a BTF with frame angle set*

$$\Theta_{\mathcal{H}} = \left\{ \tfrac{1}{m}\sqrt{m-\lambda+l(\lambda-\mu)},\ \tfrac{1}{m}\sqrt{m-\lambda} \right\}.$$

(2) *If $\mathcal{S}$ is an (n, m, λ, μ)-partial difference set for $\mathbb{G}$, then $\mathcal{H}$ is a BTF where the values of the frame angles are determined by the partial difference set's parameters (see* [**13**] *for the frame angle formulae).*

Moreover, $\mathcal{H}$ is an ETF if and only if $\mathcal{S}$ is an (n, m, λ)-difference set for $\mathbb{G}$.

Many infinite families of difference sets [**17**], divisible difference sets [**38**], and partial difference sets [**35**] are known within the combinatorial literature. Accordingly, Theorem 4.4 generates tables of infinite families of harmonic ETFs [**22**] and harmonic BTFs [**13**]. Although it is beyond the scope of this work to list all known examples here, we conclude this section by collecting three infinite families of harmonic BTFs, which we will use to construct *Steiner BTFs* in Section 5.

First, we note the existence of the underlying bidifference sets.

THEOREM 4.5. *[*[**17**, **38**]*; see also* [**7**]*]*

(1) **Simplectic difference sets**
For every $n \in \mathbb{N}$ with $n > 1$, an $(n, n-1, n-2)$-difference set exists.
(2) **Singer difference sets**
For every prime power q, a $(q^2+q+1, q+1, 1)$-difference set exists.

(3) **Picket fence sequences**
For every prime power q, a $(q^2-1, q, q-1, 1)$-relative difference set for the additive group $\mathbb{Z}_{q^2-1}$ (relative to a subgroup $\mathbb{H}$ of order $q-1$) exists.

Recalling that a relative difference set is a divisible difference set where the fourth parameter vanishes, ie $\lambda = 0$, we apply Theorem 4.4 to Theorem 4.5 to obtain the desired families of BTFs.

COROLLARY 4.6.

(1) **Simplectic ETFs**
For every $n \in \mathbb{N}$ with $n > 1$, a complex, flat, equiangular $(n, n-1)$-frame $\mathcal{H}$ exists.

(2) **Singer ETFs**
For every prime power q , a complex, flat, equiangular $(q^2+q+1, q+1)$-frame $\mathcal{H}$ exists.

(3) **Picket fence BTFs**
For every prime power q, a complex, flat, biangular (q^2-1, q)-frame $\mathcal{H}$ exists with frame angle set $\Theta_{\mathcal{H}} = \left\{\frac{1}{q}, \frac{1}{\sqrt{q}}\right\}$.

5. Steiner BTFs

In this section, we construct numerous infinite families of biangular tight frames, called **Steiner BTFs**, by exploiting the existence and "flatness" of harmonic BTFs along with well-studied combinatorial objects, called *Steiner systems* [**17**]. Our construction technique is very similar to that of the **Steiner ETFs** constructed in [**23**], involving only a slight generalization of the so-called "building blocks". For those familiar with Steiner ETFs, we simply relax the requirement that the underlying difference sets are simplectic.

We remark that some of the families of Steiner BTFs in this section were previously described in a dissertation [**26**], so we are happy to present the results here in a more formal setting; however, we also augment these results [**26**] with new examples. In particular, we introduce a second class of Steiner BTFs - distinct from those of [**26**] - which arise by passing from difference sets to certain types of divisible difference sets.

Roughly speaking, a $(2, k, v)$-*Steiner system*, $(\mathcal{V}, \mathcal{B})$, is a nonempty set, $\mathcal{V}$, of v points along with a collection, $\mathcal{B}$, of k-subsets of $\mathcal{V}$, called *blocks*, which satisfy certain incidence properties. Every Steiner system is associated to a $\{0, 1\}$-matrix [**17**], called its **incidence matrix**. For the purpose of constructing the *Steiner BTFs* in this section, we are mainly interested in the transposes of the incidence matrices of such systems. Accordingly, we find it less cumbersome to simply assign a term and formal definition to the transpose of the incidence matrix of a Steiner system - we call it a **Steiner matrix** - with the tacit understanding that the existence of a Steiner matrix is equivalent to the existence of a Steiner system [**17**], where the parameters of the Steiner matrix completely determine those of the corresponding Steiner system.

DEFINITION 5.1. Suppose k and v are positive integers, where $k \leq v$ and $\frac{v(v-1)}{k(k-1)}$ and $\frac{v-1}{k-1}$ are both integers. A (v, k)-**Steiner matrix** A is a $\{0, 1\}$-matrix of size $\frac{v(v-1)}{k(k-1)} \times v$ such that

(1) A has exactly k ones in each row,
(2) A has exactly $\frac{v-1}{k-1}$ ones in each column, and
(3) every two distinct columns of A has a dot product of one.

As with the difference sets and their generalizations discussed in the previous section, constructions of numerous infinite families of $(2, k, v)$-Steiner systems are known from the combinatorial literature [**17**]. Thus, the correspondence between Steiner systems and Steiner matrices outlined above implies the existence of numerous infinite families of corresponding Steiner matrices. Their existence is vital to the main theorem of this section.

THEOREM 5.2. *Let A be a (v, k)-Steiner matrix. To abbreviate, let $s = \frac{v-1}{k-1}$ and $m = \frac{v(v-1)}{k(k-1)}$, so that, by definition, A is an $m \times v$ binary matrix with exactly s ones in each column. Let $H_1, H_2, ..., H_v$ be the synthesis matrices of flat (t, s)-frames for $\mathbb{F}^s$ such that either* **(i)** *the corresponding frame $\mathcal{H}_j$ is equiangular for every $j \in \{1, 2, ..., v\}$ or* **(ii)** *the corresponding frame $\mathcal{H}_j$ is biangular and $\frac{1}{s} \in \Theta_{\mathcal{H}_j}$ for every $j \in \{1, 2, ..., v\}$. Furthermore, let F be the $m \times tv$ matrix constructed as follows:*

(1) *For each $j \in \{1, .., v\}$, let F_j be the $m \times t$ (block) matrix obtained by replacing each one in the jth column of A with a distinct row from H_j and replacing each zero in the jth column of A with a $1 \times t$ row of zeros.*
(2) *Concatenate to obtain the $m \times tv$ matrix $F = [F_1 F_2 \cdots F_v]$.*

In this case, F is the synthesis matrix of a biangular (tv, m)-frame, $\mathcal{F}$, for $\mathbb{F}^m$. Moreover, the frame angle set of $\mathcal{F}$ is determined as follows:

(i) *If $\mathcal{H}$ is a flat ETF, then $\Theta_{\mathcal{F}} = \{\frac{1}{s}, W_{t,s}\}$.*
(ii) *If $\mathcal{H}$ is a flat BTF with $\frac{1}{s} \in \Theta_{\mathcal{H}}$, then $\Theta_{\mathcal{F}} = \Theta_{\mathcal{H}}$.*

PROOF. First, we show that $\mathcal{F}$ is indeed a (tv, m)-frame for $\mathbb{F}^m$, noting that the following argument is independent of whether the $\mathcal{H}_j$s abide Condition **(i)** or Condition **(ii)** in the hypothesis. Since every column of F is really just a column of some H_j inflated with extra zero entries, it follows that that F's columns are unit-norm. It remains to verify that $\mathcal{F}$ is a tight frame for $\mathbb{F}^m$. Note that for any $j \in \{1, ..., v\}$, the tightness of $\mathcal{H}_j$ implies that the inner product between any two distinct rows of F_j must be zero. Since the inner product between any two distinct rows of F is the sum of the inner products between the corresponding rows of the F_js, it follows that the inner product between any two distinct rows of F is zero. Similarly, the squared norm of any row of F is the sum of the corresponding squared norms of the rows of the F_js, so the flatness of each $\mathcal{H}_j$ in conjunction with Property (1) from Definition 5.1 implies that each row of F has a squared norm of $\frac{kt}{s}$. In particular, $FF^* = \frac{kt}{s} I_m$. After noting the obvious dependence of the underlying field of $\mathcal{F}$ upon that of the $\mathcal{H}_j$s and verifying the identity $\frac{kt}{s} = \frac{tv}{m}$, we conclude that $\mathcal{F}$ is a (tv, m)-frame for $\mathbb{F}^m$.

Next, we compute the frame angle set and verify that F is a BTF. Given two distinct columns f and f' of F, there are two cases. Either Case (a) they come from different block-matrices (ie, f is a column of F_j and f' is a column of $F_{j'}$ with $j \neq j'$) or Case (b) f and f' are columns of the same block-matrix F_j for some j.

If Case (a), then the flatness of H_j and $H_{j'}$ along with Property (3) from Definition 5.1 shows that $|\langle f, f' \rangle| = \frac{1}{s}$. In particular, $\frac{1}{s} \in \Theta_{\mathcal{F}}$. Note that this is independent of whether the $\mathcal{H}_j$s satisfies Condition **(i)** or Condition **(ii)**.

Note that either **(i)** all of the $\mathcal{H}_j$s are flat ETFs or **(ii)** they are all flat BTFs with a shared frame angle of $\alpha_1 = \frac{1}{s}$, in which case the equidistributed property of BTFs along with Equation 2.1 implies that they must all agree on the second frame angle, α_2. In either case, we have

$$\Theta_{\mathcal{H}_j} = \Theta_{\mathcal{H}_{j'}} \text{ for all } j, j' \in \{1, ..., v\},$$

so let us drop the unnecessary index j and simply write $\Theta_{\mathcal{H}} = \Theta_{\mathcal{H}_j}$ for all $j \in \{1, 2, ..., v\}$.

For Case (b), it is clear by construction that $|\langle f, f' \rangle| \in \Theta_{\mathcal{H}}$, so the arbitrariness of j and j' shows that $\Theta_{\mathcal{F}} = \left\{\frac{1}{s}\right\} \cup \Theta_{\mathcal{H}}$. The claim follows by considering the two possible conditions from the hypothesis. If (**i**) each $\mathcal{H}_j$ is an ETF, then $|\langle f, f' \rangle| = W_{t,s}$, the Welch bound. Thus, $\Theta_{\mathcal{F}} = \{\frac{1}{s}, W_{t,s}\}$ and $\mathcal{F}$ is a BTF, as claimed. On the other hand, if **(ii)** each $\mathcal{H}_j$ is biangular with $\frac{1}{s} \in \Theta_{\mathcal{H}_j}$, then $\Theta_{\mathcal{F}} = \left\{\frac{1}{s}\right\} \cup \Theta_{\mathcal{H}} = \Theta_{\mathcal{H}}$, and $\mathcal{F}$ is a BTF, as claimed. □

We call any BTF constructed by the preceding theorem a **Steiner BTF**. In order to demonstrate the theorem's potency, we construct Steiner BTFs by using harmonic ETFs and certain harmonic BTFs as *building blocks*; more precisely, given a (v, k)-Steiner matrix A, we call a flat (t, s)-frame $\mathcal{H}$ a **building block** for A if $s = \frac{v-1}{k-1}$. To begin, we recall the existence of four infinite families of $(2, k, v)$-Steiner systems and, hence, their corresponding (v, k)-Steiner matrices [**17**].

THEOREM 5.3. *[*[**17**]*]*

(1) **Affine geometries**
Given a prime power q and $a \in \mathbb{N}$ with $a \geq 2$, then a (q^a, q)-Steiner matrix exists.
(2) **Projective geometries**
Given a prime power q and $a \in \mathbb{N}$ with $a \geq 2$, then a $\left(\frac{q^{a+1}-1}{q-1}, q+1\right)$-Steiner matrix exists.
(3) **Unital systems**
Given a prime power q, then a $\left(q^3 + 1, q + 1\right)$-Steiner matrix exists.
(4) **Denniston systems**
Given $a, b \in \mathbb{N}$ with $2 \leq a \leq b$, then a $\left(2^{a+b} + 2^a - 2^b, 2^a\right)$-Steiner matrix exists.

In the following subsections, we use the harmonic ETFs and BTFs described in Section 4 as building blocks for these Steiner matrices, thereby producing numerous families of Steiner BTFs.

5.1. Steiner ETFs. Using the so-called *simplectic ETFs* from Corollary 4.6 as building blocks in Theorem 5.2, we recover the plethora of **Steiner ETFs** first described in [**23**].

COROLLARY 5.4. *[*[**23**]*] If a (v, k)-Steiner matrix exists, then a complex, equiangular (n, m)-frame exists, where $m = \frac{v(v-1)}{k(k-1)}$ and $n = v(k + 1)$.*

PROOF. Let A be any (v, k)-Steiner matrix and, for brevity, write $s = \frac{v-1}{k-1}$. Let $\mathcal{H}$ be a complex, flat, equiangular $(s + 1, s)$-frame, the existence of which is assured by Corollary 4.6. Thus, $\mathcal{H}$ is an equiangular building block for A, so Theorem 5.2 yields a Steiner BTF, $\mathcal{F}$, with angle set $\Theta_{\mathcal{F}} = \{\frac{1}{s}, W_{s+1,s}\}$, so the claim follows by computing that $W_{s+1,s} = \frac{1}{s}$. □

Equiangular tight frames produced via Corollary 5.4 are called **Steiner ETFs**. The authors of [**23**] provide a thorough analysis of the Steiner ETFs produced by each family of Steiner matrices from Theorem 5.3. We defer to their work [**23**] for further details.

5.2. (Proper) Steiner BTFs from difference sets. As noted in the previous section, a complete exhibition of all known harmonic ETFs is beyond the scope of this work; accordingly, we cannot provide a complete exhibition of all Steiner BTFs generated with (non-simplectic) harmonic ETFs as building blocks. Instead, we merely demonstrate the efficacy of Theorem 5.2 by constructing three (of many possible) infinite families of Steiner BTFs via harmonic ETFs.

COROLLARY 5.5. *If q is a prime power, then*

(1) **Affine Singer (Steiner) BTFs**
a biangular (n,m)-frame $\mathcal{F}$ for $\mathbb{C}^m$ exists, where $m = q(q+1)$ and $n = q^2(q^2+q+1)$,
(2) **Projective Singer (Steiner) BTFs**
a biangular (n,m)-frame $\mathcal{F}$ for $\mathbb{C}^m$ exists, where $m = q^2+q+1$ and $n = (q^2+q+1)^2 = m^2$, and
(3) **Unital Singer (Steiner) BTFs**
if, in addition, q is a square, then a biangular (n,m)-frame $\mathcal{F}$ for $\mathbb{C}^m$ exists, where $m = \frac{q^2(q^3+1)}{q+1}$ and $n = (q+1)(q^2+q+1)$.

Moreover, in any of these three cases, the frame angle set of $\mathcal{F}$ is

$$\Theta_{\mathcal{F}} = \left\{ \frac{1}{q+1}, \frac{\sqrt{q}}{q+1} \right\}.$$

PROOF. Let q be a prime power and let $\mathcal{H}$ be a complex, flat, equiangular $(q^2+q+1, q+1)$-frame, the existence of which is assured by Corollary 4.6. By Statement (1) of Theorem 5.3, a (q^2, q) Steiner matrix A exists, by Statement (2), taking $a = 2$, a $\left(q^2+q+1, q+1\right)$-Steiner matrix B exists, and by Statement (3), if q is a square, then a $(q^3+1, q+1)$-Steiner matrix C exists. A straightforward computation shows that $\mathcal{H}$ is an equiangular building block for A, B, and C (assuming C exists). Thus, Theorem 5.2 yields a biangular (n,m)-frame $\mathcal{F}$ in each case, where the claimed values for m and n follow by elementary computation. The claimed value for the frame angle set in each case also follows from Theorem 5.2, after computing that $W_{q^2+q+1,q+1} = \frac{\sqrt{q}}{q+1}$. □

The Steiner BTFs just described - produced with harmonic ETFs as building blocks - were previously mentioned in [**26**]. We stress again that the BTFs produced in Corollary 5.5 are far from comprehensive, as there are several other families of harmonic ETFs [**17**] and Steiner matrices [**17**] that can be used to produce more Steiner BTFs. In the next subsection, we demonstrate a new class of Steiner BTFs, using certain harmonic (proper) BTFs as building blocks instead.

5.3. (Proper) Steiner BTFs from (proper) bidifference sets. As with the previous section, a complete exhibition of all known Steiner BTFs generated by (proper) harmonic BTFs as building blocks is beyond the scope of this work. In the following corollary, we use the so-called *picket fence sequences* of Corollary 4.6 to demonstrate three families of Steiner BTFs produced via (proper) bidifference sets.

COROLLARY 5.6. *If q is a prime power such that $q+1$ is also a prime power, then*

(1) **Affine picket-fence (Steiner) BTFs**
a biangular (n,m)-frame $\mathcal{F}$ for $\mathbb{C}^m$ exists, where $m = q(q+1)$ and $n = q^3(q+1)$,
(2) **Projective picket-fence (Steiner) BTFs**
a biangular (n,m)-frame $\mathcal{F}$ for $\mathbb{C}^m$ exists, where $m = q^2+q+1$ and $n = q(q+2)(q^2+q+1)$, and
(3) **Unital picket-fence (Steiner) BTFs**
if, in addition, q is a square, then a biangular (n,m)-frame $\mathcal{F}$ for $\mathbb{C}^m$ exists, where $m = \frac{q^2(q^3+1)}{q+1}$ and $n = q(q+1)(q^3+1)$.

Moreover, in any of these three cases, the frame angle set of $\mathcal{F}$ is

$$\Theta_{\mathcal{F}} = \left\{ \frac{1}{q+1}, \frac{1}{\sqrt{q+1}} \right\}.$$

PROOF. Let q be a prime power such that $q+1$ is also a prime power, and let $\mathcal{H}$ be a complex, flat, biangular $\big((q+1)^2-1, q+1\big)$-frame with frame angle set

$$\Theta_{\mathcal{H}} = \left\{ \frac{1}{q+1}, \frac{1}{\sqrt{q+1}} \right\},$$

the existence of which is assured by Corollary 4.6. By Statement (1) of Theorem 5.3, a (q^2, q) Steiner matrix A exists, by Statement (2), taking $a = 2$, a $\big(q^2+q+1, q+1\big)$-Steiner matrix B exists, and by Statement (3), if q is a square, then a $(q^3+1, q+1)$-Steiner matrix C exists. A straightforward computation shows that $\mathcal{H}$ is an biangular building block for A, B, and C (assuming C exists). Thus, Theorem 5.2 yields a biangular (n,m)-frame $\mathcal{F}$ with $\Theta_{\mathcal{F}} = \Theta_{\mathcal{H}}$ in each case, where the claimed values for m and n follow by elementary computation. □

Surprisingly, the preceding corollary along with an unintentional challenge implied by a work of Delsarte, Goethals, and Seidel [**21**] leads to a connection with an open problem from number theory.

5.3.1. *Steiner BTFs, Mersenne primes, and Fermat primes.* We were careful not to refer to the families of Steiner BTFs produced by Corollary 5.6 as *infinite families*, because the necessary condition for such a frame's existence - the existence of a primer power q such that $q+1$ is also a prime power - is equivalent to two famous open number-theoretic problems [**10**], namely the question of whether an infinite number of *Mersenne primes* or *Fermat primes* exist.

To see this, note that if a pair, q and $q+1$, of consecutive prime powers exists, then one of them must be even, and therefore a power of 2. In 2002, Mihăilescu proved [**36**] the classical conjecture of Catalan [**16**].

THEOREM 5.7. *[Catalan/Mihăilescu; [**16, 36**]] Given positive integers $a, b \in \mathbb{N}$ with $a, b \geq 2$, then the only solution to the Diophantine equation*

$$x^a + 1 = y^b$$

is $x = 2, y = 3, a = 3$ and $b = 2$.

Thus, if q and $q+1$ are a pair of consecutive prime powers, then either Case **(i)** $q = 8$ or Case **(ii)** one of the pair is an "honest" prime number while the other is a power of 2. As it turns out, Case **(ii)** boils down to q either being a **Mersenne**

prime, a prime of the form $p = 2^t - 1$, or a **Fermat prime**, a prime of the form $p = 2^t + 1$. Unfortunately, the questions of whether an infinitude of either type of primes exists have remained famously open problems for centuries [**10**]. Therefore, we cannot conclude that any of the three families of Steiner BTFs from Corollary 5.6 are infinite; nevertheless, the corollary is obviously not vacuous, since, for example, the hypothesis is satsified for $q = 2, 3, 4, 7, 8$ or any of the other known Merssenne primes [**18**] or Fermat primes [**10**].

To conclude this section, we briefly explain our "challenge" inferred from the work of Delsarte, Goethals, and Seidel [**21**] which has motivated this number-theoretic digression. In [**21**], the authors studied the upper bounds for the cardinalities of sets of unit vectors admitting few pairwise absolute inner products. Of relevance here, they showed that for any complex (n, m)-frame $\mathcal{F}$, a necessary condition for $\mathcal{F}$ to be equiangular is that $n \leq m^2$ and a necessary condition for $\mathcal{F}$ to be biangular is that $n \leq \binom{m+1}{2}^2$; however, besides a finite number of sporadic instances [**13, 29**] and the known infinite families of *maximal sets of mutually unbiased bases* (see [**13, 46**] for details), we have found it strikingly difficult to construct or locate more complex BTFs in the literature [**2, 4, 13, 29, 37, 46**] for which $n > m^2$; in other words, roughly speaking, it seems that most BTFs do not even exceed the cardinality bounds of ETFs. Thus, we pose the following question.

QUESTION 5.8. Other than maximal sets of mutually unbiased bases [**46**], do there exist other "infinite families" of complex biangular (n, m)-frames for which $n > m^2$?

In light of the preceding discussion, if an infinite number of Mersenne or Fermat primes exist, then the first and second families of Steiner BTFs produced by Corollary 5.6 are infinite families; in particular, the biangular $(q(q+2)(q^2+q+1), q^2+q+1)$-frames produced by the second family (the projective picket fence BTFs) would answer Question 5.8 in the affirmative.

6. Plücker ETFs

In this final section, we relax our emphasis from biangular tight frames to a well-studied generalization [**1, 8, 14, 34**], *(chordally) biangular tight fusion frames (BTFFs)*, with the goal of showcasing a surprising example of a BTFF which generates a Steiner ETF via the *Plücker embedding*. To begin, we present a few bare essential facts about *fusion frames* and the *Plücker embedding.*

6.1. Basics of fusion frames.

Let $l, m, n \in \mathbb{N}$, and let $\mathbb{F} = \mathbb{R}$ or $\mathbb{F} = \mathbb{C}$. A **(real or complex) tight** (n, l, m)**-fusion frame** is a set $\mathcal{F} = \{P_j\}_{j=1}^n$, where each P_j is an $m \times m$ orthogonal projection of rank l with entries over $\mathbb{F}$ that satisfies the resolution of the identity,

$$\sum_{j=1}^{n} P_j = aI_m, \tag{6.1}$$

for some $a > 0$, the fusion frame's **tightness parameter**. Recall that each P_j satisfies $P = P^2 = P^*$ and $\operatorname{tr}(P_j) = l$, so taking the trace of both sides of Equation 6.1 and solving yields the tightness parameter, $a = \frac{nl}{m}$.

Given a tight $(\boldsymbol{n}, \boldsymbol{l}, \boldsymbol{m})$-fusion frame $\mathcal{F}$, its **chordal fusion frame angles** are the elements of its **fusion frame angle set**,

$$\Theta_{\mathcal{F}} = \left\{ \sqrt{\operatorname{tr}(P_j P_{j'})} : j \neq j' \right\}.$$

We say that $\mathcal{F}$ is **chordally d-angular** if $|\Theta_{\mathcal{F}}| = d$. In particular, if $d = 1$ or $d = 2$, then we call $\mathcal{F}$ a **(chordally) equiangular tight fusion frame (ETFF)** or a **(chordally) biangular tight fusion frame (BTFF)**, respectively.

Given any (n, m)-frame for $\mathbb{F}^m$, $\{f_j\}_{j=1}^n$, then it identifies with a tight $(n, 1, m)$-fusion frame for $\mathbb{F}^m$, $\{P_j\}_{j=1}^n$, where $P_j = f_j f_j^*$ for each j. Moreover, by the identity

$$\operatorname{tr}(P_j P_{j'}) = \operatorname{tr}\left(f_j f_j^* f_{j'} f_{j'}^*\right) = |\langle f_j, f_{j'}\rangle|^2,$$

if $\{f_j\}_{j=1}^n$ is an ETF or BTF, then $\{P_j\}_{j=1}^n$ is ETFF or BTFF, respectively. In this sense, one may regard the upcoming discussion of BTTFs as a natural generalization from that of BTFs.

Finally, we recall a standard fact that identifies orthogonal projection matrices with equivalence classes of 1-tight frames.

PROPOSITION 6.1. *[**15, 26**] An $m \times m$ matrix P is an orthogonal projection of rank l over the field $\mathbb{F}$ if and only if there exists a synthesis matrix F of a 1-tight frame for $\mathbb{F}^l$ consisting of m vectors such that $P = F^* F$. Moreover, if F' is another synthesis matrix of a 1-tight frame for $\mathbb{F}^l$ consisting of m vectors such that $P = (F')^* F'$, then there exists a unitary matrix, U, such that $F = UF'$.*

In terms of the preceding proposition, we say that the frames F and F' are **generators** of the projection P, and we say they are **positively equivalent** if $\det(U) = 1$.

6.2. The Plücker embedding. The *Plücker embedding* is an algebrogeometric tool used to study the Grassmannian manifold [**25, 27**], and it has found applications in frame theory several times [**11, 12**]. For the sake of brevity, we define it in terms of frame theory.

Let $\Omega_{m,l}(\mathbb{F})$ denote the space of all $l \times m$ synthesis matrices of 1-tight frames for $\mathbb{F}^l$ consisting of m vectors. Let $\rho_{m,l} = \begin{pmatrix} m \\ l \end{pmatrix}$, the number of ways to choose distinct $l \times l$ submatrices from an $l \times m$ matrix. After fixing an ordering,

$$\{A_1, A_2, .., A_{\rho_{m,l}}\},$$

on the $m \times m$ submatrices of an arbitrary $m \times n$ matrix A, we define the **Plücker embedding** as the coordinate mapping

$$\Phi : \Omega_{m,l}(\mathbb{F}) \to \mathbb{F}^{\rho_{m,l}} : A \mapsto \left[\det(A_j)\right]_{j=1}^{\rho_{m,l}}.$$

In light of Proposition 6.1, we can lift the Plücker embedding's domain to the space of all (real or complex) $m \times m$ orthogonal projections of rank l. Let $\mathcal{G}_{m,l}(\mathbb{F})$ denote the space of all $m \times m$ orthogonal projections of rank l over the field $\mathbb{F}$. For each $P \in \mathcal{G}_{m,l}(\mathbb{F})$, select a generator $A^{(P)} \in \Omega_{m,l}(\mathbb{F})$ for P, and let $\left[\boldsymbol{A}^{(\boldsymbol{P})}\right]$ denote the class of all generators that are positively equivalent to $A^{(P)}$. We define the **lifted Plücker embedding** as the coordinate mapping

$$\overline{\Phi} : \mathcal{G}_{m,l}(\mathbb{F}) \to \mathbb{F}^{\rho_{m,l}} : P \mapsto \Phi(A), \text{ for any } A \in \left[\boldsymbol{A}^{(\boldsymbol{P})}\right].$$

This is well-defined since, if $A^{(P)} \in \left[\boldsymbol{A}^{(\boldsymbol{P})}\right]$ and $A'^{(P)} = UA \in \left[\boldsymbol{A}^{(\boldsymbol{P})}\right]$ are equivalent generators for P, then the block identity,

$$\det\left(A'^{(P)}_j\right) = \det(U)\det\left(A^{(P)}_j\right) = \det\left(A^{(P)}_j\right),$$

shows the invariance of the value of $\overline{\Phi}(P)$ with respect to the choice of generator, $A \in \left[\boldsymbol{A}^{(\boldsymbol{P})}\right]$.

6.3. A Plücker ETF. We arrive at the main purpose of this section: an example of a BTFF that "Plücker embeds" into a Steiner ETF; in particular, we construct a chordally biangular (4, 2, 16)-tight fusion frame for $\mathbb{R}^4$, $\mathcal{F} = \{P_j\}_{j=1}^{16}$, such the Plücker embedding of its elements, $\overline{\mathcal{F}} = \left\{\overline{\Phi}(P_j)\right\}_{j=1}^{16}$, forms a Steiner ETF for $\mathbb{R}^6$. To begin, let

$$A = \sqrt{\frac{2}{3}}\begin{bmatrix} 1 & -1/2 & -1/2 & 0 \\ 0 & \sqrt{3}/2 & -\sqrt{3}/2 & 0 \end{bmatrix},$$

and note that $A \in \Omega_{4,2}(\mathbb{R})$, since $FF^* = 2I_2$. Furthermore, let

$$\Gamma = \left\{ c_j = \begin{bmatrix} 0 & 1 & 0 & 0 \\ 0 & 0 & 1 & 0 \\ 0 & 0 & 0 & 1 \\ 1 & 0 & 0 & 0 \end{bmatrix}^j : j \in \{0,1,2,3\} \right\},$$

the group of 4×4 cyclic matrix permutation matrices, and let

$$\Delta = \left\{ d_0 = \begin{bmatrix} 1 & 0 & 0 & 0 \\ 0 & 1 & 0 & 0 \\ 0 & 0 & 1 & 0 \\ 0 & 0 & 0 & 1 \end{bmatrix}, d_1 = \begin{bmatrix} 1 & 0 & 0 & 0 \\ 0 & 1 & 0 & 0 \\ 0 & 0 & -1 & 0 \\ 0 & 0 & 0 & -1 \end{bmatrix}, d_2 = \begin{bmatrix} 1 & 0 & 0 & 0 \\ 0 & -1 & 0 & 0 \\ 0 & 0 & 1 & 0 \\ 0 & 0 & 0 & -1 \end{bmatrix}, d_3 = \begin{bmatrix} 1 & 0 & 0 & 0 \\ 0 & -1 & 0 & 0 \\ 0 & 0 & -1 & 0 \\ 0 & 0 & 0 & 1 \end{bmatrix} \right\},$$

a diagonal unitary representation of the noncyclic group on four elements. Next, let

$$\mathcal{A} = \left\{A_{j,k} = A\, d_j\, c_k : j,k \in \{0,1,2,3\}\right\},$$

so $\mathcal{A}$ is the orbit of A under $\Gamma \times \Delta$. Note that since the c_js and d_js are unitaries, it follows by the definition of 1-tightness that $\mathcal{A} \subset \Omega_{4,2}(\mathbb{R})$. Thus, every element of $\mathcal{A}$ is the generator of some 4×4 orthogonal projection of rank 2. Accordingly, we define the corresponding orthogonal projections,

$$\mathcal{P} = \left\{P_{j,k} = A^*_{j,k}A_{j,k} : j,k \in \{0,1,2,3\}\right\}.$$

Next, we show that $\mathcal{P}$ is a chordally biangular tight $(16,2,4)$-fusion frame. We have

$$AA^* = \begin{bmatrix} 2/3 & -1/3 & -1/3 & 0 \\ -1/3 & 2/3 & -1/3 & 0 \\ -1/3 & -1/3 & 2/3 & 0 \\ 0 & 0 & 0 & 0 \end{bmatrix}, \text{ and then, } \sum_{j=0}^{3} d_j^* A^* A d_j = \begin{bmatrix} 8/3 & 0 & 0 & 0 \\ 0 & 8/3 & 0 & 0 \\ 0 & 0 & 8/3 & 0 \\ 0 & 0 & 0 & 0 \end{bmatrix},$$

and then

$$\sum_{j=0}^{3}\sum_{k=0}^{3} P_{j,k} = \sum_{j=0}^{3}\sum_{k=0}^{3} c_k^* d_j^* A^* A d_j c_k = \sum_{k=0}^{3} c_k^* \begin{bmatrix} 8/3 & 0 & 0 & 0 \\ 0 & 8/3 & 0 & 0 \\ 0 & 0 & 8/3 & 0 \\ 0 & 0 & 0 & 0 \end{bmatrix} c_k = 8\, I_4,$$

verifying that $\mathcal{P}$ is a real, tight $(16, 2, 4)$-fusion frame. To see that $\mathcal{P}$ is chordally biangular, note that by our definition of the $P_{j,k}$s, its chordal frame angle set is

$$\Theta_{\mathcal{P}} = \left\{ \operatorname{tr}\left(c_k^* d_j^* AA^* d_j d_k c_{k'}^* d_{j'}^* AA^* d_{j'} d_{k'}\right) : j, j', k, k' \in \{0, 1, 2, 3\} \right\}.$$

Given $j, j', k, k' \in \{0, 1, 2, 3\}$, if $k = k'$ but $j \neq j'$, then a straightforward computation yields

$$\operatorname{tr}(P_{j,k} P_{j',k}) = \frac{10}{9}.$$

Otherwise, if $j \neq j'$ and $k \neq k'$, the the computation reduces to the trace inner product between two 2×2 principle submatrices,

$$\operatorname{tr}\left(P_{j,k} P_{j',k'}\right) = \operatorname{tr}\left(\begin{bmatrix} 2/3 & \pm 1/3 \\ \pm 1/3 & 2/3 \end{bmatrix} \begin{bmatrix} 2/3 & \pm 1/3 \\ \pm 1/3 & 2/3 \end{bmatrix} \right) = \frac{8}{9} \pm \frac{2}{9}.$$

Thus, $\mathcal{P}$ is a BTFF with chordal frame angle set

$$\Theta_{\mathcal{P}} = \left\{ \sqrt{10/9}, \sqrt{2/3} \right\}.$$

Finally, we show that $\mathcal{P}$ "Plücker embeds" into a Steiner ETF. Viewing $\mathcal{A}$ as a set of generators for the elements of $\mathcal{P}$, we compute the Plücker embeddings, $\mathcal{F} = \left\{\overline{\Phi}(P_j)\right\}$, and conclude that $\mathcal{F}$ is a Steiner ETF. Note that $\rho_{4,2} = 6$, and recall that the Plücker embedding requires a choice of ordering for the underlying submatrices. We use a basic dictionary ordering; that is, given an abitrary 2×4 matrix $A = [a_1\ a_2\ a_3\ a_4]$ with columns a_1, a_2, a_3 and a_4, we order the 2×2 submatrices as

$$A_1 = [a_1\ a_2], A_2 = [a_1\ a_3], A_3 = [a_1\ a_4], A_4 = [a_2\ a_3], A_5 = [a_2\ a_4], \text{ and } A_6 = [a_3\ a_4].$$

Thus, computing the Plucker embedding of $P_{j,k}$ is a simple matter of computing the determinates of its six 2×2 submatrices of its generator, $A_{j,k}$. Fixing $k = 0$, define the 6×4 (block) matrix,

$$F_0 = \left[\overline{\Phi}(P_{j,0})\right]_{j=0}^{3} = \sqrt{2} \begin{bmatrix} 1 & 1 & -1 & -1 \\ -1 & 1 & -1 & 1 \\ 0 & 0 & 0 & 0 \\ 1 & -1 & -1 & 1 \\ 0 & 0 & 0 & 0 \\ 0 & 0 & 0 & 0 \end{bmatrix}$$

Similarly, define and compute $F_k = \left[\overline{\Phi}(P_{j,k})\right]_{j=0}^{3}$ for $k = 1, 2$ and 3,

$$F_1 = \sqrt{2} \begin{bmatrix} 0 & 0 & 0 & 0 \\ 0 & 0 & 0 & 0 \\ 0 & 0 & 0 & 0 \\ 1 & 1 & -1 & -1 \\ -1 & 1 & -1 & 1 \\ 1 & -1 & -1 & 1 \end{bmatrix}, F_2 = \sqrt{2} \begin{bmatrix} 0 & 0 & 0 & 0 \\ 1 & -1 & 1 & -1 \\ -1 & 1 & 1 & -1 \\ 0 & 0 & 0 & 0 \\ 0 & 0 & 0 & 0 \\ 1 & 1 & -1 & -1 \end{bmatrix}, \text{ and } F_3 = \sqrt{2} \begin{bmatrix} 1 & -1 & -1 & 1 \\ 0 & 0 & 0 & 0 \\ -1 & -1 & 1 & 1 \\ 0 & 0 & 0 & 0 \\ 1 & -1 & 1 & -1 \\ 0 & 0 & 0 & 0 \end{bmatrix}.$$

Concatenate and rescale to obtain the 6×16 matrix

$$F = \frac{1}{\sqrt{6}} \left[F_0\ F_1\ F_2\ F_3 \right].$$

Notice that for each $j \in \{0, 1, 2, 3\}$, the 3×4 submatrix, H_j, obtained by deleting the three zero rows from F_j and rescaling appropriately, forms the synthesis matrix of a real, flat equiangular $(4, 3)$-frame. By inspection, it is clear that F

is constructed in accordance with Theorem 5.2 with respect to the $(6,2)$-Steiner matrix

$$S = \begin{bmatrix} 1 & 0 & 0 & 1 \\ 1 & 0 & 1 & 0 \\ 0 & 0 & 1 & 1 \\ 1 & 1 & 0 & 0 \\ 0 & 1 & 0 & 1 \\ 0 & 1 & 1 & 0 \end{bmatrix},$$

using the $H_j s$ as building blocks. Noting that $W_{4,3} = \frac{1}{3}$, we conclude that F is the synthesis matrix of a Steiner ETF, $\mathcal{F}$, for $\mathbb{R}^6$ consisting of 16 vectors. In light of this construction, we conclude with a natural question, the posing of which requires a definition.

DEFINITION 6.2. An equiangular (n,m)-frame for $\mathbb{F}^m$, $\mathcal{H}$, with synthesis matrix, $H = [h_1\, h_2 \ldots h_n]$, is a **Plücker ETF** if there exists a (real or complex) tight (n,l,t)-fusion frame $\mathcal{Q} = \{Q_j\}_{j=1}^n$, where $l \geq 2$, $m = \rho_{t,l}$ and $f_j = \overline{\Phi}\,(Q_j)$ for every $j \in \{1,2,...,n\}$.

Thus, the Steiner ETF, $\mathcal{F}$, that we have just constructed is a Plücker ETF. Given the volumetric nature of the Plücker embedding (ie, the determinants of the submatrices correspond to (signed) hyper-volumes of hyper-parallopipeds) and that - as an astute reader might have noticed - the generators for each P_j from our example are essentially 2-simplices living in $\mathbb{R}^4$, we find our construction strangely intuitive. Nevertheless, we are unaware of any other examples.

QUESTION 6.3. Other than the Plücker ETF, $\mathcal{F}$, constructed in this section, do Plücker ETFs with different parameters exist?

Acknowledgment

The authors would like to thank Dustin Mixon for helpful discussions regarding the affine picket-fence BTFs discussed in Section 5.

References

[1] C. Bachoc and M. Ehler, *Tight p-fusion frames*, Appl. Comput. Harmon. Anal. **35** (2013), no. 1, 1–15, DOI 10.1016/j.acha.2012.07.001. MR3053743

[2] Alexander Barg, Alexey Glazyrin, Kasso A. Okoudjou, and Wei-Hsuan Yu, *Finite two-distance tight frames*, Linear Algebra Appl. **475** (2015), 163–175, DOI 10.1016/j.laa.2015.02.020. MR3325226

[3] Alexander Barg and Oleg R. Musin, *Bounds on sets with few distances*, J. Combin. Theory Ser. A **118** (2011), no. 4, 1465–1474, DOI 10.1016/j.jcta.2011.01.002. MR2763068

[4] Alexander Barg and Wei-Hsuan Yu, *New bounds for spherical two-distance sets*, Exp. Math. **22** (2013), no. 2, 187–194, DOI 10.1080/10586458.2013.767725. MR3047911

[5] J. J. Benedetto and J. D. Kolesar, *Geometric properties of Grassmannian frames for R^2 and R^3*, EURASIP J. Appl. Signal Process. **2006** (2006), 1–17.

[6] Bruce C. Berndt, Ronald J. Evans, and Kenneth S. Williams, *Gauss and Jacobi sums*, Canadian Mathematical Society Series of Monographs and Advanced Texts, John Wiley & Sons, Inc., New York, 1998. A Wiley-Interscience Publication. MR1625181

[7] Bernhard G. Bodmann and John Haas, *Achieving the orthoplex bound and constructing weighted complex projective 2-designs with Singer sets*, Linear Algebra Appl. **511** (2016), 54–71, DOI 10.1016/j.laa.2016.09.005. MR3557826

[8] B. G. Bodmann and J. I. Haas, *Maximal Orthoplectic Fusion Frames from Mutually Unbiased Bases and Block Designs*, ArXiv e-prints (2016).

[9] Bernhard G. Bodmann, Vern I. Paulsen, and Mark Tomforde, *Equiangular tight frames from complex Seidel matrices containing cube roots of unity*, Linear Algebra Appl. **430** (2009), no. 1, 396–417, DOI 10.1016/j.laa.2008.08.002. MR2460526

[10] David M. Burton, *Elementary number theory*, 2nd ed., W. C. Brown Publishers, Dubuque, IA, 1989. MR990017

[11] Jameson Cahill, Peter G. Casazza, and Gitta Kutyniok, *Operators and frames*, J. Operator Theory **70** (2013), no. 1, 145–164, DOI 10.7900/jot.2011may10.1973. MR3085820

[12] Jameson Cahill and Nate Strawn, *Algebraic geometry and finite frames*, Finite frames, Appl. Numer. Harmon. Anal., Birkhäuser/Springer, New York, 2013, pp. 141–170, DOI 10.1007/978-0-8176-8373-3_4. MR2964009

[13] P. G. Casazza, A. Farzannia, J. I. Haas, and T. T. Tran, *Toward the Classification of Biangular Harmonic Frames*, ArXiv e-prints (2016).

[14] Peter G. Casazza and Matthew Fickus, *Minimizing fusion frame potential*, Acta Appl. Math. **107** (2009), no. 1-3, 7–24, DOI 10.1007/s10440-008-9377-1. MR2520007

[15] Peter G. Casazza and Gitta Kutyniok (eds.), *Finite frames*, Applied and Numerical Harmonic Analysis, Birkhäuser/Springer, New York, 2013. Theory and applications. MR2964005

[16] E. Catalan, *Note extraite d'une lettre adressée à l'éditeur par Mr. E. Catalan, Répétiteur à l'école polytechnique de Paris* (French), J. Reine Angew. Math. **27** (1844), 192, DOI 10.1515/crll.1844.27.192. MR1578392

[17] Charles J. Colbourn and Jeffrey H. Dinitz (eds.), *Handbook of combinatorial designs*, 2nd ed., Discrete Mathematics and its Applications (Boca Raton), Chapman & Hall/CRC, Boca Raton, FL, 2007. MR2246267

[18] W. N. Colquitt and L. Welsh Jr., *A new Mersenne prime*, Math. Comp. **56** (1991), no. 194, 867–870, DOI 10.2307/2008415. MR1068823

[19] John H. Conway, Ronald H. Hardin, and Neil J. A. Sloane, *Packing lines, planes, etc.: packings in Grassmannian spaces*, Experiment. Math. **5** (1996), no. 2, 139–159. MR1418961

[20] Ingrid Daubechies, A. Grossmann, and Y. Meyer, *Painless nonorthogonal expansions*, J. Math. Phys. **27** (1986), no. 5, 1271–1283, DOI 10.1063/1.527388. MR836025

[21] P. Delsarte, J. M. Goethals, and J. J. Seidel, *Bounds for systems of lines, and Jacobi polynomials*, Philips Research Reports **30** (1975), 91.

[22] Matthew Fickus and Dustin G. Mixon, *Tables of the existence of equiangular tight frames*, arXiv e-print, arXiv:1504.00253 (2015).

[23] Matthew Fickus, Dustin G. Mixon, and Janet C. Tremain, *Steiner equiangular tight frames*, Linear Algebra Appl. **436** (2012), no. 5, 1014–1027, DOI 10.1016/j.laa.2011.06.027. MR2890902

[24] Vivek K. Goyal, Martin Vetterli, and Nguyen T. Thao, *Quantized overcomplete expansions in* $\mathbf{R}^N$*: analysis, synthesis, and algorithms*, IEEE Trans. Inform. Theory **44** (1998), no. 1, 16–31, DOI 10.1109/18.650985. MR1486646

[25] Phillip Griffiths and Joseph Harris, *Principles of algebraic geometry*, Wiley-Interscience [John Wiley & Sons], New York, 1978. Pure and Applied Mathematics. MR507725

[26] John I. Haas, *The geometry of structured parseval frames and frame potentials*, Ph.D. thesis, University of Houston, 2015.

[27] J. W. P. Hirschfeld, *Finite projective spaces of three dimensions*, Oxford Mathematical Monographs, The Clarendon Press, Oxford University Press, New York, 1985. Oxford Science Publications. MR840877

[28] Thomas R. Hoffman and James P. Solazzo, *Complex equiangular tight frames and erasures*, Linear Algebra Appl. **437** (2012), no. 2, 549–558, DOI 10.1016/j.laa.2012.01.024. MR2921716

[29] S. G. Hoggar, *t-designs in projective spaces*, European J. Combin. **3** (1982), no. 3, 233–254, DOI 10.1016/S0195-6698(82)80035-8. MR679208

[30] Roderick B. Holmes and Vern I. Paulsen, *Optimal frames for erasures*, Linear Algebra Appl. **377** (2004), 31–51, DOI 10.1016/j.laa.2003.07.012. MR2021601

[31] John Jasper, Dustin G. Mixon, and Matthew Fickus, *Kirkman equiangular tight frames and codes*, IEEE Trans. Inform. Theory **60** (2014), no. 1, 170–181, DOI 10.1109/TIT.2013.2285565. MR3150919

[32] A. Klappenecker and M. Rotteler, *Mutually unbiased bases are complex projective 2-designs*, International Symposium on Information Theory, 2005. ISIT 2005., Sept 2005, pp. 1740–1744.

[33] C. Koukouvinos and A. L. Whiteman, *Relative difference sets*, J. Combin. Theory Ser. A **74** (1996), no. 1, 153–157, DOI 10.1006/jcta.1996.0044. MR1383512

[34] P. W. H. Lemmens and J. J. Seidel, *Equi-isoclinic subspaces of Euclidean spaces*, Nederl. Akad. Wetensch. Proc. Ser. A **76**=Indag. Math. **35** (1973), 98–107. MR0313926

[35] S. L. Ma, *A survey of partial difference sets*, Des. Codes Cryptogr. **4** (1994), no. 3, 221–261, DOI 10.1007/BF01388454. MR1277942

[36] Preda Mihăilescu, *Primary cyclotomic units and a proof of Catalan's conjecture*, J. Reine Angew. Math. **572** (2004), 167–195, DOI 10.1515/crll.2004.048. MR2076124

[37] A. Neumaier, *Graph representations, two-distance sets, and equiangular lines*, Linear Algebra Appl. **114/115** (1989), 141–156, DOI 10.1016/0024-3795(89)90456-4. MR986870

[38] Alexander Pott, *Finite geometry and character theory*, Lecture Notes in Mathematics, vol. 1601, Springer-Verlag, Berlin, 1995. MR1440858

[39] Aidan Roy and A. J. Scott, *Weighted complex projective 2-designs from bases: optimal state determination by orthogonal measurements*, J. Math. Phys. **48** (2007), no. 7, 072110, 24, DOI 10.1063/1.2748617. MR2337670

[40] Nate Strawn, *Optimization over finite frame varieties and structured dictionary design*, Appl. Comput. Harmon. Anal. **32** (2012), no. 3, 413–434, DOI 10.1016/j.acha.2011.09.001. MR2892742

[41] Thomas Strohmer and Robert W. Heath Jr., *Grassmannian frames with applications to coding and communication*, Appl. Comput. Harmon. Anal. **14** (2003), no. 3, 257–275, DOI 10.1016/S1063-5203(03)00023-X. MR1984549

[42] Mátyás A. Sustik, Joel A. Tropp, Inderjit S. Dhillon, and Robert W. Heath Jr., *On the existence of equiangular tight frames*, Linear Algebra Appl. **426** (2007), no. 2-3, 619–635, DOI 10.1016/j.laa.2007.05.043. MR2350682

[43] F. Szöllősi, *All complex equiangular tight frames in dimension 3*, arXiv preprint, arXiv 1402.6429 (2014).

[44] Richard Vale and Shayne Waldron, *Tight frames and their symmetries*, Constr. Approx. **21** (2005), no. 1, 83–112, DOI 10.1007/s00365-004-0560-y. MR2105392

[45] L. R. Welch, *Lower bounds on the maximum cross correlation of signals*, IEEE Trans. on Information Theory **20** (1974), no. 3, 397–9.

[46] William K. Wootters and Brian D. Fields, *Optimal state-determination by mutually unbiased measurements*, Ann. Physics **191** (1989), no. 2, 363–381, DOI 10.1016/0003-4916(89)90322-9. MR1003014

[47] Pengfei Xia, Shengli Zhou, and Georgios B. Giannakis, *Achieving the Welch bound with difference sets*, IEEE Trans. Inform. Theory **51** (2005), no. 5, 1900–1907, DOI 10.1109/TIT.2005.846411. MR2235693

[48] Gerhard Zauner, *Quantum designs: foundations of a noncommutative design theory*, Int. J. Quantum Inf. **9** (2011), no. 1, 445–507, DOI 10.1142/S0219749911006776. MR2931102

Department of Mathematical Science, New Mexico State University, Las Cruces, New Mexico 88003
Email address: jamesonc@nmsu.edu

Department of Mathematics, University of Missouri, Columbia, Missouri 65211-4100
Email address: casazzap@missouri.edu

Department of Mathematics, University of Missouri, Columbia, Missouri 65211-4100
Email address: haasji@missouri.edu

Department of Mathematics, University of Missouri, Columbia, Missouri 65211-4100
Email address: tremainjc@missouri.edu

Contemporary Mathematics
Volume **706**, 2018
http://dx.doi.org/10.1090/conm/706/14217

Phase retrieval by hyperplanes

Sara Botelho-Andrade, Peter G. Casazza, Desai Cheng, John Haas, Tin T. Tran, Janet C. Tremain, and Zhiqiang Xu

ABSTRACT. We show that a scalable frame does phase retrieval if and only if the hyperplanes of its orthogonal complements do phase retrieval. We then show this result fails in general by giving an example of a frame for $\mathbb{R}^3$ which does phase retrieval but its induced hyperplanes fail phase retrieval. Moreover, we show that such frames always exist in $\mathbb{R}^d$ for any dimension $d \geq 3$. We also give an example of a frame in $\mathbb{R}^3$ which fails phase retrieval but its perps do phase retrieval. We will also see that a family of hyperplanes doing phase retrieval in $\mathbb{R}^d$ must contain at least $2d-2$ hyperplanes. Finally, we provide an example of six hyperplanes in $\mathbb{R}^4$ which do phase retrieval.

1. Introduction

In some applications in engineering, the phase of a signal is lost during processing. The problem of retrieving the phase of a signal, given a set of intensity measurements, has been studied by engineers for many years. Signals passing through linear systems often result in lost or distorted phase information. This partial loss of phase information occurs in various applications including speech recognition [**3**, **11**, **12**], and optics applications such as X-ray crystallography [**2**, **9**, **10**]. The concept of *phase retrieval* for Hilbert space frames was introduced in 2006 by Balan, Casazza, and Edidin [**1**] and since then it has become an active area of research. Phase retrieval deals with recovering the phase of a signal given intensity measurements from a redundant linear system. In phaseless reconstruction the unknown signal itself is reconstructed from these measurements. In recent literature, the two terms were used interchangeably. However it is not obvious from the definitions that the two are equivalent. Recently, authors in [**4**] proved that phase retrieval is equivalent to phaseless reconstruction in both the real and complex case.

Phase retrieval has been defined for vectors as well as for projections. *Phase retrieval by projections* occur in real life problems, such as crystal twinning [**8**], where the signal is projected onto some higher dimensional subspaces and has to be recovered from the norms of the projections of the vectors onto the subspaces.

2010 *Mathematics Subject Classification.* Primary 42C15.

The first 6 authors were supported by NSF DMS 1307685; and NSF ATD 1321779; ARO W911NF-16-1-0008. Zhiqiang Xu was supported by NSFC grant (11422113, 91630203, 11331012). Part of this reseaech was carried out while the authors were visiting the Hong Kong University of Science and Technology with support from a grant from (ICERM) Institute for Computational and Experimental Research in Mathematics.

We refer the reader to [**6**] for a detailed study of phase retrieval by projections. At times these projections are identified with their target spaces. Determining when subspaces $\{W_i\}_{i=1}^n$ and $\{W_i^\perp\}_{i=1}^n$ both do phase retrieval has given way to the notion of *norm retrieval* [**5**], another important area of research.

In this paper we make a detailed study of phase retrieval by hyperplanes. We will see that it takes at least $2d-2$ hyperplanes to do phase retrieval in $\mathbb{R}^d$. We will show that scalable frames $\{\phi_i\}_{i=1}^n$ do phase retrieval if and only if their induced hyperplanes $\{\phi_i^\perp\}_{i=1}^n$ do phase retrieval. We then give examples to show this result fails in general if the frame is not scalable. In particular, we give an example of a frame for $\mathbb{R}^3$ which does phase retrieval but its induced hyperplanes fail phase retrieval. Moreover, we show that such frames always exist in $\mathbb{R}^d$ for any dimension $d \geq 3$. We also give an example of a family of hyperplanes in $\mathbb{R}^3$ which do phase retrieval but their perp vectors fail phase retrieval. Finally, we give 6 hyperplanes in $\mathbb{R}^4$ which do phase retrieval.

2. Preliminaries

In this section we will give the background material needed for the paper. We start with the definition of a frame.

DEFINITION 2.1. A family of vectors $\Phi = \{\phi_i\}_{i=1}^n$ in $\mathbb{R}^d$ is a **frame** if there are constants $0 < A \leq B < \infty$ so that for all $x \in \mathbb{R}^d$

$$A\|x\|^2 \leq \sum_{i=1}^n |\langle x, \phi_i\rangle|^2 \leq B\|x\|^2,$$

where A and B are the **lower and upper frame bounds** of the frame, respectively. The frame is called an **A-tight frame** if $A = B$ and is a **Parseval frame** if $A = B = 1$.

DEFINITION 2.2. A frame $\Phi = \{\phi_i\}_{i=1}^n$ in $\mathbb{R}^d$ is called scalable if there exist scalars $\{s_i\}_{i=1}^n$ such that $\{s_i\phi_i\}_{i=1}^n$ is a tight frame for $\mathbb{R}^d$.

The main topics here are phase retrieval and norm retrieval in $\mathbb{R}^d$.

DEFINITION 2.3. Let $\Phi = \{\phi_i\}_{i=1}^n \subset \mathbb{R}^d$ be such that for $x, y \in \mathbb{R}^d$

$$|\langle x, \phi_i\rangle| = |\langle y, \phi_i\rangle|, \text{ for all } i = 1, 2, \ldots, n.$$

Φ yields

(ii) **phaseless reconstruction** if $x = \pm y$.
(iii) **norm retrieval** if $\|x\| = \|y\|$.

REMARK 2.4. It is easy to see that $\{\phi_i\}_{i=1}^n$ does phase retrieval (norm retrieval) if and only if $\{c_i\phi_i\}_{i=1}^n$ does phase retrieval (norm retrieval), for any non-zero scalars $\{c_i\}_{i=1}^n \subset \mathbb{R}^d$.

The paper [**1**] gives the minimal number of vectors needed in $\mathbb{R}^d$ to do phase retrieval.

THEOREM 2.5 ([**1**]). *In order for a frame $\{\phi_i\}_{i=1}^n$ in $\mathbb{R}^d$ to do phase retrieval, it is necessary that $n \geq 2d-1$.*

Also [**1**] presents a fundamental classification of the frames which do phase retrieval in $\mathbb{R}^d$. For this we need a definition.

DEFINITION 2.6 ([**1**]). A frame $\Phi = \{\phi_i\}_{i=1}^n$in $\mathbb{R}^d$ satisfies the **complement property** if for all subsets $I \subset \{1, 2, \ldots, n\}$, either $\{\phi_i\}_{i \in I}$ or $\{\phi_i\}_{i \in I^c}$ spans $\mathbb{R}^d$.

A fundamental result from [**1**] is:

THEOREM 2.7 ([**1**]). *A frame Φ does phaseless reconstruction in $\mathbb{R}^d$ if and only if it has the complement property.*

It follows that if $\Phi = \{\phi_i\}_{i=1}^n$ does phase retrieval in $\mathbb{R}^d$ then $n \geq 2d - 1$. Full spark is another important notion of vectors in frame theory. A formal definition is given below:

DEFINITION 2.8. Given a family of vectors $\Phi = \{\phi_i\}_{i=1}^n$ in $\mathbb{R}^d$, the **spark** of Φ is defined as the cardinality of the smallest linearly dependent subset of Φ. When $\text{spark}(\Phi) = d + 1$, every subset of size d is linearly independent, and in that case, Φ is said to be **full spark**.

We note that from the definitions it follows that full spark frames with $n \geq 2d-1$ vectors have the complement property and hence do phaseless reconstruction. Also, if $n = 2d - 1$ then the complement property clearly implies full spark.

We will need a generalization of phase retrieval to phase retrieval by projections.

DEFINITION 2.9. A family of subspaces $\{W_i\}_{i=1}^n$ (or respectively, their induced projections $\{P_i\}_{i=1}^n$) does

(1) **phase retrieval** on $\mathbb{R}^d$ if whenever $x, y \in \mathbb{R}^d$ satisfy
$$\|P_i x\| = \|P_i y\|, \text{ for all } i = 1, 2, \ldots, n,$$
then $x = \pm y$.

(2) **norm retrieval** if $\|x\| = \|y\|$.

We will need a result from [**6**].

PROPOSITION 2.10. Let projections $\{P_i\}_{i=1}^n$ do phase retrieval on $\mathbb{R}^d$. Then $\{(I - P_i)\}_{i=1}^n$ does phase retrieval if and only if it does norm retrieval.

We note the following result from [**6**]:

THEOREM 2.11 ([**6**]). *In $\mathbb{R}^d$, for any integers $1 \leq k_i \leq d-1$, there are subspaces $\{W_i\}_{i=1}^{2d-1}$ of $\mathbb{R}^d$ with dim $W_i = k_i$ and $\{W_i\}_{i=1}^{2d-1}$ does phase retrieval.*

The major open problem in the area of real phase retrieval is:

PROBLEM 2.12. What is the least number of subspaces needed to do phase retrieval on $\mathbb{R}^d$? What are the possible dimensions of these subspaces?

For notation we will use:

NOTATION 2.13. If $\Phi = \{\phi_i\}_{i=1}^n$ is a frame in $\mathbb{R}^d$, we denote the induced hyperplanes as $\Phi^\perp = \{\phi_i^\perp\}_{i=1}^n$.

3. Phase retrieval by hyperplanes

We will need a result of Edidin [**7**], which is also generalized in [**15**].

THEOREM 3.1. *Let $\{W_i\}_{i=1}^n$ be subspaces of $\mathbb{R}^d$ with respective projections $\{P_i\}_{i=1}^n$. The following are equivalent:*

(1) $\{W_i\}_{i=1}^n$ *does phase retrieval.*
(2) *For every* $0 \neq x \in \mathbb{R}^d$, $\text{span}\{P_i x\}_{i=1}^n = \mathbb{R}^d$.

We will show that for a scalable frame Φ, both Φ and $\Phi^\perp$ do norm retrieval. For this we need a proposition.

PROPOSITION 3.2. Let $\{W_i\}_{i=1}^n$ be proper subspaces of $\mathbb{R}^d$ with respective projections $\{P_i\}_{i=1}^n$. Then for any scalars $\{a_i\}_{i=1}^n \subset \mathbb{R}$ and $0 < A \in \mathbb{R}$, the following are equivalent.

(1) For every orthonormal basis $\{u_{i,j}\}_{j=1}^{n_i}$ of W_i, the set $\{a_i u_{i,j}\}_{i=1}^n{}_{j=1}^{n_i}$ is an A-tight frame.
(2) For some orthonormal basis $\{u_{i,j}\}_{j=1}^{n_i}$ of W_i, the set $\{a_i u_{i,j}\}_{i=1}^n{}_{j=1}^{n_i}$ is an A-tight frame.
(3) $\sum_{i=1}^n a_i^2 P_i = A \cdot I$.
(4) $\sum_{i=1}^n a_i^2 (I - P_i) = \left(\sum_{i=1}^n a_i^2 - A\right) \cdot I$.
(5) For every orthonormal basis $\{v_{i,j}\}_{j=1}^{d-n_i}$ of $W_i^\perp$, the set $\{a_i v_{i,j}\}_{i=1}^n{}_{j=1}^{d-n_i}$ is a $(\sum_{i=1}^n a_i^2 - A)$-tight frame.
(6) For some orthonormal basis $\{v_{i,j}\}_{j=1}^{d-n_i}$ of $W_i^\perp$, the set $\{a_i v_{i,j}\}_{i=1}^n{}_{j=1}^{d-n_i}$ is a $(\sum_{i=1}^n a_i^2 - A)$-tight frame.

PROOF. (1) $\Rightarrow$ (2). Obvious.

(2) $\Rightarrow$ (3). Let $\{u_{i,j}\}_{j=1}^{n_i}$ be the orthonormal basis of W_i in (2). Then for any $x \in \mathbb{R}^d$,

$$a_i^2 P_i x = \sum_{j=1}^{n_i} \langle x, a_i u_{i,j} \rangle a_i u_{i,j}.$$

Hence

$$\sum_{i=1}^n a_i^2 P_i x = \sum_{i=1}^n \sum_{j=1}^{n_i} \langle x, a_i u_{i,j} \rangle a_i u_{i,j} = Ax.$$

Therefore, $\sum_{i=1}^n a_i^2 P_i = A \cdot I$.

(3) $\Rightarrow$ (1) Let $\{u_{i,j}\}_{j=1}^{n_i}$ be any orthonormal basis of W_i. Then we have

$$a_i^2 P_i x = \sum_{j=1}^{n_i} \langle x, a_i u_{i,j} \rangle a_i u_{i,j}.$$

Hence

$$Ax = \sum_{i=1}^n a_i^2 P_i x = \sum_{i=1}^n \sum_{j=1}^{n_i} \langle x, a_i u_{i,j} \rangle a_i u_{i,j}.$$

So, $\{a_i u_{i,j}\}_{i=1}^n{}_{j=1}^{n_i}$ is an A-tight frame.

(3) $\Leftrightarrow$ (4). Obvious.

Similarly, (4), (5), (6) are equivalent, but we need to see that $\sum_{i=1}^n a_i^2 - A > 0$. This follows immediately from $\sum_{i=1}^n a_i^2 P_i = A \cdot I$ and $\{W_i\}_{i=1}^n$ are proper subspaces. □

PROPOSITION 3.3. If $\{W_i\}_{i=1}^n$ satisfies one of the conditions in Proposition 3.2, then both $\{W_i\}_{i=1}^n$ and $\{W_i^\perp\}_{i=1}^n$ do norm retrieval.

PROOF. The results follow from the fact that

$$\sum_{i=1}^{n} a_i^2\|P_ix\|^2 = \sum_{i=1}^{n}\langle a_iP_ix, a_iP_ix\rangle = \langle\sum_{i=1}^{n} a_i^2P_ix, x\rangle = A\|x\|^2.$$

The other case is similar. □

COROLLARY 3.4. If $\{W_i\}_{i=1}^n$ satisfies one of the conditions in Proposition 3.2, then $\{W_i\}_{i=1}^n$ does phase retrieval if and only if $\{W_i^\perp\}_{i=1}^n$ does phase retrieval.

PROOF. This follows from Proposition 2.10 and Proposition 3.3. □

COROLLARY 3.5. If $\Phi = \{\phi_i\}_{i=1}^n$ is a scalable frame in $\mathbb{R}^d$ then Φ does phase retrieval if and only if $\Phi^\perp$ does phase retrieval.

PROOF. If P_i is the projection onto span$\{\phi_i\}$ then for any $x \in \mathbb{R}^d$,

$$\|\phi_i\|^2P_ix = \langle x, \phi_i\rangle\phi_i.$$

Since Φ is scalable then there exist scalars $\{s_i\}_{i=1}^n$ such that $\{s_i\phi_i\}_{i=1}^n$ is an A-tight frame.

Therefore, for any $x \in \mathbb{R}^d$,

$$Ax = \sum_{i=1}^{n}\langle x, s_i\phi_i\rangle s_i\phi_i = \sum_{i=1}^{n} s_i^2\|\phi_i\|^2P_ix.$$

The result follows by Corollary 3.4. □

Now we will give examples to show that Corollary 3.5 does not hold in general without the assumption the frame being scalable. First, let us examine the obvious approach to see why it fails in general. It is known that if $\{\phi_i\}_{i=1}^n$ does phase retrieval and T is an invertible operator then $\{T\phi_i\}_{i=1}^n$ does phase retrieval. If $\Phi = \{\phi_i\}_{i=1}^n$ is any frame with frame operator S which does phase retrieval, $S^{-1/2}\Phi = \{S^{-1/2}\phi_i\}_{i=1}^n$ is a Parseval frame and so

$$\left\{\left(S^{-1/2}\phi_i\right)^\perp\right\}_{i=1}^n = \left\{S^{1/2}\phi_i^\perp\right\}_{i=1}^n,$$

does phase retrieval. So we would like to apply the invertible operator $S^{-1/2}$ to our hyperplanes to conclude that $\Phi^\perp$ does phase retrieval. The problem is that it is known [**6**] the invertible operators may not take subspaces doing phase retrieval to subspaces doing phase retrieval.

EXAMPLE 3.6. There is a frame $\{\phi_i\}_{i=1}^5$ in $\mathbb{R}^3$ which does phase retrieval but the hyperplanes $\{\phi_i^\perp\}_{i=1}^5$ fail phase retrieval.

PROOF. Let $\phi_1 = (0,0,1), \phi_2 = (1,0,1), \phi_3 = (0,1,1), \phi_4 = (1, 1-\sqrt{2}, 2), \phi_5 = (1,1,1)$.

Since $\{\phi_i\}_{i=1}^5$ is a full spark frame of 5 vectors in $\mathbb{R}^3$ then it does phase retrieval.

We have,

$$\begin{aligned}
W_1 &= \phi_1^\perp = \{(x_1,x_2,x_3) \in \mathbb{R}^3 : x_3 = 0\}\\
W_2 &= \phi_2^\perp = \{(x_1,x_2,x_3) \in \mathbb{R}^3 : x_1 + x_3 = 0\}\\
W_3 &= \phi_3^\perp = \{(x_1,x_2,x_3) \in \mathbb{R}^3 : x_2 + x_3 = 0\}\\
W_4 &= \phi_4^\perp = \{(x_1,x_2,x_3) \in \mathbb{R}^3 : x_1 + (1-\sqrt{2})x_2 + 2x_3 = 0\}\\
W_5 &= \phi_5^\perp = \{(x_1,x_2,x_3) \in \mathbb{R}^3 : x_1 + x_2 + x_3 = 0\}.
\end{aligned}$$

Let P_i be the orthogonal projection onto W_i. Then $P_1(\phi_5) = (1,1,0)$; $P_2(\phi_5) = (0,1,0)$; $P_3(\phi_5) = (1,0,0)$; $P_4(\phi_5) = (1/2,(1+\sqrt{2})/2,0)$; and $P_5(\phi_5) = (0,0,0)$. Thus, $\text{span}\{P_i(\phi_5)\}_{i=1}^5 = W_1 \neq \mathbb{R}^3$. By Theorem 3.1, $\{W_i\}_{i=1}^5$ cannot do phase retrieval. □

COROLLARY 3.7. There exists $\{\phi_i\}_{i=1}^5$ in $\mathbb{R}^3$ which does phase retrieval but $\{\phi_i^\perp\}_{i=1}^5$ cannot do norm retrieval.

Now we will generalize this example to all of $\mathbb{R}^d$, $d \geq 3$. This example looks like it came from nowhere, so we first explain why this is logical by *reverse engineering* the above example in $\mathbb{R}^d$. We need a full spark set of unit vectors $\{\phi_i\}_{i=1}^{2d-1}$ (which therefore do phase retrieval on $\mathbb{R}^d$) with projections P_i onto $\text{span}\{\phi_i\}$, and a vector x so that $\{(I-P_i)x\}_{i=1}^{2d-1}$ is contained in a hyperplane. So we decide in advance that the vector x will be $x = (1,1,\ldots,1)$ and the hyperplane will be

$$H = \{(c_1, c_2, \ldots, c_{d-1}, 0) : c_i \in \mathbb{R}\}.$$

Given a $\phi = (a_1, a_2, \ldots, a_d)$ of this type,

(1) We have:

$$\|\phi\|^2 = \sum_{i=1}^d a_i^2 = 1.$$

(2) We have:

$$\begin{aligned}
(I - P_\phi)x &= (1,1,\ldots,1) - \langle x, \phi\rangle \phi\\
&= (1,1,\ldots,1) - \left(\sum_{i=1}^d a_i\right)(a_1, a_2, \ldots, a_d).
\end{aligned}$$

Since this vector is to be in the hyperplane H, we have:

$$1 = a_d \sum_{i=1}^d a_i.$$

Combining this with (1) implies:

$$a_d = \frac{\sum_{i=1}^{d-1} a_i^2}{\sum_{i=1}^{d-1} a_i}.$$

Now we can present the example:

EXAMPLE 3.8. There are vectors $\{\phi_i\}_{i=1}^{2d-1}$ in $\mathbb{R}^d$ which do phase retrieval but $\{\phi_i^\perp\}_{i=1}^{2d-1}$ does not do phase retrieval.

PROOF. Consider the set

$$A := \left\{ \left(a_1, a_2, \ldots, a_{d-1}, \frac{\sum_{i=1}^{d-1} a_i^2}{\sum_{i=1}^{d-1} a_i} \right) : a_i \in \mathbb{R}, \sum_{i=1}^{d-1} a_i \neq 0 \right\}.$$

Let $x = (1, 1, \ldots, 1) \in \mathbb{R}^d$. Let any $\phi \in A$ and denote P_ϕ the orthogonal projection onto span$\{\phi\}$. Then we have

$$(I - P_\phi)(x) = x - \langle x, \frac{\phi}{\|\phi\|} \rangle \frac{\phi}{\|\phi\|}.$$

Denote b_d the $d-$ coordinate of $(I - P_\phi)(x)$, then

$$b_d = 1 - \frac{1}{\|\phi\|^2} \left(\sum_{i=1}^{d-1} a_i + \frac{\sum_{i=1}^{d-1} a_i^2}{\sum_{i=1}^{d-1} a_i} \right) \frac{\sum_{i=1}^{d-1} a_i^2}{\sum_{i=1}^{d-1} a_i} = 1 - \frac{1}{\|\phi\|^2} \left(\sum_{i=1}^{d-1} a_i^2 + \left(\frac{\sum_{i=1}^{d-1} a_i^2}{\sum_{i=1}^{d-1} a_i} \right)^2 \right) = 0.$$

Let $\phi_1 = (1, 0, \ldots, 0, 1)$; $\phi_2 = (0, 1, \ldots, 0, 1)$; $\cdots$ $\phi_{d-1} = (0, 0, \ldots, 1, 1)$; and $\phi_d = x = (1, 1, \ldots, 1, 1)$. Then $\{\phi_i\}_{i=1}^d$ is a linearly independent set in $\mathbb{R}^d$ and $\{\phi_i\}_{i=1}^d \subset A$.

Now we will show that for any finite hyperplanes $\{W_i\}_{i=1}^k$ in $\mathbb{R}^d$, there exists a vector $\phi \in A$ such that $\phi \notin \cup_{i=1}^k W_i$. Suppose by a contradiction that $A \subset \cup_{i=1}^k W_i$.

Consider the set

$$B := \left\{ \left(x, x^2, \ldots, x^{d-2}, 1 - \sum_{i=1}^{d-2} x^i, \sum_{i=1}^{d-2} x^{2i} + \left(1 - \sum_{i=1}^{d-2} x^i \right)^2 \right) : x \in \mathbb{R} \right\},$$

then $B \subset A$.

Hence $B \subset \cup_{i=1}^k W_i$. Therefore, there exists $j \in \{1, \ldots, k\}$ such that W_j contains infinitely many vectors in B.

Let $u = (u_1, u_2, \ldots, u_d) \in W_j^\perp, u \neq 0$. Then we have

$$\langle u, \phi_x \rangle = 0$$

for infinitely many $\phi_x \in B$.

Thus,

$$\sum_{i=1}^{d-2} u_i x^i + u_{d-1} \left(1 - \sum_{i=1}^{d-2} x^i \right) + u_d \left(\sum_{i=1}^{d-2} x^{2i} + \left(1 - \sum_{i=1}^{d-2} x^i \right)^2 \right) = 0,$$

for infinitely many x.

This implies $u_1 = u_2 = \cdots = u_d = 0$, which is a contradiction.

From above, we can pick $d-1$ vectors $\{\phi_i\}_{i=d+1}^{2d-1}$ in B such that $\{\phi_i\}_{i=1}^{2d-1}$ is a full spark of vectors in $\mathbb{R}^d$. Thus, $\{\phi_i\}_{i=1}^{2d-1}$ does phase retrieval in $\mathbb{R}^d$.

Moveover, since span$\{(I - P_i)(x)\}_{i=1}^{2d-1} \neq \mathbb{R}^d$ then $\{\phi_i^\perp\}_{i=1}^{2d-1}$ cannot do phase retrieval by Theorem 3.1. □

In general, if hyperplanes $\{W_i\}_{i=1}^n$ do phase retrieval in $\mathbb{R}^d$, it does not ensure that the complement vectors do phase retrieval. The following is an example.

EXAMPLE 3.9. There are 5 vectors $\{\phi_i\}_{i=1}^5$ in $\mathbb{R}^3$ which fail phase retrieval but their induced hyperplanes $\{\phi_i^\perp\}_{i=1}^5$ do phase retrieval.

PROOF. In $\mathbb{R}^3$, let $W_1 = \text{span}\{e_2, e_3\}$; $W_2 = \text{span}\{e_1, e_3\}$; $W_3 = \text{span}\{e_1 + e_2, e_3\}$; $W_4 = \text{span}\{e_1, e_2 + e_3\}$; and $W_5 = \text{span}\{e_2, e_1 + e_3\}$.

Let P_i be the projection onto W_i. Then for any $x = (x_1, x_2, x_3)$, we have

$$P_1x = (0, x_2, x_3), \quad P_2x = (x_1, 0, x_3)$$

$$P_3x = \left(\frac{x_1 + x_2}{2}, \frac{x_1 + x_2}{2}, x_3\right), \quad P_4x = \left(x_1, \frac{x_2 + x_3}{2}, \frac{x_2 + x_3}{2}\right)$$

$$P_5x = \left(\frac{x_1 + x_3}{2}, x_2, \frac{x_1 + x_3}{2}\right).$$

For any $x \neq 0$, the rank of the matrix whose the rows are P_ix equals 3. Therefore, $\{P_ix\}_{i=1}^5$ spans $\mathbb{R}^3$. By Theorem 3.1, $\{W_i\}_{i=1}^5$ does phase retrieval in $\mathbb{R}^3$.

We also have

$$W_1^\perp = \text{span}\{e_1\}, \quad W_2^\perp = \text{span}\{e_2\}, \quad W_3^\perp = \text{span}\{u_3\},$$

$$W_4^\perp = \text{span}\{u_4\}, \quad W_5^\perp = \text{span}\{u_5\},$$

for some $u_3, u_4, u_5 \in \mathbb{R}^3$.

Since $e_1, e_2, u_3 \perp e_3$ then $\text{span}\{e_1, e_2, u_3\} \neq \mathbb{R}^3$. Thus, $\{e_1, e_2, u_3, u_4, u_5\}$ fails the complement property. Therefore, it cannot do phase retrieval. □

4. An example in $\mathbb{R}^4$

In this section we will give an example of 6 hyperplanes in $\mathbb{R}^4$ which do phase retrieval. First, we will show that this is the minimal number of hyperplanes which can do phase retrieval.

THEOREM 4.1. *If hyperplanes $\{W_i\}_{i=1}^n$ do phase retrieval in $\mathbb{R}^d$ then $n \geq 2d-2$. Moreover, if $n = 2d - 2$ then the vectors $\{W_i^\perp\}_{i=1}^{2d-2}$ are full spark.*

PROOF. Assume, by way of contradiction, that $n \leq 2d - 3$. Choose a vector

$$0 \neq x \in \cap_{i=1}^{d-1} W_i.$$

So $P_ix = x$ for all $i = 1, 2, \ldots, d-1$. It follows that the set $\{P_ix\}_{i=1}^n$ has at most $d-1$ non-zero vectors and hence cannot span $\mathbb{R}^d$, contradicting Theorem 3.1.

For the *moreover* part, we proceed by way of contradiction. Let $W_i^\perp = \text{span}\{\phi_i\}$ for $i = 1, 2, \ldots, 2d-2$ and assume there exists $I \subset [2d-2]$ with $|I| = d$ and $\{\phi_i\}_{i \in I}$ does not span the whole space. Choose $0 \neq x \perp \phi_i$ for all $i \in I$. It follows that $x \in W_i$ for all $i \in I$ and so $P_ix = x$ for all $i \in I$. But, $|I^c| = d - 2$ and so $\{P_ix\}_{i=1}^{2d-2}$ contains at most $d-1$ distinct elements and so cannot span, contradicting Theorem 3.1. □

Now we are ready for the main result of this section. In [**16**] it was shown that there are six 2-dimensional subspaces of $\mathbb{R}^4$ which do phase retrieval. We will now extend this result to hyperplanes in $\mathbb{R}^4$.

THEOREM 4.2. *Suppose that $d = 4$. There exist 6 hyperplanes $W_1, \ldots, W_6 \subset \mathbb{R}^4$ which do phase retrieval on $\mathbb{R}^4$.*

PROOF. Set

$$W_j := \phi_j^{\perp} \subset \mathbb{R}^4, \quad j = 1, \ldots, 6,$$

where

$$\begin{aligned}\phi_1 &= (2,-1,2,2)/\sqrt{13},\ \phi_2 = (2,5,4,1)/\sqrt{46},\ \phi_3 = (0,4,-1,-1)/\sqrt{18},\\ \phi_4 &= (5,4,-2,-4)/\sqrt{61},\ \phi_5 = (4,1,5.3)/\sqrt{51},\ \phi_6 = (3,-4,-4,-3)/\sqrt{50}.\end{aligned}$$

Note that $\|P_j x\|^2 = \mathrm{Tr}(A_j X)$ where $X = xx^T$, $A_j = u_j u_j^T + v_j v_j^T + \omega_j \omega_j^T$ and $u_j, v_j, \omega_j \in \mathbb{R}^4$ is an orthonormal basis of W_j. Then $W_1, \ldots, W_6$ do phase retrieval if and only if

$$\mathcal{Z} := \{Q \in \mathbb{R}^4 : Q = Q^T, \mathrm{rank}(Q) \le 2, Tr(A_j Q) = 0, j = 1, \ldots, 6\}$$

only contains zero matrix. We write Q in the form of $Q = (x_{jk})_{j,k=1}^4$ where $x_{kj} = x_{jk}$ for all $1 \le k, j \le 4$, where $x_{jk}, 1 \le j \le k \le 4$, are 10 variables. The $\mathrm{rank}(Q) \le 2$ if and only if $m_{j,k} := m_{j,k}(x_{11}, x_{12}, \ldots, x_{44}) = 0$ where $m_{j,k}$ denotes the determinant of the sub-matrix formed by deleting the jth row and kth column from the matrix Q. Noting that $A_j = I - \phi_j \phi_j^T$, we have

$$\ell_j := \ell_j(x_{11}, \ldots, x_{44}) := Tr(A_j Q) = Tr(Q) - \phi_j^T Q \phi_j.$$

The $\mathcal{Z}$ only contains the zero matrix if and only if the homogeneous polynomial system

$$\ell_1 = \cdots = \ell_6 = m_{1,1} = \cdots = m_{4,4} = 0 \tag{4.1}$$

has no non-trivial (i.e., non-zero) real solutions.

We next verify the polynomial system (4.1) only has real zero solution following the ideas of Vinzant [**14**, Theorem 1]. Using the computer algebra software `Maple`, we compute a Gröbner basis of the ideal

$$\langle \ell_1, \ldots, \ell_6, m_{1,1}, \ldots, m_{4,4} \rangle$$

and elimination (see [**13**]). The result is a polynomial $f_0 \in \mathbb{Q}[x_{34}, x_{44}]$, which is a homogeneous polynomial of degree 10:

$$\begin{aligned} f_0 &= c_{10} x_{4,4}^{10} + c_9 x_{4,4}^9 x_{3,4} + c_8 x_{4,4}^8 x_{3,4}^2 + c_7 x_{4,4}^7 x_{3,4}^3 + c_6 x_{4,4}^6 x_{3,4}^4 \\ &\quad + c_5 x_{4,4}^5 x_{3,4}^5 + c_4 x_{4,4}^4 x_{3,4}^6 + c_3 x_{4,4}^3 x_{3,4}^7 + c_2 x_{4,4}^2 x_{3,4}^8 + c_1 x_{4,4} x_{3,4}^9 + c_0 x_{3,4}^{10}, \end{aligned}$$

where

$c_{10} = 6158368146944401257559417502053559572598060554305329739568779 00,$

$c_9 = -88497259445238795884856247314424179703069776451922820509818 3524,$

$c_8 = +375495105627626896030324796105779806146849701151805087612126 02923,$

$c_7 = -2617842892452520683425111576738689980030770359229357584545688 69970,$

$c_6 = +13186463610143742038055954937168015374620839229188399654359011 51518,$

$c_5 = +23236725037290134712712186115418226060873143131038552222668872 57194,$

$c_4 = +8410996559292025399905068706483499389429274202255882749684672 86492,$

$c_3 = +24531184661387436242724764944997332563822672346953985098573154 58204,$

$c_2 = +26867026353615602035620126806679118345824764445881244783119660 09776,$

$c_1 = +5987247506697840627080058242507159240327313046306355233940526 2912,$

$c_0 = +9504840500329006177437937293743836329176142273561737549053687 87200.$

We can verify that the univariate polynomial $f_0(1, x_{4,4})$ has no real zero roots using Sturm sequence, and hence

$$\{(x_{3,4}, x_{4,4}) \in \mathbb{R}^2 : f_0(x_{3,4}, x_{4,4}) = 0\} = \{(0,0)\},$$

which implies that if $(x_{1,1}, x_{1,2}, \ldots, x_{3,4}, x_{4,4})$ is a real solution of (4.1) then $x_{3,4} = x_{4,4} = 0$. By computing a Gröbner basis of the ideal, we obtain that

$$1 \in \langle x_{3,4}, x_{4,4}, x_{j,k} - 1, \ell_1, \ldots, \ell_6, m_{1,1}, \ldots, m_{4,4} \rangle, \quad 1 \leq j \leq k \leq 4$$

which means that (4.1) does not have nonzero real roots with $x_{3,4} = x_{4,4} = 0$. The maple code for these computation is posted at http://lsec.cc.ac.cn/~xuzq/phase.htm. Combining results above, we obtain that (4.1) has no non-trivial real solutions. □

COROLLARY 4.3. There are six hyperplanes $\{W_i\}_{i=1}^6$ doing phase retrieval on $\mathbb{R}^4$ but $\{W_i^\perp\}_{i=1}^6$ does not do phase retrieval.

What happens in higher dimensional spaces? One already knows, when $d = 5$, it takes at least $2d - 1 = 9$ hyperplanes to do phase retrieval in $\mathbb{R}^5$ [**7**]. Hence, the next dimension that we need to consider is $d = 6$. To extend the proof of Theorem 4.2 to the case where $d = 6$, we need to use Maple to compute a Grobner basis of an ideal which is generated by 410 polynomials (when $d = 4$, the ideal is generated by 22 polynomials). We tried to do that but Maple does not work for such a large system. So we have not been able so far to extend this example past $d = 6$.

References

[1] Radu Balan, Pete Casazza, and Dan Edidin, *On signal reconstruction without phase*, Appl. Comput. Harmon. Anal. **20** (2006), no. 3, 345–356, DOI 10.1016/j.acha.2005.07.001. MR2224902

[2] R. H. Bates and D. Mnyama, *The status of practical Fourier phase retrieval*, Advances in Electronics and Electron Physics, **67** (1986), 1-64.

[3] C. Becchetti and L. P. Ricotti, *Speech recognition theory and C++ implementation*, Wiley (1999).

[4] Sara Botelho-Andrade, Peter G. Casazza, Hanh Van Nguyen, and Janet C. Tremain, *Phase retrieval versus phaseless reconstruction*, J. Math. Anal. Appl. **436** (2016), no. 1, 131–137, DOI 10.1016/j.jmaa.2015.11.045. MR3440085

[5] Saeid Bahmanpour, Jameson Cahill, Peter G. Casazza, John Jasper, and Lindsey M. Woodland, *Phase retrieval and norm retrieval*, Trends in harmonic analysis and its applications, Contemp. Math., vol. 650, Amer. Math. Soc., Providence, RI, 2015, pp. 3–14. MR3441730

[6] Jameson Cahill, Peter G. Casazza, Jesse Peterson, and Lindsey Woodland, *Phase retrieval by projections*, Houston J. Math. **42** (2016), no. 2, 537–558. MR3529968

[7] Dan Edidin, *Projections and phase retrieval*, Appl. Comput. Harmon. Anal. **42** (2017), no. 2, 350–359, DOI 10.1016/j.acha.2015.12.004. MR3591243

[8] J. Drenth, *Principles of protein x-ray crystallography*, Springer, 2010.

[9] J. R. Fienup, *Reconstruction of an object from the modulus of its fourier transform*, Optics Letters, **3** (1978), 27-29.

[10] J. R. Fienup, *Phase retrieval algorithms: A comparison*, Applied Optics, **21** (15) (1982), 2758-2768.

[11] L. Rabiner, and B. H. Juang, *Fundamentals of speech recognition*, Prentice Hall Signal Processing Series (1993).

[12] Joseph M. Renes, Robin Blume-Kohout, A. J. Scott, and Carlton M. Caves, *Symmetric informationally complete quantum measurements*, J. Math. Phys. **45** (2004), no. 6, 2171–2180, DOI 10.1063/1.1737053. MR2059685

[13] Bernd Sturmfels, *What is ... a Gröbner basis?*, Notices Amer. Math. Soc. **52** (2005), no. 10, 1199–1200. MR2186903

[14] Cynthia Vinzant, *A small frame and a certificate of its injectivity*, Sampling Theory and Applications (SampTA) Conference Proceedings. (2015)197-200.

[15] Yang Wang and Zhiqiang Xu, *Generalized phase retrieval: measurement number, matrix recovery and beyond*, Available online: arXiv:1605.08034.

[16] Z. Xu, *The minimal measurement number for low-rank matrix recovery*, Appl. Comput. Harmon. Anal. (2017), http://dx.doi.org/10.1016/j.acha.2017.01.005 .

Department of Mathematics, University of Missouri, Columbia, Missouri 65211-4100
Email address: sandrade102087@gmail.com

Department of Mathematics, University of Missouri, Columbia, Missouri 65211-4100
Email address: Casazzap@missouri.edu

Department of Mathematics, University of Missouri, Columbia, Missouri 65211-4100
Email address: chengdesai@gmail.com

Department of Mathematics, University of Missouri, Columbia, Missouri 65211-4100
Email address: terraformthedreamscape@gmail.com

Department of Mathematics, University of Missouri, Columbia, Missouri 65211-4100
Email address: tinmizzou@gmail.com

Department of Mathematics, University of Missouri, Columbia, Missouri 65211-4100
Email address: Tremainjc@missouri.edu

LSEC, Inst. Comp. Math., Academy of Mathematics and System Science, Chinese Academy of Sciences, Beijing, 100091, China
Email address: xuzq@lsec.cc.ac.cn

Contemporary Mathematics
Volume **706**, 2018
http://dx.doi.org/10.1090/conm/706/14210

Tight and full spark Chebyshev frames with real entries and worst-case coherence analysis

David Ellis, Eric Hayashi, and Shidong Li

ABSTRACT. Constructions of tight and full spark Chebyshev frames from truncations of Vandermonde-like matrices of orthogonal polynomials are presented. These are frames with real entries. Worst case coherence analysis is also carried out. We show that for sufficiently high degree, the minimum angle between distinct frame vectors is bounded below by about 44 degrees. In a related study, we also provide a worst case coherence analysis for equal norm tight $M \times N$ frames from truncated DFT matrix. The cosine of the smallest angle between these distinct frame elements is asymptotic to $\frac{\sin(\pi\alpha)}{\pi\alpha}$ where $\alpha = M/N$.

1. Introduction

A set of vectors $\{x_n\}_{n=1}^N$ in an M-dimensional inner product space $\mathbb{H}$, with $N \geq M$, is called a *frame* for $\mathbb{H}$ if there exist $0 < A \leq B < \infty$ such that

$$\forall f \in \mathbb{H}, \quad A\|f\|^2 \leq \sum_{n=1}^{N} |\langle f, x_n\rangle|^2 \leq B\|f\|^2.$$

Studies and constructions of finite frames with particular optimality criteria turn out to be exceedingly deep and broad. Tight and equal norm frames may have been the start of such efforts, exemplified by the work of Benedetto and Fickus on the notion of frame potentials and the role of uniform norm tight frames as their minimizers [**4**]. This was followed by many works, e.g., [**12**], [**13**], [**14**], [**18**], [**21**], and applications of tight frames.

Applications of tight frames range from optimal signal reconstructions with erasures, for instance, [**5**], [**11**], [**18**], [**20**], [**22**], [**23**], [**24**], and [**25**], to optimal coding schemes in communications, e.g., [**2**], [**3**].

The study of *full spark frames* (or *maximally robust frames*) is yet another thread perhaps going back to the work of [**27**]. Full spark frames play vital roles in applications such as sparse representation [**15, 26**] and compressed sensing [**7**–**10**, **16**]. The term of *spark* of a frame matrix $\mathbb{D}$ of (column) vectors was first introduced

2010 *Mathematics Subject Classification.* Primary 94A12, 42C15.

Key words and phrases. Frames, tight frames, equal norm tight frames, full spark frames, coherence, Vandermonde-like matrices, Chebyshev polynomials.

The third author was partially supported by the NSF grant DMS-1010058, DMS-1313490, AFOSR grant AF9550-11-1-02, and National key R&D grant of China, 2016YEE0200400, Shaanxi key R&D grant 2017KW-ZD-12.

by Donoho and Elad in [**17**] as the smallest number of linearly dependent columns of $\mathbb{D}$. It was shown in [**17**] that the sparsity of x must be less than one half of the spark of $\mathbb{D}$ in order to have the unique sparsest solution in the linear system $y = \mathbb{D}x$. As a result, the larger the spark is of a system/frame matrix $\mathbb{D}$, the easier it is to find a unique sparsest solution x to the system $y = \mathbb{D}x$. Thus, full spark frames comprise an important class of frames.

Subsequent studies of sparse solutions to underdetermined systems $y = \mathbb{D}x$ considered the notion of low *worst-case coherence*, e.g., [**6**]. The worst-case coherence μ of a frame matrix $\mathbb{D}$ is defined as the largest correlation between normalized (column) frame elements: $\mu \equiv \max_{i \neq j} |\langle d_i, d_j \rangle|$. For example it was shown in [**28**] that low coherence is instrumental in delivering the unique sparsest solution by ℓ_1-minimization procedures.

The threads of study mentioned above combine to make a case for the construction of frames that simultaneously are tight, equal normed, full spark, (see [**27**]) and have low worst-case coherence (see [**1**]). The methods of construction are circumscribed by the fact that testing a general frame matrix for the full spark property is NP-hard (see [**1**]).

A simple method to construct a full spark frame is to take the columns of a truncated non-singular Vandermonde matrix. In [**1**], it was shown that when $N \geq 2M$, the only equal norm tight $M \times N$ frames arising from truncated Vandermonde matrices are those with nodes on the unit circle.

In a related study, we observed that Vandermonde-like matrices of orthogonal polynomials can be used in the construction of real, tight and full spark frames through truncations. Particular attention is also given to the real-valued frames, as such constructions are still scarce to the best of our knowledge. To understand the properties of these classes of tight and full spark frames, we carry out coherence analyses for several classes of real tight frames. First we show that certain $M \times 2M$ truncates of Vandermonde-like frames associated with Chebyshev polynomials are tight and have moderate worst-case coherence that compare favorably with complex $M \times 2M$ Vandermonde frames. For sufficiently high degree, the minimum angle between distinct frame vectors is bounded below by about 44 degrees. And although the norms of the frame vectors are not equal, their ratios lie in intervals of the form $\left[c^{-1} - \varepsilon, c + \varepsilon\right]$ where $c = \sqrt{\frac{\pi+2}{\pi-2}} \approx 2.1222$ and ε can be made arbitrarily small for suitably large M. Next we give a coherence analysis for $M \times N$ frames obtained from truncated discrete Fourier transform (DFT) matrices. We show that the cosine of the smallest angle between distinct frame elements is asymptotic to $\frac{\sin(\pi\alpha)}{\pi\alpha}$ where $\alpha = M/N$. Finally we consider two constructions of real equal norm tight $M \times N$ frames that exhibit the aforementioned asymptotic worst-case coherence.

2. Real Vandermonde-like matrices

Following Gautschi [**19**], we define a Vandermonde-like matrix to be a square matrix of the form

$$P_n(x_1, x_2, ..., x_n) = \begin{bmatrix} p_0(x_1) & p_0(x_2) & \cdots & p_0(x_n) \\ p_1(x_1) & p_1(x_2) & \cdots & p_1(x_n) \\ \vdots & \vdots & & \vdots \\ p_{n-1}(x_1) & p_{n-1}(x_2) & \cdots & p_{n-1}(x_n) \end{bmatrix}$$

where p_0, p_1, p_2, $\cdots$ is a sequence of orthogonal polynomials with respect to some weight on an interval $[a,b]$, and $\{x_1, x_2, ..., x_n\}$ are the zeros of p_n. For $k = 0, 1, 2, \cdots$, we can write

$$p_k(x) = \sum_{j=0}^{k} a_{kj} x^j$$

with $a_{kk} \neq 0$. Applying elementary row operations, it is easily seen that

$$\det P_n(x_1, x_2, ..., x_n) = \left(\prod_{0 \le k < n} a_{kk} \right) \det \begin{bmatrix} 1 & 1 & \cdots & 1 \\ x_1 & x_2 & \cdots & x_n \\ x_1^2 & x_2^2 & \cdots & x_n^2 \\ \vdots & \vdots & & \vdots \\ x_1^{n-1} & x_2^{n-1} & \cdots & x_n^{n-1} \end{bmatrix}$$

which is a non-zero multiple of a standard Vandermonde determinant. Thus when the nodes are distinct, $\det P_n$ is non-zero. The following proposition is an immediate corollary of this observation

PROPOSITION 1. *Let $\{p_k\}_{k=0}^n$ be a sequence of orthogonal polynomials. Let $x_1, x_2, ..., x_n$ be the zeroes of p_n. Fix $m \le n$ and define the $m \times n$ matrix D by*

$$D = \begin{bmatrix} p_0(x_1) & p_0(x_2) & \cdots & p_0(x_n) \\ p_1(x_1) & p_1(x_2) & \cdots & p_1(x_n) \\ \vdots & \vdots & & \vdots \\ p_{m-1}(x_1) & p_{m-1}(x_2) & \cdots & p_{m-1}(x_n) \end{bmatrix}.$$

Then the columns of D form a full spark frame on $\mathbb{R}^m$, that is, every $m \times m$ submatrix of D is non-singular.

We consider the case where $\{p_k\}$ are normalized with respect to weight w on $[a,b]$. Let $\{x_k\}_{k=1}^n$ denote the zeroes of p_n. It is well-known that these zeroes are simple, and lie in the interval $[a,b]$. Let $\{\lambda_k\}$ denote the corresponding sequence of Christoffel numbers. Then

$$\sum_{s=1}^{n} \lambda_s p_j(x_s) p_k(x_s) = \delta_{jk} \text{ when } 0 \le j, k \le n-1.$$

In the case that $\{p_k\}$ are the normalized Chebyshev polynomials of the first kind, (with weight $w(x) = (1-x^2)^{-1/2}$ on $[-1,1]$), $\lambda_1 = \lambda_2 = \cdots = \lambda_n$, which implies that the rows of D are orthogonal and have equal norm. This means that the columns of D form a tight frame. Another benefit of using Chebyshev polynomials is that $p_j(x_k)$, the entries of D, are easily calculated.

2.1. The tight and full spark Chebyshev frame. Let $M \ge 1, N = 2M$, and

$$T_m(x) := \sum_{k=0}^{\lfloor m/2 \rfloor} \binom{m}{2k} \left(x^2 - 1\right)^k x^{m-2k}$$

be the Chebyshev polynomial of degree m of the first kind.

We replace T_0 by $1/\sqrt{2}$ so that T_m for $m = 0, 1, 2, ..$ have the same norm with respect to the weight $\frac{1}{\sqrt{1-x^2}}$. It is well-known that

$$T_m(\cos(\theta)) = \begin{cases} \cos(m\theta), & \text{for } m = 1, 2, 3, ... \\ \frac{1}{\sqrt{2}}, & \text{for } m = 0. \end{cases}$$

Let $x_k = x_{k,N} = \cos\left[\frac{\pi(2k-1)}{2N})\right]$, the kth zero of T_N, $k = 1, 2, \ldots, N$. For convenience, we write $\theta_k = \frac{\pi}{2N}(2k-1)$. Define the real Chebyshev frame to be, in a matrix form whose columns are frame vectors,

$$D(M,N) = \begin{bmatrix} T_0(x_1) & T_0(x_2) & \cdots & T_0(x_N) \\ T_1(x_1) & T_1(x_2) & \cdots & T_1(x_N) \\ \vdots & \vdots & & \vdots \\ T_{M-1}(x_1) & T_{M-1}(x_2) & \cdots & T_{M-1}(x_N) \end{bmatrix}.$$

Then
(2.1)

$$D(M,N) = \begin{bmatrix} 1/\sqrt{2} & 1/\sqrt{2} & \cdots & 1/\sqrt{2} \\ \cos(\theta_1) & \cos(\theta_2) & \cdots & \cos(\theta_N) \\ \cos(2\theta_1) & \cos(2\theta_2) & & \cos(2\theta_N) \\ \vdots & \vdots & & \vdots \\ \cos((M-1)\theta_1) & \cos((M-1)\theta_2) & \cdots & \cos((M-1)\theta_N) \end{bmatrix}.$$

2.2. Worst case coherence in the $M \times 2M$ case. Let

$$R_{j,k} = \sum_{m=0}^{M-1} T_m(x_j)T_m(x_k),\ 1 \le j \le k \le N.$$

THEOREM 1.

$$\lim_{N\to\infty} \max\left\{\frac{R_{j,k}^2}{R_{j,j}R_{k,k}},\ 1 \le j < k \le N\right\} \le \frac{4}{\left(\pi - \frac{2}{3}\right)\pi} \approx 0.71726$$

PROOF.

$$\begin{aligned} R_{j,k} &= \sum_{m=0}^{M-1} T_m(x_j)T_m(x_k), 1 \le j \le k \le N. \\ &= \tfrac{1}{2} + \sum_{m=1}^{M-1} \cos(m\theta_j)\cos(m\theta_k) \\ &= \tfrac{1}{2} + \tfrac{1}{2}\sum_{m=1}^{M-1} [\cos(m(\theta_j+\theta_k)) + \cos(m(\theta_j-\theta_k))] \\ &= -\tfrac{1}{2} + \tfrac{1}{2}\mathrm{Re}\sum_{m=0}^{M-1}\left[e^{im(\theta_j+\theta_k)} + e^{im(\theta_j-\theta_k)}\right] \\ &= -\tfrac{1}{2} + \tfrac{1}{2}\mathrm{Re}\left\{\frac{1-e^{iM(\theta_j+\theta_k)}}{1-e^{i(\theta_j+\theta_k)}} + \frac{1-e^{iM(\theta_j-\theta_k)}}{1-e^{i(\theta_j-\theta_k)}}\right\}. \end{aligned}$$

Standard manipulations give

$$\begin{aligned} R_{j,k} &= -\tfrac{1}{2} + \tfrac{1}{2}\mathrm{Re}\left\{ \frac{\sin\left[\frac{M}{2}(\theta_j+\theta_k)\right]}{\sin\left[\frac{1}{2}(\theta_j+\theta_k)\right]} e^{\left(\frac{M}{2}-\frac{1}{2}\right)i(\theta_j+\theta_k)} \right. \\ &\quad \left. + \frac{\sin\left(\frac{M}{2}(\theta_j-\theta_k)\right)}{\sin\left[\frac{1}{2}(\theta_j-\theta_k)\right]} e^{\left(\frac{M}{2}-\frac{1}{2}\right)i(\theta_j-\theta_k)} \right\} \\ &= -\tfrac{1}{2} + \tfrac{1}{2}\frac{\sin\left[\frac{M}{2}(\theta_j+\theta_k)\right]}{\sin\left[\frac{1}{2}(\theta_j+\theta_k)\right]} \cos\left[\tfrac{M-1}{2}(\theta_j+\theta_k)\right] \\ &\quad + \tfrac{1}{2}\frac{\sin\left[\frac{M}{2}(\theta_j-\theta_k)\right]}{\sin\left[\frac{1}{2}(\theta_j-\theta_k)\right]} \cos\left[\tfrac{M-1}{2}(\theta_j-\theta_k)\right] \\ &= \tfrac{1}{4}\sin(M(\theta_j+\theta_k))\cot\left[\tfrac{\theta_j+\theta_k}{2}\right] - \tfrac{1}{4}\cos\left(M(\theta_j+\theta_k)\right) \\ &\quad + \tfrac{1}{4}\sin(M(\theta_j-\theta_k))\cot(\frac{1}{2}(\theta_j-\theta_k)) - \tfrac{1}{4}\cos(M(\theta_j-\theta_k)). \end{aligned}$$

Substituting $\theta_j = \frac{\pi}{2N}(2j-1)$ yields

$$\begin{aligned} R_{j,k} &= \tfrac{1}{4}\sin(\frac{\pi}{2}(j+k-1))\cot(\frac{\pi}{2N}(j+k-1)) - \tfrac{1}{4}\cos(\frac{\pi}{2}(j+k-1)) \\ &\quad + \frac{1}{4}\sin(\frac{\pi}{2}(j-k))\cot(\frac{\pi}{2N}(j-k)) - \tfrac{1}{4}\cos(\frac{\pi}{2}(j-k)). \end{aligned} \tag{2.2}$$

For $k = j$

$$\begin{aligned} R_{j,j} &= \tfrac{1}{4}\sin\left[\frac{\pi}{2}(2j-1)\right]\cot\left[\frac{\pi}{2N}(2j-1)\right] \\ &\quad - \tfrac{1}{4}\cos\left[\tfrac{\pi}{2}(2j-1)\right] + \tfrac{N}{4} - \tfrac{1}{4} \\ &= \tfrac{1}{4}(N-1) + \tfrac{1}{4}\sin\left[\frac{\pi}{2}(2j-1)\right]\cot\left[\frac{\pi}{2N}(2j-1)\right]. \end{aligned}$$

Thus

$$R_{j,j} = \tfrac{N-1}{4} + \tfrac{1}{4}(-1)^{j+1}\cot\left[\tfrac{\pi}{2N}(2j-1)\right]. \tag{2.3}$$

In finding the minimum angle between columns, we can assume WLOG that $1 \le j < k \le M$. To see this, suppose $1 \le j < k \le N$. Let

$$j' = N - j + 1 \text{ and } k' = N - k + 1. \tag{2.4}$$

Since N is even, j and j' always have opposite parity, namely, if one is even, the other must be odd. It is easily checked that for $1 \le j \le N$,

$$\begin{aligned} 4R_{j,j} &= (-1)^{j'+1}\cot\left(\left(\frac{\pi}{2N}(2j'-1)\right)\right) + N - 1 \\ &= 4R_{j',j'}. \end{aligned} \tag{2.5}$$

Also, using the identity $\cot(t+\pi/2) = -\tan(t)$, we get

$$\begin{aligned} 4R_{j,k} &= (-1)^{M+1}\sin\left(\frac{\pi}{2}(j-k')\right)\tan\left(\frac{\pi}{2N}(j-k')\right) - (-1)^M\cos\left(\frac{\pi}{2}(j-k')\right) + \\ &\quad (-1)^{M+1}\sin\left(\frac{\pi}{2}(j+k'-1)\right)\tan\left(\left(\frac{\pi}{2N}(j+k'-1)\right)\right) - \\ &\quad (-1)^M\cos\left(\frac{\pi}{2}(j+k'-1)\right). \end{aligned} \tag{2.6}$$

With the two replacements in (2.4), we get:

$$\begin{aligned}4R_{j,k} &= (-1)^{2M}\sin\left(\frac{\pi}{2}(j'-k')\right)\cot\left(\frac{\pi}{2N}(j'-k')\right)-(-1)^{2M}\cos\left(\frac{\pi}{2}(j'-k')\right)+\\ &\quad (-1)^{2M}\sin\left(\frac{\pi}{2}(j'+k'-1)\right)\cot\left(\left(\frac{\pi}{2N}(j'+k'-1)\right)\right)-\\ &\quad (-1)^{2M}\cos\left(\frac{\pi}{2}(j'+k'-1)\right)=4R_{j',k'}.\end{aligned}$$

So if j and k both exceed $M-1$, we replace them by j' and k' without changing the angle. If $1\le j<M\le k\le N$, replace k by k'. Thus, we have in all cases (noting that either $\sin\left(\frac{\pi}{2}(j+k-1)\right)$ or $\sin\left(\frac{\pi}{2}(j-k)\right)$ must vanish) that

$$|4R_{j,k}|\le\cot\left(\frac{\pi}{2N}\right)+o(N).$$

Case 1: j and k both odd. Then, by (2.3),

$$16R_{j,j}R_{k,k}\ge(N+o(N))^2$$

so

$$\sqrt{\frac{R_{j,k}^2}{R_{jj}R_{kk}}}\le\frac{\cot(\pi/(2N))+o(N)}{(N+o(N))}\sim\frac{2}{\pi}\doteq 0.636\,62.$$

Case 2: j and k both even. Then $2\le|j-k|,\ |j+k-1|\le 2M-3=N-3$. Thus

$$|4R_{j,k}|\le\cot\left(\frac{\pi}{N}\right)+o(N).$$

Moreover, by (2.3),

$$16R_{j,j}R_{k,k}\ge\left(N-\cot\left(\frac{3\pi}{2N}\right)+o(N)\right)\left(N-\cot\left(\frac{5\pi}{2N}\right)+o(N)\right)$$

hence

$$\begin{aligned}\sqrt{\frac{R_{j,k}^2}{R_{jj}R_{kk}}} &\le \sqrt{\frac{\left(\cot\left(\frac{\pi}{N}\right)+o(N)\right)^2}{\left(N-\cot\left(\frac{3\pi}{2N}\right)+o(N)\right)\left(N-\cot\left(\frac{5\pi}{2N}\right)+o(N)\right)}}\\ &\sim \sqrt{\frac{1}{\left(\pi-\frac{2}{3}\right)\left(\pi-\frac{2}{5}\right)}}\\ &\doteq 0.383\,90\end{aligned}$$

Case 3: j (even) $<k$ (odd).

$$|4R_{j,k}|\le\cot\left(\frac{\pi}{2N}\right)+o(N)$$

and

$$16R_{j,j}R_{k,k}\ge\left(N-\cot\frac{3\pi}{2N}+o(N)\right)(N+o(N))$$

so

$$\begin{aligned}
\sqrt{\frac{R_{j,k}^2}{R_{jj}R_{kk}}} &\leq \sqrt{\frac{\left(\cot\left(\frac{\pi}{2N}\right)+o(N)\right)^2}{\left(N-\cot\left(\frac{3\pi}{2N}\right)+o(N)\right)(N+o(N))}} \\
&\sim \sqrt{\frac{\left(\frac{2N}{\pi}\right)^2}{\left(N-\frac{2N}{3\pi}\right)(N)}} \\
&= \sqrt{\frac{4}{\left(\pi-\frac{2}{3}\right)\pi}} \\
&\doteq 0.71726.
\end{aligned}$$

Case 4: j (odd) $< k$ (even)

$$|4R_{j,k}| \leq \cot\left(\frac{\pi}{2N}\right) + o(N)$$

and

$$16R_{j,j}R_{k,k} \geq (N+o(N))\left(N-\cot\left(\frac{5\pi}{2N}\right)+o(N)\right)$$

so

$$\begin{aligned}
\sqrt{\frac{R_{j,k}^2}{R_{jj}R_{kk}}} &\leq \sqrt{\frac{\left(\cot\left(\frac{\pi}{2N}\right)+o(N)\right)^2}{(N+o(N))\left(N-\cot\left(\frac{5\pi}{2N}\right)+o(N)\right)}} \\
&\sim \sqrt{\frac{\left(\frac{2N}{\pi}\right)^2}{N\left(N-\frac{2N}{5\pi}\right)}} \\
&= \sqrt{\frac{4}{\pi\left(\pi-\frac{2}{5}\right)}} \\
&\doteq 0.68148.
\end{aligned}$$

□

Thus, we see that the asymptotic minimum angle between columns is at least

$$\arccos\left(\sqrt{\frac{4}{\left(\pi-\frac{2}{3}\right)\pi}}\right) \doteq 44.171 \text{ degrees.}$$

2.3. Eccentricity of $M \times N$ Chebyshev frames. For positive integers M and N with $M = \alpha N$ with $\alpha < 1$ let $D(M,N)$ denote the $M \times N$ Chebyshev frame defined by 2.1. Let f_j denote the jth column. Although $D(M,N)$ is not an equal norm frame, for large N the relative lengths of its columns are modulated by the parameter α. More precisely

THEOREM 2. *Let $\varepsilon > 0$. Then for sufficiently large N,*

$$\sqrt{\frac{\pi\alpha-\sin(\pi\alpha)}{\pi\alpha+\sin(\pi\alpha)}} - \varepsilon < \frac{\|f_j\|}{\|f_k\|} < \sqrt{\frac{\pi\alpha+\sin(\pi\alpha)}{\pi\alpha-\sin(\pi\alpha)}} + \varepsilon$$

for $1 \leq j,k \leq N$.

PROOF. From above,

$$4R_{j,j} = \sin(\pi\alpha(2j-1))\cot\left(\frac{\pi}{2N}(2j-1)\right) - \cos(\pi\alpha(2j-1)) + 2\alpha N - 1.$$

Also,

$$R_{j,j} = R_{j',j'}$$

where $j' = N - j + 1$, so we can assume WLOG that $1 \leq j,k \leq \frac{N}{2}$. We also have the inequality

$$|\sin t\beta| \leq |t|\,|\sin\beta|$$

for all integers t. Thus

$$|4R_{j,j}| \leq (2j-1)\sin(\pi\alpha)\cot\left(\frac{\pi}{2N}(2j-1)\right) + 2\alpha N.$$

Since $x\cot x$ decreases on the interval $0 < x < \pi/2$, we get

$$\begin{aligned} |4R_{j,j}| &\leq \sin(\pi a)\cot\left(\frac{\pi}{2N}\right) + 2\alpha N \\ &= \sin(\pi\alpha)\frac{2N}{\pi} + 2\alpha N + o(N) \\ &= 2\alpha N\left(1 + \frac{\sin(\pi\alpha)}{\pi\alpha}\right) + o(N). \end{aligned}$$

Similarly,

$$\begin{aligned} |4R_{j,j}| &\geq 2\alpha N - \sin(\pi a)\cot\left(\frac{\pi}{2N}\right) - 2 \\ &= 2\alpha N\left(1 - \frac{\sin(\pi\alpha)}{\pi\alpha}\right) + o(N). \end{aligned}$$

Combining these inequalities we get

$$\sqrt{\frac{2\alpha N\left(1 - \frac{\sin(\pi\alpha)}{\pi\alpha}\right) + o(N)}{2\alpha N\left(1 + \frac{\sin(\pi\alpha)}{\pi\alpha}\right) + o(N)}} \leq \frac{\|f_j\|}{\|f_k\|} \leq \sqrt{\frac{2\alpha N\left(1 + \frac{\sin(\pi\alpha)}{\pi\alpha}\right) + o(N)}{2\alpha N\left(1 - \frac{\sin(\pi\alpha)}{\pi\alpha}\right) + o(N)}}.$$

Thus

$$\sqrt{\frac{1 - \frac{\sin(\pi\alpha)}{\pi\alpha} + o(1)}{1 + \frac{\sin(\pi\alpha)}{\pi\alpha} + o(1)}} \leq \frac{\|f_j\|}{\|f_k\|} \leq \sqrt{\frac{1 + \frac{\sin(\pi\alpha)}{\pi\alpha} + o(1)}{1 - \frac{\sin(\pi\alpha)}{\pi\alpha} + o(1)}}$$

which yields the desired result. □

In the case $N = 2M$ we can substitute $\alpha = \frac{1}{2}$ into the above formula to get the ratios of lengths of frame elements between $c^{-1} - \varepsilon$ and $c + \varepsilon$ with $c = \sqrt{\frac{\pi+2}{\pi-2}}$.

2.4. Cosine-sine frames seeded from Chebyshev frames. Since the frames $D(M,N)$ are not equal norm frames, the coherence analysis is difficult except in special cases. To remedy this, we follow the technique used in [**19**] to form real equal norm tight (ENT) full spark frames from a DFT matrix. This involves augmenting (stacking) the frame matrix to make cosine-sine pairs in each column. The resulting equal-norm property allows for a simple analysis of the worst case coherence. However some care must be exercised to maintain tightness. For each $M \le \frac{1}{2}N$ we define a stacked cosine-sine matrix as follows. In the odd case $M = 2p+1$,

$$F(M,N) := \begin{bmatrix} 1/\sqrt{2} & 1/\sqrt{2} & \cdots & 1/\sqrt{2} \\ \cos(2\theta_1) & \cos(2\theta_2) & \cdots & \cos(2\theta_N) \\ \vdots & \vdots & & \vdots \\ \cos(2p\theta_1) & \cos(2p\theta_2) & \cdots & \cos(2p\theta_n) \\ \sin(2\theta_1) & \sin(2\theta_2) & \cdots & \sin(2\theta_N) \\ \vdots & \vdots & & \vdots \\ \sin(2p\theta_1) & \sin(2p\theta_2) & \cdots & \sin(2p\theta_N) \end{bmatrix}.$$

In the even case $M = 2p$ define $F(M,N)$ by

$$F(M,N) := \begin{bmatrix} \cos(\theta_1) & \cos(\theta_2) & \cdots & \cos(\theta_N) \\ \cos(3\theta_1) & \cos(3\theta_2) & \cdots & \cos(3\theta_N) \\ \vdots & \vdots & & \vdots \\ \cos((2p-1)\theta_1) & \cos((2p-1)\theta_2) & \cdots & \cos((2p-1)\theta_n) \\ \sin(\theta_1) & \sin(\theta_2) & \cdots & \sin(\theta_N) \\ \sin(3\theta_1) & \sin(3\theta_2) & \cdots & \sin(3\theta_N) \\ \vdots & \vdots & & \vdots \\ \sin((2p-1)\,\theta_1) & \sin((2p-1)\,\theta_2) & \cdots & \sin((2p-1)\,\theta_N) \end{bmatrix}.$$

The use of purely even coefficients in the first case is required to preserve tightness. The use of purely odd coefficients in the second case is required to preserve tightness as well as the full spark property.

THEOREM 3. *The frames $F(M,N)$ as defined above are full spark, equal norm and tight. Moreover, for fixed α, the worst-case coherence coefficient tends to $\frac{\sin(\pi\alpha)}{\pi\alpha}$ as $N \to \infty$.*

We defer the proof until the final section. It turns out that these frames are unitary multiples of frames related to *DFT* matrices that were introduced in [**27**]. So this result becomes a corollary to the results proved in the next section.

3. Frames seeded from DFT matrices

For postitive M and N with $M = \alpha N$ and $0 < \alpha \le \frac{1}{2}$, let

$$DFT(M,N) := \left[e^{ijx_k}\right]_{0\le j<M-1,\ 1\le k\le N}$$

where $x_k := \frac{2\pi k}{N}$. (For convenience, 0 is the starting index for the rows.) In [**1**] Alexeev, Cahill, and Mixon proved:

THEOREM 4. *The only $M \times N$ Vandermonde matrices that are equal norm and tight have bases in the complex unit circle. Among these, the frames with the smallest worst-case coherence are of the form $DFT(M,N)$, provided $N \ge 2M$.*

Our contribution is the following:

THEOREM 5. *For fixed $\alpha \le 1/2$, the asymptotic worst-case coherence coefficient for $DFT(M,N)$ equals $\frac{\sin(\pi\alpha)}{\pi\alpha}$. That is, the cosine of the minimum angle between columns of $DFT(M,N)$ approaches $\frac{\sin(\pi\alpha)}{\pi\alpha}$ as $N \to \infty$ with $M/N = \alpha$.*

PROOF. Note that each column has length $\sqrt{M}$. Therefore $C_{j,k}$, the inner product of columns j and k, which measures the angle between those two vectors, is given by

$$\begin{aligned} C_{j,k} &= \frac{1}{M}\sum_{l=0}^{M-1} e^{li(x_j-x_k)}e^{(\pi(M-1)i(j-k')/N)} \\ &= \frac{1}{M}\sum_{l=0}^{M-1} e^{\frac{2\pi}{N}li(j-k)}e^{(\pi(M-1)i(j-k')/N)} \\ &= \frac{1}{M}\frac{\sin\left(\frac{M\pi}{N}(j-k)\right)}{\sin\left(\frac{\pi}{N}(j-k)\right)}e^{(\pi(M-1)i(j-k')/N)} \\ &= \frac{1}{N\alpha}\frac{\sin\left(\pi\alpha(j-k)\right)}{\sin\left(\frac{\pi}{N}(j-k)\right)}e^{(\pi(M-1)i(j-k')/N)}. \end{aligned}$$

Thus

$$\begin{aligned} |C_{1,2}| &= \frac{1}{N\alpha}\frac{\sin(\pi\alpha)}{\sin\left(\frac{\pi}{N}\right)} \\ &= \frac{\sin(\pi\alpha)}{\pi\alpha} + o(1) \end{aligned}$$

as $N\to\infty$. Consequently,

$$\lim_{N\to\infty}\sup_{1\le j<k\le N}|C_{j,k}| \ge \frac{\sin(\pi\alpha)}{\pi\alpha}.$$

To verify the reverse inquality, first consider $1\le j<k\le N$ with $x := |j-k|$. We have

$$\begin{aligned} |C_{j,k}| &= \frac{1}{N\alpha}\left|\frac{\sin\left(M\frac{\pi}{N}x\right)}{\sin\left(\frac{\pi}{N}x\right)}\right| \\ &= \frac{1}{N\alpha}\left|\frac{\sin\left((M-1)\frac{\pi}{N}x\right)\cos\left(\frac{\pi}{N}x\right)+\cos\left(\frac{\pi}{N}x\right)\sin\left(\frac{\pi}{N}x\right)}{\sin\left(\frac{\pi}{N}x\right)}\right| \\ &\le \frac{1}{N\alpha}\left[\left|\sin\left((M-1)\frac{\pi}{N}x\right)\cot\left(\frac{\pi}{N}x\right)\right|+1\right]. \end{aligned}$$

First consider the case $1\le x\le N/2$. Note that $|\sin(n\theta)|\le|n|\,|\sin\theta|$ for all integers n. Thus

$$|C_{j,k}| \le \frac{1}{N\alpha}x\sin\left((M-1)\frac{\pi}{N}\right)\cot\left(\frac{\pi}{N}x\right)+\frac{1}{N\alpha}.$$

Note that the function $t\cot t$ decreases for $0<t\le\frac{\pi}{2}$ and that in the present case, $\frac{\pi}{N}\le\frac{\pi}{N}x\le\frac{\pi}{2}-\frac{\pi}{N}$. Thus the right hand side of the inequality is maximized by setting $x=1$, which yields

$$|C_{j,k}| \le \frac{1}{N\alpha}\sin\left((M-1)\frac{\pi}{N}\right)\cot\left(\frac{\pi}{N}\right)+\frac{1}{N\alpha}.$$

The same inequality holds when $N/2\le x\le N-1$ because the function $f(x) := \frac{\sin\left(M\frac{\pi}{N}x\right)}{\sin\left(\frac{\pi}{N}x\right)}$ satisfies

$$\left|f\left(\frac{N}{2}-x\right)\right| = \left|f\left(\frac{N}{2}+x\right)\right|$$

when $0 < x < \frac{N}{2}$. Now letting $N \to \infty$ shows that

$$\lim_{N\to\infty} \sup_{1\le j<k\le N} |C_{j,k}| \le \frac{\sin(\pi\alpha)}{\pi\alpha}.$$

□

3.1. Real frames seeded from DFT matrices. The following equal norm tight full spark frames were introduced by Puschel and Kovacevic in [**27**]. We denote them by $Q(M,N)$. In the odd case $M = 2p+1$, let

$$Q(M,N) := \begin{bmatrix} 1/\sqrt{2} & 1/\sqrt{2} & \cdots & 1/\sqrt{2} \\ \cos x_1 & \cos x_2 & \cdots & \cos x_N \\ \cos 2x_1 & \cos 2x_2 & \cdots & \cos 2x_N \\ \vdots & \vdots & & \vdots \\ \cos(px_1) & \cos(px_2) & \cdots & \cos(px_N) \\ \sin x_1 & \sin x_2 & \cdots & \sin x_N \\ \sin 2x_1 & \sin 2x_2 & \cdots & \sin 2x_N \\ \vdots & \vdots & & \vdots \\ \sin(px_1) & \sin(px_2) & \cdots & \sin(px_N) \end{bmatrix}.$$

Note that

$$\begin{bmatrix} \sqrt{2} & & \\ & I & iI \\ & I & -iI \end{bmatrix} Q(M,N) = DFT(M,N)$$

where I is a properly dimensioned identity matrix. The block matrix is a scalar multiple of a unitary matrix and

$$Q(M,N) = \begin{bmatrix} \frac{1}{\sqrt{2}} & & \\ & \frac{1}{2}I & \frac{1}{2}I \\ & -\frac{i}{2}I & \frac{i}{2}I \end{bmatrix} DFT(M,N). \tag{3.1}$$

In the even case $M = 2p$, let J denote the $p\times p$ permutation matrix whose nonzero entries are ones along the cross-diagonal. Define

$$Q(M,N) := \begin{bmatrix} \frac{1}{2}J & \frac{1}{2}I \\ -\frac{i}{2}J & \frac{1}{2}I \end{bmatrix} DFT(M,N)\,W \tag{3.2}$$

with $W := diag\left[w^{-\frac{(2p-1)}{2}}, w^{-2\frac{(2p-1)}{2}}, \ldots, w^{-N\frac{(2p-1)}{2}}\right]$ where $w = e^{\frac{2\pi}{N}i}$. Then

$$Q(M,N) = \begin{bmatrix} \cos\left(\frac{1}{2}x_1\right) & \cos\left(\frac{1}{2}x_2\right) & \cdots & \cos\left(\frac{1}{2}x_N\right) \\ \cos\left(\frac{3}{2}x_1\right) & \cos\left(\frac{3}{2}x_2\right) & \cdots & \cos\left(\frac{3}{2}x_N\right) \\ \vdots & \vdots & & \vdots \\ \cos\left(\frac{2p-1}{2}x_1\right) & \cos\left(\frac{2p-1}{2}x_2\right) & \cdots & \cos\left(\frac{2p-1}{2}x_N\right) \\ \sin\left(\frac{1}{2}x_1\right) & \sin\left(\frac{1}{2}x_2\right) & \cdots & \sin\left(\frac{1}{2}x_N\right) \\ \sin\left(\frac{3}{2}x_1\right) & \sin\left(\frac{3}{2}x_2\right) & \cdots & \sin\left(\frac{3}{2}x_N\right) \\ \vdots & \vdots & & \vdots \\ \sin\left(\frac{2p-1}{2}x_1\right) & \sin\left(\frac{2p-1}{2}x_2\right) & \cdots & \sin\left(\frac{2p-1}{2}x_N\right) \end{bmatrix}.$$

Since the transformations relating $Q(M,N)$ with $DFT(M,N)$ are essentially unitary, the equal norm, tightness, and full spark properties of $DFT(M,N)$ carry through to $Q(M,N)$. This proves:

THEOREM 6. *(Puschel and Kovacevich) In both even and odd cases, $Q(M,N)$ is an equal-norm, tight, full spark frame.*

Furthermore, the coherence properties of the DFT matrices carry over.

THEOREM 7. *For fixed $\alpha \in (0, \frac{1}{2}]$, consider integers M and N with $M = N\alpha$. The cosine of the minimum angle between columns of $Q(M,N)$ approaches $\frac{\sin(\pi\alpha)}{\pi\alpha}$ as $N \to \infty$.*

PROOF. This follows immediately from 3.1 and 3.2. □

3.2. Relation to Chebyshev frames. Let $M \leq \frac{1}{2}N$. Following previous notation let $\theta_k = \theta_{N,k} = \frac{\pi}{2N}(2k-1)$ and $x_k = x_{N,k} = \frac{2\pi}{N}k$ for $1 \leq k \leq N$. When $M = 2p$ it is easily checked that

$$\begin{bmatrix} I & iI \\ I & -iI \end{bmatrix} F(M,N) = \Lambda \begin{bmatrix} I & iI \\ I & -iI \end{bmatrix} Q(M,N)$$

where

$$\Lambda := diag\left[e^{-\pi i/2N}, e^{-3\pi i/2N}, ..., e^{-(2p-1)\pi i/2N}; e^{\pi i/2N}, e^{3\pi i/2N}, ..., e^{(2p-1)\pi i/2N}\right].$$

In the odd case $M = 2p-1$,

$$\begin{bmatrix} \sqrt{2} & & \\ & I & iI \\ & I & -iI \end{bmatrix} F(M,N) = \Lambda' \begin{bmatrix} \sqrt{2} & & \\ & I & iI \\ & I & -iI \end{bmatrix} Q(M,N)$$

with

$$\Lambda' := diag\left[1, e^{-\pi i/N}, e^{-2\pi i/N}, ..., e^{-p\pi i/N}; e^{\pi i/N}, e^{2\pi i/N}, ..., e^{p\pi i/N}\right].$$

Thus $F(M,N)$ is a unitary multiple of $Q(M,N)$ for both even and odd cases of M. As a result, $F(M,N)$ shares all the same properties as Q.

References

[1] Boris Alexeev, Jameson Cahill, and Dustin G. Mixon, *Full spark frames*, J. Fourier Anal. Appl. **18** (2012), no. 6, 1167–1194, DOI 10.1007/s00041-012-9235-4. MR3000979

[2] John J. Benedetto, Alexander M. Powell, and Özgür Yılmaz, *Sigma-Delta ($\Sigma\Delta$) quantization and finite frames*, IEEE Trans. Inform. Theory **52** (2006), no. 5, 1990–2005, DOI 10.1109/TIT.2006.872849. MR2234460

[3] John J. Benedetto, Alexander M. Powell, and Özgür Yılmaz, *Second-order sigma-delta* ($\Sigma\Delta$) *quantization of finite frame expansions*, Appl. Comput. Harmon. Anal. **20** (2006), no. 1, 126–148, DOI 10.1016/j.acha.2005.04.003. MR2200933

[4] John J. Benedetto and Matthew Fickus, *Finite normalized tight frames*, Adv. Comput. Math. **18** (2003), no. 2-4, 357–385, DOI 10.1023/A:1021323312367. Frames. MR1968126

[5] Bernhard G. Bodmann and Vern I. Paulsen, *Frames, graphs and erasures*, Linear Algebra Appl. **404** (2005), 118–146, DOI 10.1016/j.laa.2005.02.016. MR2149656

[6] Emmanuel Candès and Justin Romberg, *Sparsity and incoherence in compressive sampling*, Inverse Problems **23** (2007), no. 3, 969–985, DOI 10.1088/0266-5611/23/3/008. MR2329927

[7] Emmanuel J. Candès, Justin Romberg, and Terence Tao, *Robust uncertainty principles: exact signal reconstruction from highly incomplete frequency information*, IEEE Trans. Inform. Theory **52** (2006), no. 2, 489–509, DOI 10.1109/TIT.2005.862083. MR2236170

[8] Emmanuel J. Candès, Justin K. Romberg, and Terence Tao, *Stable signal recovery from incomplete and inaccurate measurements*, Comm. Pure Appl. Math. **59** (2006), no. 8, 1207–1223, DOI 10.1002/cpa.20124. MR2230846

[9] Emmanuel J. Candes and Terence Tao, *Decoding by linear programming*, IEEE Trans. Inform. Theory **51** (2005), no. 12, 4203–4215, DOI 10.1109/TIT.2005.858979. MR2243152

[10] Emmanuel J. Candes and Terence Tao, *Near-optimal signal recovery from random projections: universal encoding strategies?*, IEEE Trans. Inform. Theory **52** (2006), no. 12, 5406–5425, DOI 10.1109/TIT.2006.885507. MR2300700

[11] Peter G. Casazza and Jelena Kovačević, *Equal-norm tight frames with erasures*, Adv. Comput. Math. **18** (2003), no. 2-4, 387–430, DOI 10.1023/A:1021349819855. Frames. MR1968127

[12] P. Cassaza. Custom building finite frames. *Contemporary Mathematics.*

[13] Peter G. Casazza and Matthew Fickus, *Minimizing fusion frame potential*, Acta Appl. Math. **107** (2009), no. 1-3, 7–24, DOI 10.1007/s10440-008-9377-1. MR2520007

[14] Peter G. Casazza and Gitta Kutyniok, *A generalization of Gram-Schmidt orthogonalization generating all Parseval frames*, Adv. Comput. Math. **27** (2007), no. 1, 65–78, DOI 10.1007/s10444-005-7478-1. MR2317921

[15] Scott Shaobing Chen, David L. Donoho, and Michael A. Saunders, *Atomic decomposition by basis pursuit*, SIAM Rev. **43** (2001), no. 1, 129–159, DOI 10.1137/S003614450037906X. Reprinted from SIAM J. Sci. Comput. **20** (1998), no. 1, 33–61 (electronic) [MR1639094 (99h:94013)]. MR1854649

[16] David L. Donoho, *Compressed sensing*, IEEE Trans. Inform. Theory **52** (2006), no. 4, 1289–1306, DOI 10.1109/TIT.2006.871582. MR2241189

[17] David L. Donoho and Michael Elad, *Optimally sparse representation in general (nonorthogonal) dictionaries via l^1 minimization*, Proc. Natl. Acad. Sci. USA **100** (2003), no. 5, 2197–2202, DOI 10.1073/pnas.0437847100. MR1963681

[18] De-Jun Feng, Long Wang, and Yang Wang, *Generation of finite tight frames by Householder transformations*, Adv. Comput. Math. **24** (2006), no. 1-4, 297–309, DOI 10.1007/s10444-004-7637-9. MR2222273

[19] Walter Gautschi, *The condition of Vandermonde-like matrices involving orthogonal polynomials*, Linear Algebra Appl. **52/53** (1983), 293–300, DOI 10.1016/0024-3795(83)80020-2. MR709357

[20] Vivek K. Goyal, Jelena Kovačević, and Jonathan A. Kelner, *Quantized frame expansions with erasures*, Appl. Comput. Harmon. Anal. **10** (2001), no. 3, 203–233, DOI 10.1006/acha.2000.0340. MR1829801

[21] Deguang Han, *Tight frame approximation for multi-frames and super-frames*, J. Approx. Theory **129** (2004), no. 1, 78–93, DOI 10.1016/j.jat.2004.04.012. MR2070181

[22] Jinsong Leng and Deguang Han, *Optimal dual frames for erasures II*, Linear Algebra Appl. **435** (2011), no. 6, 1464–1472, DOI 10.1016/j.laa.2011.03.043. MR2807163

[23] Jerry Lopez and Deguang Han, *Optimal dual frames for erasures*, Linear Algebra Appl. **432** (2010), no. 1, 471–482, DOI 10.1016/j.laa.2009.08.031. MR2566493

[24] Roderick B. Holmes and Vern I. Paulsen, *Optimal frames for erasures*, Linear Algebra Appl. **377** (2004), 31–51, DOI 10.1016/j.laa.2003.07.012. MR2021601

[25] Jinsong Leng, Deguang Han, and Tingzhu Huang, *Optimal dual frames for communication coding with probabilistic erasures*, IEEE Trans. Signal Process. **59** (2011), no. 11, 5380–5389, DOI 10.1109/TSP.2011.2162955. MR2883607

[26] B. K. Natarajan, *Sparse approximate solutions to linear systems*, SIAM J. Comput. **24** (1995), no. 2, 227–234, DOI 10.1137/S0097539792240406. MR1320206

[27] Markus Puschel and Jelena Kovacevic. Real, tight frames with maximal robustness to erasures. In *Data Compression Conference, 2005. Proceedings. DCC 2005*, pages 63–72. IEEE, 2005.

[28] H. Rauhut. Compressive sensing and structured random matrices. *in Theoretical Foundations and Numerical Methods for Sparse Recovery.*

Department of Mathematics, San Francisco State University, 1600 Holloway Avenue, San Francisco, California 94132

Email address: dellis@sfsu.edu

Department of Mathematics, San Francisco State University, 1600 Holloway Avenue, San Francisco, California 94132

Email address: hayashi@sfsu.edu

Department of Mathematics, San Francisco State University, 1600 Holloway Avenue, San Francisco, California 94132

Email address: shidong@sfsu.edu

Contemporary Mathematics
Volume **706**, 2018
http://dx.doi.org/10.1090/conm/706/14220

Fusion frames and distributed sparsity

Roza Aceska, Jean-Luc Bouchot, and Shidong Li

Abstract. We analyze the problem of recovering signals from low quality sensing devices, and implement a combination of compressed sampling and distributed sensing to model so-called fusion frames structured signals. Within our framework, it is possible to recover signals with high accuracy by increasing the total number of measurements, while keeping the cheap sensors at hand. We show that, under suitable constraints, nonsparse signals can be recovered with high probability. Moreover, we show that, with the use of a few linear measurements, cheap sensors are sufficient when combined with the fusion frame methodology. Using our new method, we show that it is possible, and sometimes necessary, to split a signal via local projections for accurate, stable, and robust estimation.

1. Introduction

In applications of sampling theory, a good sensor network design ensures that recovery from sufficiently many samples of the signal of interest $\mathbf{x} \in \mathbb{K}^N$ is possible. Here $\mathbb{K}^N$ denotes a finite dimensional Hilbert space over the field of real or complex numbers. When anticipating that the signal $\mathbf{x} \in \mathbb{K}^N$ is sparse, we can start with scarce, noisy measurements $\mathbf{y} = \mathbf{A}\mathbf{x} + \mathbf{e} \in \mathbb{K}^m$, $m \ll N$, and recover a sparse $\hat{\mathbf{x}} \in \mathbb{K}^N \approx \mathbf{x}$. The recovery is a result of solving the mathematical program

$$(\boldsymbol{\ell}^0\text{-min}) \qquad \hat{\mathbf{x}} := \operatorname{argmin} \|\mathbf{z}\|_0, \quad \text{subject to } \|\mathbf{A}\mathbf{z} - \mathbf{y}\|_2 \leq \eta.$$

The ($\boldsymbol{\ell}^0$-min) problem is NP-Hard, and its solution is approximated by the solution of its convex relaxation, known as the Basis Pursuit denoising

$$(\text{BPDN}) \qquad \hat{\mathbf{x}} := \operatorname{argmin} \|\mathbf{z}\|_1, \quad \text{subject to } \|\mathbf{A}\mathbf{z} - \mathbf{y}\|_2 \leq \eta.$$

For a given *sparsity* s of $\mathbf{x}$, the number of random subgaussian linear measurements needs to grow as $m \gtrsim s\log(N/s)$ for $\widehat{\mathbf{x}}$ to be a good enough approximation to $\mathbf{x}$, which means, we have a stable and robust recovery from relatively few measurements.

2010 *Mathematics Subject Classification.* Primary: 42C15, 68M10, 68Q85, 94A12.

Key words and phrases. Fusion frames, distributed sparsity.

The first author was partially supported by the BSU Aspire Research Grant "Frame Theory and Modern Sampling Strategies". The second author thanks the support of the European Research Council through the grant StG 258926 and the organizors as well as the Hausdorff Center for Mathematics for his participation in the Hausdorff Trimester Program on Mathematics of Signal Processing. The third author was partially supported by US NSF grants USA (DMS-1313490, DMS-1010058).

We analyze in this paper the problem of signal sampling and recovery when the quality of sensors is constrained, hereby limiting the number of measurements m, independently of the sparsity or complexity of the signal of interest. The sensors constraints can be due to many reasons such as cost (e.g. using sensors at a coarser resolution is cheaper than one at the finest), legal regulation (for instance in nuclear medicine where one must not expose a patient to high amounts of radiations at once). The quality of recovery is not ensured when the signals being sampled are *not sparse enough*, given the constraints of the sensors used.

In this paper, we assume certain constraints on the sensor design which are fixed due to some outside reasons. Under these assumptions, we take on the following challenge: split the information carried by the signal in a clever way so that a mathematical recovery is possible; that is, we apply the fusion frame theory to the signal recovery problem. We show that it is possible to handle very complex signals in an efficient and stable manner.

1.1. Motivation: Unavailability of high quality observation devices. A time-invariant bounded linear operator is represented by a circulant matrix $\mathbf{A}$. Suppose $\mathbf{A}$ represents a sensing device whose number of rows m is physically limited by the sampling rate (or resolution) of the device. Suppose also that the sparsity of the sampled signal $\mathbf{x}$ is substantially larger than what the standard compressed sensing techniques could handle when working with $\mathbf{Ax}$.

In this context, the limitations on the sensing devices combined with the (potentially) high number of non-zeros in the signals makes it impossible for a state-of-the-art algorithm to recover the unknown signal $\mathbf{x}$ accurately. As illustrated in Figure 1, we suggest to apply n such devices in parallel after pre-filtering. The fused compressed sensing technique introduced here allows to resolve the problem that otherwise a single device can not.

We present an example of such a real-life scenario to elaborate on the necessity of the proposed fused compressed sensing technique: Suppose an application

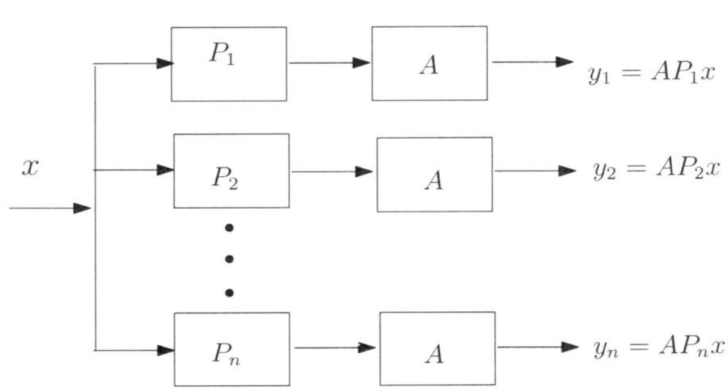

FIGURE 1. Sparse linear array geometry

requires a sophisticated sensing device $\tilde{\mathbf{A}}$ of very high sensing rate. Such a device may be very expensive, or not available on the market, which is why we combine 10 parallel projections or filtering operations $\{P_j\}_{j=1}^{10}$ prior to measuring using 10 lower rate devices, each with measurement matrix $\mathbf{A}$. The (sparse) signal $\mathbf{x}$ is then subsequently recovered by various techniques in each subspace/channel and, through the theory of fusion frames [**10**, **12**, **13**], merged into a single vector. Such a fusion operation is possible if $\{P_j\}_{j=1}^{10}$ are the projections or any filtering operations satisfying $A\mathbf{I} \leq \sum_i P_j^* P_j \leq B\mathbf{I}$ for some $0 < A \leq B < \infty$.

Here is why such a pre-filtering/projection mechanism manages to distribute the sparsity. Since the signal has a 20GHz bandwidth, the non-zeros can be assumed to be distributed within the entire bandwidth. Otherwise, the bandwidth could be assumed to be smaller.

In order for the fusion operator to be bounded away from 0, the projections P_i are certainly bandpass in nature. Namely, P_i's are distributed over adjacent and overlapping bands so that there won't be "holes" in the frequency domain. Consequently, each P_i represents different bandpass filtering with different bands and carries a "different part of sparsity"; that is, the sparsity is distributed in different bands.

Evidently, if we design such a fused compressed sensing technique, by enabling the subdivision of sparsity of $\mathbf{x}$ into individual subspaces, the sparse recovery problem becomes a feasible one and can be resolved by multiple sensing devices with sampling rates lower than 2GHz, which are not only widely available but also economical.

The value of fused compressed sensing techniques presented in this paper is clearly reflected in this situation where an otherwise too expensive or impossible problem can now be resolved by using a number of lower resolution/sampling rate devices and by making a *reasonable* number of observations, and processed by the fusion frame theory.

1.2. Contributions. In this paper we combine mathematical tools from compressed sensing and fusion frame theory to break the limitations on the signal complexity induced by traditional recovery methods. Our Theorem 3.11 states the following: Given a fusion frame system $\mathcal{W} = (W_i, P_i)_{i=1}^n$ for $\mathbb{K}^N$ with frame operator S and frame bounds $0 < C \leq D < \infty$, and a matrix $\mathbf{A} \in \mathbb{K}^{m\times N}$ satisfying a Partial RIP condition (see Definition 3.4), we can recover a $\mathbf{s}$ distributed sparse vector $\mathbf{x} \in \mathbb{K}^N$ (see Definition 3.1) as

$$\widehat{\mathbf{x}} = S^{-1} \sum_{i=1}^{n} \widehat{\mathbf{x}^{(i)}},$$

where the $\widehat{\mathbf{x}^{(i)}}$ are obtained as solutions to the n local s-sparse recovery problems

$$\widehat{\mathbf{x}^{(i)}} := \operatorname{argmin} \|\mathbf{z}\|_0, \qquad \text{subject to } \|\mathbf{A}P_i\mathbf{z} - \mathbf{y}^{(i)}\|_2 \leq \eta.$$

Moreover, assuming certain uniformity over all subspaces, the error of solution approximation $\|\widehat{\mathbf{x}} - \mathbf{x}\|_2$ is well-controlled by inequality (3.5).

In Section 2 we review the basic tools from compressed sensing and fusion frame theory. We extend these tools to the case of recovery with assumed local redundancy in the fusion frame decomposition in Section 3. In particular, we derive a mathematical theory allowing to work with fusion frames where sparsity is exploited along a subspace decomposition. Our theoretical results are illustrated on numerical examples where we try to recover a signal with dense Fourier spectrum from cheap sensors in Section 4.

2. General tools and models

2.1. Compressed sensing tools. Compressed sensing (CS) relies ([**17, 19**] and references within) on the inherent sparsity of natural signals $\mathbf{x}$ for their recovery from seemingly few measurements $\mathbf{y} = A\mathbf{x} + \mathbf{e}$ for some linear measurement matrix $\mathbf{A} \in \mathbb{K}^{m\times N}$, with $m \ll N$. Here the vector $\mathbf{e} \in \mathbb{K}^m$ represents the measurement noise, and is assumed to be bounded, $\|\mathbf{e}\|_2 \leq \eta$. With the sparsity assumption, CS aims at finding (approximate) solutions to the (ℓ^0-min) problem. General recovery guarantees ensure that the recovery is stable and robust, that is, the solution $\widehat{\mathbf{x}}$ satisfies

$$\|\widehat{\mathbf{x}} - \mathbf{x}\|_2 \leq \frac{C}{\sqrt{s}}\sigma_s(\mathbf{x})_1 + D\eta,$$
$$\|\widehat{\mathbf{x}} - \mathbf{x}\|_1 \leq C\sigma_s(\mathbf{x})_1 + D\sqrt{s}\eta,$$

where $\sigma_s(\mathbf{x})_1 := \min_{\mathbf{z}:\|\mathbf{z}\|_0\leq s} \|\mathbf{x} - \mathbf{z}\|_1$ is called the best s-term approximation of $\mathbf{x}$.

Plethora of conditions (some of which we look closer in Section 3) on $\mathbf{A}$ have been derived to ensure that the previous (or similar) estimates hold. These are based on restricted isometry constants ($\delta_{2s} < \sqrt{2} - 1$ [**11**] or $\delta_{2s} < 4/\sqrt{41}$ [**19**], $\delta_s < 1/3$), null space properties ($\mathbf{A}$ fulfills the robust and stable NSP(s, ρ, τ) if $\|\mathbf{v}_S\|_1 \leq \rho\|\mathbf{v}_{\overline{S}}\|_1 + \tau\|\mathbf{A}\mathbf{v}\|_2$ for any vector $\mathbf{v} \in \mathbb{K}^N$, and S any index set with $|S| \leq s$), coherence ($\mu_1(s) + \mu_1(s-1) < 1$), quotient property, and so on. Similar conditions and bounds can be found when greedy and thresholding algorithms are used to approximate (ℓ^0-min) (see [**5–7, 18, 21, 23**] for some relatively recent results in this direction).

There is still a need for building adequate sensing matrices $\mathbf{A}$, which ensure good recovery bounds. So far, no deterministic matrices $\mathbf{A}$ can be built with reasonably few rows, and we have to rely on randomness to build measurement matrices. It has been shown, for instance, that matrices with independent random subgaussian entries fulfill the RIP(s, δ)under certain constraints on the number of measurements [**4, 22**].

2.2. Frames, fusion frames, and distributed signal processing. By definition, a sequence $\mathcal{F} = \{\mathbf{f}_i\}_{i\in I}$ in a Hilbert space $\mathcal{H}$ is a frame [**14**] for $\mathcal{H}$ if there exist $0 < A \leq B < \infty$ (lower and upper frame bounds) such that

$$A\|\mathbf{f}\|^2 \leq \sum_{i\in I} |\langle \mathbf{f}, \mathbf{f}_i\rangle|^2 \leq B\|\mathbf{f}\|^2 \text{ for all } \mathbf{f} \in \mathcal{H}.$$

The representation space associated with $\mathcal{F}$ is $\ell^2(I)$ and its analysis and synthesis operators are respectively given by $T(\mathbf{f}) = \{\langle \mathbf{f}, \mathbf{f}_i\rangle\}_{i\in I}$ and $T^*(\{c_i\}_{i\in I}) = \sum_{i\in I} c_i\mathbf{f}_i$, for all $\mathbf{f} \in \mathcal{H}$ and $\{c_i\}_{i\in I} \in \ell^2(I)$. It shows that the frame operator $S := T^*T$ is a positive, self-adjoint and invertible operator; this means that recovery of any $\mathbf{f} \in \mathcal{H}$ is possible, if Sf is known; however, computing S^{-1} can be computationally challenging. Each frame $\mathcal{F}$ is accompanied by at least one so-called dual frame $\mathcal{G} = \{\mathbf{g}_i\}_{i\in I}$, which satisfies

$$\mathbf{f} = \sum_{i\in I}\langle \mathbf{f}, \mathbf{f}_i\rangle \mathbf{g}_i = \sum_{i\in I}\langle \mathbf{f}, \mathbf{g}_i\rangle \mathbf{f}_i \text{ for all } \mathbf{f} \in \mathcal{H}.$$

Whenever a frame is A-tight ($A = B$), the problem of function reconstruction is simplified, since in this case the frame operator is given by $S = A\mathbf{I}$, with $\mathbf{I}$ denoting the identity operator.

Fusion frames have been initially created [**12, 13**] to model the setting of a wireless sensor network. A large sensor network is typically split into redundant sub-networks; the local measurements within each sub-network are sent to a local sub-station, which submits the gathered information further to a central processing station for final reconstruction. Each sub-network is related to a frame for a subspace of a Hilbert space. The subspaces have to satisfy certain overlapping properties. The theory of fusion frames derives a number of means to fuse together subspace information for a complete signal reconstruction.

DEFINITION 2.1 (Fusion frames). Given an index set I, let $\mathcal{W} := \{W_i \mid i \in I\}$ be a family of closed subspaces in $\mathcal{H}$. We denote by P_i the orthogonal projections onto W_i. Then $\mathcal{W}$ is a fusion frame, if there exist $C, D > 0$ such that

$$C\|\mathbf{f}\|^2 \leq \sum_{i\in I} \|P_i(\mathbf{f})\|^2 \leq D\|\mathbf{f}\|^2 \text{ for all } \mathbf{f} \in \mathcal{H}.$$

REMARK 2.2. Fusion frames are often accompanied by respective weights. In the weighted case, the frame condition reads $C\|\mathbf{f}\|^2 \leq \sum_{i\in I} v_i^2 \|P_i(\mathbf{f})\|^2 \leq D\|\mathbf{f}\|^2$ for all $\mathbf{f} \in \mathcal{H}$ and some positive weights $(v_i)_{i\in I}$. All (unweighted) results derived in this paper apply mutatis mutandis to the weighted case.

The following theorem [**12**] illustrates the relationship between the local and global properties of a fusion frame:

THEOREM 2.3. *For each $i \in I$, let W_i be a closed subspace of $\mathcal{H}$, and let $\mathcal{F}_i = \{\mathbf{f}_{ij} \mid j \in J_i\}$ be a frame for W_i, with frame bounds A_i, B_i. If $0 < A = \inf_{i\in I} A_i \leq \sup_{i\in I} B_i = B < \infty$, then the following conditions are equivalent:*

- *$\cup_{i\in I}\{\mathbf{f}_{ij} \mid j \in J_i\}$ is a frame for $\mathcal{H}$.*
- *$\{(W_i, \mathcal{F}_i)\}_{i\in I}$ is a fusion frame for $\mathcal{H}$.*

In particular, if $\{(W_i, \mathcal{F}_i)\}_{i\in I}$ is a fusion frame system for $\mathcal{H}$ with frame bounds C and D, then $\cup_{i\in I}\{\mathbf{f}_{ij} \mid j \in J_i\}$ is a frame for $\mathcal{H}$ with frame bounds AC and BD.

In fusion frame theory, an input signal is represented by a collection of vector coefficients that represent the projection onto each subspace. The representation space used in this setting is

$$\left(\sum_{i\in I} \oplus W_i\right)_{\ell^2} = \{\{\mathbf{f}_i\}_{i\in I} \mid \mathbf{f}_i \in W_i\}$$

with $\{\|\mathbf{f}_i\|\}_{i\in I} \in \ell^2(I)$. The analysis operator is given by $T(\mathbf{f}) := \{P_i(\mathbf{f})\}_{i\in I}$ for all $\mathbf{f} \in \mathcal{H}$, while its adjoint is the synthesis operator $T^* : \left(\sum_{i\in I} \oplus W_i\right)_{\ell^2} \to \mathcal{H}$, defined by

$$T^*(\mathbf{f}) = \sum_i \mathbf{f}_i, \text{ where } \mathbf{f} = \{\mathbf{f}_i\}_{i\in I} \in \left(\sum_{i\in I} \oplus W_i\right)_{\ell^2}.$$

The fusion frame operator $S = T^*T$ is given by $S(\mathbf{f}) = \sum_{i\in I} P_i(\mathbf{f})$. It is easy to verify that S is a positive and invertible operator on $\mathcal{H}$. A distributed fusion processing is feasible in an elegant way, since the reconstruction formula for all $\mathbf{f} \in \mathcal{H}$ is

$$\mathbf{f} = \sum_{i \in I} S^{-1} P_i(\mathbf{f}) = S^{-1} \sum_{i \in I} P_i(\mathbf{f}).$$

2.3. Signal recovery in fusion frames. This work describes an approach for sensing and reconstructing signals in a fusion frame structure. As presented above, given some local information $\mathbf{x}^{(i)} := P_i(\mathbf{x})$, for $1 \leq i \leq n$, a vector can easily be reconstructed by applying the inverse fusion frame operator

$$\mathbf{x} := S^{-1}S(\mathbf{x}) = S^{-1} \sum_{i=1}^{n} P_i(\mathbf{x}) = S^{-1} \left(\sum_{i=1}^{n} \mathbf{x}^{(i)} \right). \tag{2.1}$$

Throughout this work, we assume that the projected vectors are sampled independently from one another, with n devices modeled by the same sensing matrix $\mathbf{A}$. The task at hand is as follows: *Use the measurements* $\mathbf{y}^{(i)} = \mathbf{A}\mathbf{x}^{(i)} + \mathbf{e}^{(i)}$ *to compute an estimate* $\widehat{\mathbf{x}^{(i)}}$ *of the local component* $\mathbf{x}^{(i)}$*; then compute the approximation* $\widehat{\mathbf{x}}$ *of the signal* $\mathbf{x}$ *via* (2.1).

From this point on, there are two ways of thinking about the problem. In a first scenario, the signals are measured in the subspace and the recovery of each $\mathbf{x}^{(i)}$ is done locally. In other words, it accounts for solving n (ℓ^0-min) problems (or their approximations via (BPDN) for instance) in the subspaces, and then transmitting the estimated local signals to a central unit taking care of the fusion via an analogue of (2.1).The other approach consists of transmitting the local observations $\mathbf{y}^{(i)} = \mathbf{A}P_i\mathbf{x} + \mathbf{e}^{(i)}$ to a central processing unit which takes care of the whole reconstruction process. In this case, the vector $\mathbf{x}$ can be recovered by solving a unique (ℓ^0-min) problem directly with, letting $\mathbf{I}_n$ denote the n dimensional identity matrix,

$$\mathbf{y} = \begin{pmatrix} \mathbf{y}^{(1)} \\ \vdots \\ \mathbf{y}^{(n)} \end{pmatrix} = \mathbf{I}_n \otimes \mathbf{A} \begin{bmatrix} P_1 \\ \vdots \\ P_n \end{bmatrix} \mathbf{x} \tag{2.2}$$

While the latter case is interesting on its own (see for instance [**16**]), we investigate here some results for the first case. Our results can be generalized further, integrating ideas where the measurements matrix (here, the sensors) vary locally, as is the case in [**16**] or driven with some structured acquisition (see for instance [**9**]).

We would like to put our work in context. This paper is not the first one to describe the use of fusion frames in sparse signal recovery. However, it is inherently different from previous works in [**2**, **8**]. In [**8**] the authors provide a framework for recovering sparse fusion frame coefficients. In other words, given a fusion frame system $\{(W_i, P_i)\}_{i=1}^n$ a vector is represented on this fusion frame as a set of n vectors $(\mathbf{x}^{(i)})_{i=1}^n$ where each of the $\mathbf{x}^{(i)}$ corresponds to the coefficient vector in subspace W_i. The main idea of the authors is that the original signal may only lie in few of the n subspaces, implying that most of the $\mathbf{x}^{(i)}$ should be 0. To rephrase the problem, we can say that the vector has to lie in a sparse subset of the original fusion frame. In terms of the band-pass filtering discussed in the introduction, this would be equivalent to the coefficients to 0 within a whole band. While [**8**] is concerned with some recovery guarantees under (fusion-type) RIP and average case analysis, the paper [**2**] gives uniform results for subgaussian measurements and derives results on the minimum number of vector measurements required for robust and stable recovery.

3. An extension of traditional compressed sensing

Our goal is to ensure robust and stable recovery of complex signals by smartly combining local information. For this purpose, we extend the tools of traditional CS via the use of fusion frames. We note that the authors of [**15**, **16**] introduce a model which corresponds to the the idea of solving a single sparse recovery problem. They analyze RIP conditions to ensure stable and robust recovery of $\mathbf{x}$ via a single matrix similar to Equation (2.2).

We start with a fusion frame $\mathcal{W} = (W_i, P_i)_{i=1}^n$ for $\mathbb{K}^N$, and assume that our signal of interest is $\mathbf{x} \in \mathbb{K}^N$. We aim at signal recovery by means of (a) CS in local subspaces and (b) fusion processing; that is, the local pieces of information are computed as solutions to the problems

$$(\mathcal{P}_{1,\eta}) \qquad \min_{\mathbf{x}\in\mathbb{K}^N} \|\mathbf{z}\|_1 \text{ s.t. } \|\mathbf{A}P_i\mathbf{z} - \mathbf{y}^{(i)}\|_2 \le \eta,$$

where P_i is the projection operator related to subspace W_i. In the noiseless case, the problem is solved by the basis pursuit

$$(\mathcal{P}_{1,0}) \qquad \min_{\mathbf{x}\in\mathbb{K}^N} \|\mathbf{z}\|_1 \text{ s.t. } \mathbf{A}P_i\mathbf{z} = \mathbf{y}^{(i)}.$$

Numerically, solving a set of smaller subproblems is typically more efficient than solving a single large system. In addition, the compressed sensing step in local subspaces can be done in a parallel fashion, greatly improving the computational efficiency.

DEFINITION 3.1. Let $s > 0$ and set $\mathbf{s} = (s, \dots, s)$ be the n-tuple with identical entries s. A signal $\mathbf{x} \in \mathbb{K}^N$ is said to be $\mathbf{s}$-uniformly distributed sparse (or $\mathbf{s}$-UD sparse) with respect to a fusion frame $\mathcal{W} = (W_i, P_i)_{i=1}^n$, if $\|P_i(\mathbf{x})\|_0 \le s$, for every $1 \le i \le n$. We also refer to $\mathbf{s}$ as a *uniform sparsity pattern* of $\mathbf{x}$ with respect to $\mathcal{W}$.

REMARK 3.2. It is important to understand what it means for a signal to be distributed sparse. In particular, an s sparse signal need not be distributed $\mathbf{s}$ sparse. Conversely, an $\mathbf{s}$ dibstributed sparse signal might also be dense.

An important particular case, is when dealing with projections parallel to the original axis of the space. In this case, an s sparse vector is also $\mathbf{s}$ distributed sparse, where the worst case scenario is if all the s non zero terms fall within a single subspace. An optimal design of the fusion frame is dependent on the signal, and plays a great role in our approach.

By $\Sigma_{\mathbf{s}}^{(\mathcal{W})}$ we denote the set of all $\mathbf{s}$-UD sparse vectors with respect to the family of subspaces $\mathcal{W}$. Since the sparsity of the signal is uniformly distributed among the subspaces, the usual CS recovery guarantees ensure that

$$m \asymp s \log(N/s)$$

(subgaussian) observations per subspace are required for a stable and robust recovery of $\mathbf{x}$. This accounts for a total number of measurements scaling as $m_T \asymp ns\log(N/s)$. Most of the computations can be easily carried in a distributed setting, where only pieces of the information are available. Only the fusion process requires all the local information to compute the final estimation of a signal.

Note that, for $p > 0$ and a UD sparsity pattern $\mathbf{s} = (s, \cdots, s)$ with $s \in \mathbb{N}^+$, the $\boldsymbol{\ell}^p$ errors of best $\mathbf{s}$-term approximations are understood in a distributed setting, and are embedded in the vector

$$\sigma_{\mathbf{s}}^{\mathcal{W}}(\mathbf{x})_p := \left(\sigma_s(P_1\mathbf{x})_p, \sigma_s(P_2\mathbf{x})_p, \cdots, \sigma_s(P_n\mathbf{x})_p\right)^T.$$

3.1. Partial properties. The *null space property* (NSP) is used in the CS literature as a necessary and sufficient condition for the sparse recovery problem via (ℓ^0-min). A matrix $\mathbf{A}$ is said to satisfy the (robust) null space property with parameters $\rho \in (0,1)$ and $\tau > 0$ relative to a set $S \subset \{1, \cdots, N\}$ if

$$\|\mathbf{v}_S\|_1 \leq \rho \|\mathbf{v}_{\overline{S}}\|_1 + \tau \|\mathbf{A}\mathbf{v}\|_2, \quad \text{for all } \mathbf{v} \in \mathbb{K}^N.$$

More generally, we say that the matrix $\mathbf{A}$ satisfies the NSP of order s if it satisfies the NSP relative to all sets S such that $|S| \leq s$. Here we talk about a sparsity pattern and ask that the NSP property be valid for all local subspaces up to a certain (local) sparsity level.

DEFINITION 3.3 (Uniformly distributed partial null space property (UDP-NSP)). Let $\mathcal{W} = (W_i, P_i)_{i=1}^n$ be a fusion frame for $\mathbb{K}^N$, and let $\mathbf{s}$ be a UD sparsity pattern (with entries s). A sensing matrix $\mathbf{A} \in \mathbb{K}^{m \times N}$ is said to fulfill the UDP-NSP with UD pattern $\mathbf{s}$ with respect to $\mathcal{W}$ and uniform constants $\rho \in (0,1)$ and $\tau > 0$ if

$$\|(P_i\mathbf{v})_{S_i}\|_1 \leq \rho \|(P_i\mathbf{v})_{\overline{S_i}}\|_1 + \tau \|\mathbf{A}\mathbf{v}\|_2,$$

for all $\mathbf{v} \in \mathbb{K}^N, 1 \leq i \leq n, S_i \subset W_i,$ and $|S_i| \leq s$.

Definition 3.3 is reminiscent of the work on sparse recovery with partially known support [**3**]. Note that in our case, there is no need to enforce a condition on the vector $\mathbf{v}$ to lie in the range of the other subspaces. In a sense, this is taken care of by the fusion process and the fact that we have multiple measurement vectors.

In CS, a stronger condition is often used to ensure recovery: the *Restricted Isometry Property* (RIP). A matrix is said to satisfy the $RIP(s, \delta)$ if it behaves like an isometry (up to a constant δ) on every s-sparse vector $\mathbf{v} \in \Sigma_s$. Formally speaking, $\mathbf{A} \in \mathbb{K}^{m \times N}$ satisfies $RIP(s, \delta)$, for some $s \geq 2$ and $\delta \in (0,1)$ if

$$(1-\delta)\|\mathbf{v}\|_2^2 \leq \|\mathbf{A}\mathbf{v}\|_2^2 \leq (1+\delta)\|\mathbf{v}\|_2^2, \quad \text{for every } \mathbf{v} \in \Sigma_s. \tag{3.1}$$

The lowest δ satisfying the inequalities is called the *restricted isometry constant.* We want to derive similar properties on our sensing matrix for the distributed sparse signal model.

DEFINITION 3.4 (Uniform Partial RIP (UP-RIP)). Let $\mathcal{W} = (W_i, P_i)_{i=1}^n$ be a fusion frame, and let $\mathbf{A} \in \mathbb{K}^{m \times N}$. Assume that $\mathbf{A}P_i$ satisfies the $RIP(s, \delta)$ on each W_i, with $\delta \in (0,1)$, $i \in I = \{1, \cdots, n\}$. Then, we say that $\mathbf{A} \in \mathbb{K}^{m \times N}$ satisfies the UP-RIP with respect to $\mathcal{W}$, with uniform bound δ and uniform sparsity pattern $\mathbf{s}$.

REMARK 3.5. Definition 3.4 reduces to the definition of the classical RIP when $n = 1$ (one projection, one subspace).

The UP-RIP is characterized by an inequality similar to Equation (3.1):

PROPOSITION 1. *Let $\mathcal{W} = (W_i)_{i=1}^n$ be a fusion frame for $\mathbb{K}^N$, with frame bounds $0 < C \leq D < \infty$. Let $\mathbf{A} \in \mathbb{K}^{m \times N}$ satisfy the UP-RIP with respect to $\mathcal{W}$, with uniform bound δ and sparsity pattern $\mathbf{s} = (s, \cdots, s)$, and let $C_o = C(1-\delta)$, $D_o = D(1+\delta)$. Then, for any $\mathbf{v} \in \Sigma_{\mathbf{s}}^{\mathcal{W}}$.*

$$C_o\|\mathbf{v}\|_2^2 \leq \sum_{i=1}^n \|\mathbf{A}\mathbf{v}_i\|_2^2 \leq D_o\|\mathbf{v}\|_2^2.$$

PROOF. Using the fusion frame inequality, and inequalities (3.1) on each W_i, $1 \leq i \leq n$, we obtain

$$\begin{aligned} C_o\|\mathbf{v}\|_2^2 &\leq (1-\delta)\sum_i \|\mathbf{v}_i\|_2^2 \leq \sum_i (1-\delta)\|\mathbf{v}_i\|_2^2 \leq \sum_i \|\mathbf{A}\mathbf{v}_i\|_2^2 \\ &\leq \sum_i (1+\delta)\|\mathbf{v}_i\|_2^2 \leq (1+\delta)\sum_i \|\mathbf{v}_i\|_2^2 \leq D_o\|\mathbf{v}\|_2^2. \end{aligned}$$

□

We recall a standard RIP result for subgaussian matrices (Thm9.2 in [**19**]):

THEOREM 3.6. *Let $\varepsilon > 0$. Let $\mathbf{A}$ be an $m \times N$ subgaussian random matrix. Then there exists a constant $C > 0$ (depending only on subgaussian parameters β, k) such that the RIP constant δ_s of $\frac{1}{\sqrt{m}}\mathbf{A}$ satisfies $\delta_s \leq \delta$ with probability at least $1-\varepsilon$, if*

$$m \geq C\delta^{-2}\left(s\ln(eN/s) + \ln(2\varepsilon^{-1})\right).$$

It is relatively easy to show that:

THEOREM 3.7. *Let $\varepsilon > 0$. Let $\mathcal{W} = (W_i, P_i)_{i=1}^n$ be a fusion frame for $\mathbb{K}^N$. Let $\mathbf{A} \in \mathbb{K}^{m\times N}$ be a subgaussian matrix with parameters β, k. Then, there exists a constant $C = C_{\beta,k}$ such that the UP-RIP constants of $\frac{1}{\sqrt{m}}\mathbf{A}$ satisfy $\delta_s \leq \delta$, with probability at least $1-\varepsilon$, provided*

$$m \geq C\frac{1}{\delta^2}\left(s\ln(eN/s) + \ln(2\varepsilon^{-1}n)\right).$$

REMARK 3.8. All Gaussian and Bernoulli random matrices are subgaussian random matrices, so Theorem 3.7 holds true at least for Gaussian and Bernoulli random matrices.

3.2. Recovery in general fusion frames settings. Using the tools introduced in Subsection 3.1, we show that any signals with uniform sparsity pattern $\mathbf{s}$ can be recovered in a stable and robust manner via a fusion frame approach.

THEOREM 3.9. *Let $\mathbf{A} \in \mathbb{K}^{m\times N}$ and $\mathcal{W} = (W_i, P_i)_{i=1}^n$ a fusion frame with frame bounds $0 < C \leq D < \infty$ and frame operator S. Let $\left(\mathbf{y}^{(i)}\right)_{i=1}^n$ be the linear measurements $\mathbf{y}^{(i)} = \mathbf{A}P_i\mathbf{x} + \mathbf{e}^{(i)}$, $1 \leq i \leq n$ for some uniformly bounded noise vectors $\mathbf{e}^{(i)}$ such that $\|\mathbf{e}^{(i)}\|_2 \leq \eta$. Denote by $\widehat{\mathbf{x}^{(i)}}$ the solution to the local Basis pursuit problems $(\mathcal{P}_{1,\eta})$.*

If the matrix $\mathbf{A}$ satisfies the UDP-NSP with sparsity pattern $\mathbf{s}$ with constants $0 < \rho < 1$ and $\tau > 0$ with respect to $\mathcal{W}$, then $\widehat{\mathbf{x}} = S^{-1}\sum_i \widehat{\mathbf{x}^{(i)}}$ approximates $\mathbf{x}$ in the following sense:

$$\|\widehat{\mathbf{x}} - \mathbf{x}\|_2 \leq \|\widehat{\mathbf{x}} - \mathbf{x}\|_1 \leq \frac{2}{C}\left(\frac{1+\rho}{1-\rho}\sum_{i=1}^n \sigma_{\mathbf{s}}(P_i\mathbf{x}) + \frac{2n\tau\eta}{1-\rho}\right). \tag{3.2}$$

PROOF. The solution is given by the fusion process $\widehat{\mathbf{x}} = S^{-1}\left(\sum_{i=1}^n \widehat{\mathbf{x}^{(i)}}\right)$ with $\widehat{\mathbf{x}^{(i)}}$ the solutions to the local problems $(\mathcal{P}_{1,\eta})$. It holds

$$\|\mathbf{x} - \widehat{\mathbf{x}}\|_2 = \left\|S^{-1}\left(\sum_{i=1}^n P_i\mathbf{x} - \sum_{i=1}^n \widehat{\mathbf{x}^{(i)}}\right)\right\|_2 \leq C^{-1}\sum_{i=1}^n \left\|P_i\mathbf{x} - \widehat{\mathbf{x}^{(i)}}\right\|_2.$$

Thus

$$\|\mathbf{x} - \widehat{\mathbf{x}}\|_2 \leq C^{-1} \sum_{i=1}^{n} \left\| P_i\mathbf{x} - \widehat{\mathbf{x}^{(i)}} \right\|_1 .$$

For each $i \in \{1, \cdots, n\}$, we estimate the error on subspace W_i in the $\boldsymbol{\ell}^1$ sense: if $\mathbf{v} := P_i\mathbf{x} - \widehat{\mathbf{x}^{(i)}}$ and $S_i \subset W_i$ the set of best s_i components of $\mathbf{x}$ supported on W_i, then the robust null space property yields

$$\|(P_i\mathbf{v})_{S_i}\|_1 \leq \rho \left\|(P_i\mathbf{v})_{\overline{S_i}}\right\|_1 + \tau\|\mathbf{A}\mathbf{v}\|_2.$$

Combining with [**19**, Lemma 4.15] stating

$$\left\|(P_i\mathbf{v})_{\overline{S_i}}\right\|_1 \leq \left\|P_i\widehat{\mathbf{x}^{(i)}}\right\|_1 - \|P_i\mathbf{x}\|_1 + \left\|(P_i\mathbf{v})_{S_i}\right\|_1 + 2\left\|(P_i\mathbf{x})_{\overline{S_i}}\right\|_1 .$$

Summing both inequalities, we arrive at

$$(1-\rho)\left\|(P_i\mathbf{v})_{\overline{S_i}}\right\|_1 \leq \|P_i\widehat{\mathbf{x}^{(i)}}\|_1 - \|P_i\mathbf{x}\|_1 + 2\left\|(P_i\mathbf{x})_{\overline{S_i}}\right\|_1 + \tau\|\mathbf{A}\mathbf{v}\|_2.$$

Applying once again the local robust NSP, it holds

$$\begin{aligned}
\|P_i\mathbf{v}\|_1 = \|(P_i\mathbf{v})_{S_i}\|_1 + \left\|(P_i\mathbf{v})_{\overline{S_i}}\right\|_1 &\leq \rho\left\|(P_i\mathbf{v})_{\overline{S_i}}\right\|_1 + \tau\|\mathbf{A}\mathbf{v}\|_2 + \left\|(P_i\mathbf{v})_{\overline{S_i}}\right\|_1 \\
&\leq (1+\rho)\left\|(P_i\mathbf{v})_{\overline{S_i}}\right\|_1 + \tau\|\mathbf{A}\mathbf{v}\|_2 \\
&\leq \frac{1+\rho}{1-\rho}\left(\|P_i\widehat{\mathbf{x}^{(i)}}\|_1 - \|P_i\mathbf{x}\|_1 + 2\left\|(P_i\mathbf{x})_{\overline{S_i}}\right\|_1\right) + \frac{4\tau}{1-\rho}\|\mathbf{A}\mathbf{v}\|_2.
\end{aligned}$$

Noticing that $P_i\widehat{\mathbf{x}^{(i)}} = \widehat{\mathbf{x}^{(i)}}$, and $\widehat{\mathbf{x}^{(i)}}$ being the optimal solution to $(\mathcal{P}_{1,\eta})$, it is clear that $\|\widehat{\mathbf{x}^{(i)}}\|_1 \leq \|P_i\mathbf{x}\|_1$ from what we can conclude that

$$\left\|P_i\mathbf{x} - \widehat{\mathbf{x}^{(i)}}\right\|_1 = \|P_i\mathbf{v}\|_1 \leq 2\frac{1+\rho}{1-\rho}\sigma_{\mathbf{s}}^{\mathcal{W}}(\mathbf{x})_{1,i} + \frac{4\tau}{1-\rho}\|\mathbf{A}\mathbf{v}\|_2. \tag{3.3}$$

Remembering that $\|\cdot\|_2 \leq \|\cdot\|_1$ and summing up the contributions for all i in $\{1, \cdots, n\}$ and applying the inverse frame operator finishes the proof. □

3.2.1. *UP-RIP based recovery.* We will now demonstrate that the UP-RIP is sufficient for stable and robust recovery; the following result, showing the existence of random matrices satisfying the UDP-NSP, is crucial to our argument.

THEOREM 3.10. *Let $\mathbf{A} \in \mathbb{K}^{m\times N}$ be a matrix satisfying the UP-RIP($2\mathbf{s}$,δ), with $\mathbf{s} = (s, \cdots, s)$ and $\delta < 4/\sqrt{41}$ on all W_i, $1 \leq i \leq n$. Then, $\mathbf{A}$ satisfies the UDP-NSP with constants*

$$\begin{aligned}
\rho &:= \tfrac{\delta}{\sqrt{1-\delta^2}-\delta/4} < 1 \\
\tau &:= \tfrac{\sqrt{1+\delta}\sqrt{s}}{\sqrt{1-\delta^2}-\delta/4}.
\end{aligned} \tag{3.4}$$

PROOF. The proof of this result consists in simply applying [**19**, Theorem 6.13] to every subspace independently. □

By Theorem 3.9 and Theorem 3.10, we have a first recovery result under UP-RIP:

THEOREM 3.11. *Let $\mathcal{W} = (W_i, P_i)_{i=1}^n$ be a fusion frame for $\mathbb{K}^N$ with frame operator S and frame bounds $0 < C \leq D < \infty$. Let $\mathbf{A} \in \mathbb{K}^{m\times N}$ be a matrix satisfying the UP-RIP($2\mathbf{s}$, δ) where $\delta < 4/\sqrt{41}$.*

Then any uniformly distributed sparse vector $\mathbf{x} \in \Sigma_{\mathbf{s}}^{(\mathcal{W})}$ *can be recovered by solving* n (BPDN) *problems, and the recovery is*

$$\widehat{\mathbf{x}} = S^{-1} \sum \widehat{\mathbf{x}^{(i)}}.$$

In addition, let the noise in each (BPDN) *problem be controlled by* $\|\mathbf{e}^{(i)}\|_2 \leq \eta$, $1 \leq i \leq n$. *Then*

$$\|\widehat{\mathbf{x}} - \mathbf{x}\|_2 \leq \frac{2\,(2\delta+1)\,C^{-1}}{1-\frac{41}{16}\delta^2} \left(\sum_{i=1}^{n} \sigma_s(P_i\mathbf{x})_1 + 2n\eta\sqrt{s}\sqrt{1+\delta} \right). \tag{3.5}$$

PROOF. By (3.4), it follows

$$\frac{1+\rho}{1-\rho} = \frac{1-\delta^2/16+2\delta\sqrt{1-\delta^2}}{1-\frac{41}{16}\delta^2}$$

$$\frac{\tau}{1-\rho} = \frac{\sqrt{s}\sqrt{1+\delta}\left(\sqrt{1-\delta^2}+\frac{5}{4}\delta\right)}{1-\frac{41}{16}\delta^2}$$

Due to inequality (3.2), $\|\widehat{\mathbf{x}} - \mathbf{x}\|_2$ is bounded from above by

$$\frac{2}{C}\sum_{i=1}^{n} \frac{\left(1-\delta^2/16+2\delta\sqrt{1-\delta^2}\right)\sigma_s(P_i\mathbf{x})_1 + 2\sqrt{1+\delta}\left(\sqrt{1-\delta^2}+\frac{5}{4}\delta\right)\sqrt{s}\eta}{1-\frac{41}{16}\delta^2} \tag{3.6}$$

Notice that $1-\delta^2/16+2\delta\sqrt{1-\delta^2} \leq 2\delta+1$ when $\delta \in (0, 4/\sqrt{41})$; also, $\sqrt{1-\delta^2}+\frac{5}{4}\delta \leq 1+\frac{5}{4}\delta < 1+2\delta$. Then, (3.6) implies

$$\|\widehat{\mathbf{x}} - \mathbf{x}\|_2 \leq \frac{2}{C}\sum_{i=1}^{n} \frac{(2\delta+1)\,\sigma_s(P_i\mathbf{x})_1 + 2\sqrt{1+\delta}\,(2\delta+1)\,\sqrt{s}\eta}{1-\frac{41}{16}\delta^2}$$

i.e.

$$\|\widehat{\mathbf{x}} - \mathbf{x}\|_2 \leq \frac{2\,(2\delta+1)\,C^{-1}}{1-\frac{41}{16}\delta^2} \left(\sum_{i=1}^{n} \sigma_s(P_i\mathbf{x})_1 + 2n\sqrt{s}\eta\sqrt{1+\delta} \right).$$

□

REMARK 3.12. Traditional results in CS theory show a decay of the ℓ^2 error in $s^{-1/2}\sigma_s(\mathbf{x})_1$. This is not the case here. However, such a bound can be achieved, when working with extensions of the NSP to the ℓ^q robust null space properties (see [**19**, 4.21] for the single subspace definition). This is actually a passage implicit in the proof of Theorem 3.10 (see for instance [**19**, Theorem 4.22] with the appropriate changes). We choose not to detail these results to ease the presentation.

4. Application to dense spectrum estimation

We derive here a simple, yet convincing, example of the application of our approach to the estimation of a dense Fourier spectrum from relatively cheap sensors.

We artificially construct a signal whose Fourier spectrum is sparse. To this end, we consider an $2N+1$-spectrum with $N = 2500$. 60% of these frequencies (i.e. 3000 indices) are selected at random and the magnitude is set uniformly at random in $[0, 1]$. An example of such a random spectrum is illustrated as the blue dots on Figure 2b.

We then split the spectrum in $n = 60$ equally sized frequency bands, such that each frequency is present in at least 3 bands. The set of projections in the fusion frame approach is then simply a direct band-pass filter. The associated frame

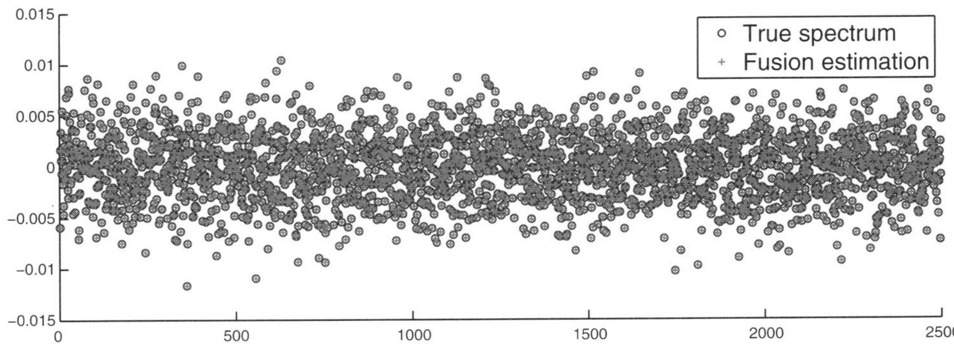

(A) Signal generated with a dense Fourier spectrum

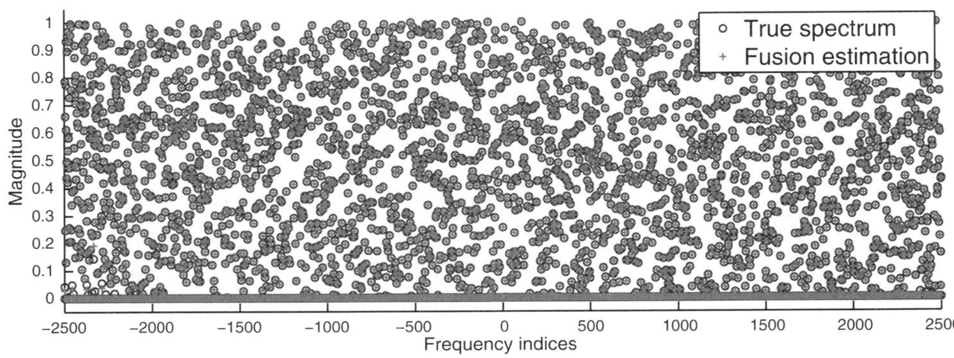

(B) Fourier spectrum: the support and the magnitudes are generated uniformly at random.

FIGURE 2. Noiseless recovery via the fused compressed sensing approach introduced. Blue circles are true values, while red crosses represent our estimations.

operator is hence $S = 3\mathbf{I}$. The measurements are taken via a random subsampled Fourier matrix. Consider the unscaled Fourier matrix $\mathcal{F}_N := \left(e^{\frac{\imath 2\pi k t}{N}}\right)_{-N\leq t,k\leq N}$ and extract m rows indexed by $\Omega \subset \{1, \cdots, 2N+1\}$ at random and construct the sensing matrix as $\mathbf{A} = \frac{1}{\sqrt{m}} \left(\mathcal{F}_N(i,k)\right)_{i\in\Omega, 1\leq k\leq 2N+1}$. While the optimal bound known so far for RIPs for Fourier matrices [**20**] suggest that m should scale as $m_{\text{theoretical}} \geq Cs\log(s)^2\log(N)$, we have here set

$$m = 2s\log(2N+1),$$

which is believed to be closer to the optimal bounds, and still numerically performs well. It corresponds in this situation to $m = 1197$ measurements per sensor. The sparse recovery is approximated by means of the HTP algorithm [**18**] to speed up slightly the calculations. The recovered spectrum after n local recovery and a fusion of the data can be seen as the red crosses on Figure 2b. This corresponds to the signal depicted in Figure 2a, where, once again, the blue circles corresponds to the true signal and the red crosses to the one generated after a Fast Fourier Transform on the recovered spectrum.

Lastly Figure 3 shows the result of our fused compressed sensing approach when independent additive Gaussian noise is included in the measurement process. As

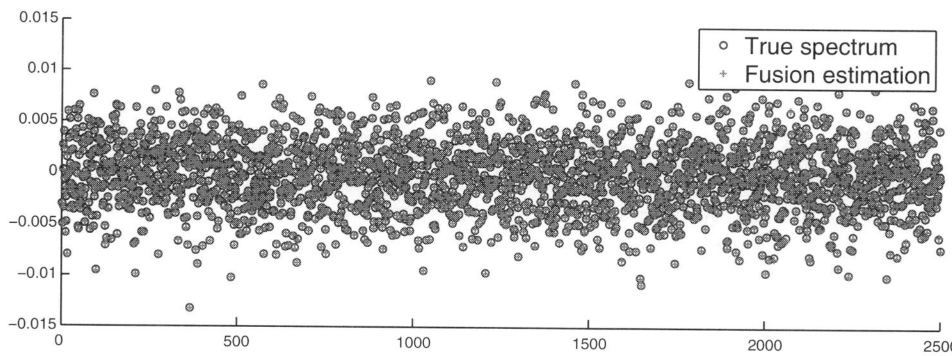

(A) Noisy signal recovery with additive Gaussian noise with variance 0.01.

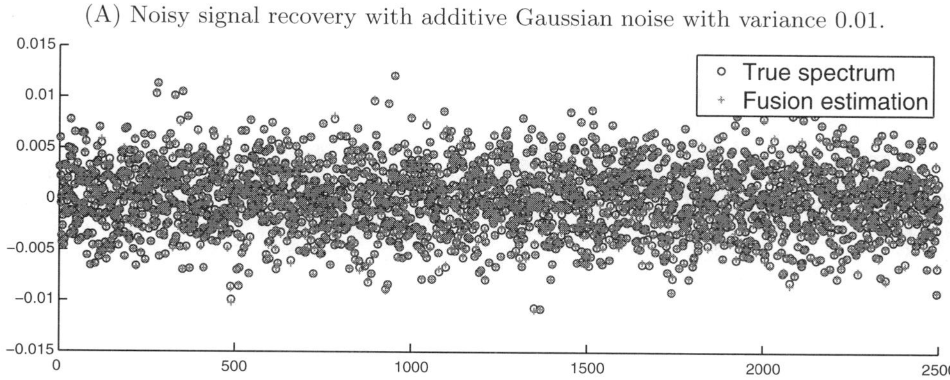

(B) Noisy signal recovery with additive Gaussian noise with variance 0.02.

FIGURE 3. Noisy signal recovery via our fused compressed sensing approach. Top corresponds to .01 additive Gaussian random noise to each sensor, while the bottom one corresponds to .02 additive Gaussian noise.

can be justified by our theorems, the approach is robust, even when fairly strong Gaussian noise is added. Figure 3b includes noise with 0.02 variance indepently at each sensor and yet yields almost perfect recovery of the signal. It is important to notice here that the sensors can be considered *very cheap* in the sense that they account for around 40% of the whole signal size, which in turns means that they also represent only 2/3 of the number of non-zeroes in the spectrum. In this particular case, there is no hope to recover the underlying spectrum without multiple measurements, and the use of the Fusion Frame approach allows for efficient computations and recovery without facing further problems.

5. Conclusion

In applications, it is beneficial to increase the number of non-zeros that can be recovered, and the related number of measurements scales with the sparsity/density. To avoid this explosion in sizes, we have combined ideas from distributed compressed sensing and fusion frames reconstruction methods to recover a signal $\mathbf{x} \in \mathbb{K}^N$ from the (noisy) measurements $\mathbf{y}^{(i)} = \mathbf{A}P_i\mathbf{x}+\mathbf{e}^{(i)}$, where the $\{P_i\}_{i=1}^n$ is a family of n projections in $\mathbb{K}^N$. We have solved this high-dimensional problem $\mathbf{y} = \mathbf{Ax}$ by combining results obtained from n localized problems, by means of fusion frames.

Each localized problem is significantly easier to solve. We have shown that our approach extends classical theories in compressed sensing to a distributed sampling setting. In particular, it allows to gives a strong mathematical foundations to the engineering problems that may require approaches similar to the ones we described here. One can assume non-uniform sparsity patterns, and nonuniform constants in order to expand Definitions 3.1, 3.3 and 3.4; such generalizations are explored in [**1**]. In particular, by fitting the projections appropriately to the sought to recover signal, one can easily analyze the block sparse model and the t-separation model. One may also be interested in ℓ^1 analysis models, which are investigated in [**1**] . In particular, this becomes handy when dealing with expansions in coherent frames.

References

[1] Roza Aceska, Jean-Luc Bouchot, and Shidong Li, *Local sparsity and recovery of fusion frames structured signals*, arXiv preprint arXiv:1604.00424, 2017.

[2] Ulaş Ayaz, Sjoerd Dirksen, and Holger Rauhut, *Uniform recovery of fusion frame structured sparse signals*, Appl. Comput. Harmon. Anal. **41** (2016), no. 2, 341–361, DOI 10.1016/j.acha.2016.03.006. MR3534442

[3] Afonso S. Bandeira, Katya Scheinberg, and Luis Nunes Vicente, *On partial sparse recovery*, arXiv preprint arXiv:1304.2809, 2013.

[4] Richard Baraniuk, Mark Davenport, Ronald DeVore, and Michael Wakin, *A simple proof of the restricted isometry property for random matrices*, Constr. Approx. **28** (2008), no. 3, 253–263, DOI 10.1007/s00365-007-9003-x. MR2453366

[5] Thomas Blumensath and Mike E. Davies, *Iterative hard thresholding for compressed sensing*, Appl. Comput. Harmon. Anal. **27** (2009), no. 3, 265–274, DOI 10.1016/j.acha.2009.04.002. MR2559726

[6] Jean-Luc Bouchot, *A generalized class of hard thresholding algorithms for sparse signal recovery*, Approximation theory XIV: San Antonio 2013, Springer Proc. Math. Stat., vol. 83, Springer, Cham, 2014, pp. 45–63, DOI 10.1007/978-3-319-06404-8_4. MR3218570

[7] Jean-Luc Bouchot, Simon Foucart, and Pawel Hitczenko, *Hard thresholding pursuit algorithms: number of iterations*, Appl. Comput. Harmon. Anal. **41** (2016), no. 2, 412–435, DOI 10.1016/j.acha.2016.03.002. MR3534445

[8] Petros Boufounos, Gitta Kutyniok, and Holger Rauhut, *Sparse recovery from combined fusion frame measurements*, IEEE Trans. Inform. Theory **57** (2011), no. 6, 3864–3876, DOI 10.1109/TIT.2011.2143890. MR2817059

[9] Claire Boyer, Jérémie Bigot, and Pierre Weiss, *Compressed sensing with structured sparsity and structured acquisition*, Applied and Computational Harmonic Analysis (2017).

[10] Jameson Cahill, Peter G. Casazza, and Shidong Li, *Non-orthogonal fusion frames and the sparsity of fusion frame operators*, J. Fourier Anal. Appl. **18** (2012), no. 2, 287–308, DOI 10.1007/s00041-011-9200-7. MR2898730

[11] Emmanuel J. Candès, *The restricted isometry property and its implications for compressed sensing* (English, with English and French summaries), C. R. Math. Acad. Sci. Paris **346** (2008), no. 9-10, 589–592, DOI 10.1016/j.crma.2008.03.014. MR2412803

[12] Peter G. Casazza and Gitta Kutyniok, *Frames of subspaces*, Wavelets, frames and operator theory, Contemp. Math., vol. 345, Amer. Math. Soc., Providence, RI, 2004, pp. 87–113, DOI 10.1090/conm/345/06242. MR2066823

[13] Peter G. Casazza, Gitta Kutyniok, and Shidong Li, *Fusion frames and distributed processing*, Appl. Comput. Harmon. Anal. **25** (2008), no. 1, 114–132, DOI 10.1016/j.acha.2007.10.001. MR2419707

[14] Ole Christensen, *An introduction to frames and Riesz bases*, Applied and Numerical Harmonic Analysis, Birkhäuser Boston, Inc., Boston, MA, 2003. MR1946982

[15] Il Yong Chun and Ben Adcock, *Uniform recovery from subgaussian multi-sensor measurements*, 2016.

[16] Il Yong Chun and Ben Adcock, *Compressed sensing and parallel acquisition*, IEEE Trans. Inform. Theory **63** (2017), no. 8, 4860–4882. MR3683541

[17] Massimo Fornasier and Holger Rauhut, *Compressive sensing*, Handbook of Mathematical Methods in Imaging (2011), 187–228.

[18] Simon Foucart, *Hard thresholding pursuit: an algorithm for compressive sensing*, SIAM J. Numer. Anal. **49** (2011), no. 6, 2543–2563, DOI 10.1137/100806278. MR2873246

[19] Simon Foucart and Holger Rauhut, *A mathematical introduction to compressive sensing*, Applied and Numerical Harmonic Analysis, Birkhäuser/Springer, New York, 2013. MR3100033

[20] Ishay Haviv and Oded Regev, *The restricted isometry property of subsampled Fourier matrices*, Proceedings of the Twenty-Seventh Annual ACM-SIAM Symposium on Discrete Algorithms, ACM, New York, 2016, pp. 288–297, DOI 10.1137/1.9781611974331.ch22. MR3478398

[21] D. Needell and J. A. Tropp, *CoSaMP: iterative signal recovery from incomplete and inaccurate samples*, Appl. Comput. Harmon. Anal. **26** (2009), no. 3, 301–321, DOI 10.1016/j.acha.2008.07.002. MR2502366

[22] Holger Rauhut, *Compressive sensing and structured random matrices*, Theoretical foundations and numerical methods for sparse recovery, Radon Ser. Comput. Appl. Math., vol. 9, Walter de Gruyter, Berlin, 2010, pp. 1–92, DOI 10.1515/9783110226157.1. MR2731597

[23] Tong Zhang, *Sparse recovery with orthogonal matching pursuit under RIP*, IEEE Trans. Inform. Theory **57** (2011), no. 9, 6215–6221, DOI 10.1109/TIT.2011.2162263. MR2857968

Department of Mathematical Sciences, Ball State University, Muncie, Indiana 47306
Email address: `raceska@bsu.edu`

Chair for Mathematics C (Analysis), RWTH Aachen University, Aachen, Germany
Email address: `bouchot@mathc.rwth-aachen.de`

Department of Mathematics, San Francisco State University, San Francisco, California 94132
Email address: `shidong@math.sfsu.edu`

Contemporary Mathematics
Volume **706**, 2018
http://dx.doi.org/10.1090/conm/706/14218

The Kadison-Singer problem

Marcin Bownik

Abstract. We give a self-contained presentation of results related to the Kadison-Singer problem, which was recently solved by Marcus, Spielman, and Srivastava [**35**]. This problem connects with an unusually large number of research areas including: operator algebras (pure states), set theory (ultrafilters), operator theory (paving), random matrix theory, linear and multilinear algebra, algebraic combinatorics (real stable polynomials), algebraic curves, frame theory, harmonic analysis (Fourier frames), and functional analysis.

1. Introduction

The goal of this paper is to give a self-contained presentation of mathematics involved in Kadison-Singer problem [**33**]. This problem was shown to be equivalent to a large number of problems such as: Anderson paving conjecture [**3**–**5**], Bourgain–Tzafriri restricted invertibility conjecture [**10**–**12**], Akemann–Anderson projection paving conjecture [**1**], Feichtinger conjecture [**15, 16, 26**], R_ϵ conjecture [**21**], and Weaver conjecture [**46**]. The breakthrough solution of the Weaver conjecture [**46**] by Marcus, Spielman, and Srivastava [**35**] has validated all of these conjectures.

While a lot has been written about Kadison-Singer problem before it was solved [**12, 15**–**19, 21**] and after its solution [**23, 36, 42**–**45**], we believe that there is still a space and an interest for yet another presentation which would present consequences of the solution of Kadison-Singer problem in an optimized form. We have aimed at presenting the material in a modular form as a sequence of implications which are largely independent of each other. This paper is the result of the author's attempts at achieving this goal. It is based on a series of lectures on the subject given by the author at University of Oregon, University of Gdańsk, Institute of Mathematics of Polish Academy of Sciences, and Tel Aviv University. The author is grateful for the interest and hospitality of these institutions.

The general outline of the paper can be described by the following diagram:

$$\begin{array}{ccccccccccccccccc}
(KS) & \Leftarrow & (PB) & \Leftarrow & (PS) & \Leftarrow & (PR) & \Leftarrow & (PP_{\frac{1}{2}}) & \Leftarrow & (KS_r^\infty) & \Leftarrow & (KS_r) & \Leftarrow & (MSS) & \Leftarrow & (MCP) \\
 & & & & & & & & & & \Downarrow & & \Downarrow & & & & \\
 & & & & & & & & (PP_\delta) & \Leftarrow & (KS_2^\infty) & & (FEI) & \Rightarrow & (R_\epsilon) & \Rightarrow & (BT)
\end{array}$$

2010 *Mathematics Subject Classification.* Primary 42C15, 47B15.

The author was partially supported by NSF grant DMS-1665056 and by a grant from the Simons Foundation #426295.

The above symbols represent abbreviations used throughout the paper. The most important are: the original Kadison-Singer problem (KS), Weaver's conjecture (KS_r), Marcus-Spielman-Srivastava solution (MSS), and its mixed characteristic polynomial formulation (MCP). In this paper we will prove all of the above implications including the proof of the core statement (MCP).

1.1. Notes. The existing literature discussing the solution of the Kadison-Singer problem is quite varied. It gives a deeper appreciation for the many areas of mathematics this problem has touched. Tao [**43**] has written a beautiful exposition containing a simplified proof of the solution. Tanbay [**42**] has given a nice entertaining historical perspective on the Kadison-Singer problem. Matheron [**36**] gives a long and exhaustive exposition (in French), primarily from the viewpoint of operator algebras. Valette [**45**] has written a Bourbaki exposition (in French). Casazza, who has worked and popularized the Kadison-Singer problem, has written a joint paper with Tremain [**23**] discussing consequences of the solution. Timotin [**44**] gives another presentation of the proof of the Kadison-Singer problem. Harvey [**30**] gives a gentle introduction aimed at readers without much background in functional analysis or operator theory. Finally, the book by Stevens [**40**] contains a relatively elementary and self-contained account of the Kadison-Singer problem and its proof.

2. From Kadison-Singer problem to Weaver's conjecture

2.1. Kadison-Singer problem. We start with the Kadison-Singer problem [**33**], now a theorem, which was originally formulated in 1959.

DEFINITION 2.1. Let $\mathcal{D} \subset \mathcal{B}(\ell^2(\mathbb{N}))$ be the algebra of diagonal matrices. A state $s : \mathcal{D} \to \mathbb{C}$ is a positive bounded linear functional ($A \geq 0 \implies s(A) \geq 0$) such that $s(\mathbf{I}) = 1$. A state is pure if it is not a convex combination of other states.

THEOREM (KS). *Let $s : \mathcal{D} \to \mathbb{C}$ be a pure state. Then, there exists a unique state $\tilde{s} : \mathcal{B}(\ell^2(\mathbb{N})) \to \mathbb{C}$ that extends s.*

The original formulation of Kadison-Singer problem involves the concept of a maximal abelian self-adjoint (MASA) subalgebra $\mathcal{A}$ of $\mathcal{B}(\mathcal{H})$, where $\mathcal{H}$ is an infinite-dimensional separable Hilbert space. Kadison and Singer [**33**] have shown that every MASA $\mathcal{A}$ decomposes into discrete and continuous parts. More precisely, there exists an orthogonal decomposition $\mathcal{H} = \mathcal{H}_d \oplus \mathcal{H}_c$ with P denoting the orthogonal projection of $\mathcal{H}$ onto $\mathcal{H}_d$ such that:

- $\mathcal{A}_d = \{PA|_{\mathcal{H}_d} : A \in \mathcal{A}\}$ is a discrete MASA, i.e., $\mathcal{A}_d$ is the commutant of the set of its minimal projections,
- $\mathcal{A}_c = \{(\mathbf{I} - P)A|_{\mathcal{H}_c} : A \in \mathcal{A}\}$ is a continuous MASA, i.e., $\mathcal{A}_c$ contains no minimal projections.

An example of a continuous MASA $\mathcal{A}_c$ are multiplication operators on $L^2[0,1]$ by functions in $L^\infty[0,1]$. Kadison and Singer have shown that as long as $\mathcal{H}_c \neq \{0\}$, there exists a pure state on $\mathcal{A}$, which has non-unique state extensions on $\mathcal{B}(\mathcal{H})$. They have hinted [**33**, §5] that the same might hold in general though they were careful to state this in a form of a question, rather than a conjecture.

PROBLEM (Kadison-Singer). *Let $\mathcal{H}$ be an infinite-dimensional separable Hilbert space and let $\mathcal{A}$ be a discrete maximal abelian self-adjoint subalgebra (MASA) of $\mathcal{B}(\mathcal{H})$. Does every pure state on $\mathcal{A}$ extend to a unique pure state on $\mathcal{B}(\mathcal{H})$?*

One can show that every discrete MASA $\mathcal{A}$ is unitarily equivalent with the diagonal matrix algebra $\mathcal{D}$ in Theorem (KS). That is, there exists a unitary $U : \mathcal{H} \to \ell^2(\mathbb{N})$ such that $\mathcal{A} = U^*\mathcal{D}U$. Hence, Theorem (KS) gives an affirmative answer to the Kadison-Singer problem.

Clearly, the diagonal matrix algebra $\mathcal{D}$ is isometrically isomorphic with $\ell^\infty(\mathbb{N})$. That is, $x \in \ell^\infty(\mathbb{N})$ corresponds to diagonal operator $\operatorname{diag}(x)$ with sequence x on the main diagonal. Pure states on $\ell^\infty(\mathbb{N})$ can be described in terms of the ultrafilters.

DEFINITION 2.2. Let $\mathcal{F}$ be a collection of non-empty subsets of $\mathbb{N}$. We say that $\mathcal{F}$ is a filter if:

(i) if $F_1, \dots, F_n \in \mathcal{F}$, $n \geq 1$, then $F_1 \cap \dots \cap F_n \in \mathcal{F}$,
(ii) if $F \in \mathcal{F}$ and $F \subset G \subset \mathbb{N}$, then $G \in \mathcal{F}$.

We say that $\mathcal{U}$ is an ultrafilter, if it is a maximal filter with respect to the inclusion partial order. Equivalently,

(iii) for any $A \subset \mathbb{N}$, either $A \in \mathcal{U}$ or $\mathbb{N} \setminus A \in \mathcal{U}$.

Given an ultrafiter $\mathcal{U}$, we can define the concept of a limit of a bounded sequence on $\mathbb{N}$.

DEFINITION 2.3. Fix an ultrafilter $\mathcal{U}$. Let $x = (x_j)_{j\in\mathbb{N}} \in \ell^\infty(\mathbb{N})$. For any subset $A \subset \mathbb{N}$, define

$$C_A = \overline{\{x_j : j \in A\}} \subset \mathbb{C}.$$

We define

$$\lim_{\mathcal{U}} x = x_0 \iff \{x_0\} = \bigcap_{A\in\mathcal{U}} C_A.$$

It is easy to see that the above limit is always well-defined. The intersection of any finite family of compact sets C_A, $A \in \mathcal{U}$ is non-empty. Hence, the entire intersection is non-empty as well. Moreover, it consists of exactly one point. On the contrary, if it contained two points $x_0 \neq \tilde{x}_0$, then we would consider the set

$$A = \{j \in \mathbb{N} : |x_j - x_0| < |x_0 - \tilde{x}_0|/2\}.$$

Then by the ultrafilter property (iii), we have two possibilities. Either $A \in \mathcal{U}$, which forces $\tilde{x}_0$ outside the intersection $\bigcap_{A\in\mathcal{U}} C_A$, or $\mathbb{N} \setminus A \in \mathcal{U}$, which forces $\tilde{x}_0$ outside. Either way, the above intersection must be a singleton.

LEMMA 2.1. *There is one-to-one correspondence between pure state on the algebra $\mathcal{D}$ of diagonal matrices and ultrafilters on $\mathbb{N}$. More precisely, each pure state s on $\mathcal{D} \cong \ell^\infty(\mathbb{N})$ is of the form*

$$s(\operatorname{diag}(x)) = \lim_{\mathcal{U}} x \qquad \text{for all } x \in \ell^\infty(\mathbb{N}) \tag{2.1}$$

for some unique ultrafilter $\mathcal{U}$.

PROOF. Suppose that s is a pure state on $\ell^\infty(\mathbb{N})$. For $A \subset \mathbb{N}$, let P_A be the orthogonal projection of $\ell^2(\mathbb{N})$ onto $\overline{\operatorname{span}}\{e_j : j \in A\}$, where $\{e_j\}_{j\in\mathbb{N}}$ is a standard o.n. basis of $\ell^2(\mathbb{N})$. Define

$$\mathcal{U} = \{A \subset \mathbb{N} : s(P_A) = 1\}.$$

We claim that $\mathcal{U}$ is an ultrafilter. This can be shown in two ways. The Čech-Stone compactification of $\mathbb{N}$ is a maximal compact Hausdorff space $\beta\mathbb{N}$, which contains $\mathbb{N}$ as a dense subset. By the universal property of $\beta\mathbb{N}$, the space $\ell^\infty(\mathbb{N})$ is isometrically isomorphic with $C(\beta\mathbb{N})$. By the Riesz representation theorem, positive functionals on $C(\beta\mathbb{N})$ are identified with positive regular Borel measures. Hence, s corresponds to a probabilistic measure on $\beta\mathbb{N}$. In addition, since s is pure, this measure must be a point measure on $\beta\mathbb{N}$. In particular, s is commutative. Hence, $s(P_A) = s(P_A)^2$, which implies that $s(P_A) \in \{0,1\}$. Likewise,

$$s(P_{A\cap B}) = s(P_A)s(P_B) \qquad \text{for any } A, B \subset \mathbb{N}, \tag{2.2}$$

which implies that $\mathcal{U}$ is an ultrafilter.

This can also be seen by a direct argument as follows. Since P_A and $\mathbf{I} - P_A = P_{\mathbb{N}\setminus A}$ are both positive, we have $0 \le s(P_A) \le s(\mathbf{I}) = 1$. Suppose that $s(P_A) = \theta$ for some $0 < \theta < 1$. Then, we can write $s = \theta s_1 + (1-\theta)s_2$, where $s_1(T) = \frac{1}{\theta}s(P_A T)$ and $s_2(T) = \frac{1}{1-\theta}s((\mathbf{I} - P_A)T)$ for $T \in \mathcal{D}$. It is easy to show that s_1 and s_2 are states on $\mathcal{D}$, which contradicts that s is a pure state. Consequently, $s(P_A) \in \{0,1\}$. By the positivity of s, it is clear that (2.2) holds if either $s(P_A) = 0$ or $s(P_B) = 1$. Now, if $s(P_A) = s(P_B) = 1$, then

$$s(P_{\mathbb{N}\setminus(A\cap B)}) = s(P_{(\mathbb{N}\setminus A)\cup(\mathbb{N}\setminus B)}) \le s(P_{\mathbb{N}\setminus A}) + s(P_{\mathbb{N}\setminus B}) = 0,$$

This shows (2.2), which again implies that $\mathcal{U}$ is an ultrafilter.

Every $x \in \ell^\infty(\mathbb{N})$ can be approximated in norm by simple functions, i.e., finite linear combinations of indicator functions $\mathbf{1}_{A_i}$ for disjoint subsets $A_i \subset \mathbb{N}$, $i = 1, \dots, n$. By definition

$$s(\operatorname{diag}(\mathbf{1}_{A_i})) = s(P_{A_i}) = \lim_{\mathcal{U}} \mathbf{1}_{A_i} = \begin{cases} 1 & A_i \in \mathcal{U}, \\ 0 & \text{otherwise.} \end{cases}$$

Thus, (2.1) holds for indicator functions and by the density argument for all $x \in \ell^\infty(\mathbb{N})$. This implies that two distinct pure states must correspond to distinct ultrafilters, which shows one-to-one correspondence. $\square$

2.2. Paving conjectures. To crack Theorem (KS) Anderson has proposed the concept of paving. We will adopt the following definition.

DEFINITION 2.4. Let $T \in \mathcal{B}(\ell^2(I))$, where I is at most countable. We say that T has (r,ε)-paving if there exists a partition $\{A_1, \dots, A_r\}$ of I such that

$$||P_{A_j} T P_{A_j}|| \le \varepsilon ||T|| \qquad \text{for } j = 1, \dots, r. \tag{2.3}$$

Here, for $A \subset I$, let P_A be the orthogonal projection of $\ell^2(I)$ onto $\overline{\operatorname{span}}\{e_i : i \in A\}$, where $\{e_i\}_{i\in I}$ is a standard o.n. basis of $\ell^2(I)$.

The following result states the paving conjecture for bounded operators with zero diagonal.

THEOREM (PB). *For every $\varepsilon > 0$, there exists $r = r(\varepsilon)$ such that every $T \in \mathcal{B}(\ell^2(I))$ with zero diagonal can be (r,ε)-paved.*

We are now ready to establish the first implication in our scheme.

LEMMA 2.2. *$(PB) \implies (KS)$.*

PROOF. Let $\mathbb{E} : \mathcal{B}(\ell^2(\mathbb{N})) \to \mathcal{D}$ be the non-commutative conditional expectation which erases all off-diagonal entries. That is, for any $T \in \mathcal{B}(\ell^2(\mathbb{N}))$, let $\mathbb{E}(T)$ be the diagonal operator which has the same diagonal entries as T. Let s be any pure state on $\mathcal{D}$. It is easy to show that $\tilde{s}(T) = s(\mathbb{E}(T))$, $T \in \mathcal{B}(\ell^2(\mathbb{N}))$, defines a state extending s. Hence, the difficult part is showing the uniqueness.

Let $\tilde{s} : \mathcal{B}(\ell^2(\mathbb{N})) \to \mathbb{C}$ be any state extending s. Since

$$\tilde{s}(T) = \tilde{s}(T - \mathbb{E}(T)) + s(\mathbb{E}(T)),$$

it suffices to show that

$$\tilde{s}(T) = 0 \qquad \text{for all } T \in \mathcal{B}(\ell^2(\mathbb{N})) \text{ with } \mathbb{E}(T) = \mathbf{0}. \tag{2.4}$$

By (PB) for any $\varepsilon > 0$ we can find $A_1, \dots, A_r$ such that (2.3) holds. By the ultrafilter property, there exists $j_0 \in [r] := \{1, \dots, r\}$ such that

$$s(P_{A_j}) = \delta_{j,j_0} \qquad \text{for } j \in [r].$$

One can easily verify that

$$\langle T_1, T_2 \rangle := \tilde{s}(T_1 T_2^*) \qquad T_1, T_2 \in \mathcal{B}(\ell^2(\mathbb{N})),$$

defines a semidefinite inner product on $\mathcal{B}(\ell^2(\mathbb{N}))$. In particular, by the Cauchy-Schwarz inequality we have

$$|\tilde{s}(T_1 T_2^*)|^2 \le \tilde{s}(T_1 T_1^*) \tilde{s}(T_2 T_2^*).$$

Thus, for any $j \neq j_0$ and $R \in \mathcal{B}(\ell^2(\mathbb{N}))$ we have

$$0 = \tilde{s}(R P_{A_j}) = \tilde{s}(P_{A_j} R).$$

We conclude that

$$\tilde{s}(T) = \sum_{j=1}^{r} \tilde{s}(T P_{A_j}) = \tilde{s}(T P_{A_{j_0}}) = \tilde{s}(P_{A_{j_0}} T P_{A_{j_0}}).$$

Thus, $|\tilde{s}(T)| \le ||P_{A_{j_0}} T P_{A_{j_0}}|| \le \varepsilon ||T||$. Since $\varepsilon > 0$ is arbitrary, this shows (2.4). □

Paving conjectures can be formulated for smaller classes of operators than bounded (PB) such as: self-adjoint operators (PS), reflections (PR), and projections $(PP_{\frac{1}{2}})$.

THEOREM (PS). *For every $\varepsilon > 0$, there exists $r = r(\varepsilon)$ such that every self-adjoint operator S on $\ell^2(I)$ with zero diagonal can be (r, ε)-paved.*

THEOREM (PR). *For every $\varepsilon > 0$, there exists $r = r(\varepsilon)$ such that every reflection R on $\ell^2(I)$, i.e., $R = R^*$ and $R^2 = \mathbf{I}$, with zero diagonal can be (r, ε)-paved.*

THEOREM $(PP_{\frac{1}{2}})$. *For every $\varepsilon > 0$, there exists $r = r(\varepsilon)$ such that every projection P on $\ell^2(I)$, i.e., $P = P^*$ and $P^2 = P$, with all diagonal entries equal to $\frac{1}{2}$ can be $(r, \frac{1+\varepsilon}{2})$-paved.*

While the implication $(PS) \implies (PB)$ is trivial, we need to show the converse implication. At the same time we shall keep track how the paving parameters (r, ε) are affected by these implications.

LEMMA 2.3. *(PS) holds for (r, ε) $\implies$ (PB) holds for $(r^2, 2\varepsilon)$.*

PROOF. Take any $T \in \mathcal{B}(\ell^2(I))$ with $\mathbb{E}(T) = \mathbf{0}$. We decompose it as sum of self-adjoint and skew-adjoint operators

$$T = S_1 + S_2 \qquad \text{where } S_1 = \frac{T + T^*}{2}, \ S_2 = \frac{T - T^*}{2}.$$

By the paving property (PB) for S_1 and $iS_2 = (iS_2)^*$ we can find partitions $\{A_1, \dots, A_r\}$ and $\{B_1, \dots, B_r\}$ such that

$$||P_{A_i \cap B_j} S_1 P_{A_i \cap B_j}|| \le ||P_{A_i} S_1 P_{A_i}|| \le \varepsilon ||S_1|| \le \varepsilon ||T|| \qquad i, j \in [r].$$

Since the same estimate holds for S_2 we have

$$||P_{A_i\cap B_j}TP_{A_i\cap B_j}|| \le ||P_{A_i\cap B_j}S_1P_{A_i\cap B_j}||+||P_{A_i\cap B_j}S_2P_{A_i\cap B_j}|| \le 2\varepsilon||T|| \quad i,j\in[r].$$

Hence, the partition $\{A_i \cap B_j\}_{i,j\in[r]}$ yields $(r^2, 2\varepsilon)$ paving of T. □

LEMMA 2.4. *(PR) holds for (r,ε) $\implies$ (PS) holds for (r,ε).*

PROOF. Take any $S = S^* \in \mathcal{B}(\ell^2(I))$ with $\mathbb{E}(S) = \mathbf{0}$. Without loss of generality assume that $||S|| = 1$. Consider an operator R on $\ell^2(I) \oplus \ell^2(I)$ given by

$$\begin{bmatrix} S & \sqrt{\mathbf{I}-S^2} \\ \sqrt{\mathbf{I}-S^2} & -S \end{bmatrix}.$$

A direct calculation shows that $R^2 = \mathbf{I}$, $R = R^*$, and $\mathbb{E}(R) = \mathbf{0}$. That is, R is a reflection on $\ell^2(I \cup I')$, where I' is a copy of the index set I. By (PR) there exists a partition of $I \cup I'$ which yields (r,ε) paving of R. Restricting this partition to I yields (r,ε) paving of S. □

LEMMA 2.5. *$(PP_{\frac{1}{2}})$ holds for $(r, \frac{1+\varepsilon}{2})$ $\implies$ (PR) holds for (r^2,ε).*

PROOF. Take any reflection $R \in \mathcal{B}(\ell^2(I))$ with $\mathbb{E}(R) = \mathbf{0}$. Define $Q = (\mathbf{I} + R)/2$. Then, Q is a projection $Q = Q^* = Q^2$ with $\mathbb{E}(Q) = \frac{1}{2}\mathbf{I}$. Suppose that for some $A \subset I$ we have

$$||P_AQP_A|| \le \beta := \frac{1+\varepsilon}{2}.$$

Since Q is positive this can be phrased in terms of the partial order $\le$ on self-adjoint operators

$$\mathbf{0} \le P_AQP_A \le \beta P_A.$$

Since $R = 2Q - \mathbf{I}$ we obtain

$$-P_A \le P_ARP_A \le (2\beta - 1)P_A = \varepsilon P_A. \tag{2.5}$$

We repeat the same for a projection $Q_1 = (\mathbf{I} - R)/2$. Assuming that for some $B \subset I$ we have

$$||P_BQ_1P_B|| \le \beta := \frac{1+\varepsilon}{2},$$

yields

$$-P_B \le P_B(-R)P_B \le (2\beta - 1)P_B = \varepsilon P_B. \tag{2.6}$$

Taking $C = A \cap B$ and combining (2.5) and (2.6) yields

$$-\varepsilon P_C = -\varepsilon P_CP_BP_C \le P_CRP_C \le \varepsilon P_CP_AP_C = \varepsilon P_C. \tag{2.7}$$

Hence, $||P_CRP_C|| \le \varepsilon$. By the paving property $(PP_{\frac{1}{2}})$, we can find partitions $\{A_1,\dots,A_r\}$ and $\{B_1,\dots,B_r\}$ which produce $(r, \frac{1+\varepsilon}{2})$-paving of Q and Q_1, resp. By (2.7) their common refinement partition $\{C_{i,j} = A_i \cap B_j\}_{i,j\in[r]}$ yields (r^2,ε)-paving of R. □

2.3. Weaver's conjecture. Next, we will show that paving conjectures follow from Weaver's KS_r conjecture, which was verified by Marcus, Spielman, and Srivastava [**35**]. We state it in a general infinite dimensional form (KS_r^∞) and later deduce it from its finite dimensional counterpart (KS_r). We start with the standard definition in frame theory.

DEFINITION 2.5. A family of vectors $\{u_i\}_{i\in I}$ in a Hilbert space $\mathcal{H}$ is called a *frame* for $\mathcal{H}$ if there are constants $0 < A \le B < \infty$ (called *lower and upper frame bounds*, respectively) so that

$$A\|u\|^2 \le \sum_{i\in I} |\langle u, u_i\rangle|^2 \le B\|u\|^2 \qquad \text{for all } u \in \mathcal{H}. \tag{2.8}$$

If we only have the right hand inequality in (2.8), we call $\{u_i\}_{i\in I}$ a *Bessel sequence* with Bessel bound B. If $A = B$, $\{u_i\}_{i\in I}$ is called a *tight frame* and if $A = B = 1$, it is called a *Parseval frame.*

THEOREM (KS_r^∞). *Let I be at most countable index set and let $\mathcal{H}$ be a separable Hilbert space. Let $\{u_i\}_{i\in I} \subset \mathcal{H}$ be a Bessel sequence with bound 1,*

$$\sum_{i\in I} |\langle u, u_i\rangle|^2 \le 1 \quad \textit{for all } ||u|| = 1 \qquad \textit{and} \qquad ||u_i||^2 \le \delta \quad \textit{for all } i. \tag{2.9}$$

Then for any positive integer r, there exists a partition $\{I_1, \ldots, I_r\}$ of I such that each $\{u_i\}_{i\in I_k}$, $k = 1, \ldots, r$, is a Bessel sequence with the following bound

$$\sum_{i\in I_k} |\langle u, u_i\rangle|^2 \le \left(\frac{1}{\sqrt{r}} + \sqrt{\delta}\right)^2 \qquad \textit{for all } ||u|| = 1. \tag{2.10}$$

Next we show how (KS_r^∞) implies projection paving.

LEMMA 2.6. $(KS_r^\infty) \implies (PP_{\frac{1}{2}})$.

PROOF. Let Q be an arbitrary projection on $\ell^2(I)$ with $\mathbb{E}(Q) = \frac{1}{2}\mathbf{I}$. Define vectors $u_i = Qe_i$, $i \in I$, where $\{e_i\}_{i\in I}$ is a standard o.n. basis of $\ell^2(I)$. Then, Q is represented by the Gram matrix of $\{u_i\}_{i\in I}$

$$Q = (\langle Qe_i, Qe_j\rangle)_{i,j\in I} = (\langle u_i, u_j\rangle)_{i,j\in I} \qquad \text{and} \qquad ||u_i||^2 = \frac{1}{2} = \delta.$$

The Gram matrix $Q = TT^*$ is a composition of the analysis operator

$$T : \mathcal{H} \to \ell^2(I) \qquad \text{where } Tu = (\langle u, u_i\rangle)_{i\in I} \quad \text{for } u \in \mathcal{H},$$

with the synthesis operator

$$T^* : \ell^2(I) \to \mathcal{H} \qquad \text{where } Ta = \sum_{i\in I} a_i u_i \quad \text{for } a = (a_i)_{i\in I} \in \ell^2(I).$$

The frame operator is a composition of these operators, but in a reverse order

$$S : \mathcal{H} \to \mathcal{H} \qquad S = T^*T = \sum_{i\in I} u_i \otimes u_i,$$

where $u_i \otimes u_i : \mathcal{H} \to \mathcal{H}$ is a rank one positive operator given by

$$(u_i \otimes u_i)(u) = \langle u, u_i\rangle u_i \quad \text{for } u \in \mathcal{H}.$$

By (KS_r^∞), for any $r \in \mathbb{N}$, there exists a partition $\{I_1, \dots, I_r\}$ such that

$$||P_{I_k} Q P_{I_k}|| = ||(\langle u_i, u_j \rangle)_{i,j \in I_k}|| = \left\| \sum_{i \in I_k} u_i \otimes u_i \right\| \le \left(\frac{1}{\sqrt{r}} + \frac{1}{\sqrt{2}} \right)^2 < \frac{1}{2} + \frac{3}{\sqrt{r}}.$$

The second equality is the consequence of the fact the norms of Gram and frame operator are the same $||TT^*|| = ||T^*T||$. Thus, Q can be $(r, \frac{1+\varepsilon}{2})$-paved for $r = 36/\varepsilon^2$. □

Marcus, Spielman, and Srivastava [**35**] have shown the following version of Weaver's conjecture. The key feature of (KS_r) is independence of Bessel bound on a number of vectors m and a dimension d.

THEOREM (KS_r). *Let $\{u_i\}_{i \in [m]} \subset \mathbb{C}^d$ be a Parseval frame*

$$\sum_{i=1}^m |\langle u, u_i \rangle|^2 = 1 \quad \text{for all } ||u|| = 1 \qquad \text{and} \qquad ||u_i||^2 \le \delta \quad \text{for all } i. \tag{2.11}$$

Then for any positive integer r, there exists a partition $\{I_1, \dots, I_r\}$ of $[m]$ such that each $\{u_i\}_{i \in I_k}$, $k = 1, \dots, r$, is a Bessel sequence with bound $\left(\frac{1}{\sqrt{r}} + \sqrt{\delta}\right)^2$, i.e., (2.10) *holds.*

To deduce Theorem (KS_r^∞) from (KS_r) we will use the following fact, which is sometimes referred to as a pinball principle. Its proof is essentially a combination of diagonal argument with pigeonhole principle.

LEMMA 2.7. *Fix a natural number r and assume for every natural number n, we have a partition $\{I_i^n\}_{i=1}^r$ of $[n]$. Then there are natural numbers $\{n_1 < n_2 < \cdots\}$ so that if $j \in I_i^{n_j}$ for some $i \in [r]$, then $j \in I_i^{n_k}$ for all $k \ge j$. For any $i \in [r]$ define $I_i = \{j : j \in I_i^{n_j}\}$. Then,*

(i) *$\{I_i\}_{i=1}^r$ is a partition of $\mathbb{N}$.*
(ii) *If $I_i = \{j_1 < j_2 < \cdots\}$, then for every natural number k we have*

$$\{j_1, j_2, \dots, j_k\} \subset I_i^{n_{j_k}}.$$

Instead of giving a separate proof of Lemma 2.7, we include its justification in the proof of Lemma 2.8.

LEMMA 2.8. $(KS_r) \implies (KS_r^\infty)$.

PROOF. First, observe that the Parseval frame assumption (2.11) can be weakened by the Bessel condition. Indeed, suppose that $\{u_i\}_{i \in [m]}$ is merely a Bessel sequence with bound 1 and $||u_i||^2 \le \delta$. Define $d \times d$ matrix T as

$$T = \mathbf{I} - \sum_{i=1}^m u_i \otimes u_i.$$

Since T is positive semidefinite, we can find vectors $\{u_i\}_{i=m+1}^{m'}$, $m' > m$, such that

$$T = \sum_{i=m+1}^{m'} u_i \otimes u_i \qquad \text{and} \qquad ||u_i||^2 \le \delta \text{ for } i \ge m+1.$$

Indeed, it suffices to choose vectors u_i to be appropriately scaled eigenvectors of T. Consequently, $\{u_i\}_{i \in [m']}$ becomes a Parseval frame for $\mathbb{C}^d$ and by (KS_r) we can find a partition $\{I_1, \dots, I_r\}$ of $[m']$ such that corresponding subsets $\{u_i\}_{i \in I_k}$ have required Bessel bounds. Restricting these partition to $[m]$ yields the same conclusion for $\{u_i\}_{i \in I_k \cap [m]}$, $k = 1, \dots, r$.

Now suppose $\{u_i\}_{i \in I}$ is an infinite Bessel sequence in a Hilbert space $\mathcal{H}$ as in (2.9). Since I is countable, we may assume $I = \mathbb{N}$. For any $n \in \mathbb{N}$ we can apply (KS_r) to the initial sequence $\{u_i\}_{i \in [n]}$. Hence, for each $n \in \mathbb{N}$ we have a partition $\{I_1^n, \dots, I_r^n\}$ of $[n]$, which yields required Bessel bounds. To show the existence of a global partition of $\{I_1, \dots, I_r\}$ of $\mathbb{N}$ satisfying (2.10), it suffices to apply Lemma 2.7. This boils down to repeated applications of pigeonhole principle. The first vector u_1 must land infinitely many times to one of the slots $I_{j_1}^n$ for some $j_1 = 1, \dots, r$. Let $N_1 \subset \mathbb{N}$ be the collection of such n. Then, we repeat the same argument to the second vector u_2 for partitions of $[n]$, where $n \in N_1$. Again, we can find a slot $I_{j_2}^n$, where the second vector u_2 lands for infinitely many $n \in N_2 \subset N_1$. Repeating this process yields a nested sequence of infinite subsets $N_1 \supset N_2 \supset \dots$ and indices $j_1, j_2, \dots$ in $[r]$ such that the initial vectors $u_1, \dots, u_m$, $m \in \mathbb{N}$, all land to the same respective slots $I_{j_1}^n, \dots, I_{j_m}^n$ for all $n \in N_m$. This yields a global partition of $\mathbb{N}$ by $I_k = \{i \in \mathbb{N} : j_i = k\}$, $k \in [r]$. Thus, (2.10) holds when I_k replaced by $I_k \cap [m]$. Letting $m \to \infty$ shows the required Bessel bound (2.10). □

An interesting special case of Weaver's conjecture (KS_r) happens when $r = 2$.

THEOREM (KS_2^∞). *Let I be at most countable index set and let $\mathcal{H}$ be a separable Hilbert space. Let $\{u_i\}_{i \in I} \subset \mathcal{H}$ be a Parseval frame and $\|u_i\|^2 \le \delta$ for all i. Then, there exists a partition $\{I_1, I_2\}$ of I such that each $\{u_i\}_{i \in I_k}$, $k = 1, 2$ is a frame with bounds*

$$\frac{1}{2} - O(\sqrt{\delta}) \le \sum_{i \in I_k} |\langle u, u_i \rangle|^2 \le \frac{1}{2} + O(\sqrt{\delta}) \qquad \text{for all } \|u\| = 1. \tag{2.12}$$

LEMMA 2.9. $(KS_r^\infty) \implies (KS_2^\infty)$.

PROOF. (KS_r) for $r = 2$ yields partition $\{I_1, I_2\}$ such that

$$\sum_{i \in I_k} |\langle u, u_i \rangle|^2 \le \frac{1}{2} + \sqrt{2\delta} + \delta \qquad \text{for } \|u\| = 1, \ k = 1, 2.$$

Subtracting the equality (2.11) yields the lower bound in (2.12). □

REMARK 2.1. Note that the bound (2.12) produces something non-trivial only for $0 < \delta < (2 + \sqrt{2})^{-2} \approx 0.0857864$. Casazza, Marcus, Speegle, and the author [**13**] have shown the improved bound in (KS_2). For $0 < \delta < \frac{1}{4}$, the bound (2.12) holds where $O(\sqrt{\delta})$ is replaced by $\sqrt{2\delta(1 - 2\delta)}$. Therefore, by a variant of Lemma 2.6 we have the following variant of paving for projections.

THEOREM (PP_δ). *Let $0 < \delta < 1/4$. Every projection P on $\ell^2(I)$, i.e., $P = P^*$ and $P^2 = P$, with all diagonal entries $\le \delta$ can be $(2, \frac{1}{2} + \sqrt{2\delta(1 - 2\delta)})$-paved.*

2.4. Notes. Another well-known equivalent of the Kadison-Singer problem, which we didn't discuss here, is a relative Dixmier property studied by Berman, Halpern, Kaftal, and Weiss [**7, 27–29**]. Every bounded operator $T \in \mathcal{B}(\ell^2(I))$ satisfies

$$\mathbb{E}(T) \in \overline{\text{conv}}\{UTU^* : U \text{ is a diagonal unitary on } \ell^2(I)\}.$$

The connection of the Kadison-Singer problem with paving was investigated by Anderson and Akemann [**1**,**3**–**5**]. A streamlined presentation of paving implications presented here has been shown by Casazza, Edidin, Kalra, and Paulsen [**18**]. A pinball principle, Lemma 2.7, was shown in [**16**].

For each of the classes of matrices/operators considered above, such as:

(B) bounded matrices with zero diagonal,
(S) self-adjoint matrices with zero diagonal,
($P_{\frac{1}{2}}$) projections with $\frac{1}{2}$ on diagonal,

we can ask for the smallest $r \in \mathbb{N}$ such that all matrices in this class have (r,ε)-paving for some $\varepsilon < 1$. By keeping track of the values in Lemma 2.6, we have shown $(12,\varepsilon)$ paving for $(P_{\frac{1}{2}})$ and thus $(12^2,\varepsilon)$ paving for (S) for some $\varepsilon < 1$. This was recently improved by Ravichandran [**38**] who has shown $(4,\varepsilon)$ paving for $(P_{\frac{1}{2}})$ and thus $(16,\varepsilon)$ paving for (S). It is known that $(2,\varepsilon)$ paving does not work for $(P_{\frac{1}{2}})$, see [**20**]. Does $(3,\varepsilon)$ paving work for $(P_{\frac{1}{2}})$? Likewise, we can ask for largest δ such that $(2,\varepsilon)$ paving works for all projections with δ on diagonal. Paving property can be formulated for other operator norms and matrices with zero diagonal. However, paving remains an open problem for operator ℓ^p norms, $p \neq 2$, though Schechtman [**41**] has recently shown paving for the Schatten C_p class norm for $2 < p < \infty$ extending earlier results of Berman, Halpern, Kaftal, and Weiss [**8**].

3. Proof of Weaver's conjecture

Weaver's conjecture is a consequence of the following probabilistic result due to Marcus, Spielman, and Srivastava [**35**]. The special case was shown by Casazza, Marcus, Speegle, and the author [**13**].

THEOREM (MSS). *Let $\epsilon > 0$. Suppose that $v_1, \dots, v_m$ are jointly independent random vectors in $\mathbb{C}^d$, which take finitely many values and satisfy*

$$\sum_{i=1}^{m} \mathbb{E}\left[v_i v_i^*\right] = \mathbf{I} \qquad \text{and} \qquad \mathbb{E}\left[\|v_i\|^2\right] \le \epsilon \quad \text{for all } i. \tag{3.1}$$

Then,

$$\mathbb{P}\left(\left\| \sum_{i=1}^{m} v_i v_i^* \right\| \le (1+\sqrt{\epsilon})^2 \right) > 0. \tag{3.2}$$

In the special case when $v_1, \dots, v_m$ take at most two values and $\epsilon < 1/4$, we have

$$\mathbb{P}\left(\left\| \sum_{i=1}^{m} v_i v_i^* \right\| \le 1 + 2\sqrt{\epsilon}\sqrt{1-\epsilon} \right) > 0.$$

LEMMA 3.1. $(MSS) \implies (KS_r)$.

PROOF. Assume $\{u_i\}_{i\in[m]} \subset \mathbb{C}^d$ satisfies (2.11). For any $r \in \mathbb{N}$, let $v_1, \dots, v_m$ be independent random vectors in $(\mathbb{C}^d)^{\oplus r} = \mathbb{C}^{rd}$ such that each vector v_i takes r values

$$\begin{bmatrix} \sqrt{r}u_i \\ 0 \\ \vdots \\ 0 \end{bmatrix}, \dots, \begin{bmatrix} 0 \\ \vdots \\ 0 \\ \sqrt{r}u_i \end{bmatrix}$$

each with probability $\frac{1}{r}$. Then,

$$\sum_{i=1}^{m} \mathbb{E}\left[v_i v_i^*\right] = \begin{bmatrix} \sum_{i=1}^m u_i u_i^* & & \\ & \ddots & \\ & & \sum_{i=1}^m u_i u_i^* \end{bmatrix} = \begin{bmatrix} \mathbf{I}_d & & \\ & \ddots & \\ & & \mathbf{I}_d \end{bmatrix} = \mathbf{I}_{dr},$$

and

$$\mathbb{E}\left[||v_i||^2\right] = r||u_i||^2 \le \epsilon := r\delta.$$

Hence, we can apply (MSS) to deduce (3.1). Choose an outcome for which the bound in (3.2) happens. For this outcome define

$$I_k = \{i \in [m] : v_i \text{ is non-zero in } k^{\text{th}} \text{ entry}\}, \qquad \text{for } k = 1, \dots, r.$$

Thus, the block diagonal matrix

$$\sum_{i=1}^{m} v_i v_i^* = \begin{bmatrix} r\sum_{i\in I_1} u_i u_i^* & & \\ & \ddots & \\ & & r\sum_{i\in I_r} u_i u_i^* \end{bmatrix}$$

has norm bounded by $(1+\sqrt{\epsilon})^2$. This implies that each block has norm

$$\left\| \sum_{i\in I_k} u_i u_i^* \right\| \le \frac{1}{r}(1+\sqrt{r\delta})^2 = \left(\frac{1}{\sqrt{r}} + \sqrt{\delta}\right)^2.$$

Since a rank one operator $u_i \otimes u_i$ on $\mathbb{C}^d$ is represented by the $d \times d$ matrix $u_i u_i^*$, we obtain (KS_r). $\square$

3.1. Mixed characteristic polynomial. The main result of this section involves the concept of a mixed characteristic polynomial (MCP).

DEFINITION 3.1. Let $A_1, \dots, A_m$ be $d \times d$ matrices. The mixed characteristic polynomial is defined as for $z \in \mathbb{C}$ by

$$\mu[A_1, \dots, A_m](z) = \left(\prod_{i=1}^{m}(1-\partial_{z_i})\right) \det\left(z\mathbf{I} + \sum_{i=1}^{m} z_i A_i\right)\Bigg|_{z_1=\dots=z_m=0}.$$

By determinant expansion one can show that $\det\left(z\mathbf{I} + \sum_{i=1}^m z_i A_i\right)$ is a polynomial in $\mathbb{C}[z, z_1, \dots, z_m]$ of degree $\le d$. Hence, $\mu[A_1, \dots, A_m](z)$ is a polynomial in $\mathbb{C}[z]$ of degree $\le d$. These polynomials satisfy a number of interesting properties if $A_1, \dots, A_m$ are positive definite.

THEOREM (MCP). *Let $\epsilon > 0$. Suppose $A_1, \dots, A_m$ are $d \times d$ positive semidefinite matrices satisfying*

$$\sum_{i=1}^{m} A_i = \mathbf{I} \quad \text{and} \quad \mathrm{Tr}(A_i) \le \epsilon \quad \text{for all } i. \tag{3.3}$$

Then, all roots of the mixed characteristic polynomial $\mu[A_1, \dots, A_m]$ are real and the largest root is at most $(1+\sqrt{\epsilon})^2$.

It remains to accomplish two major tasks: prove the implication $(MCP) \implies (MSS)$ and then show (MCP). Before doing this we need to show a few basic properties of μ.

LEMMA 3.2. *For a fixed $z \in \mathbb{C}$, the mixed characteristic polynomial mapping*

$$\mu : M_{d\times d}(\mathbb{C}) \times \ldots \times M_{d\times d}(\mathbb{C}) \to \mathbb{C}$$

is multi-affine and symmetric. That is, μ affine in each variable and its value is the same for any permutation of its arguments $A_1, \ldots, A_m$.

PROOF. The fact the μ is symmetric is immediate from the definition. We claim that for any $d \times d$ matrix B, a function

$$f : M_{d\times d}(\mathbb{C}) \to \mathbb{C}, \qquad f(A_1) = (1-\partial_{z_1}) \det(B+z_1A_1)|_{z_1=0} \qquad \text{for } A_1 \in M_{d\times d}(\mathbb{C})$$

is affine. Indeed, if B is invertible, then by Jacobi's formula

$$f(A_1) = \det(B) - \det(B)\partial_{z_1} \det(\mathbf{I} + z_1B^{-1}A_1)|_{z_1=0} = \det(B)(1 - \operatorname{tr}(B^{-1}A_1)).$$

Since invertible matrices are dense in the set of all matrices, by continuity we deduce the general case. Thus, for any choice of matrices $A_2, \ldots, A_m$, a mapping

$$(M_{d\times d}(\mathbb{C}), \mathbb{C}^{m-1}) \ni (A_1, z_2, \ldots, z_m) \mapsto (1 - \partial_{z_1}) \det\left(z\mathbf{I} + \sum_{i=1}^m z_iA_i\right)\Big|_{z_1=0}$$

is affine in the A_1 variable and a polynomial of degree $\le d$ in $z_2, \ldots, z_m$ variables. Applying linear operators, such as partial differential operators with constant coefficients $(1 - \partial_{z_i})$, $i = 2, \ldots, m$, preserves this property. Consequently, the mapping $A_1 \mapsto \mu[A_1, \ldots, A_m](z)$ is affine. By symmetry, μ is multi-affine. □

LEMMA 3.3. *If $A_1, \ldots, A_m$ are rank one $d \times d$ matrices, then the mixed characteristic polynomial is a characteristic polynomial of the sum $A = A_1 + \ldots + A_m$,*

$$\mu[A_1, \ldots, A_m](z) = \det(z\mathbf{I} - A) \qquad z \in \mathbb{C}. \tag{3.4}$$

PROOF. Any rank one matrix is of the from uv^* for some $u, v \in \mathbb{C}^d$. By the Sylvester determinant identity $\det(\mathbf{I} + tuv^*) = 1 + tv^*u$ for any $t \in \mathbb{C}$. Hence, for any $d \times d$ matrix B the mapping

$$\mathbb{C} \ni t \mapsto \det(B + tuv^*) = b_0 + b_1t, \qquad b_0, b_1 \in \mathbb{C},$$

is affine. If B is invertible, then this follows by factoring out B, which reduces to the case $B = \mathbf{I}$. Since invertible matrices are dense in the set of all matrices, by continuity we deduce the general case. This implies that for fixed $z \in \mathbb{C}$, the polynomial

$$p(z_1, \ldots, z_m) := \det\left(z\mathbf{I} + \sum_{i=1}^m z_iA_i\right) = b + \sum_{1\le i_1<\ldots<i_j\le m} a_{i_1} \ldots a_{i_j} z_{i_1} \ldots z_{i_j}$$

is affine multilinear in $z_1, \ldots, z_m$. Hence, we can recover values of p by the following formula

$$p(t_1, \ldots, t_m) = \left(\prod_{i=1}^m (1 + t_i\partial_{z_i})\right) p(z_1, \ldots, z_m)\Big|_{z_1=\ldots=z_m=0}.$$

Taking $t_1 = \ldots = t_m = -1$ yields

$$\mu[A_1, \ldots, A_m](z) = p(-1, \ldots, -1) = \det(z\mathbf{I} - A).$$

□

LEMMA 3.4. *Let $X_1, \dots, X_m$ be $d \times d$ jointly independent random matrices, which take finitely many values. Then,*

$$\mathbb{E}\left[\mu[X_1, \dots, X_m](z)\right] = \mu[\mathbb{E}[X_1], \dots, \mathbb{E}[X_m]](z) \qquad z \in \mathbb{C}. \tag{3.5}$$

In addition, if random matrices X_i, $i = 1, \dots, m$, are rank one, then

$$\mathbb{E}\left[\det\left(z\mathbf{I} - \sum_{i=1}^{m} X_i\right)\right] = \mu[\mathbb{E}[X_1], \dots, \mathbb{E}[X_m]](z) \qquad z \in \mathbb{C}. \tag{3.6}$$

PROOF. By Lemma 3.2 for any matrices $B_1, \dots, B_n$ and $A_2, \dots, A_m$ and coefficients $p_1, \dots, p_n$ satisfying $\sum_{i=1}^n p_i = 1$, we have

$$\mu\left[\sum_{i=1}^{n} p_i B_i, A_2, \dots, A_m\right](z) = \sum_{i=1}^{n} p_i \mu[B_i, A_2, \dots, A_m].$$

Then the joint independence of $X_1, \dots, X_m$ yields (3.5). Combining (3.5) with Lemma 3.3 yields (3.6). $\square$

3.2. Real stable polynomials. The proof of $(MCP) \implies (MSS)$ relies on the concept of a real stable polynomial.

DEFINITION 3.2. Let $\mathbb{C}_+ = \{z \in \mathbb{C} : \operatorname{Im}(z) > 0\}$ be the upper half plane. We say that a polynomial $p \in \mathbb{C}[z_1, \dots, z_m]$ is *stable* if $p(z_1, \dots, z_m) \neq 0$ for every $(z_1, \dots, z_m) \in \mathbb{C}_+^m$. A polynomial is called *real stable* if it is stable and all of its coefficients are real.

Note that a univariate polynomial is real stable if and only if all its roots are real. We will show a few basic properties of real stable polynomials.

LEMMA 3.5. *If $A_1, \dots, A_m$ are positive semidefinite hermitian $d \times d$ matrices, then*

$$p(z, z_1, \dots, z_m) = \det\left(z\mathbf{I} + \sum_{i=1}^{m} z_i A_i\right) \in \mathbb{C}[z, z_1, \dots, z_m]$$

is a real stable polynomial.

PROOF. If inputs $z, z_1, \dots, z_m$ are real, then the values $p(z, z_1, \dots, z_m)$ are also real, since a determinant of a hermitian matrix is real. Hence, $p \in \mathbb{R}[z, z_1, \dots, z_m]$. On the contrary suppose that $p(z, z_1, \dots, z_m) = 0$ for some $(z, z_1, \dots, z_m) \in \mathbb{C}_+^{m+1}$. That is,

$$\left(z\mathbf{I} + \sum_{i=1}^{m} z_i A_i\right) v = 0 \qquad \text{for some } 0 \neq v \in \mathbb{C}^{m+1}.$$

Hence,

$$0 = \operatorname{Im}\left\langle \left(z\mathbf{I} + \sum_{i=1}^{m} z_i A_i\right) v, v \right\rangle = \operatorname{Im}(z)||v||^2 + \sum_{i=1}^{m} \operatorname{Im}(z_i)\langle A_i v, v\rangle \geq \operatorname{Im}(z)||v||^2 > 0,$$

which is a contradiction. $\square$

LEMMA 3.6. *Suppose that $p \in \mathbb{R}[z_1, \dots, z_m]$ is stable.*

- **(restriction)** *for fixed $t \in \mathbb{R}$, polynomial $p(t, z_2, \dots, z_m) \in \mathbb{R}[z_2, \dots, z_m]$ is stable unless it is identically zero.*
- **(differentiation)** *if $t \in \mathbb{R}$, then $(1 + t\partial_{z_1})p$ is real stable.*

PROOF. By Hurwitz's theorem, if a sequence of non-vanishing holomorphic functions $\{f_n\}_{n\in\mathbb{N}}$ on an open connected domain $\Omega \subset \mathbb{C}^m$ converges uniformly on compact sets, then its limit f is either non-vanishing or $f \equiv 0$. Let $\Omega = \mathbb{C}_+^{m-1}$. Define

$$f_n(z_1, \dots, z_{m-1}) = p(t + i/n, z_1, \dots, z_{m-1}) \qquad \text{for } (z_1, \dots, z_{m-1}) \in \Omega, \ n \in \mathbb{N}.$$

Letting $n \to \infty$, Hurwitz's theorem implies the restriction property.

To show differentiation property we can assume that $t \neq 0$. Fix $z_2, \dots, z_m \in \Omega$. By definition $q(z) = p(z, z_2, \dots, z_m) \in \mathbb{C}[z]$ is stable. Hence, we can write $q(z) = c\prod_{i=1}^d (z - w_i)$ for some roots $w_1, \dots, w_d \in \mathbb{C} \setminus \mathbb{C}_+$. Then,

$$q(z) + tq'(z) = c\prod_{i=1}^d (z - w_i)\Big(1 + \sum_{i=1}^d \frac{t}{z - w_i}\Big).$$

Take any $z \in \mathbb{C}_+$. Since $\operatorname{Im}(w_i) \le 0$, we have $z - w_i \in \mathbb{C}_+$, and hence $\operatorname{Im}(1/(z - w_i)) < 0$ for all $i = 1, \dots, d$. Hence, $\sum_{i=1}^d \frac{t}{z-w_i}$ has non-zero imaginary part. This implies that $q(z) + tq'(z) \neq 0$ for any $z \in \mathbb{C}_+$. Since $z_2, \dots, z_m \in \Omega$ are arbitrary, $(1 + t\partial_{z_1})p$ is stable. □

As a corollary of Lemma 3.5 and 3.6 we have:

COROLLARY 3.7. *If $A_1, \dots, A_m$ are positive semidefinite hermitian $d \times d$ matrices, then the mixed characteristic polynomial $\mu[A_1, \dots, A_m]$ is real, stable, and monic of degree d.*

The following elementary lemma plays a key role in our arguments. Recall that for any $p \in \mathbb{R}[z]$, p is stable $\iff$ p has all real roots. Let $\operatorname{maxroot}(p)$ be the largest root of p.

LEMMA 3.8. *Let $p, q \in \mathbb{R}[z]$ be stable monic polynomials of the same degree. Suppose that every convex combination $(1-t)p + tq$, $0 \le t \le 1$, is also stable. Then for any $0 \le t_0 \le 1$, $\operatorname{maxroot}(((1-t_0)p + t_0q)$ lies between $\operatorname{maxroot}(p)$ and $\operatorname{maxroot}(q)$.*

PROOF. Without loss of generality we can assume $\operatorname{maxroot}(p) \le \operatorname{maxroot}(q)$ and $0 < t_0 < 1$. Our goal is to show that

$$m_p := \operatorname{maxroot}(p) \le \operatorname{maxroot}(((1-t_0)p + t_0q) \le \operatorname{maxroot}(q) =: m_q.$$

For $x > m_q$, both $p(x)$ and $q(x)$ are positive, and hence $((1-t_0)p + t_0q)(x) > 0$. This shows the second inequality.

We shall prove the first inequality by contradiction. Suppose that $(1-t_0)p+t_0q$ has no roots $[m_p, m_q]$. This implies that $(1-t_0)p + t_0q > 0$ for all $x \ge m_p$. In particular, $q(m_p) > 0$. Hence, q must have at least 2 roots (counting multiplicity) to the right of m_p. Let D be an open disk in $\mathbb{C}$ centered at $\frac{m_p+m_q}{2}$ and radius $\frac{m_q-m_p}{2}$. We claim that

$$((1-t)p + tq)(z) \neq 0 \qquad \text{for all } z \in \partial D \text{ and } t_0 \le t \le 1.$$

Indeed, since $(1-t)p + tq$ is stable, this is easily verified at $z = m_p$ and $z = m_q$. By compactness

$$\inf_{(z,t)\in\partial D\times[t_0,1]} |((1-t)p + tq)(z)| > 0.$$

By Rouche's theorem, polynomials $(1-t)p+tq$ have the same number of zeros in D for all $t_0 \le t \le 1$. This is a contradiction with the hypothesis that $(1-t_0)p+t_0q$ has no roots in D, but q has at least 2 roots in D. □

Lemma 3.8 can be generalized to control other roots, such as second largest, third largest, etc. This leads to a concept of an interlacing family of polynomials, which plays a fundamental role in the arguments of Marcus-Spielman-Srivastava [**34**, **35**].

3.3. Interlacing family of polynomials. We shall not give the formal definition of this concept. Instead, following Tao [**43**] we will use the following lemma.

LEMMA 3.9. *Let $p_1, \dots, p_n \in \mathbb{R}[z]$ be stable monic polynomials of the same degree. Suppose that every convex combination*

$$\sum_{i=1}^n t_i p_i, \qquad \text{where } \sum_{i=1}^n t_i = 1,\ t_i \ge 0$$

is a stable polynomial. Then, for any such convex combination there exists $1 \le i_0 \le n$ such that

$$\operatorname{maxroot}(p_{i_0}) \le \operatorname{maxroot}\left(\sum_{i=1}^n t_i p_i\right). \tag{3.7}$$

PROOF. Using Lemma 3.8 we can easily show (3.7) by induction on the number of polynomials $p_1, \dots, p_n$. □

LEMMA 3.10. *Let X be a random rank one positive semidefinite $d \times d$ matrix. Let $A_1, \dots, A_m$ be $d \times d$ deterministic positive semidefinite matrices. Then with positive probability we have*

$$\operatorname{maxroot}(\mu[X, A_1, \dots, A_m]) \le \operatorname{maxroot}(\mu[\mathbb{E}[X], A_1, \dots, A_m]). \tag{3.8}$$

PROOF. Suppose that X takes values $B_1, \dots, B_n$. Define polynomials $p_i = \mu[B_i, A_1, \dots, A_m]$, $i = 1, \dots, m$. By Corollary 3.7 these are real, stable, and monic polynomials of degree d. Moreover, any convex combination $\sum_{i=1}^n t_i p_i$ is also stable. Indeed, consider a random variable Y taking values $B_1, \dots, B_n$ with probabilities $t_1, \dots, t_n$, resp. By Lemma 3.4 and Corollary 3.7

$$\sum_{i=1}^n t_i p_i = \mathbb{E}[\mu[Y, A_1, \dots, A_m]] = \mu[\mathbb{E}[Y], A_1, \dots, A_m]$$

is a real stable polynomial. Hence, Lemma 3.9 yields (3.8). □

Iterating Lemma 3.10 gives the required control on roots of mixed characteristic polynomials. This is an essence of the method of interlacing family of polynomials.

LEMMA 3.11. *Suppose that $X_1, \dots, X_m$ are jointly independent random rank one positive semidefinite $d \times d$ matrices which take finitely many values. Then with positive probability*

$$\operatorname{maxroot}(\mu[X_1, \dots X_m]) \le \operatorname{maxroot}(\mu[\mathbb{E}[X_1], \dots, \mathbb{E}[X_m]]). \tag{3.9}$$

PROOF. By Lemma 3.10 a random matrix X_1 takes some value A_1 (with positive probability) such that

$$\operatorname{maxroot}(\mu[\mathbb{E}[X_1], \dots, \mathbb{E}[X_m]]) \ge \operatorname{maxroot}(\mu[A_1, \mathbb{E}[X_2], \dots, \mathbb{E}[X_m]]).$$

By the independence assumption, if we condition the probability space to the event $X_1 = A_1$, then random variables $X_2, \dots, X_m$ have the same joint distribution. Again by Lemma 3.10, X_2 takes some value A_2 (with positive probability) such that

$$\operatorname{maxroot}(\mu[A_1, \mathbb{E}[X_2], \dots, \mathbb{E}[X_m]]) \geq \operatorname{maxroot}(\mu[A_1, A_2, \mathbb{E}[X_3], \dots, \mathbb{E}[X_m]]).$$

Conditioning on the event $X_1 = A_1$ and $X_2 = A_2$, and repeating this argument for remaining random variables yields (3.9). □

Finally, we can complete the proof of the main implication.

LEMMA 3.12. $(MCP) \implies (MSS)$.

PROOF. Take random vectors $v_1, \dots, v_m$ as in Theorem (MSS). Define rank one positive semidefinite random matrices $X_i = v_i v_i^*$, $i = 1, \dots, m$. The assumption (3.1) translates into the assumption (3.3) for $A_i = \mathbb{E}[X_i]$. Since X_i are hermitian, by Lemma 3.3

$$\left\| \sum_{i=1}^m v_i v_i^* \right\| = \left\| \sum_{i=1}^m X_i \right\| = \operatorname{maxroot}\left(\det\left(z\mathbf{I} - \sum_{i=1}^m X_i \right) \right) = \operatorname{maxroot}(\mu[X_1, \dots X_m]).$$

By Lemma 3.11, the bound on $\operatorname{maxroot}(\mu[A_1, \dots A_m])$ in the conclusion of Theorem (MCP) yields the same bound on $\|\sum_{i=1}^m v_i v_i^*\|$ with positive probability. □

3.4. Multivariate barrier argument. The proof of Theorem (MCP) hinges on a multivariate barrier argument.

DEFINITION 3.3. Let $p \in \mathbb{R}[z_1, \dots, z_m]$. We say that $x = (x_1, \dots, x_m) \in \mathbb{R}^m$ is *above the roots* of p if

$$p(x + t) > 0 \qquad \text{for all } t \in [0, \infty)^m.$$

A *barrier function* of p in direction of z_i, $i = 1, \dots, m$, is defined for such x as

$$\Phi_p^i(x) = \partial_{z_i} \log p(x) = \frac{\partial_{z_i} p(x)}{p(x)}.$$

We need the following result about zeros of real stable polynomials in two variables, which is illustrated in Figure 1.

LEMMA 3.13. *Let $p \in \mathbb{R}[z, w]$ be a stable polynomial. Then for all but finitely many $x \in \mathbb{R}$, a polynomial $p(x, w) \in \mathbb{R}[w]$ has all real roots and constant degree $d \in \mathbb{N}$. Let $y_1(x) \leq \ldots \leq y_d(x)$ be its roots counting multiplicity. Then, for each $i \in [d]$, $x \mapsto y_i(x)$ is non-increasing.*

PROOF. We can write $p(z, w) = \sum_{i=0}^d w^i q_i(z)$, where each $q_i \in \mathbb{R}[z]$. By Lemma 3.6 for fixed $x \in \mathbb{R}$, a polynomial $p(x, w) \in \mathbb{R}[w]$ has all real roots. Its degree equals d if and only if $q_d(x) \neq 0$. The fundamental theorem of algebra implies the existence of roots $y_1(x) \leq \ldots \leq y_d(x)$. It remains to show that $x \mapsto y_i(x)$ is non-increasing, where $i \in [d]$.

We claim that for every real root $p(x, y) = 0$, $(x, y) \in \mathbb{R}^2$, we have

$$\partial_z p(x, y) \leq 0 \qquad \text{and} \qquad \partial_w p(x, y) \leq 0. \tag{3.10}$$

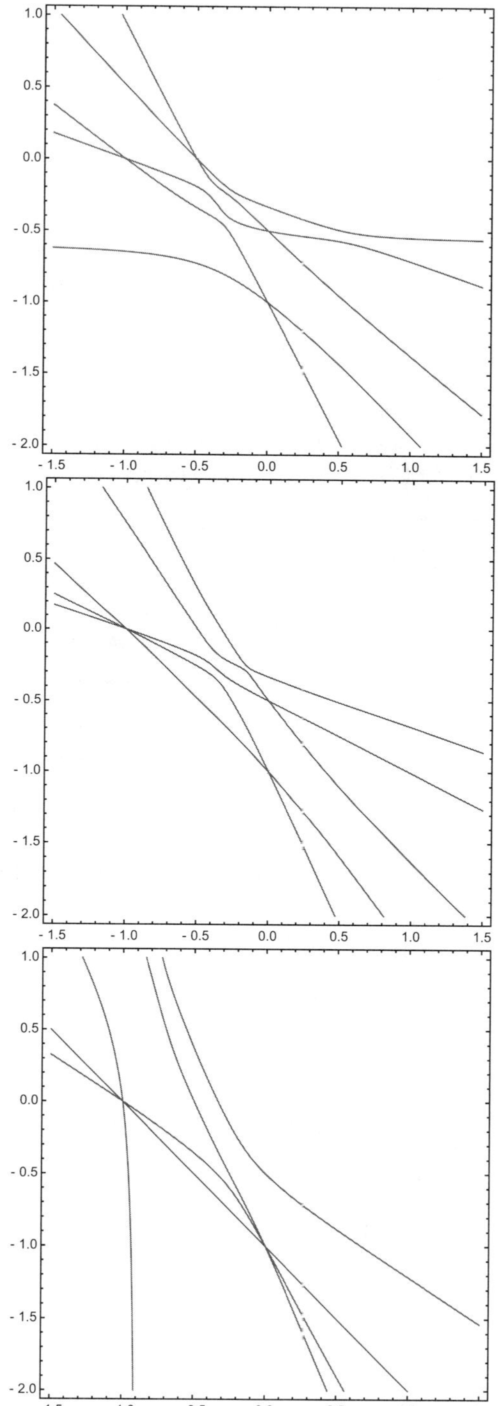

FIGURE 1. Examples of zero sets of real stable polynomials on the plane.

On the contrary, suppose that $\alpha = \partial_w p(x,y) > 0$. By the implicit function theorem for holomorphic functions [**25**, Theorem I.7.6], there exists complex neighborhoods $U_x, U_y \subset \mathbb{C}$ of x and y, resp., and a holomorphic function $h : U_x \to U_y$ such that

$$\{(z,w) \in U_x \times U_y : p(z,w) = 0\} = \{(z, h(z)) : z \in U_x\}.$$

Taking $z = x + \varepsilon i$ and $h(z) \approx y + h'(x)\varepsilon i = y + \alpha\varepsilon i$ for small $\varepsilon > 0$ produces a root of p with positive imaginary parts, which contradicts stability of p. By symmetry we deduce (3.10).

To finish the proof, it is convenient to use a basic fact about algebraic curves

$$\{(x,y) \in \mathbb{R}^2 : p(x,y) = 0\}.$$

Every algebraic curve decomposes as a finite union of *branches* connected by some points and a finite number of vertical lines. A branch is the graph of a smooth and monotone function defined on an open (possibly unbounded) interval in the x-axis. Hence, a branch is the graph of some $y_i(x)$ restricted to an appropriate open interval. Differentiating $p(x, y_i(x)) = 0$ with respect to x yields

$$\partial_z p(x, y_i(x)) + \partial_w p(x, y_i(x)) y_i'(x) = 0.$$

By (3.10) we have $y_i'(x) \le 0$. □

LEMMA 3.14. *Let $p \in \mathbb{R}[z_1, \dots, z_m]$ be stable. Let $1 \le i, j \le m$. Then, for any $k \in \mathbb{N}_0$, partial derivatives of the barrier function of p satisfy*

$$(-1)^k \partial_{z_j}^k \Phi_p^i(x) \ge 0 \qquad \text{if } x \in \mathbb{R}^n \text{ is above the roots of } p.$$

In particular, $t \mapsto \Phi_p^i(x + te_j)$ is non-negative, non-increasing, and convex function of $t \ge 0$.

PROOF. First suppose that $i = j$. Freezing all variables except $z_i = z_j$, by Lemma 3.6 (restriction) we can assume that $m = 1$. Suppose that $x \in \mathbb{R}$ is above all roots of a stable polynomial $p \in \mathbb{R}[z]$. The stability of p implies that p has all real roots. Hence, $p(z) = c\prod_{j=1}^d (z - y_j)$, where $y_j \in \mathbb{R}$, and

$$\Phi_p(x) = \left(\frac{p'}{p}\right)(x) = \sum_{j=1}^d \frac{1}{x - y_j}.$$

By a direct calculation

$$(-1)^k (\Phi_p)^{(k)}(x) = k! \sum_{j=1}^d \frac{1}{(x - y_j)^{k+1}}.$$

Since x is above the roots of p, we have $x > \max(y_j)$ and all of the above terms are positive. The above argument also covers trivially the case $k = 0$.

It remains to deal with the case $i \ne j$ and $k \ge 1$. By Lemma 3.6 and symmetry, we can assume that $m = 2$, $i = 1$, and $j = 2$. Suppose that $x = (x_1, x_2)$ is above the roots of a stable polynomial $p \in \mathbb{R}[z_1, z_2]$. Since

$$(-1)^k \partial_{z_2}^k \Phi_p^1(x) = (-1)^k \partial_{z_2}^k \partial_{z_1} \log p(x) = \partial_{z_1}((-1)^k \partial_{z_2}^k \log p)(x)$$

it suffices to show that $x_1 \mapsto (-1)^k \partial_{z_2}^k \log p(x_1, x_2)$ is non-decreasing.

By Lemma 3.13 we can write

$$p(x_1, x_2) = c(x_1) \prod_{i=1}^d (x_2 - y_i(x_1)).$$

Hence,

$$x_1 \mapsto (-1)^k \partial_{z_2}^k \log p(x_1, x_2) = -(k-1)! \sum_{i=1}^{d} \frac{1}{(x_2 - y_i(x_1))^k}.$$

Since x is above the roots of p, we have $x_2 > \max(y_i(x_1))$ and by Lemma 3.13, the above function is non-decreasing. □

The following lemma provides the crucial control of the barrier function of $(1 - \partial_{z_j})q$ in terms of the barrier function of q.

LEMMA 3.15. *Let $q \in \mathbb{R}[z_1, \dots, z_m]$ be stable. Suppose that $x \in \mathbb{R}^m$ lies above the roots of q and*

$$\Phi_q^j(x) \le 1 - \frac{1}{\delta} \qquad \text{for some } j \in [m] \text{ and } \delta > 0. \tag{3.11}$$

Then, $x + \delta e_j$ lies above the roots of $(1 - \partial_{z_j})q$ and

$$\Phi_{(1-\partial_{z_j})q}^i(x + \delta e_j) \le \Phi_q^i(x) \qquad \text{for all } i \in [m]. \tag{3.12}$$

PROOF. Take any $y \in \mathbb{R}^m$ above x, that is, $y_i \ge x_i$ for all $i \in [m]$. By Lemma 3.14 (monotonicity), we have $\Phi_q^j(y) \le \Phi_q^j(x) < 1$. Hence,

$$(1 - \partial_{z_j})q(y) = q(y)(1 - \Phi_q^j(y)) > 0. \tag{3.13}$$

In particular, $x + \delta e_j$ is above the roots of $(1 - \partial_{z_j})q$. By (3.13)

$$\log((1 - \partial_{z_j})q)(y) = \log q(y) + \log(1 - \Phi_q^j)(y).$$

Applying ∂_{z_i}, $i \in [m]$, shows

$$\Phi_{(1-\partial_{z_j})q}^i(y) = \Phi_q^i(y) - \frac{\partial_{z_i}\Phi_q^j(y)}{1 - \Phi_q^j(y)}.$$

Since

$$\partial_{z_i}\Phi_q^j(y) = \partial_{z_i}\partial_{z_j}\log q(y) = \partial_{z_j}\partial_{z_i}\log q(y) = \partial_{z_j}\Phi_q^i(y),$$

the required bound (3.12) is equivalent with the inequality

$$\frac{\partial_{z_j}\Phi_q^i(x+\delta e_j)}{1 - \Phi_q^j(x+\delta e_j)} = \Phi_q^i(x+\delta e_j) - \Phi_{(1-\partial_{z_j})q}^i(x+\delta e_j) \ge \Phi_q^i(x+\delta e_j) - \Phi_q^i(x). \tag{3.14}$$

By Lemma 3.14 (convexity) and (monotonicity), we have

$$\Phi_q^i(x+\delta e_j) - \Phi_q^i(x) \le \delta\partial_{z_j}\Phi_q^i(x+\delta e_j) \le 0.$$

Hence, (3.14) is implied by multiplying the inequality

$$\frac{1}{1 - \Phi_q^j(x+\delta e_j)} \le \delta \tag{3.15}$$

by $\partial_{z_j}\Phi_q^i(x+\delta e_j)$. Finally, (3.15) holds true as a consequence of Lemma 3.14 (monotonicity) and (3.11)

$$\Phi_q^j(x+\delta e_j) \le \Phi_q^j(x) \le 1 - \frac{1}{\delta}.$$

This shows (3.14) and consequently (3.12). □

Applying inductively Lemma 3.15 yields the crucial corollary.

COROLLARY 3.16. *Let $q \in \mathbb{R}[z_1, \dots, z_m]$ be stable. Suppose that $x \in \mathbb{R}^m$ lies above the roots of q and for some $\delta > 0$ we have*

$$\Phi_q^j(x) \le 1 - \frac{1}{\delta} \qquad \text{for all } j \in [m].$$

Then, $x + (\delta, \dots, \delta)$ lies above the roots of $\prod_{i=1}^m (1 - \partial_{z_i}) q$.

PROOF. For $k = 0, \dots, m$, define

$$y_k = x + \delta \sum_{i=1}^k e_i, \qquad q_k = \prod_{i=1}^k (1 - \partial_{z_i}) q.$$

Then, using Lemma 3.15, we show inductively that $y_k \in \mathbb{R}^m$ lies above the roots of q_k for all $k \in [m]$. □

Finally, we are ready to give the proof of (MCP).

PROOF OF THEOREM (MCP). Define

$$p(z_1, \dots, z_m) = \det\left(\sum_{i=1}^m z_i A_i\right) \in \mathbb{R}[z_1, \dots, z_m] \tag{3.16}$$

By Lemmas 3.5 and 3.6, p is a real stable polynomial. By Jacobi's formula for any $j \in [m]$,

$$\begin{aligned}\partial_{z_j} p(x_1, \dots, x_m) &= \partial_t \det\left(\sum_{i=1}^m x_i A_i + t A_j\right)\Bigg|_{t=0} \\ &= \det\left(\sum_{i=1}^m x_i A_i\right) \operatorname{tr}\left(\left(\sum_{i=1}^m x_i A_i\right)^{-1} A_j\right).\end{aligned}$$

Hence, by (3.3)

$$\Phi_p^j(t, \dots, t) = \operatorname{tr}(t^{-1} A_j) \le \frac{\epsilon}{t} \qquad \text{for } t > 0.$$

Moreover, $x = (t, \dots, t)$ lies above the roots of p for any $t > 0$. Take any $t, \delta > 0$ such that

$$\frac{\epsilon}{t} + \frac{1}{\delta} \le 1. \tag{3.17}$$

By Corollary 3.16, $(t + \delta, \dots, t + \delta)$ lies above the roots of $\prod_{i=1}^m (1 - \partial_{z_i}) p$. Since

$$\prod_{i=1}^m (1 - \partial_{z_i}) p(z, \dots, z) = \mu[A_1, \dots, A_m](z) \qquad \text{for any } z \in \mathbb{C},$$

the largest root of $\mu[A_1, \dots, A_m]$ is $\le t + \delta$. Minimizing $t + \delta$ under the constraint (3.17) yields $t = \sqrt{\epsilon} + \epsilon$ and $\delta = 1 + \sqrt{\epsilon}$. Hence, the largest root of $\mu[A_1, \dots, A_m]$ is bounded by $(1 + \epsilon)^2$. □

3.5. Notes. The strategy of the proofs of (MSS) and (MCP) follows the original proof of Marcus, Spielman, and Srivastava [**35**] with strong influence by Tao's blog article [**43**]. The main difference is in the proof of Lemma 3.14. Tao uses more elementary properties of real stable polynomials in the form of Lemma 3.13, whereas the original proof uses Helton-Vinnikov's theorem [**9, 31**]. This result states that every real stable polynomial in two variables of degree d has a determinantal representation $p(x, y) = \pm \det(xA + yB + C)$ for some $d \times d$ positive semidefinite matrices A, B and a symmetric matrix C.

The proof of the special case of Theorem (MSS), for random variables taking at most two values, is more technical and it can be found in [**13**]. It relies on a variant of Theorem (MCP) for matrices $A_1, \ldots, A_m$ of rank ≤ 2. This corresponds to a determinantal polynomial (3.16) which is quadratic with respect to each variable $z_1, \ldots, z_m$. Amazingly, such deceptively simple polynomial encodes all the information about roots of the mixed characteristic polynomial $\mu[A_1, \ldots, A_m]$, which is needed for showing (MCP).

4. Applications of Weaver's conjecture

In this section we show applications of the solution of the Kadison-Singer problem which are outside of the main sequence of implications $(KS) \Leftarrow \ldots \Leftarrow (MCP)$. Our main goal is to show quantitative bounds in Feichtinger's conjecture. To achieve this we need some background about Naimark's dilation theorem.

4.1. Naimark's complements of frame partitions. We start with well-known Naimark's dilation theorem.

LEMMA 4.1. *Let $\{u_i\}_{i\in I}$ be a Parseval frame in a Hilbert space $\mathcal{H}$. Then there exists a larger Hilbert space $\mathcal{K} \supset \mathcal{H}$ and an o.n. basis $\{e_i\}_{i\in I} \subset \mathcal{K}$ such that*

$$u_i = Pe_i \qquad \text{for all } i \in I, \text{ where } P \text{ is an orthogonal projection of } \mathcal{K} \text{ onto } \mathcal{H}.$$

Conversely, if P is a projection of $\mathcal{K}$ onto a closed subspace $\mathcal{H}$, then $\{Pe_i\}_{i\in I}$ is a Parseval frame in $\mathcal{H}$.

PROOF. Consider the analysis $T : \mathcal{H} \to \ell^2(I)$ as in Lemma 2.6. Since $\{u_i\}_{i\in I}$ is a Parseval frame, T is an isometry of $\mathcal{H}$ onto $T(\mathcal{H}) \subset \ell^2(I)$. Let Q be the orthogonal projection of $\ell^2(I)$ onto $T(\mathcal{H})$. Let $\{e_i\}_{i\in I}$ be the standard o.n. basis of $\ell^2(I)$. Since T is an isometry, it suffices to show the conclusion for Parseval frame $\{Tu_i\}_{i\in I}$ in $T(\mathcal{H})$. In turn, this is a consequence of the following calculation. Since $\{Tu_i\}_{i\in I}$ is a Parseval frame

$$Qa = \sum_{i\in I} \langle a, Tu_i\rangle Tu_i \qquad \text{for all } a \in \ell^2(I).$$

Thus, for any $i \in I_0$,

$$Qe_{i_0} = \sum_{i\in I} \langle e_{i_0}, Tu_i\rangle Tu_i = \sum_{i\in I} \overline{\langle u_i, u_{i_0}\rangle} Tu_i = T\Big(\sum_{i\in I} \langle u_{i_0}, u_i\rangle u_i\Big) = Tu_{i_0}.$$

□

Lemma 4.1 leads to the concept of a Naimark's complement. This is a Parseval frame $\{(\mathbf{I} - P)e_i\}_{i\in I}$ in $\mathcal{K} \ominus \mathcal{H}$, where $\mathbf{I}$ is the identity on $\mathcal{K} \cong \ell^2(I)$. Recall the definition of a Riesz sequence.

DEFINITION 4.1. A family of vectors $\{u_i\}_{i\in I}$ in a Hilbert space $\mathcal{H}$ is a *Riesz sequence* if there are constants $A, B > 0$ so that for all $\{a_i\} \in \ell^2(I)$ we have

$$A\sum_{i\in I}|a_i|^2 \le \left\|\sum_{i\in I} a_i u_i\right\|^2 \le B\sum_{i\in I}|a_i|^2. \tag{4.1}$$

We call A, B *lower and upper Riesz bounds* for $\{u_i\}_{i\in I}$.

Note that it suffices to verify (4.1) only for sequences $\{a_i\}$ with finitely many non-zero coefficients, since a standard convergence argument yields the same bounds (4.1) for all infinitely supported sequences $\{a_i\} \in \ell^2(I)$. In general we do not require that frame, Bessel, and Riesz bounds in Definitions 2.5 and 4.1 are optimal. In particular, a Bessel sequence with bound B is automatically a Bessel sequence with bound $B' \ge B$.

LEMMA 4.2. *Let $P : \ell^2(I) \to \ell^2(I)$ be the orthogonal projection onto a closed subspace $\mathcal{H} \subset \ell^2(I)$. Then, for any subset $J \subset I$ and $\delta > 0$, the following are equivalent:*

(i) *$\{Pe_i\}_{i\in J}$ is a Bessel sequence with bound $1-\delta$,*
(ii) *$\{(\mathbf{I}-P)e_i\}_{i\in J}$ is a Riesz sequence with lower bound δ, where $\mathbf{I}$ is the identity on $\ell^2(I)$.*

PROOF. Note that for any sequence of coefficients $\{a_i\} \in \ell^2(J)$,

$$\sum_{i\in J}|a_i|^2 = \left\|\sum_{i\in J} a_i Pe_i\right\|^2 + \left\|\sum_{i\in J} a_i(\mathbf{I}-P)e_i\right\|^2. \tag{4.2}$$

Thus,

$$\left\|\sum_{i\in J} a_i Pe_i\right\|^2 \le (1-\delta)\sum_{i\in J}|a_i|^2 \iff \left\|\sum_{i\in J} a_i(\mathbf{I}-P)e_i\right\|^2 \ge \delta\sum_{i\in J}|a_i|^2. \tag{4.3}$$

Observe that the inequality in the left hand side of (4.3) is equivalent to (i). This follows from the well-known fact that $||T|| = ||T^*||$, where T is the analysis operator

$$T : \mathcal{H} \to \ell^2(I), \qquad T\phi = \{\langle u, Pe_i\rangle\}_{i\in J}, \quad u \in \mathcal{H},$$

and its adjoint is the synthesis operator

$$T^* : \ell^2(I) \to \mathcal{H}, \qquad T^*(\{a_i\}_{i\in J}) = \sum_{i\in J} a_i Pe_i, \quad \{a_i\}_{i\in J} \in \ell^2(J).$$

This yields the equivalence of (i) and (ii). □

4.2. The Feichtinger conjecture. We are now ready to formulate the quantitative version of the Feichtinger conjecture which was shown by Casazza, Marcus, Speegle, and the author [**13**].

THEOREM (FEI). *Let I be at most countable set and let $\mathcal{H}$ be a separable Hilbert space. Suppose $\{u_i\}_{i\in I}$ is a Bessel sequence in $\mathcal{H}$ with bound 1 that consists of vectors of norms $\|u_i\|^2 \ge \varepsilon$, where $\varepsilon > 0$. Then there exists a universal constant $C > 0$, such that I can be partitioned into $r \le C/\varepsilon$ subsets $I_1, \dots, I_r$, such that each subfamily $\{u_i\}_{i\in I_j}$, $j = 1, \dots, r$, is a Riesz sequence.*

In the proof of Theorem (FEI) we shall use the following adaptation of the Schur-Horn theorem for Riesz sequences.

LEMMA 4.3. *Let S be a positive semi-defnite $M \times M$ matrix with eigenvalues $\lambda_1 \geq \ldots \geq \lambda_M \geq 0$. Let $d_1 \geq \ldots \geq d_M \geq 0$ be such that*

$$\sum_{i=1}^{M} d_i = \sum_{i=1}^{M} \lambda_i \quad \text{and} \quad \sum_{i=1}^{k} d_i \leq \sum_{i=1}^{k} \lambda_i \quad \text{for all } 1 \leq k \leq M. \tag{4.4}$$

Then there exists a collection of vectors $\{v_i\}_{i=1}^M$ in $\mathbb{C}^M$ such its frame operator is S and $\|v_i\|^2 = d_i$ for all $i = 1, \ldots, M$.

Lemma 4.3 has a converse, which states that the norms of $\{v_i\}_{i=1}^M$ and eigenvalues of its frame operator must satisfy (4.4). Since we will not need this, we simply omit the proof of the converse result.

PROOF. By the Schur-Horn theorem, there exists a hermitian matrix $\tilde{S}$ with eigenvalues $\lambda_1 \geq \ldots \geq \lambda_M$ and diagonal $d_1 \geq \ldots \geq d_M$. Since S and $\tilde{S}$ are unitarily equivalent, there exists a unitary $M \times M$ matrix U such that $\tilde{S} = U^* S U$. Define vectors $v_i = S^{1/2} U e_i$, where e_i, $i \in [M]$, are standard basis vectors in $\mathbb{C}^M$. Then,

$$||v_i||^2 = \langle S^{1/2} U e_i, S^{1/2} U e_i \rangle = \langle S U e_i, U e_i \rangle = \langle \tilde{S} e_i, e_i \rangle = d_i.$$

Moreover, the frame operator of $\{v_i\}_{i=1}^M$ satisfies for $v \in \mathbb{C}^M$,

$$\begin{aligned} \left(\sum_{i=1}^{M} v_i \otimes v_i \right)(v) &= \sum_{i=1}^{M} (S^{1/2} U e_i) \otimes (S^{1/2} U e_i)(v) = \sum_{i=1}^{M} \langle v, S^{1/2} U e_i \rangle S^{1/2} U e_i \\ &= S^{1/2} \left(\sum_{i=1}^{M} \langle S^{1/2} v, U e_i \rangle U e_i \right) = S^{1/2} S^{1/2} v = S v. \end{aligned}$$

The penultimate step follows from the fact that $\{U e_i\}_{i=1}^M$ is an o.n. basis. □

We start from the special case of Theorem (FEI) and then show its increasingly general versions.

LEMMA 4.4. *Theorem (FEI) holds under the additional assumption that I is finite and $\varepsilon = 0.92$. In this case, a Bessel sequence $\{u_i\}_{i \in I}$ with bound 1 and $\|u_i\|^2 \geq \varepsilon$ can be partitioned into two Riesz sequences with lower bound 0.02.*

In light of Remark 2.1, the value of 0.92 can be replaced by any number $> 3/4$, but we are not after best constants here.

PROOF. Assume momentarily that $\{u_i\}_{i \in I}$ is a Parseval frame in a finite dimensional Hilbert space $\mathcal{H}$. By Lemma 4.1 we can imbed $\mathcal{H}$ into $\ell^2(I)$ such that $u_i = P e_i$, $i \in I$, where P is an orthogonal projection of $\ell^2(I)$ onto $\mathcal{H}$. Then, vectors $v_i = (\mathbf{I} - P) e_i$, $i \in I$, form a Parseval frame in $\ell^2(I) \ominus \mathcal{H}$. Since $||u_i||^2 \geq \varepsilon$, we have $||v_i||^2 \leq 1 - \varepsilon < \delta := 0.08$. By Theorem (KS_2^∞) we can find a subset $J \subset I$, such that both $\{v_i\}_{i \in J}$ and $\{v_i\}_{i \in I \setminus J}$ are Bessel sequences with bound $\frac{1}{2} + \sqrt{2\delta} + \delta = 0.98$. Thus, by Lemma 4.2, both $\{u_i\}_{i \in J}$ and $\{u_i\}_{i \in I \setminus J}$ are Riesz sequences with lower bound $1 - 0.98 = 0.02$.

Assume now that $\{u_i\}_{i \in I}$ is a Bessel sequence with bound 1 and $\|u_i\|^2 \geq \varepsilon$. Since I is finite, we can assume that $I = [n]$ and $\mathcal{H} = \mathbb{C}^d$. By increasing the dimension of the ambient space, we claim that a Bessel sequence $\{u_i\}_{i \in [n]}$ can be extended to a Parseval frame by adjoining some collection of vectors $\{u_{n+i}\}_{i=1}^{d+N}$ in $\mathbb{C}^{d+N}$ satisfying $||u_i|| \geq \varepsilon$, where N is sufficiently large.

Indeed, suppose that the frame operator of $\{u_i\}_{i=1}^n$, which acts on $\mathbb{C}^d$, has eigenvalues $1 \ge \lambda_1 \ge \lambda_2 \ge \ldots \ge \lambda_d \ge 0$. For a fixed N, consider an operator on $\mathbb{C}^{d+N}$,

$$\tilde{S} = \mathbf{I}_{d+N} - S \oplus \mathbf{0}_N, \qquad \text{where } \mathbf{0}_N \text{ is the zero operator on } \mathbb{C}^N.$$

Then, $\tilde{S}$ has the following eigenvalues listed in decreasing order

$$\underbrace{1, \ldots, 1}_{N}, 1 - \lambda_d, \ldots, 1 - \lambda_1. \tag{4.5}$$

Thus, we need to show the existence of vectors $\{u_{n+i}\}_{i=1}^{d+N}$ in $\mathbb{C}^{d+N}$ such that:

(i) its frame operator is $\tilde{S}$, and
(ii) $||u_{n+i}||^2 = C$ for all $i = 1, \ldots, d+N$ for some constant $C \in [\varepsilon, 1]$.

By Lemma 4.3, this is possible provided eigenvalue sequence (4.5) majorizes, in the sense of (4.4), the sequence

$$\underbrace{C, \ldots, C}_{d+N}. \tag{4.6}$$

However, the majorization (4.5) is automatic for the constant sequence (4.6) provided that

$$(d+N)C = N + \sum_{i=1}^{d} (1 - \lambda_i).$$

Thus, by choosing sufficiently large N, we have $C \ge \varepsilon$, which shows the claim. Now, we apply the previous argument for a Parseval frame $\{u_i\}_{i=1}^{n+d+N}$. Hence, we can find a partition into two Riesz sequences. Restricting this partition to the original sequence $\{u_i\}_{i=1}^n$ yields the same conclusion. □

LEMMA 4.5. *Theorem (FEI) holds under the assumption that I is finite and $||u_i||^2 = \varepsilon > 0$ for all i. In this case, a Bessel sequence $\{u_i\}_{i \in I}$ with bound 1 can be partitioned into two Riesz sequences with bounds $\varepsilon/50$ and $\varepsilon/0.92$.*

PROOF. By scaling Lemma 4.4 yields the following result: any finite tight frame $\{w_i\}$ with constant B and with norms $||w_i||^2 \ge 0.92B$ can be partitioned into two Riesz sequences with bounds $B/50$ and B.

Now, suppose that $\{u_i\}_{i \in I}$ is a Bessel sequence with bound 1 and $||u_i||^2 = \varepsilon$. By Theorem (KS_r) for each r we can find a partition $\{\tilde{I}_j\}_{j=1}^r$ of I such that each $\{u_i\}_{i \in \tilde{I}_j}$ is a Bessel sequence with bound

$$B = \left(\frac{1}{\sqrt{r}} + \sqrt{\varepsilon} \right)^2.$$

Now we choose large enough r such that

$$||u_i||^2 = \varepsilon \ge 0.92B = 0.92 \left(\frac{1}{\sqrt{r}} + \sqrt{\varepsilon} \right)^2. \tag{4.7}$$

A simple calculation shows that the above inequality simplifies to

$$\frac{2}{\sqrt{r\varepsilon}} + \frac{1}{r\varepsilon} \le \frac{0.08}{0.92}.$$

Hence, it suffices to choose

$$r \geq \frac{9}{\varepsilon}\left(\frac{0.92}{0.08}\right)^2.$$

By Lemma 4.4, each $\{u_i\}_{i\in\tilde{I}_j}$ can be partitioned into two Riesz sequences with lower bound $B/50 \geq \varepsilon/50$ and upper bound $B \leq \varepsilon/0.92$. This gives the required partition of size $2r$ and completes the proof of Lemma 4.5. □

Theorem (FEI) is now a consequence of Lemmas 2.7 and 4.5.

Proof of Theorem (FEI). Suppose $\{u_i\}_{i\in I}$ is an infinite Bessel sequence in a Hilbert space $\mathcal{H}$ satisfying $||u_i||^2 \geq \varepsilon$. Without loss of generality we can assume that $||u_i||^2 = \varepsilon$ for all $i \in I$. Indeed, $\{\sqrt{\varepsilon}\frac{u_i}{||u_i||}\}_{i\in I}$ is also Bessel sequence with bound 1. Applying (FEI) for this sequence yields the same conclusion for the original Bessel sequence $\{u_i\}_{i\in I}$.

Since I is countable, we may assume $I = \mathbb{N}$. For any $n \in \mathbb{N}$, we apply Lemma 4.5 to the initial sequence $\{u_i\}_{i\in[n]}$. Hence, we find a partition $\{I_1^n, \dots, I_r^n\}$ of $\{u_i\}_{i\in[n]}$ into Riesz sequences with uniform lower and upper bounds of $\varepsilon/50$ and $\varepsilon/0.92$, resp. To show the existence of a global partition of $\{I_1, \dots, I_r\}$ of $\{u_i\}_{i\in\mathbb{N}}$ into Riesz sequences, it suffices to apply Lemma 2.7. This done in the same way as in the proof of Lemma 2.8. □

4.3. Casazza-Tremain conjecture. A stronger variant of the Feichtinger conjecture, called R_ε conjecture, was studied by Casazza and Tremain [**22**]. This result states that Bessel sequences consisting of unit norm vectors can be partitioned into almost orthogonal sequences.

Theorem (R_ε). *Suppose that $\{u_i\}_{i\in I}$ is a unit norm Bessel sequence with bound B in a separable Hilbert space $\mathcal{H}$. Then for any $\varepsilon > 0$ there exists a partition $\{I_1, \dots, I_r\}$ of I of size $r = O(B/\varepsilon^4)$, such that each $\{u_i\}_{i\in I_j}$, $j = 1, \dots, r$, is a Riesz sequence with bounds $1-\varepsilon$ and $1+\varepsilon$.*

In the proof of Theorem (R_ε) we will use the following lemma. The case when $J = I$ is a well-known fact, see [**24**, Section 3.6]. For the sake of completeness we will give the proof of Lemma 4.6.

Lemma 4.6. *Suppose $\{u_i\}_{i\in I}$ is a Riesz basis in a Hilbert space $\mathcal{H}$. Let $\{u_i^*\}_{i\in I}$ be its unique biorthogonal (dual) Riesz basis, i.e.,*

$$\langle u_i, u_j^* \rangle = \delta_{i,j} \qquad \text{for all } i, j \in I.$$

Let $J \subset I$ be any subset. Then, $\{u_i\}_{i\in J}$ is a Riesz sequence with bounds A and B $\iff$ $\{u_i^*\}_{i\in J}$ is a Riesz sequence with bounds $1/B$ and $1/A$.

Proof. Suppose that $\{u_i\}_{i\in J}$ has upper Riesz bound B. This is equivalent to the Bessel condition

$$\sum_{i\in J} |\langle u, u_i \rangle|^2 \leq B||u||^2 \qquad \text{for all } u \in \mathcal{H}. \tag{4.8}$$

For any sequence $\{a_i\}_{i\in J} \in \ell^2$, there exists a unique $u \in \mathcal{H}$ such that

$$\langle u, u_i \rangle = \begin{cases} a_i & i \in J, \\ 0 & \text{otherwise.} \end{cases}$$

Since $u = \sum_{i\in J} a_i u_i^*$, by (4.8) we have

$$\Big\| \sum_{i\in J} a_i u_i^* \Big\|^2 = ||u||^2 \geq \frac{1}{B} \sum_{i\in J} |\langle u, u_i\rangle|^2 = \frac{1}{B} \sum_{i\in J} |a_i|^2.$$

Conversely, if $\{u_i^*\}_{i\in J}$ has lower Riesz bound $1/B$, then (4.8) holds and $\{u_i\}_{i\in J}$ has upper Riesz bound B. By symmetry, $\{u_i^*\}_{i\in J}$ has upper Riesz bound $1/A$ if and only if $\{u_i\}_{i\in J}$ has lower Riesz bound A, which completes the proof of the lemma. □

LEMMA 4.7. $(FEI) \implies (R_\varepsilon)$.

PROOF. In the first step we apply a scaled version of Theorem (FEI) to find a partition of $\{u_i\}_{i\in I}$ of size $O(B)$ into Riesz sequences with uniform lower and upper bounds. By Lemma 4.5, these bounds are $1/50$ and $1/0.92$, resp.

Suppose that $\{u_i\}_{i\in I'}$, $I' \subset I$ is one of these unit-norm Riesz sequences. In the next step we need to tighten these bounds as follows. Let $\{u_i^*\}_{i\in I'}$ be the unique biorthogonal (dual) Riesz basis to $\{u_i\}_{i\in I'}$ in its closed linear span $\mathcal{H}' = \overline{\operatorname{span}}\{u_i : i \in I'\}$. By Lemma 4.6 the upper Riesz bound of $\{u_i^*\}_{i\in I'}$ is 50. Applying Theorem (KS_r) to both $\{u_i\}_{i\in I'}$ and $\{u_i^*\}_{i\in I'}$, we can find partitions into Riesz sequences, which reduce upper bounds to $1+\varepsilon$. A calculation shows that this requires partitions each of size $O(1/\varepsilon^2)$. Taking common refinement of all of these partitions produces a partition of $\{u_i\}_{i\in I}$ of size $O(B/\varepsilon^4)$. Let $\{u_i\}_{i\in J}$ be any element of of this partition. Then, both $\{u_i\}_{i\in J}$ and $\{u_i^*\}_{i\in J}$ are Riesz sequences with upper bounds $1+\varepsilon$. Lemma 4.6 implies that $\{u_i\}_{i\in J}$ has lower bound $1/(1+\varepsilon) \geq 1-\varepsilon$. □

4.4. Bourgain-Tzafriri conjecture. Theorem (R_ε) yields automatically the Bourgain–Tzafriri restricted invertibility conjecture.

THEOREM (BT). *Let $\{e_i\}_{i\in I}$ be an orthonormal basis of a separable Hilbert space $\mathcal{H}$. Let $T : \mathcal{H} \to \mathcal{H}$ be a bounded linear operator with norm $\|T\|^2 \leq B$ and $\|Te_i\| = 1$ for all $i \in I$, where $B > 1$. Then, for any $\varepsilon > 0$, there exists a partition $\{I_1, \dots, I_r\}$ of I of size $r = O(B/\varepsilon^4)$, such that T is $(1+\varepsilon)$-isometry when restricted to each orthogonal subspace*

$$\mathcal{H}_j = \overline{\operatorname{span}}\{e_i : i \in I_j\}.$$

That is, for all $j = 1, \dots, r$,

$$(1-\varepsilon)||f||^2 \leq ||Tf||^2 \leq (1+\varepsilon)||f||^2 \qquad \text{for all } f \in \mathcal{H}_j. \tag{4.9}$$

LEMMA 4.8. $(R_\varepsilon) \implies (BT)$.

PROOF. Define vectors $u_i = Te_i$, $i \in I$. By our hypothesis $\{u_i\}_{i\in I}$ is a unit norm Bessel sequence with bound B. By Theorem (R_ε), there exists a partition $\{I_1, \dots, I_r\}$ of I of size $r = O(B/\varepsilon^4)$ such that each collection $\{u_i\}_{i\in I_j}$, $j \in [r]$, is a Riesz sequence with bounds $1-\varepsilon$ and $1+\varepsilon$. Translating this back for the property of T yields (4.9). □

A classical application of the results studied in this section involves Fourier frames. If $E \subset [0,1]$ has positive Lebesgue measure, then the collection of functions $\phi_n(t) = e^{2\pi int}\chi_E(t)$, $n \in \mathbb{Z}$, is a Parseval frame for $L^2(E)$, often called a Fourier frame. Since this is an equal norm frame, i.e., $||\phi_n||^2 = |E|$ for all $n \in \mathbb{Z}$, Theorem (R_ε) yields the following corollary.

COROLLARY 4.9. *There exists a universal constant $c > 0$ such that for any $\varepsilon > 0$ and any subset $E \subset [0,1]$ with positive measure, the corresponding Fourier frame $\{\phi_n\}_{n\in\mathbb{Z}}$ can be decomposed as the union of $r \le c\varepsilon^{-4}|E|^{-1}$ Riesz sequences with bounds $1 \pm \varepsilon$.*

4.5. Notes. The proof of (FEI) and (R_ε) follows the approach in [**13**] with some minor simplifications. One can show (FEI) with less effort by deducing it from Theorem (PB) as in [**16**, Proposition 3.1], but with worse bounds on the partition size r. The bound on r in Theorem (FEI) is asymptotically optimal as $\varepsilon \to 0$. This can be seen by considering a union of $\lfloor 1/\varepsilon \rfloor$ o.n. bases scaled by the factor $\sqrt{\varepsilon}$. A more general version of Lemma 4.3 for frames can be found in [**6, 14**]. Ravichandran and Srivastava [**39**] have recently shown an asymptotically optimal multi-paving result, which can be used to improve the size of partition from $r = O(B/\varepsilon^4)$ to $r = O(B/\varepsilon^2)$ in (R_ε) and (BT).

Lawton [**32**] and Paulsen [**37**] have shown that the partition subsets of $\mathbb{Z}$ in Corollary 4.9 can be chosen to be syndetic sets, i.e., subsets of $\mathbb{Z}$ with bounded gaps. The study of the Feichtinger conjecture for Fourier frames is connected with the problem of paving for Laurent operators. A Laurent operator $L_\varphi : L^2[0,1] \to L^2[0,1]$ is given by $L_\varphi f = \varphi f$ for $f \in L^2[0,1]$, where the symbol $\varphi \in L^\infty[0,1]$. The problem of paving for Laurent operators was studied by Halpern, Kaftal, and Weiss [**27**]. It was continued by Bourgain and Tzafriri [**12**], who have shown that every Fourier frame has a Riesz sequence indexed by a subset $\Lambda \subset \mathbb{Z}$ of positive upper density at least $c|E|$. This is the consequence of their celebrated restricted invertibility theorem [**10, 11**], which also holds for ℓ^p spaces. While the Bourgain-Tzafriri restricted invertibility conjecture, Theorem (BT), holds for ℓ^p when $p = 2$, it is an open problem for $p \neq 2$.

Akemann and Weaver [**2**] have shown an interesting generalization of (KS_r) in the form of Lyapunov's theorem.

THEOREM (AW). *Suppose $\{u_i\}_{i\in I}$ is a Bessel family with bound 1 in a separable Hilbert space $\mathcal{H}$, which consists of vectors of norms $\|u_i\|^2 \le \delta$, where $\delta > 0$. Suppose that $0 \le \tau_i \le 1$ for all $i \in I$. Then, there exists a subset of indices $I_0 \subset I$ such that*

$$\left\| \sum_{i\in I_0} u_i \otimes u_i - \sum_{i\in I} \tau_i u_i \otimes u_i \right\| \le C\delta^{1/8}, \tag{4.10}$$

where $C > 0$ is a universal constant.

The proof of Theorem (AW) relies solely on (KS_r), and hence we could have added another implication to our diagram $(KS_r) \implies (AW)$. However, we will stop here and instead invite the reader to explore other interesting consequences of the breakthrough solution of the Kadison-Singer problem.

Acknowledgments

The author is grateful for useful conversations on the Kadison-Singer problem with Bernhard Bodmann, Jean Bourgain, Pete Casazza, Bill Johnson, Adam Marcus, Gideon Schechtman, Darrin Speegle, and participants of the AIM workshop "Beyond Kadison-Singer: paving and consequences" in December 2014 and the MSRI workshop "Hot Topics: Kadison-Singer, Interlacing Polynomials, and Beyond" in March 2015.

References

[1] C. A. Akemann and J. Anderson, *Lyapunov theorems for operator algebras*, Mem. Amer. Math. Soc. **94** (1991), no. 458, iv+88, DOI 10.1090/memo/0458. MR1086563

[2] C. Akemann and N. Weaver, *A Lyapunov-type theorem from Kadison-Singer*, Bull. Lond. Math. Soc. **46** (2014), no. 3, 517–524, DOI 10.1112/blms/bdu005. MR3210706

[3] J. Anderson, *Extreme points in sets of positive linear maps on $\mathcal{B}(\mathcal{H})$*, J. Funct. Anal. **31** (1979), no. 2, 195–217, DOI 10.1016/0022-1236(79)90061-2. MR525951

[4] J. Anderson, *Extensions, restrictions, and representations of states on C^*-algebras*, Trans. Amer. Math. Soc. **249** (1979), no. 2, 303–329, DOI 10.2307/1998793. MR525675

[5] J. Anderson, *A conjecture concerning the pure states of $B(H)$ and a related theorem*, Topics in modern operator theory (Timişoara/Herculane, 1980), Operator Theory: Adv. Appl., vol. 2, Birkhäuser, Basel-Boston, Mass., 1981, pp. 27–43. MR672813

[6] J. Antezana, P. Massey, M. Ruiz, and D. Stojanoff, *The Schur-Horn theorem for operators and frames with prescribed norms and frame operator*, Illinois J. Math. **51** (2007), no. 2, 537–560. MR2342673

[7] K. Berman, H. Halpern, V. Kaftal, and G. Weiss, *Matrix norm inequalities and the relative Dixmier property*, Integral Equations Operator Theory **11** (1988), no. 1, 28–48, DOI 10.1007/BF01236652. MR920733

[8] K. Berman, H. Halpern, V. Kaftal, and G. Weiss, *Some C_4 and C_6 norm inequalities related to the paving problem*, Operator theory: operator algebras and applications, Part 2 (Durham, NH, 1988), Proc. Sympos. Pure Math., vol. 51, Amer. Math. Soc., Providence, RI, 1990, pp. 29–41. MR1077417

[9] J. Borcea and P. Brändén, *Multivariate Pólya-Schur classification problems in the Weyl algebra*, Proc. Lond. Math. Soc. (3) **101** (2010), no. 1, 73–104, DOI 10.1112/plms/pdp049. MR2661242

[10] J. Bourgain and L. Tzafriri, *Invertibility of "large" submatrices with applications to the geometry of Banach spaces and harmonic analysis*, Israel J. Math. **57** (1987), no. 2, 137–224, DOI 10.1007/BF02772174. MR890420

[11] J. Bourgain and L. Tzafriri, *Restricted invertibility of matrices and applications*, Analysis at Urbana, Vol. II (Urbana, IL, 1986), London Math. Soc. Lecture Note Ser., vol. 138, Cambridge Univ. Press, Cambridge, 1989, pp. 61–107. MR1009186

[12] J. Bourgain and L. Tzafriri, *On a problem of Kadison and Singer*, J. Reine Angew. Math. **420** (1991), 1–43. MR1124564

[13] M. Bownik, P. Casazza, A. Marcus, D. Speegle, *Improved bounds in Weaver and Feichtinger conjectures*, J. Reine Angew. Math. (to appear).

[14] M. Bownik and J. Jasper, *Existence of frames with prescribed norms and frame operator*, Excursions in harmonic analysis. Vol. 4, Appl. Numer. Harmon. Anal., Birkhäuser/Springer, Cham, 2015, pp. 103–117. MR3411093

[15] M. Bownik and D. Speegle, *The Feichtinger conjecture for wavelet frames, Gabor frames and frames of translates*, Canad. J. Math. **58** (2006), no. 6, 1121–1143, DOI 10.4153/CJM-2006-041-3. MR2270922

[16] P. G. Casazza, O. Christensen, A. M. Lindner, and R. Vershynin, *Frames and the Feichtinger conjecture*, Proc. Amer. Math. Soc. **133** (2005), no. 4, 1025–1033, DOI 10.1090/S0002-9939-04-07594-X. MR2117203

[17] P. G. Casazza and D. Edidin, *Equivalents of the Kadison-Singer problem*, Function spaces, Contemp. Math., vol. 435, Amer. Math. Soc., Providence, RI, 2007, pp. 123–142, DOI 10.1090/conm/435/08371. MR2359423

[18] P. Casazza, D. Edidin, D. Kalra, and V. I. Paulsen, *Projections and the Kadison-Singer problem*, Oper. Matrices **1** (2007), no. 3, 391–408, DOI 10.7153/oam-01-23. MR2344683

[19] P. G. Casazza, M. Fickus, J. C. Tremain, and E. Weber, *The Kadison-Singer problem in mathematics and engineering: a detailed account*, Operator theory, operator algebras, and applications, Contemp. Math., vol. 414, Amer. Math. Soc., Providence, RI, 2006, pp. 299–355, DOI 10.1090/conm/414/07820. MR2277219

[20] P. G. Casazza, M. Fickus, D. G. Mixon, and J. C. Tremain, *The Bourgain-Tzafriri conjecture and concrete constructions of non-pavable projections*, Oper. Matrices **5** (2011), no. 2, 351–363, DOI 10.7153/oam-05-25. MR2830604

[21] P. G. Casazza and J. C. Tremain, *The Kadison-Singer problem in mathematics and engineering*, Proc. Natl. Acad. Sci. USA **103** (2006), no. 7, 2032–2039, DOI 10.1073/pnas.0507888103. MR2204073

[22] P. G. Casazza and J. C. Tremain, *Revisiting the Bourgain-Tzafriri restricted invertibility theorem*, Oper. Matrices **3** (2009), no. 1, 97–110, DOI 10.7153/oam-03-04. MR2500595

[23] P. G. Casazza and J. C. Tremain, *Consequences of the Marcus/Spielman/Srivastava solution of the Kadison-Singer problem*, New trends in applied harmonic analysis, Appl. Numer. Harmon. Anal., Birkhäuser/Springer, Cham, 2016, pp. 191–213. MR3497721

[24] O. Christensen, *An introduction to frames and Riesz bases*, Applied and Numerical Harmonic Analysis, Birkhäuser Boston, Inc., Boston, MA, 2003. MR1946982

[25] K. Fritzsche and H. Grauert, *From holomorphic functions to complex manifolds*, Graduate Texts in Mathematics, vol. 213, Springer-Verlag, New York, 2002. MR1893803

[26] K. Gröchenig, *Localized frames are finite unions of Riesz sequences*, Adv. Comput. Math. **18** (2003), no. 2-4, 149–157, DOI 10.1023/A:1021368609918. Frames. MR1968117

[27] H. Halpern, V. Kaftal, and G. Weiss, *Matrix pavings and Laurent operators*, J. Operator Theory **16** (1986), no. 2, 355–374. MR860353

[28] H. Halpern, V. Kaftal, and G. Weiss, *The relative Dixmier property in discrete crossed products*, J. Funct. Anal. **69** (1986), no. 1, 121–140, DOI 10.1016/0022-1236(86)90110-2. MR864758

[29] H. Halpern, V. Kaftal, and G. Weiss, *Matrix pavings in* $B(H)$, Operators in indefinite metric spaces, scattering theory and other topics (Bucharest, 1985), Oper. Theory Adv. Appl., vol. 24, Birkhäuser, Basel, 1987, pp. 201–214, DOI 10.1147/sj.262.0201. MR903073

[30] N. Harvey, *An introduction to the Kadison-Singer Problem and the Paving Conjecture*, preprint available at `https://www.cs.ubc.ca/ nickhar/Publications/KS/KS.pdf`.

[31] J. W. Helton and V. Vinnikov, *Linear matrix inequality representation of sets*, Comm. Pure Appl. Math. **60** (2007), no. 5, 654–674, DOI 10.1002/cpa.20155. MR2292953

[32] W. Lawton, *Minimal sequences and the Kadison-Singer problem*, Bull. Malays. Math. Sci. Soc. (2) **33** (2010), no. 2, 169–176. MR2666420

[33] R. V. Kadison and I. M. Singer, *Extensions of pure states*, Amer. J. Math. **81** (1959), 383–400, DOI 10.2307/2372748. MR0123922

[34] A. W. Marcus, D. A. Spielman, and N. Srivastava, *Interlacing families I: Bipartite Ramanujan graphs of all degrees*, Ann. of Math. (2) **182** (2015), no. 1, 307–325, DOI 10.4007/annals.2015.182.1.7. MR3374962

[35] A. W. Marcus, D. A. Spielman, and N. Srivastava, *Interlacing families II: Mixed characteristic polynomials and the Kadison-Singer problem*, Ann. of Math. (2) **182** (2015), no. 1, 327–350, DOI 10.4007/annals.2015.182.1.8. MR3374963

[36] É. Matheron, *Le problème de Kadison-Singer* (French), Ann. Math. Blaise Pascal **22** (2015), no. S2, 151–265. MR3453286

[37] V. I. Paulsen, *Syndetic sets, paving and the Feichtinger conjecture*, Proc. Amer. Math. Soc. **139** (2011), no. 3, 1115–1120, DOI 10.1090/S0002-9939-2010-10600-7. MR2745663

[38] M. Ravichandran, *Mixed Determinants and the Kadison-Singer problem*, preprint (2016) available at `arXiv:1609.04195`.

[39] M. Ravichandran and N. Srivastava, *Asymptotically optimal multi-paving*, preprint available (2017) at `arXiv:1706.03737`.

[40] M. Stevens, *The Kadison-Singer property*, SpringerBriefs in Mathematical Physics, vol. 14, Springer, Cham, 2016. With a foreword by Klaas Landsman. MR3616518

[41] G. Schechtman, *Three observations regarding Schatten p classes*, J. Operator Theory **75** (2016), no. 1, 139–149, DOI 10.7900/jot.2014dec07.2048. MR3474100

[42] B. Tanbay, *A letter on the Kadison-Singer problem*, Rev. Roumaine Math. Pures Appl. **59** (2014), no. 2, 293–302. MR3299507

[43] T. Tao, *Real stable polynomials and the Kadison-Singer problem*, blog entry available at `https://terrytao.wordpress.com/tag/kadison-singer-problem/`

[44] D. Timotin, *The solution to the Kadison–Singer Problem: yet another presentation*, preprint `arXiv:1501.00464`.

[45] A. Valette, *Le problème de Kadison-Singer (d'après A. Marcus, D. Spielman et N. Srivastava)* (French, with French summary), Astérisque **367-368** (2015), Exp. No. 1088, x, 451–476. MR3363599

[46] N. Weaver, *The Kadison-Singer problem in discrepancy theory*, Discrete Math. **278** (2004), no. 1-3, 227–239, DOI 10.1016/S0012-365X(03)00253-X. MR2035401

Department of Mathematics, University of Oregon, Eugene, Oregon 97403–1222
Current address: Institute of Mathematics, Polish Academy of Sciences, ul. Wita Stwosza 57, 80–952 Gdańsk, Poland
Email address: mbownik@uoregon.edu

Contemporary Mathematics
Volume **706**, 2018
http://dx.doi.org/10.1090/conm/706/14212

Spectral properties of an operator polynomial with coefficients in a Banach algebra

Anatoly G. Baskakov and Ilya A. Krishtal

ABSTRACT. We use the method of similar operators and recent results on Wiener's lemma extensions to study the spectral properties of a polynomial of a closed linear operator with coefficients in some Banach algebra of bounded linear operators. For example, we show that if the closed operator generates a bounded C_0-group and the Banach algebra satisfies some mild conditions then the spectrum of the operator polynomial in the algebra of all bounded linear operators coincides with its spectrum in the algebra of coefficients.

1. Introduction

This paper is motivated, in part, by the question of when the solutions of abstract differential equations inherit the smoothness or other properties of the right-hand-side of the equation. There is a great multitude of results of this nature that are available in the literature; we mention [**3**, **20**, **32**, **33**] to name just a few. We decided to shed some new light on this question through a general study of the spectral properties of an operator polynomial

$$\mathscr{A} = A^n + B_1 A^{n-1} + B_2 A^{n-2} + \ldots + B_n, \tag{1.1}$$

where the coefficients $B_1, \ldots, B_n$ belong to a Banach algebra (or, more generally, a Banach module) $\mathfrak{B}$ of operators in the Banach algebra $B(\mathcal{X})$ of all bounded operators on a complex Banach space $\mathcal{X}$, and $A : D(A) \subset \mathcal{X} \to \mathcal{X}$ is a closed linear operator. We are mainly interested in the case when the operator iA generates a bounded C_0-group $\{T(t),\ t \in \mathbb{R}\}$ of operators in $B(\mathcal{X})$, for example, when $A = \frac{d}{dt}$ with a suitable domain. The set $\mathfrak{B}$ is assumed to contain all operators that, in some sense, almost commute with A. We decided to call the sets with the required property substantial with respect to the operator A (see Definition 2.15). Usually, the set $\mathfrak{B}$ will be an algebra of operators in $B(\mathcal{X})$. In certain cases, however, it may be difficult or impractical to verify that a given class of operators is an algebra. To circumvent this restriction, we introduce a more general notion of an admissible module (see Definition 3.2), which is just as suitable for our purposes.

2010 *Mathematics Subject Classification.* Primary 47A10, 46H25.

Key words and phrases. Operator polynomials, spectral analysis, method of similar operators.

The first author is supported in part by the Ministry of Education and Science of the Russian Federation in the frameworks of the project part of the state work quota (Project No 1.3464.2017/4.6). The second author is supported in part by NSF grant DMS-1322127.

The marquee result of this paper can be stated as follows. In its formulation we denote by

$$\sigma(\mathscr{A}) = \mathbb{C} \setminus \{\lambda \in \mathbb{C} : (\mathscr{A} - \lambda I)^{-1} \in B(\mathcal{X})\}$$

the *spectrum* of the operator $\mathscr{A}$ and by

$$\sigma_{\mathfrak{B}}(\mathscr{A}) = \mathbb{C} \setminus \{\lambda \in \mathbb{C} : (\mathscr{A} - \lambda I)^{-1} \in \mathfrak{B}\}$$

the spectrum of $\mathscr{A}$ in the Banach algebra (or module) $\mathfrak{B}$. The corresponding resolvent sets will be denoted by $\rho(\mathscr{A})$ and $\rho_{\mathfrak{B}}(\mathscr{A})$.

THEOREM 1.1. *Let $A : D(A) \subset \mathcal{X} \to \mathcal{X}$ be such that the operator iA is a generator of a bounded C_0-group of operators in $B(\mathcal{X})$, $\mathfrak{B}$ be an admissible Banach module of operators in $B(\mathcal{X})$ which is substantial with respect to A, and $\mathscr{A}$ be a polynomial of A given by* (1.1). *Then $\sigma(\mathscr{A}) = \sigma_{\mathfrak{B}}(\mathscr{A})$.*

Substantial sets of operators in $B(\mathcal{X})$ often have a corresponding class or classes of elements in $\mathcal{X}$ that are invariant for operators in the set. Thus, if y is in such a class, an operator $\mathscr{A}$ given by (1.1) is invertible, and $\mathscr{A}x = y$, the above theorem guarantees that x is in the same class with y.

Our proof of Theorem 1.1 is, in our opinion, fairly direct and elegant. It is, however, based on some sophisticated machinery that was developed in [**6**–**8**, **11**, **15**, **16**]. We first use the method of similar operators [**7**] and a Wiener's lemma extension [**16**] to prove Theorem 3.4, which is a special case of Theorem 1.1 for a polynomial $\mathscr{A} = A + B$ of degree 1. We then complete the proof of Theorem 1.1 using the standard reduction techniques of higher order differential equations to first order equations.

The paper is organized as follows. Section 2 draws on [**15**, **16**] to introduce the basic notions of the spectral theory of Banach modules, memory decay of linear operators on Banach modules, and the relevant extension of Wiener's lemma. It concludes with a definition of substantial classes of operators, numerous examples of which are presented in Section 5. The basic ideas of the method of similar operators [**6**, **7**] are outlined in Section 3. The method is the key ingredient of the proof of Theorem 3.4. The following Section 4 furnishes the remainder of the proof of Theorem 1.1 via the reduction of the general case to the one proved in Theorem 3.4. Finally, in Section 6 we illustrate some of the classes defined in Section 5 with examples based on matrices as well as differential and integral operators.

2. Banach modules and memory decay

We begin by introducing the basic notions of the spectral theory of Banach modules. We refer to [**11**, **15**, **16**] and references therein for more details. In this section, as well as in the remainder of the paper, by $\mathcal{X}$ we denote a complex Banach space and by $T : \mathbb{R}^d \to B(\mathcal{X})$ a bounded group representation. Via a standard renormalization [**23**] of $\mathcal{X}$ which does not affect our results, we may assume that the representation is, in fact, isometric. We shall also assume that $\mathcal{X}$ is a Banach $L^1(\mathbb{R}^d)$-module, that is, there is a well-defined map $(f, x) \mapsto fx : L^1(\mathbb{R}^d) \times \mathcal{X} \to \mathcal{X}$ which satisfies the usual associativity, distributivity and other standard properties (see, e.g., [**11**, **15**]). We especially emphasize the inequality

$$\|fx\|_{\mathcal{X}} \le \|f\|_{L^1(\mathbb{R}^d)} \|x\|_{\mathcal{X}}, \quad f \in L^1(\mathbb{R}^d), x \in \mathcal{X}. \tag{2.1}$$

A subspace $\mathcal{X}_0 \subseteq \mathcal{X}$ is a *submodule* of $\mathcal{X}$ if $fx \in \mathcal{X}_0$ for all $x \in \mathcal{X}_0$ and all $f \in L^1(\mathbb{R}^d)$. An important class of submodules is defined using the notion of the

Beurling spectrum. In the definition below and elsewhere in the paper, we denote by $\hat{f}$ the Fourier transform of a function $f \in L^1(\mathbb{R}^d)$:

$$\hat{f}(\lambda) = \int_{\mathbb{R}^d} f(t) e^{-i\lambda \cdot t} dt.$$

DEFINITION 2.1. *Given $x \in \mathcal{X}$, its* Beurling spectrum $\Lambda(x)$ *is a subset of $\mathbb{R}^d$ defined by $\Lambda(x) = \{\lambda \in \mathbb{R}^d : fx = 0 \text{ implies } \hat{f}(\lambda) = 0,\ f \in L^1(\mathbb{R}^d)\}$. Given $\sigma \subseteq \mathbb{R}^d$, the* spectral submodule $\mathcal{X}_\sigma$ *is given by $\mathcal{X}_\sigma = \{x \in \mathcal{X} : \Lambda(x) \subseteq \sigma\}$.*

DEFINITION 2.2. *We say that the Banach module structure of $\mathcal{X}$ is* associated with a representation T *if*

$$T(t)(fx) = f_t x = f(T(t)x),\ t \in \mathbb{R}^d,\ f \in L^1(\mathbb{R}^d),\ x \in \mathcal{X}, \tag{2.2}$$

where $f_t = f(\cdot - t)$. The structure is said to be non-degenerate *if $fx = 0$ for all $f \in L^1(\mathbb{R}^d)$ implies $x = 0$.*

The same Banach space may have several module structures. We shall use the notation $(\mathcal{X}, T)$ if we need to refer to the structure associated with the representation T.

DEFINITION 2.3. *Given $x \in \mathcal{X}$, we define the map $x_T : \mathbb{R}^d \to \mathcal{X}$ by $x_T(t) = T(t)x$. The submodule $\mathcal{X}_T$ of T-*continuous *vectors in $\mathcal{X}$ is then defined by*

$$\mathcal{X}_T = \{x \in \mathcal{X} : \text{ the map } x_T \text{ is continuous}\}.$$

With a slight abuse of notation we let

$$T(f)x = \int_{\mathbb{R}^d} f(t) T(-t) x dt,\ f \in L^1(\mathbb{R}^d),\ x \in \mathcal{X}_T. \tag{2.3}$$

The following lemma is then fairly easy to prove.

LEMMA 2.4. [**15**, Lemma 2.7] *Assume that the Banach module structure of $\mathcal{X}$ is non-degenerate and associated with the representation T. Then $T(f)x = fx$ for all $x \in \mathcal{X}_T$.*

In the following lemma we collect the spectral properties of Banach modules which we frequently use throughout the paper (sometimes, without mention).

LEMMA 2.5. [**16**, Lemma 2.2]. *Let $\mathcal{X}$ be a non-degenerate Banach L^1-module with the structure associated with a representation T. Then*

(i) *$\Lambda(x)$ is closed for every $x \in \mathcal{X}$ and $\Lambda(x) = \emptyset$ if and only if $x = 0$;*
(ii) *$\Lambda(Xx + Yy) \subseteq \Lambda(x) \cup \Lambda(y)$ for all $X, Y \in B(\mathcal{X})$ that commute with all operators $T(f)$, $f \in L^1$;*
(iii) *$\Lambda(fx) \subseteq (\operatorname{supp} \hat{f}) \cap \Lambda(x)$ for all $f \in L^1$ and $x \in \mathcal{X}$;*
(iv) *$fx = 0$ if $(\operatorname{supp} \hat{f}) \cap \Lambda(x) = \emptyset$, where $f \in L^1$ and $x \in \mathcal{X}$;*
(v) *$fx = x$ if $\Lambda(x)$ is a compact set, and $\hat{f} \equiv 1$ in some neighborhood of $\Lambda(x)$, $f \in L^1$, $x \in \mathcal{X}$.*

As we showed in [**15, 16**], Banach module structures are especially suitable for describing the spectral (memory) decay of elements of $\mathcal{X}$. In [**16**], we defined several kinds of spectral decay and provided various useful characterizations. To prove Theorem 1.1, we need only one kind – exponential. A few of the other kinds of spectral decay appear among the examples in Section 5.

Among several equivalent ways of defining exponential spectral decay (see [**16**, Proposition 3.4]), we choose one that doesn't require any auxiliary definitions.

DEFINITION 2.6. *A vector $x \in \mathcal{X}$ has* exponential spectral decay *with respect to the representation T if there is a $\delta > 0$ such that the function $x_T : \mathbb{R}^d \to \mathcal{X}$ given by $x_T(t) = T(t)x$, $t \in \mathbb{R}^d$, admits a holomorphic extension to the strip*

$$\mathbb{C}^d(-\delta, \delta) = \{z = (z_1, \dots, z_d) \in \mathbb{C}^d : -\delta < \Im m\ z_i < \delta,\ 1 \le i \le d\},$$

which is continuous and bounded in $\overline{\mathbb{C}^d(-\delta, \delta)}$.

To provide an equivalent definition mentioned above, we need to introduce an extended module structure as we did in [**16**, Section 2.3]. This structure will also allow us to define other useful kinds of spectral decay in Section 5.

We shall make use of the spaces $\widehat{L^1}(\mathbb{R}^d) = \{\hat\varphi : \varphi \in L^1\}$ and $\widehat{L^1_{loc}}(\mathbb{R}^d) = \{h : \mathbb{R}^d \to \mathbb{C}$ such that $h\hat\varphi \in \widehat{L^1}(\mathbb{R}^d)$ for all $\varphi \in L^1(\mathbb{R}^d)$ with $\operatorname{supp}\hat\varphi$ compact$\}$. Clearly, $\widehat{L^1}(\mathbb{R}^d) \subset \widehat{L^1_{loc}}(\mathbb{R}^d)$.

An $L^1(\mathbb{R}^d)$-module structure on $\mathcal{X}$ can equivalently be viewed as an $\widehat{L^1}(\mathbb{R}^d)$-module structure with $\hat\varphi \diamond x = \varphi x$, $\varphi \in L^1(\mathbb{R}^d)$, $x \in \mathcal{X}$. We extend this structure to $\widehat{L^1_{loc}}(\mathbb{R}^d)$ by defining a (possibly unbounded) linear operator $h\diamond = \check{T}(h)$ in the following way. For $x = \varphi x$ with $\operatorname{supp}\hat\varphi$ compact, we let

$$\check{T}(h)x = h \diamond x := (h\hat\varphi) \diamond x = (h\hat\varphi)^\vee x = T((h\hat\varphi)^\vee)x.$$

As we showed in [**16**], this definition is unambiguous since $h \diamond x$ does not depend on the choice of $\varphi \in L^1(\mathbb{R}^d)$ such that $x = \varphi x$ and $\operatorname{supp}\hat\varphi$ is compact. We have also showed that for each $h \in \widehat{L^1_{loc}}(\mathbb{R}^d)$ this operator is closable and we use the same notation, $\check{T}(h) = h\diamond$, to denote its closure. Thus, $\check{T}(h)$ is a well-defined closed linear operator and its domain $D(\check{T}(h)) = D(h\diamond)$ is a subset of $\mathcal{X}_T$.

The following two types of functions in $\widehat{L^1_{loc}}(\mathbb{R}^d)$ are especially useful. For $\lambda = (\lambda_1, \dots, \lambda_d) \in \mathbb{R}^d$ and $\alpha = (\alpha_1, \dots, \alpha_d) \in \mathbb{R}^d$ we let

$$\mathbf{e}_\alpha(\lambda) = e^{\alpha\cdot\lambda} = \exp\left(\sum_{k=1}^d \alpha_k \lambda_k\right) \tag{2.4}$$

and

$$\mathbf{h}_\alpha(\lambda) = e^{\alpha\cdot\tilde\lambda} = \exp\left(\sum_{k=1}^d \alpha_k |\lambda_k|\right), \tag{2.5}$$

where $\tilde\lambda = (|\lambda_1|, \dots, |\lambda_d|)$.

LEMMA 2.7. [**16**, Proposition 3.4]. *A vector $x \in \mathcal{X}_T$ has exponential spectral decay if and only if $x \in D(\boldsymbol{h}_\alpha\diamond)$ for some $\alpha \in \mathbb{R}^d_+$.*

We refer to [**15**] for numerous examples of Banach modules and associated representations.

To introduce the notion of a generator of a Banach $L^1(\mathbb{R})$-module, we use the family of functions $f_\lambda \in L^1(\mathbb{R})$, $\lambda \in \mathbb{C} \setminus \{i\mathbb{R}\}$, given by

$$f_\lambda(t) = e^{\lambda t}\chi_{(-\infty,0]}(t),\ \Re e\lambda > 0, \text{ and } f_\lambda(t) = -e^{\lambda t}\chi_{[0,\infty)}(t),\ \Re e\lambda < 0, \tag{2.6}$$

where χ_S is the characteristic function of the set S.

DEFINITION 2.8. *We say that a closed linear operator A is the* generator *of the Banach $L^1(\mathbb{R})$-module $\mathcal{X}$ if the resolvent of the operator iA satisfies $(\lambda I - iA)^{-1} = T(f_\lambda)$, $\lambda \in \mathbb{C} \setminus \{i\mathbb{R}\}$, where f_λ is given by (2.6).*

LEMMA 2.9. *Let $(\mathcal{X}, T)$ be a non-degenerate Banach $L^1(\mathbb{R})$-module. Then there exists a closed linear operator A that is its generator.*

PROOF. First, we observe that the function $\lambda \mapsto T(f_\lambda)$ satisfies Hilbert's resolvent identity, that is

$$T(f_\lambda) - T(f_\mu) = (\mu - \lambda)T(f_\lambda)T(f_\mu), \ \lambda, \mu \in \mathbb{C} \setminus \{i\mathbb{R}\}.$$

The above equality easily follows since $\hat{f}_\lambda - \hat{f}_\mu = (\mu - \lambda)\hat{f}_\lambda \hat{f}_\mu$. Hence, this function is a pseudo-resolvent, and, by [**12**, Theorem 1], there exists a linear relation A such that $(\lambda I - iA)^{-1} = T(f_\lambda)$, $\lambda \in \mathbb{C} \setminus \{i\mathbb{R}\}$. The same theorem implies that A can be chosen to be a closed linear operator, provided that all operators $T(f_\lambda)$, $\lambda \in \mathbb{C} \setminus \{i\mathbb{R}\}$, are injective.

Assume $\lambda \in \mathbb{C} \setminus \{i\mathbb{R}\}$, $x \in X$, and $T(f_\lambda)x = 0$. Let $I_\lambda \subseteq L^1(\mathbb{R})$ be the closed two-sided ideal in $L^1(\mathbb{R})$ generated by the translates of the function f_λ. On one hand, the definition of the Banach module structure together with (2.1) and (2.2) ensure that $T(f)x = 0$ for all $f \in I_\lambda$. On the other hand, since $\hat{f}_\lambda \neq 0$, Wiener's Theorem [**22**, XI.4.7] implies that $I_\lambda = L^1(\mathbb{R})$. Since the Banach module is non-degenerate, we conclude that $x = 0$. □

REMARK 2.10. *It is not hard to see ([**12**]) that if A is the generator of a (non-degenerate) Banach module $(\mathcal{X}, T)$ and the group $\{T(t),\ t \in \mathbb{R}\}$ is a C_0-group, then the operator iA is the infinitesimal generator of the group.*

In this paper, we are primarily interested in Banach modules that are related to the operator A in (1.1). Obviously, the first module that is of interest to us is the module $(\mathcal{X}, T)$ generated by the operator A. Since the operators $T(t)$, $t \in \mathbb{R}$, commute with the generator A, they also define a bounded strongly continuous representation on the Banach space $\mathcal{Y}$ defined as the set $D(A)$ equipped with the graph norm. Thus, the second Banach module we consider is $(\mathcal{Y}, T)$; we shall use the same symbol T to denote the representations on both $\mathcal{X}$ and $\mathcal{Y}$ as it will be clear from the context which one is meant. We shall also use several Banach module structures on the spaces of operators.

In the remainder of the section $\mathcal{X} = (\mathcal{X}, T)$ is a Banach module generated by an operator A and $\mathcal{Y}$ is the space $D(A)$ equipped with the graph norm. We shall denote by $L(\mathcal{X}, \mathcal{Y})$ and $L(\mathcal{Y}, \mathcal{X})$ the Banach spaces of all bounded linear operators from $\mathcal{X}$ to $\mathcal{Y}$ and from $\mathcal{Y}$ to $\mathcal{X}$, respectively. As in [**15, 16**], we also define Banach module structures on the space of operators $\mathcal{B} \in \{B(\mathcal{X}), L(\mathcal{X}, \mathcal{Y}), L(\mathcal{Y}, \mathcal{X})\}$. With another slight abuse of notation, we let $\widetilde{T} : \mathbb{R} \to B(\mathcal{B})$ be given by $\widetilde{T}(t)X = T(t)XT(-t)$, $X \in \mathcal{B}$, and write

$$\widetilde{T}(f)X = \int_{\mathbb{R}} f(t)\widetilde{T}(-t)X dt, \ f \in L^1(\mathbb{R}), \ X \in \mathcal{B}, \tag{2.7}$$

where the convergence of the integral is understood in the strong operator topology. We remark that the module $(\mathcal{B}, \widetilde{T})$ is generated by the commutator $\widetilde{A} : D(\widetilde{A}) \subseteq \mathcal{B} \to \mathcal{B}$ given by $\widetilde{A}(X) = AX - XA$, $B \in D(\widetilde{A}) \subset \mathcal{B}$ [**23**]. Moreover, if $\mathcal{B} = B(\mathcal{X})$ and $h_n(\lambda) = \lambda^n$, $n \in \mathbb{N}$, one easily sees that

$$\check{\widetilde{T}}(h_n) = \widetilde{A}^n, \ n \in \mathbb{N}. \tag{2.8}$$

We also note the following very useful spectral properties of the representation $\widetilde{T}$ (see [**16**, Lemma 4.1]):

(2.9) $$\Lambda(XY,\widetilde{T}) \subseteq \overline{\Lambda(X,\widetilde{T}) + \Lambda(Y,\widetilde{T})} \text{ and } \Lambda(Xx,T) \subseteq \overline{\Lambda(X,\widetilde{T}) + \Lambda(x,T)}.$$

To define some of the operator classes in Section 5, we shall also use an $L^1(\mathbb{R}^2)$-module structure on $\mathcal{B} \in \{B(\mathcal{X}), L(\mathcal{X},\mathcal{Y}), L(\mathcal{Y},\mathcal{X})\}$. We introduce the representation $\bar{T} : \mathbb{R}^2 \to B(\mathcal{X})$ given by $\bar{T}(t_1,t_2)B = T(t_1)BT(t_2)$, $B \in \mathcal{B}$, and define the structure via

(2.10) $$\bar{T}(f)B = \iint f(t_1,t_2)\bar{T}(-t_1,-t_2)Bdt_1dt_2,\ f \in L^1(\mathbb{R}^2),\ B \in \mathcal{X},$$

where, again, the convergence of the integral is understood in the strong operator topology.

The following spectral property of the representation $\bar{T}$ will be useful for us.

PROPOSITION 2.11. *Assume that* $X, Y \in B(\mathcal{X})$ *satisfy* $\Lambda(X,\bar{T}) \subset S \times \mathbb{R}$ *and* $\Lambda(Y,\bar{T}) \subset \mathbb{R} \times S$ *for a compact set* S *such that* $S \subset (0,\infty)$ *or* $S \subset (-\infty,0)$. *Then, if the generator* A *is invertible,*

$$\|A^{-1}X\| \le \sup_{s\in S}|s|^{-1}\|X\| \text{ and } \|YA^{-1}\| \le \sup_{s\in S}|s|^{-1}\|Y\|.$$

PROOF. Let $f \in L^1(\mathbb{R})$ be such that $\hat{f}(s) = 1/s$ on S. Then $A^{-1}X = T(f)X$ and $YA^{-1} = YT(f)$. Choosing f as in the proof of [**14**, Theorem 1] yields $\|f\| = \sup_{s\in S}|s|^{-1}$, and the desired inequalities follow. □

Next, we apply Definition 2.6 to describe exponential spectral decay of operators. For reasons explained in [**16**], we prefer to call it memory decay.

DEFINITION 2.12. *Let* $\mathcal{B} \in \{B(\mathcal{X}), L(\mathcal{X},\mathcal{Y}), L(\mathcal{Y},\mathcal{X})\}$. *An operator* $X \in \mathcal{B}$ *has* exponential memory decay *with respect to the generator* A *if it has exponential spectral decay with respect to the representation* $\widetilde{T} : \mathbb{R} \to B(\mathcal{B})$, *i.e., there is a* $\delta > 0$, *such that the function* $X_{\widetilde{T}} : \mathbb{R} \to \mathcal{B}$ *given by* $X_{\widetilde{T}}(t) = \widetilde{T}(t)X$, $t \in \mathbb{R}$, *admits a holomorphic extension to the strip*

$$\mathbb{C}(-\delta,\delta) = \{z \in \mathbb{C} : -\delta < \Im m z < \delta\},$$

which is continuous and bounded in $\overline{\mathbb{C}(-\delta,\delta)}$ *(in the norm of* $\mathcal{B}$*).*

REMARK 2.13. *Assume that* $X \in L(\mathcal{X},\mathcal{Y})$. *Then* X *can also be considered as an element of* $B(\mathcal{X})$ *with* $\|X\|_{B(\mathcal{X})} \le \|X\|_{L(\mathcal{X},\mathcal{Y})}$. *It follows that exponential memory decay of* $X \in L(\mathcal{X},\mathcal{Y})$ *implies that of* $X \in B(\mathcal{X})$. *Similarly, an operator* $X \in B(\mathcal{X})$ *can also be considered as an element of* $L(\mathcal{Y},\mathcal{X})$ *with* $\|X\|_{B(\mathcal{X})} \ge \|X\|_{L(\mathcal{Y},\mathcal{X})}$. *It follows that exponential memory decay of* $X \in B(\mathcal{X})$ *implies that of* $X \in L(\mathcal{Y},\mathcal{X})$.

It is not too hard to observe that exponential memory decay of an operator X implies that the commutators $XT(t) - T(t)X$, $t \in \mathbb{R}$, are in some sense small. If all the commutators are 0, then the operator X is called *memoryless* (see [**15**]). Clearly, memoryless operators have exponential memory decay.

The following result is a crucial component of the proof of our main theorem.

THEOREM 2.14 ([**16**, Theorem 6.5]). *Assume that an operator* $X \in L(\mathcal{Y},\mathcal{X})$ *is invertible and has exponential memory decay. Then* $X^{-1} \in L(\mathcal{X},\mathcal{Y})$ *also has exponential memory decay.*

We conclude this section with the definition of a substantial set of operators.

DEFINITION 2.15. *A set* $\mathfrak{B}$ *of operators in* $B(\mathcal{X})$ *is called* substantial *with respect to the generator* A *(or simply* A-substantial*) if* $\mathfrak{B}$ *contains all operators with exponential memory decay with respect to the generator* A.

We remark that a substantial Banach algebra does not have to be inverse closed in $B(\mathcal{X})$, even though many of such algebras have this property [**10**, **16**, **24**, **28**–**31**, **35**]. Another obvious but useful observation is that substantial sets of operators necessarily contain the identity operator $I \in B(\mathcal{X})$ and the operators $T(t)$, $t \in \mathbb{R}^2$, which are all memoryless.

3. The method of similar operators and a special case of the main result

The method of similar operators [**7**, **8**] is designed to study the spectral properties of an operator $A - B$ where A is a closed linear operator with well-understood spectral properties and B is a perturbation that is in some sense small with respect to A. The idea of the method is to find another perturbation B_0 of A such that $A - B_0$ is similar to $A - B$ and its spectrum can be easily estimated because of the special structure of B_0. We refer to the recent papers [**13**, **17**, **18**] illustrating the effective use of the method for spectral estimation.

In this paper, we are interested in the case when iA is a generator of an isometric C_0-group of operators $\{T(t), t \geq 0\}$ in $B(\mathcal{X})$ and B belongs to an A-substantial Banach module $\mathfrak{B}$. Accordingly, we modify the standard definitions and theorems of the method of similar operators in the following way.

DEFINITION 3.1. *Let* $\mathcal{X}_i$ *and* $\mathcal{Y}_i$, $i = 1, 2$, *be complex Banach spaces. Two closed linear operators* $A_i : D(A_i) \subseteq \mathcal{X}_i \to \mathcal{Y}_i$, $i = 1, 2$, *are called* similar *if there exists an invertible operator* $U \in L(\mathcal{X}_2, \mathcal{X}_1)$ *such that* $UD(A_2) = D(A_1)$ *and* $A_1 U x = U A_2 x$ *for all* $x \in D(A_2)$. *The operator* U *is called the* similarity transform operator *of* A_1 *into* A_2.

To introduce the class of admissible Banach modules, we make use of two families of transforms: $J_\alpha, \Gamma_\alpha \in B(B(\mathcal{X}))$, $\alpha > 0$. These families of transforms, as well as many of the operator classes in Section 5, are defined using a bounded uniform partition of unity (BUPU). To avoid another abstract definition, we choose to use a specific BUPU generated by the Fejér kernel. Thus, we let $\phi \in L^1(\mathbb{R})$ be defined by its Fourier transform:

$$\hat{\phi}(\lambda) = (1 - |\lambda|)\chi_{[-1,1]}(\lambda), \ \lambda \in \mathbb{R}. \tag{3.1}$$

We then let ϕ_a, $a \in \mathbb{R}$, be given by $\hat{\phi}_a(\lambda) = \hat{\phi}(\lambda - a)$, so that the family $(\phi_n)_{n\in\mathbb{Z}}$ forms a BUPU according to [**16**, Definition 3.2]. By $\phi_{a,\alpha}$, $a \in \mathbb{R}$, $\alpha > 0$, we denote the functions defined by $\hat{\phi}_{a,\alpha} = \hat{\phi}_a(\cdot/\alpha)$, and let

$$\hat{\varphi}_\alpha = \hat{\phi}_{-1,\alpha} + \hat{\phi}_{0,\alpha} + \hat{\phi}_{1,\alpha}, \ \alpha > 0. \tag{3.2}$$

Observe that for each $\alpha > 0$ we have $\varphi_\alpha \in L^1(\mathbb{R})$ and $\|\varphi_\alpha\|_1$ is independent of α. We define the transforms $J_\alpha \in B(B(\mathcal{X}))$ by

$$J_\alpha X = \widetilde{T}(\varphi_\alpha)X = \int_{\mathbb{R}} \varphi_\alpha(t) T(-t) X T(t) dt,$$

$X \in B(\mathcal{X})$, $\alpha > 0$. Since the representation $\widetilde{T}$ is isometric, our observation about $\|\varphi_\alpha\|_1$ and (2.7) imply that $\{J_\alpha, \alpha > 0\}$ is a bounded subset of $B(B(\mathcal{X}))$. Additionally, we observe that since $\operatorname{supp}\hat{\varphi}_\alpha = [-2\alpha, 2\alpha]$, we have $\Lambda(J_\alpha X) \subseteq [-2\alpha, 2\alpha]$, $X \in B(\mathcal{X})$, by [**16**, Lemma 2.2]. Therefore, [**16**, Corollary 2.13] implies that $J_\alpha X$, $X \in B(\mathcal{X})$, has exponential memory decay with respect to A.

The transforms Γ_α, $\alpha > 0$, are defined via

$$\Gamma_\alpha X = \widetilde{T}(\psi_\alpha)(X - JX) = \widetilde{T}(\psi_\alpha - \psi_\alpha * \varphi_\alpha)X,$$

where $\psi_\alpha \in L^1(\mathbb{R})$ is such that $\hat{\psi}_\alpha(\lambda) = \frac{1}{\lambda}$ in a neighborhood of $\mathbb{R}\setminus[-\alpha, \alpha]$. Observe that $\|\psi_\alpha - \psi_\alpha * \varphi_\alpha\|_1 = \frac{\pi}{2\alpha}$ (see [**5, 11**] and references therein). By [**16**, Lemma 2.2], we have $\Lambda(\Gamma_\alpha X) \subseteq \mathbb{R} \setminus (-\alpha, \alpha)$ and $\|\Gamma_\alpha X\| \leq \frac{\pi}{2\alpha}\|X\|$, $X \in B(\mathcal{X})$. Moreover, using the same argument as in the proof of [**14**, Theorem 1], we deduce that

$$\|\Gamma_\alpha(\widetilde{T}(\phi_{a,\beta})X)\| = \|\widetilde{T}(\phi_{a,\beta})(\Gamma_\alpha X)\| \leq \frac{\|\widetilde{T}(\phi_{a,\beta})X\|}{|a| - \beta}, \ |a| > \alpha + \beta, \tag{3.3}$$

for all $X \in B(\mathcal{X})$ and $\alpha, \beta > 0$. Additionally, basic facts about the functional calculus introduced by (2.7) imply [**7**, §21] that $(\Gamma_\alpha X)D(A) \subseteq D(A)$, the commutator $\widetilde{A}(\Gamma_\alpha X) = A\Gamma_\alpha X - (\Gamma_\alpha X)A$ extends continuously to a bounded operator in $B(\mathcal{X})$, and

$$\widetilde{A}(\Gamma_\alpha X) = A\Gamma_\alpha X - (\Gamma_\alpha X)A = X - J_\alpha X \text{ for all } X \in B(\mathcal{X}), \alpha > 0. \tag{3.4}$$

Formula (3.4) represents the key equation of the method of similar operators.

DEFINITION 3.2. *A Banach module $\mathfrak{B}$ of the $L^1(\mathbb{R})$-module $(B(\mathcal{X}), \widetilde{T})$ is called* admissible *for A if the following properties hold:*

(1) *The module $\mathfrak{B}$ is continuously embedded into $B(\mathcal{X})$, i.e. there is a constant $C_1 > 0$ such that $\|X\|_{B(\mathcal{X})} \leq C_1\|X\|_{\mathfrak{B}}$ for all $X \in \mathfrak{B}$.*
(2) *The representation $\widetilde{T}$ satisfies*

$$\|\widetilde{T}(t)X\|_{\mathfrak{B}} \leq C_2\|X\|_{\mathfrak{B}}, \ t \in \mathbb{R}, \ X \in \mathfrak{B},$$

for some $C_2 > 0$ that is independent of X.
(3) *$X\Gamma_\alpha Y$, $(\Gamma_\alpha X)Y \in \mathfrak{B}$ for all $X, Y \in \mathfrak{B}$ and $\alpha > 0$. Moreover, for each $\alpha > 0$ there is $\gamma_\alpha > 0$ such that*

$$\max\{\|X\Gamma_\alpha Y\|_{\mathfrak{B}}, \|(\Gamma_\alpha X)Y\|_{\mathfrak{B}}\} \leq \gamma_a\|X\|_{\mathfrak{B}}\|Y\|_{\mathfrak{B}}, \ X, Y \in \mathfrak{B},$$

and $\gamma_\alpha \to 0$ as $\alpha \to \infty$.

We call an admissible module strictly admissible *if also $T(t)X \in \mathfrak{B}$ for all $t \in \mathbb{R}$ and $X \in \mathfrak{B}$.*

We note that Definition 3.2(2) implies that the restrictions of the transforms J_α and Γ_α to an admissible module $\mathfrak{B}$ belong to $B(\mathfrak{B})$. We shall use the same notation to denote these restrictions. Additionally, we remark that if $\mathfrak{B}$ is a substantial algebra that satisfies the first two conditions in Definition 3.2, then it is automatically a strictly admissible Banach module. We mention that in strictly admissible modules, we have $XT(t) \in \mathfrak{B}$ and $X(\lambda I - A)^{-1} \in \mathfrak{B}$, $t \in \mathbb{R}$, $X \in \mathfrak{B}$, $\lambda \in \rho(A)$. We also mention that, since iA is a generator of an isometric C_0-group of operators, we have that for every $X \in B(\mathcal{X})$ and $\varepsilon > 0$ there is $\lambda_\varepsilon \in \rho(A)$ such that $\|X(\lambda_\varepsilon I - A)^{-1}\|_{B(\mathcal{X})} \leq \varepsilon$, which is one of the requirements in the general scheme of the method of similar operators.

The main theorem of the method of similar operators for the operator $A - B$ is formulated using the non-linear functions $\Phi_\alpha : \mathfrak{B} \to \mathfrak{B}$ which are defined in the following way:

$$\Phi_\alpha(X) = B\Gamma_\alpha X - (\Gamma_\alpha X)J_\alpha X + B, \ \alpha > 0. \tag{3.5}$$

THEOREM 3.3. *Let iA be a generator of a bounded C_0-group of operators in $B(\mathcal{X})$ and B belong to an admissible Banach module $\mathfrak{B}$. Let $\alpha > 0$ be such that $\gamma_\alpha \|J_\alpha\|_{B(\mathfrak{B})}\|B\|_\mathfrak{B} < \frac{1}{6}$, where γ_α is defined in Definition 3.2(3). Then the operator $A - B$ is similar to the operator $A - J_\alpha X$, where $X \in \mathfrak{B}$ is the only fixed point of the function Φ_α given by (3.5). Moreover, in this case, the similarity transform operator of $A - B$ into $A - J_\alpha X$ is given by $I + \Gamma_\alpha X$ and $\|\Gamma_\alpha X\|_\mathfrak{B} < 1$.*

We refer to [**7**, **8**] for the proof of all the statements in the above theorem.

We are now in position to prove our main theorem for the case of polynomials of first degree.

THEOREM 3.4. *Let iA be a generator of a bounded C_0-group of operators in $B(\mathcal{X})$ and B belong to an admissible Banach module $\mathfrak{B}$ which is substantial with respect to A. Then $\sigma(A - B) = \sigma_\mathfrak{B}(A - B)$.*

PROOF. Let $J = J_\alpha$ and $\Gamma = \Gamma_\alpha$ be two transforms in Theorem 3.3. Applying the theorem to the operator $A - B$, we get

$$(A - B)(I + \Gamma X)x = (I + \Gamma X)(A - JX)x$$

for all $x \in D(A)$. It follows that for each $\lambda \in \rho(A - B)$ we have

$$(A - B - \lambda I)^{-1} = (I + \Gamma X)^{-1}(A - JX - \lambda I)^{-1}(I + \Gamma X) \in B(\mathcal{X}).$$

Since $X \in \mathfrak{B}$, we have $(I + \Gamma X) \in \mathfrak{B}$. Next, we recall that by construction $JX \in \mathfrak{B}$ has exponential memory decay with respect to A. Remark 2.13 implies that the assertion remains valid for $JX \in L(\mathcal{Y}, \mathcal{X})$, where, as before, $\mathcal{Y}$ is the Banach space $D(A)$ equipped with the graph norm. Hence, $A - JX - \lambda I \in L(\mathcal{Y}, \mathcal{X})$ also has exponential memory decay with respect to A. From Theorem 2.14 we derive that $(A - JX - \lambda I)^{-1} \in L(\mathcal{X}, \mathcal{Y})$ has the same property and from Remark 2.13 it follows that so does $(A - JX - \lambda I)^{-1} \in B(\mathcal{X})$. Using the assumption that $\mathfrak{B}$ is substantial, we derive that $(A - JX - \lambda I)^{-1} \in \mathfrak{B}$. Finally, since $X \in \mathfrak{B}$ and $\|\Gamma X\|_\mathfrak{B} < 1$, we may use the Neumann series representation of $(I + \Gamma X)^{-1}$ and Definition 3.2(3) to conclude that

$$(A - B - \lambda I)^{-1} = (I + \Gamma X)^{-1}(A - JX - \lambda I)^{-1}(I + \Gamma X) \in \mathfrak{B}.$$

Therefore, $\lambda \in \rho_\mathfrak{B}(A - B)$, and the theorem is proved. □

REMARK 3.5. *In certain cases, the result of Theorem 3.4 will remain valid for admissible modules that are not necessarily substantial. One only needs to check that $(A - JX - \lambda I)^{-1}$ belongs to the module. For example, if one considers almost periodic operators defined in* [**4**], *the operator $(A - JX - \lambda I)^{-1}$ may or may not be almost periodic depending on the choice of A. The same conclusion follows for $(A - B - \lambda I)^{-1}$.*

REMARK 3.6. *The proof of Theorem 3.4 is much easier in the case when $\mathfrak{B}$ is an inverse closed subalgebra. Indeed, with this assumption it essentially suffices to observe that $(A - B - \lambda I)^{-1} = (A - \lambda I)^{-1}(I - B(A - \lambda I)^{-1})^{-1} \in \mathfrak{B}$. We point out, however, that even in this case we believe the result to be new.*

4. Proof of the main result

In this section we shall prove Theorem 1.1. Clearly, it will suffice to show that if an operator $\mathscr{A}$ of the form (1.1) is invertible, that is $\mathscr{A}^{-1} \in B(\mathcal{X})$, then $\mathscr{A}^{-1}$ is, in fact, in $\mathfrak{B}$. We shall show this by reducing invertibility of $\mathscr{A}$ to that of an operator studied in the previous section. We employ an adjustment of the classical technique of reduction of higher order differential equations to a system of first order equations.

By $\mathcal{X}^n$ we shall denote the complex Banach space which is a product of n copies of the complex Banach space $\mathcal{X}$ with the max norm. The operators in $B(\mathcal{X}^n)$ will be represented by $n \times n$ operator matrices with entries in $B(\mathcal{X})$. Similar convention will be used for unbounded operators with the same domain. The Banach module structure of $\mathcal{X}$ defined via the representation T generated by the operator A extends to $\mathcal{X}^n$ in the following way. The operator

$$(4.1)\qquad \mathbb{A} = \begin{pmatrix} A & 0 & 0 & \dots & 0 \\ 0 & A & 0 & \dots & 0 \\ 0 & 0 & A & \dots & 0 \\ \vdots & \vdots & \vdots & \ddots & \vdots \\ 0 & 0 & 0 & \dots & A \end{pmatrix} : [D(A)]^n \subseteq \mathcal{X}^n \to \mathcal{X}^n$$

generates a bounded representation $\mathbb{T} : \mathbb{R} \to B(\mathcal{X}^N)$ given by

$$(4.2)\qquad \mathbb{T}(t) = \begin{pmatrix} T(t) & 0 & 0 & \dots & 0 \\ 0 & T(t) & 0 & \dots & 0 \\ 0 & 0 & T(t) & \dots & 0 \\ \vdots & \vdots & \vdots & \ddots & \vdots \\ 0 & 0 & 0 & \dots & T(t) \end{pmatrix} \in B(\mathcal{X}^n),\ t \in \mathbb{R},$$

which, in turn, equips $\mathcal{X}^n$ with a Banach module structure via an analog of (2.3). The representation $\widetilde{\mathbb{T}}$ and the module structure on $B(\mathcal{X}^n)$ are then defined as in Section 2 via (2.7). From these definitions, it is easily seen that an operator in $B(\mathcal{X}^n)$ has exponential memory decay with respect to $\mathbb{A}$ if and only if each entry of its matrix has exponential memory decay with respect to A. Accordingly, given a class $\mathfrak{B}$ of operators in $B(\mathcal{X})$ we let $\mathfrak{B}_n$ be the class of operators in $B(\mathcal{X}^n)$ whose matrices have entries in $\mathfrak{B}$. Among the equivalent norms in $\mathfrak{B}_n$ we choose the max of the $\mathfrak{B}$-norms of the entries (rescaled as in Remark 5.1, if necessary). With this definition, the class $\mathfrak{B}_n$ is $\mathbb{A}$-substantial whenever $\mathfrak{B}$ is A-substantial.

Next, we proceed to define the operator $\mathbb{B}$ such that invertibility of the operator $\mathscr{A}$ given by (1.1) is equivalent to that of the operator $\mathbb{A}-\mathbb{B}$. Guided by the standard reduction scheme, we let

$$(4.3)\qquad \mathbb{B} = \begin{pmatrix} 0 & I & 0 & \dots & 0 \\ 0 & 0 & I & \dots & 0 \\ \vdots & \vdots & \vdots & \ddots & \vdots \\ 0 & 0 & 0 & \dots & I \\ -B_n & -B_{n-1} & -B_{n-2} & \dots & -B_1 \end{pmatrix} \in \mathfrak{B}_n.$$

To complete the proof of Theorem 1.1 it remains to state the following result and apply Theorem 3.4 to the operator $\mathbb{A} - \mathbb{B}$.

THEOREM 4.1. *Assume that an operator iA is a generator of a bounded C_0-group of operators in $B(\mathcal{X})$ and the operator $\mathscr{A}$ is defined via (1.1), where the coefficients B_n, $n = 1, \dots, n-1$, are in $B(\mathcal{X})$. Then the operator $\mathscr{A}$ is invertible if and only if the operator $\mathbb{A} - \mathbb{B}$ is invertible. Moreover, if $\mathscr{A}$ is invertible, then the matrix of $(\mathbb{A} - \mathbb{B})^{-1}$ has the form*

$$(4.4)\quad \begin{aligned} (\mathbb{A}-\mathbb{B})^{-1} &= \begin{pmatrix} * & * & * & \dots & \mathscr{A}^{-1} \\ * & * & * & \dots & * \\ * & * & * & \dots & * \\ \vdots & \vdots & \vdots & \ddots & \vdots \\ * & * & * & \dots & * \end{pmatrix} \\ &= \begin{pmatrix} A^{-1}-D_1 & A^{-2}-D_2 & \dots & A^{1-n}-D_{n-1} & \mathscr{A}^{-1} \\ -AD_1 & A^{-1}-AD_2 & \dots & A^{2-n}-AD_{n-1} & A\mathscr{A}^{-1} \\ -A^2D_1 & -A^2D_2 & \dots & A^{3-n}-A^2D_{n-1} & A^2\mathscr{A}^{-1} \\ \vdots & \vdots & \vdots & \vdots & \vdots \\ -A^{n-1}D_1 & -A^{n-1}D_2 & \dots & -A^{n-1}D_{n-1} & A^{n-1}\mathscr{A}^{-1} \end{pmatrix}, \end{aligned}$$

where

$$D_j = \mathscr{A}^{-1} \sum_{k=0}^{j} B_{k-j+n} A^{-k}, \quad 1 \le j \le n-1.$$

PROOF. The statement of the theorem is easily verified via a straightforward (block) matrix multiplication. □

For strictly admissible modules the reduction to polynomials of first degree can be done in a much easier way. For an invertible operator $\mathscr{A}$, we have

$$\mathscr{A} = (A + B_1 + B_2 A^{-1} + \dots + B_n A^{1-n}) A^{n-1} = (A + B_0) A^{n-1},$$

and $B_0 = B_1 + B_2 A^{-1} + \dots + B_n A^{1-n} \in \mathfrak{B}$ by definition of strict admissibility. Therefore, the following enhancement of Theorem 1.1 is valid.

THEOREM 4.2. *Let $iA : D(A) \subset \mathcal{X} \to \mathcal{X}$ be a generator of a bounded C_0-group of operators in $B(\mathcal{X})$, $\mathfrak{B}$ be a strictly admissible Banach module of operators in $B(\mathcal{X})$ which is substantial with respect to A, and $\mathscr{A}$ be a polynomial of A given by (1.1). Then, for $\lambda \in \rho(\mathscr{A})$, we have $\lambda \in \rho_{\mathfrak{B}}(\mathscr{A})$, and*

$$(4.5)\quad (\mathscr{A} - \lambda I)^{-1} = (A - \lambda I)^{1-n} L,$$

for some $L \in \mathfrak{B}$.

5. Various classes of operators with memory decay

This section serves as a small catalog of operator classes with various types of memory decay. In several cases, we provide conditions which make the class a Banach algebra or an admissible Banach module. Some of the examples in this section may seem rather arcane, but they do cover a wide variety of cases that are of interest. We also point out that our main result appears to be new even for many of the simplest examples of admissible modules. As before, in this section, $\mathcal{X}$ is a Banach module with the structure generated by an operator A.

REMARK 5.1. *Typically, the definition of a Banach algebra $\mathfrak{B}$ requires that $\|XY\| \le \|X\|\|Y\|$ for any two elements of the algebra. For some of the examples in this section, however, we will provide a norm that satisfies a weaker requirement, namely, that $\|XY\| \le C\|X\|\|Y\|$ for some constant $C > 0$ that is independent of X and Y. Then an equivalent norm $|||\cdot|||$ satisfying the more stringent condition can be defined by letting*

$$|||X||| = \sup_{\|Y\|\le 1} \|XY\|, \ X, Y \in \mathfrak{B}.$$

As mentioned in Section 3, many of the classes below are defined using a bounded uniform partition of unity (BUPU). We shall make use of the BUPU defined by (3.1) and remark that using any other BUPU would typically result in defining the same classes (see [**16**] for more details and references on BUPU).

Another notion that we use to define some of the classes below is that of a weight. In this context, weights are typically used to measure the relative importance of the corresponding parts of an operator, for example, to quantify the off-diagonal decay of a matrix. We shall make use of two kinds of weights – weights on $\mathbb{R}$ and weights on $\mathbb{R}^2$, the latter will be referred to as bi-weights.

DEFINITION 5.2. *A* weight *is an even measurable function $v : \mathbb{R} \to \mathbb{R}$ such that $v(t) \ge 1$ for all $t \in \mathbb{R}$. A* bi-weight *is a measurable function $w : \mathbb{R}^2 \to \mathbb{R}$ that satisfies*

$$w(s,t) = w(t,s) \ge 1 \quad \textit{and} \quad \sup_{s\in\mathbb{R}} w(s, s+t) < \infty, \ s, t \in \mathbb{R}.$$

We remark that for a bi-weight w the function $v_w : \mathbb{R} \to \mathbb{R}$ given by

$$v_w(t) = \sup_{s\in\mathbb{R}} w(s, s+t), \ t \in \mathbb{R},$$

is a weight. We shall refer to v_w as the *majorante* of w.

To describe the properties of the various classes of operators defined below, we proceed to introduce various conditions on weights. We cite the survey [**26**], where the role of some of these conditions is carefully explained.

DEFINITION 5.3. *A weight v is*

- submultiplicative, *if $v(s+t) \le v(s)v(t)$, $s, t \in \mathbb{R}$;*
- subconvolutive, *if $v^{-1} \in L^1(\mathbb{R})$ and $v^{-1} * v^{-1} \le Cv^{-1}$ (as a pointwise inequality) for some $C > 0$;*
- quasisubconvolutive, *if $u^{-1} = ((1+|\cdot|)v)^{-1} \in L^1(\mathbb{R})$, and*

$$\int_{|t|\ge\alpha} \frac{v(s)}{u(t)v(s-t)} dt \to 0 \tag{5.1}$$

  as $\alpha \to \infty$ uniformly in $s \in \mathbb{R}$;
- balanced *if there exist $a, b \in (0, \infty)$ such that*

$$a \le \inf_{t\in[0,1]} \frac{v(s+t)}{v(s)} \le \sup_{t\in[0,1]} \frac{v(s+t)}{v(s)} \le b, \ s \in \mathbb{R};$$

- subexponential, *if for every $\gamma > 1$ there is $M > 0$ such that $v(t) \le M\gamma^{|t|}$ for all $t \in \mathbb{R}$;*
- *a* GRS-weight [**25**], *if it is submultiplicative and $\lim_{n\to\infty} [v(nt)]^{1/n} = 1$.*

Each of the above notions applies to a bi-weight w if it applies to its majorante v_w.

EXAMPLE 5.4. *A typical weight is given by*

$$v(t) = e^{a|t|^b}(1+|t|)^p. \tag{5.2}$$

The above weight is balanced if $0 \le b \le 1$ *and submultiplicative if also* $a, p \ge 0$*; under these conditions it is a GRS-weight if and only if* $0 \le b < 1$*, in which case it is also subexponential. The weight* (5.2) *is subconvolutive if* $a = 0$ *and* $p > 1$*. It is quasisubconvolutive if* $a = 0$ *and* $p > 0$*; in particular, we note that the weight* $v(t) = 1 + |t|$ *is quasisubconvolutive, but not subconvolutive.*

5.1. $\widetilde{T}$-continuous operators. The algebra B_c of $\widetilde{T}$-continuous operators is defined by $B_c = (B(\mathcal{X}))_{\widetilde{T}} = \{X \in B(\mathcal{X}) :$ the map $t \mapsto \widetilde{T}(t)X : \mathbb{R} \to B(\mathcal{X})$ is continuous$\}$. The norm of B_c is inherited from $B(\mathcal{X})$. We refer to [**15**, Lemma 4.3] for several useful characterizations of the algebra B_c. Most of the operator classes below are contained in B_c.

5.2. Wiener type. Let v be a weight. The Wiener-type class $\mathcal{W}_v$ is defined by

$$\mathcal{W}_v = \{X \in B(\mathcal{X}) : \|X\|_{\mathcal{W}_v} = \int_{\mathbb{R}} \|\widetilde{T}(\phi_a)X\| v(a)da < \infty\},$$

where ϕ_a is given following (3.1). In case of the trivial weight $v \equiv 1$, we typically write $\mathcal{W}$ instead of $\mathcal{W}_v$. The classical Wiener algebra of periodic functions with summable Fourier coefficients can be written in this form.

In [**16**], we have shown that $\mathcal{W}$ has an equivalent collection of norms. The following lemma extends the result to the weighted case.

LEMMA 5.5. *Let v be a balanced weight. Then*

$$\|X\|_{v,\alpha} = \sum_{n\in\mathbb{Z}} \|\widetilde{T}(\phi_{n,\alpha})X\| v(n), \ \hat{\phi}_{n,\alpha}(\lambda) = \hat{\phi}(\lambda/\alpha - n), \ \alpha > 0,$$

is an equivalent norm for $\mathcal{W}_v$.

PROOF. We will show the equivalence for $\alpha = 1$. The general case is proved in a similar manner.

Let $k = k(a) = \lfloor a \rfloor$ be the floor of a. Observe that

$$\phi_a = \phi_a * (\phi_{k-1} + \phi_k + \phi_{k+1} + \phi_{k-2}). \tag{5.3}$$

Therefore, since $\|\widetilde{T}(\phi_a)\| = 1$, we have

$$\begin{aligned}
\|X\|_{\mathcal{W}_v} &= \sum_{k\in\mathbb{Z}} \int_k^{k+1} \|\widetilde{T}(\phi_a)X\| v(a)da \\
&= \sum_{k\in\mathbb{Z}} \int_k^{k+1} \|\widetilde{T}(\phi_a)\widetilde{T}(\phi_{k-1} + \phi_k + \phi_{k+1} + \phi_{k-2})X\| v(a)da \\
&\le \sum_{k\in\mathbb{Z}} \left(\sum_{j=-1}^{2} \|\widetilde{T}(\phi_{k+j})X\| \right) \int_k^{k+1} v(a)da \\
&= \sum_{k\in\mathbb{Z}} \|\widetilde{T}(\phi_k)X\| \int_{k-2}^{k+2} v(a)da \le C \sum_{k\in\mathbb{Z}} \|\widetilde{T}(\phi_k)X\| v(k),
\end{aligned}$$

where the last inequality holds for some $C > 0$ because the weight is balanced. The other desired inequality follows similarly from

$$\|\widetilde{T}(\phi_n)X\| = \int_n^{n+1} \left\| \widetilde{T}(\phi_n) \left(\sum_{j=-2}^{2} \widetilde{T}(\phi_{a+n-j}) \right) X \right\| d(a),$$

and the fact that the weight v is balanced. □

REMARK 5.6. *The above proof indicates how balanced weights are used to essentially reduce the study of various classes of operators to a discrete case. If the representation T is periodic, the operators $E_k = T(\phi_{k,\alpha})$, $k \in \mathbb{Z}$, are disjunctive idempotents that form a resolution of the identity (for a proper choice of α that depends on the period). Then an operator $X \in B(\mathcal{X})$ can be identified with the operator matrix with entries E_kXE_j, $j,k \in \mathbb{Z}$, and the operators $\widetilde{T}(\phi_{n,\alpha})X = \sum_{k\in\mathbb{Z}} \sum_{j-k=n} E_kXE_j$ represent the diagonals of the matrix. In this case, it is easier to study the algebraic properties of the classes because then*

$$\widetilde{T}(\phi_{n,\alpha})(XY) = \sum_{k\in\mathbb{Z}} \sum_{j-k=n} (\widetilde{T}(\phi_{k,\alpha})X)(\widetilde{T}(\phi_{j,\alpha})Y).$$

In [**16**], we have also shown that $\mathcal{W}$ is a Banach algebra. A similar argument applies in the case of a submultiplicative weight as well (we note that submultiplicative weights are automatically balanced). Additionally, in the same paper we showed that the algebra $\mathcal{W}$ is inverse closed. We expect that the algebra $\mathcal{W}_v$ is inverse closed if v is a submultiplicative GRS-weight.

The algebra $\mathcal{W}_v$ will be substantial as long as the submultiplicative weight v is also subexponential.

A similar kind of algebra have been studied by many different authors in various guises. We cite [**10, 19, 29, 30, 34–36**] to name just a few.

5.3. Jaffard type. Let v be a weight and w be a bi-weight. We shall define two kinds of Jaffard-type classes using these weights.

The Jaffard-1 class $\mathfrak{J}_w^1$ is defined by

$$\mathfrak{J}_w^1 = \{X \in B(\mathcal{X}) : \|X\|_{\mathfrak{J}_w^1} = \|X\|_{B(\mathcal{X})} + |X|_{\mathfrak{J}_w^1} < \infty\},$$

where

$$|X|_{\mathfrak{J}_w^1} = \sup_{a,b\in\mathbb{R}} \|\bar{T}(\phi_a \otimes \phi_b)X\| w(a,b),$$

ϕ_a is given following (3.1), and $(\phi_a \otimes \phi_b)(s,t) = \phi_a(s)\phi_b(t)$.

Similarly, the Jaffard-2 class $\mathfrak{J}_v^2$ is defined by

$$\mathfrak{J}_v^2 = \{X \in B(\mathcal{X}) : \|X\|_{\mathfrak{J}_v^2} = \|X\|_{B(\mathcal{X})} + |X|_{\mathfrak{J}_v^2} < \infty\},$$

where

$$|X|_{\mathfrak{J}_v^2} = \sup_{a\in\mathbb{R}} \|\widetilde{T}(\phi_a)X\| v(a),$$

and ϕ_a is given following (3.1).

Clearly, $\mathfrak{J}_w^1$ and $\mathfrak{J}_v^2$ are modules of operators in $B(\mathcal{X})$ satisfying the first two conditions of Definition 3.2.

LEMMA 5.7. *Assume that a weight v and a bi-weight w are balanced and subconvolutive. Then $\mathfrak{J}_w^1$ and $\mathfrak{J}_v^2$ are Banach algebras.*

PROOF. We outline the proof in case of $\mathfrak{F}_v^2$; the other case can be proved along the same lines.

Using (5.3) once again, we get

$$\begin{aligned}\|\widetilde{T}(\phi_a)(XY)\| &= \left\|\widetilde{T}(\dot{\varphi}_a)\left(\left(\sum_{k\in\mathbb{Z}}\widetilde{T}(\phi_k)X\right)\left(\sum_{m\in\mathbb{Z}}\widetilde{T}(\phi_m)Y\right)\right)\right\| \\ &\le \sum_{k\in\mathbb{Z}}\sum_{\substack{m\in\mathbb{Z}:\\ m+k-a\le 2}}\left\|\widetilde{T}(\phi_k)X\right\|\left\|\widetilde{T}(\phi_m)Y\right\| \\ &\le |X|_{\mathfrak{F}_v^2}|Y|_{\mathfrak{F}_v^2}\sum_{k\in\mathbb{Z}}\sum_{\substack{m\in\mathbb{Z}:\\ m-a\le 2}} v^{-1}(k)v^{-1}(m-k).\end{aligned}$$

The desired Banach algebra inequality follows from the above since the weight v is balanced and subconvolutive. □

Conditions on the weights can be relaxed if we only require admissibility of the modules $\mathfrak{F}_w^1$ and $\mathfrak{F}_v^2$.

LEMMA 5.8. *Assume that a weight v and a bi-weight w are balanced and quasisubconvolutive. Then $\mathfrak{F}_w^1$ and $\mathfrak{F}_v^2$ are strictly admissible Banach modules.*

PROOF. Again, we only outline the proof in case of $\mathfrak{F}_v^2$.

We use (3.3) to modify the key inequalities in the proof of Lemma 5.7:

$$\begin{aligned}\|\widetilde{T}(\phi_a)((\Gamma_\alpha X)Y)\| &= \left\|\widetilde{T}(\dot{\varphi}_a)\left(\left(\sum_{k\in\mathbb{Z}}\widetilde{T}(\phi_k)(\Gamma_\alpha X)\right)\left(\sum_{m\in\mathbb{Z}}\widetilde{T}(\phi_m)Y\right)\right)\right\| \\ &\le \sum_{k\in\mathbb{Z}}\sum_{\substack{m\in\mathbb{Z}:\\ m+k-a\le 2}}\left\|\widetilde{T}(\phi_k)(\Gamma_\alpha X)\right\|\left\|\widetilde{T}(\phi_m)Y\right\| \\ &\le |X|_{\mathfrak{F}_v^2}|Y|_{\mathfrak{F}_v^2}\sum_{|k|\ge\alpha-1}\sum_{\substack{m\in\mathbb{Z}:\\ m-a\le 2}}\frac{v^{-1}(k)v^{-1}(m-k)}{|k|-1} \\ &\le Cv^{-1}(a)|X|_{\mathfrak{F}_v^2}|Y|_{\mathfrak{F}_v^2}\sum_{|k|\ge\alpha-1}\frac{v(a)}{(1+|k|)v(k)v(a-k)}\end{aligned}$$

for some $C>0$, since the weight v is balanced. Similar estimates hold for

$$\|\widetilde{T}(\phi_a)(X(\Gamma_\alpha Y))\|,$$

and the admissibility of $\mathfrak{F}_v^2$ follows since the weight v is balanced and quasisubconvolutive. The defining property of strict admissibility is obvious from the definition of $|\cdot|_{\mathfrak{F}_v^2}$. □

The class $\mathfrak{F}_w^1$ is especially amenable for application of Theorem 4.2. Indeed, the following result is immediate from the theorem and Proposition 2.11.

THEOREM 5.9. *Let $iA : D(A)\subset\mathcal{X}\to\mathcal{X}$ be a generator of a bounded C_0-group of operators in $B(\mathcal{X})$, and $\mathscr{A}$ be a polynomial of A given by (1.1) with $\mathfrak{B}=\mathfrak{F}_w^1$ for some balanced quasisubconvolutive bi-weight w. Then, for $\lambda\in\rho(\mathscr{A})$, we have $(\mathscr{A}-\lambda I)^{-1}\in\mathfrak{F}_{\bar{w}}^1$, where $\bar{w}(a,b)=(1+|a|)^{n-1}w(a,b)$.*

5.4. Schur type. Let w be a bi-weight. The Schur-type class $\mathcal{S}_{p,w}$ is defined by

$$\mathcal{S}_{p,w} = \{X \in B(\mathcal{X}) : \|X\|_{\mathcal{S}_{p,w}} < \infty\},$$

where

$$\|X\|_{\mathcal{S}_{p,w}} = \sup_{s\in\mathbb{R}} \left(\int_{a\in\mathbb{R}} [\|T(\phi_a)X\| w(a,s)]^p da \right)^{\frac{1}{p}} + \sup_{s\in\mathbb{R}} \left(\int_{a\in\mathbb{R}} [\|XT(\phi_a)\| w(s,a)]^p da \right)^{\frac{1}{p}}$$

and ϕ_a is given following (3.1). As before, we shall omit the symbol w in the notation if $w \equiv 1$.

A version of this class of operators was studied in [**36**], where it was shown that, under suitable conditions on w, it is an inverse closed subalgebra of $B(\mathcal{X})$. In [**37**] it is shown that, in general, a Schur-type algebra need not be inverse closed. It is also not hard to see that $\mathcal{S}$ is not a subset of B_c.

5.5. Multiplicators. To introduce operator classes defined by a multiplicator, we use the extended module structure from Section 2. For $h \in \widehat{L^1_{loc}}(\mathbb{R})$, we define the first class $\mathscr{M}^1_h$ to be the set $D(h\diamond) = D(\check{\widetilde{T}}(h))$ equipped with the graph norm. Similarly, for $h \in \widehat{L^1_{loc}}(\mathbb{R}^2)$, we define the second class $\mathscr{M}^2_h$ to be the set $D(h\diamond) = D(\check{\widetilde{T}}(h))$, also equipped with the graph norm.

The following lemma follows immediately from the above definitions.

LEMMA 5.10. *Assume that $h_1, h_2 \in \widehat{L^1_{loc}}(\mathbb{R}^d)$, $d \in \{1,2\}$, are such that h_1/h_2 is a Fourier transform of a finite complex Borel measure. Then $\mathscr{M}^d_{h_2}$ embeds continuously into $\mathscr{M}^d_{h_1}$.*

COROLLARY 5.11. *Assume that $h \in \widehat{L^1_{loc}}(\mathbb{R}^d)$, $d \in \{1,2\}$, is such that for every $\alpha > 0$ the function $\boldsymbol{h}_\alpha/h$ is a Fourier transform of a finite complex Borel measure, where $\boldsymbol{h}_\alpha$ is defined by (2.5). Then the class $\mathscr{M}^d_h$ is A-substantial.*

It seems difficult to provide admissibility conditions for a general function h. We restrict our attention to the case when $h \in \widehat{L^1_{loc}}(\mathbb{R})$ is an entire function, that is

$$h(\lambda) = \sum_{n=0}^{\infty} \frac{h_n}{n!}\lambda^n, \ \lim_{n\to\infty} \left(\frac{h_n}{n!}\right)^{1/n} = 0. \tag{5.4}$$

We shall also assume that $h_n \geq 0$ for all $n \geq 0$ and $h_0 = 1$. Recalling (2.8), we see that for such h we have

$$\check{\widetilde{T}}(h) = \sum_{n=0}^{\infty} \frac{h_n}{n!}\widetilde{A}^n, \ \widetilde{A}X = AX - XA,$$

and we let

$$\widetilde{\mathscr{M}_h} = \left\{ X \in B(\mathcal{X}) : \|X\|_h = \sum_{n=0}^{\infty} \frac{h_n}{n!} \|\widetilde{A}^n X\| < \infty \right\}.$$

Clearly, $\widetilde{\mathscr{M}_h} \subseteq \mathscr{M}^1_h$ in general, and $\widetilde{\mathscr{M}_h} = \mathscr{M}^1_h$ when h is a polynomial.

LEMMA 5.12. *Assume that the sequence of coefficients* (h_n) *in* (5.4) *is submultiplicative, i.e.* $h_{m+k} \leq h_m h_k$, $m, k \geq 0$. *Then the class* $\mathscr{M}_h$ *is a Banach algebra with respect to* $\|\cdot\|_h$.

PROOF. Observe that the commutator $\widetilde{A}$ obeys the Leibniz formula for differentiation of a product, i.e. $\widetilde{A}(XY) = (\widetilde{A}X)Y + X(\widetilde{A}Y)$. Hence,

$$\begin{aligned}\|XY\|_h &= \sum_{n=0}^{\infty} \frac{h_n}{n!} \left\|\widetilde{A}^n(XY)\right\| = \sum_{n=0}^{\infty} \frac{h_n}{n!} \left\|\sum_{k=0}^{n} \binom{n}{k} (\widetilde{A}^k X)(\widetilde{A}^{n-k} Y)\right\| \\ &\leq \sum_{k=0}^{\infty} \sum_{m=0}^{\infty} \frac{h_{m+k}}{m!k!} \left\|\widetilde{A}^k X\right\| \left\|\widetilde{A}^m Y\right\| \\ &\leq \sum_{k=0}^{\infty} \sum_{m=0}^{\infty} \frac{h_k}{k!} \frac{h_m}{m!} \left\|\widetilde{A}^k X\right\| \left\|\widetilde{A}^m Y\right\| = \|X\|_h \|Y\|_h,\end{aligned}$$

and the assertion of the lemma follows. □

5.6. Approximation schemes. For $\sigma_a = (-a, a)$ consider the family of spectral submodules $(B(\mathcal{X}))_{\sigma_a}$ of the Banach module $(B(\mathcal{X}), \widetilde{T})$. This family constitutes an *approximation scheme* [**1, 2, 27**] in the Banach algebra $B(\mathcal{X})$, that is $(B(\mathcal{X}))_{\sigma_0} = \{0\}$, $(B(\mathcal{X}))_{\sigma_a} \subseteq (B(\mathcal{X}))_{\sigma_b}$ for $a \leq b$, and $(B(\mathcal{X}))_{\sigma_a} \cdot (B(\mathcal{X}))_{\sigma_b} \subseteq (B(\mathcal{X}))_{\sigma_{a+b}}$. The a-error of approximation $E_a(X)$ for $X \in B(\mathcal{X})$ is defined via

$$E_a(X) = \inf_{Y \in (B(\mathcal{X}))_{\sigma_a}} \|X - Y\|.$$

It is well known [**15, 27**] that

$$B_c = \{X \in B(\mathcal{X}) : \lim_{a \to \infty} E_a(X) = 0\}.$$

By specifying a rate of decay for $E_a(X)$ as $a \to \infty$, we may define approximation spaces $\mathcal{E}_r^p$ in $B(\mathcal{X})$ by the norms

$$\|X\|_{\mathcal{E}_r^p}^p = \sum_{k=0}^{\infty} E_k(X)^p \frac{(1+k)^{rp}}{1+k},$$

for $1 \leq p < \infty$ with the obvious change for $p = \infty$, $r > 0$. We cite [**27**] and references therein for the proof of the fact that these spaces are, in fact, inverse closed Banach algebras in $B(\mathcal{X})$. We remark that an equivalent norm on $\mathcal{E}_r^p$ is given by

$$\|X\|_{\mathcal{E}_r^p}^p \simeq \|X\|_{B(\mathcal{X})}^p + \int_1^{\infty} E_a(X)^p \frac{(1+a)^{rp}}{1+a} da,$$

for $1 \leq p < \infty$ with the obvious change for $p = \infty$.

Alternatively, given a weight v that is increasing on $[0, \infty)$, we may consider the Banach module $\widetilde{\mathcal{E}}_v$ defined by the norm

$$\|X\|_{\widetilde{\mathcal{E}}_v} = \|X\|_{B(\mathcal{X})} + \sup_{\alpha > 0} E_\alpha(X) v(\alpha).$$

A minor adjustment of the proof of [**27**, Proposition 4.2] shows that $\widetilde{\mathcal{E}}_v$ is an admissible Banach algebra provided that there is a constant $C > 0$ such that $v(2\alpha) \leq C v(\alpha)$ for all $\alpha \in [0, \infty)$.

6. Examples

In this section, we describe for a few specific types of operators what it means to belong to some of the classes in the previous section.

6.1. Matrices. Here, we let $\mathcal{X} = \ell^p(\mathbb{Z})$, $p \in [1,\infty)$, be the space of p-summable sequences, or $\mathcal{X} = c_0(\mathbb{Z})$ – the space of sequences vanishing at infinity. We shall identify operators on $\mathcal{X}$ with their matrices in the standard basis. In particular, the generator A will be given by a diagonal matrix with entries $a_{nn} = n$, $n \in \mathbb{Z}$, and the group $\{T(\tau)\}$ will be the 2π-periodic modulation given by the diagonal matrices with entries $t_{nn}(\tau) = e^{in\tau}$, $n \in \mathbb{Z}$. In this case, for $x \in \mathcal{X}$ and a matrix $X = (x_{mk}) \in B(\mathcal{X})$, we have $\Lambda(x,T) = \{n \in \mathbb{Z} : x(n) \neq 0\}$, $\Lambda(X,\widetilde{T}) = \{n \in \mathbb{Z} :$ there exist $m,k \in \mathbb{Z}$ such that $n = m-k$ and $x_{mk} \neq 0\}$, and $\Lambda(X,\bar{T}) = \{(m,k) \in \mathbb{Z}^2 : x_{mk} \neq 0\}$. Hence, the classes introduced in Section 5 appear in their most familiar form.

In particular, if v is a weight and w is a bi-weight, we have, for $X = (x_{mk}) \in B(\mathcal{X})$,

$$\|X\|_{W_v} \sim \sum_{n\in\mathbb{Z}} \left[v(n) \sup_{m-k=n} |x_{mk}| \right],$$

$$\|X\|_{\mathfrak{F}^1_w} \sim \|X\|_{B(\mathcal{X})} + \sup_{m,k\in\mathbb{Z}} \left[w(m,k)|x_{mk}| \right],$$

$$\|X\|_{\mathfrak{F}^2_v} \sim \|X\|_{B(\mathcal{X})} + \sup_{n\in\mathbb{Z}} \left[v(n) \sup_{m-k=n} |x_{mk}| \right],$$

$$\|X\|_{\mathcal{S}_{p,w}} = \sup_{k\in\mathbb{Z}} \left(\sum_{m\in\mathbb{Z}} \left[w(m,k) \sup_{j\in\mathbb{Z}} |x_{mj}| \right]^p \right)^{\frac{1}{p}} + \sup_{m\in\mathbb{Z}} \left(\sum_{k\in\mathbb{Z}} \left[w(m,k) \sup_{j\in\mathbb{Z}} |x_{jk}| \right]^p \right)^{\frac{1}{p}},$$

where $\sim$ denotes the equivalence of norms.

Next, we observe that the commutator $Y = AX - XA$ has entries $y_{mk} = (m-k)x_{mk}$. Therefore, for a multiplier h given by (5.4) and $X = (x_{mk}) \in B(\mathcal{X})$, we have

$$\|X\|_h = \sum_{n=0}^{\infty} \frac{h_n}{n!} \|Y^{(n)}\|,$$

where $Y^{(0)} = X$ and each $Y^{(n)}$, $n \in \mathbb{N}$, has entries $y^{(n)}_{mk} = (m-k)^n x_{mk}$.

REMARK 6.1. *More generally, one can take $(a_{nn})_{n\in\mathbb{Z}}$ to be any sequence of real numbers. This would lead to almost periodic operators such as the ones studied in* [**4**] *and some of the references therein.*

6.2. Differential and integral operators. In this subsection, we let $\mathcal{X} = C_{b,u} = C_{b,u}(\mathbb{R}, E)$ be the space of bounded uniformly continuous functions with values in a Banach space E. We also let $A = d/dt$ with $D(A) = C^1_{b,u}(\mathbb{R}, E)$ – the space of continuously differentiable functions with derivative in $C_{b,u}$. In this case, the representation T generated by A is the translation: $T(t)x(\tau) = x(t+\tau)$, $t \in \mathbb{R}$, $x \in C_{b,u}$.

First, we consider multiplication operators given by

$$(Bx)(t) = b(t)x(t), \ t \in \mathbb{R},$$

where $b \in C_{b,u}(\mathbb{R}, B(E))$ are continuous operator-valued functions. Computing the representation $\widetilde{T}$ we get

$$((\widetilde{T}(\tau)B)x)(t) = b(\tau + t)x(t).$$

Thus, $\Lambda(B, \widetilde{T}) = \Lambda(b, T) = \operatorname{supp} \hat{b}$, and, therefore, memory decay of the operator B is determined by the decay of the Fourier transform of the function b. Therefore, for a weight v and a multiplier h given by (5.4), we have

$$\|B\|_{W_v} = \int_{\mathbb{R}} \|\phi_a * b\|_{C_{b,u}} v(a)da,$$

$$\|B\|_{\mathfrak{F}_v^2} = \|b\|_{C_{b,u}} + \sup_{a\in\mathbb{R}} \|\phi_a * b\|_{C_{b,u}} v(a),$$

$$\|B\|_h = \sum_{m=0}^{\infty} \frac{h_m}{m!} \|b^{(m)}\|_{C_{b,u}}.$$

If $\mathscr{A} = d/dt + B$, the inverse operator $\mathscr{A}^{-1}$ is defined via Green's function:

$$(\mathscr{A}^{-1}x)(t) = \int_{\mathbb{R}} G(t,s)x(s)ds,$$

where the function $G : \mathbb{R}^2 \to B(E)$ is infinitely many times differentiable on $\{(t,s) \in \mathbb{R}^2, t \neq s\}$ and $\|G(t,s)\| \leq Me^{\gamma|t-s|}$, $t,s \in \mathbb{R}^2$, for some $M, \gamma > 0$ [**9, 21**]. A straightforward computation shows that

$$((\bar{T}(u_1,u_2)\mathscr{A}^{-1})x)(t) = \int_{\mathbb{R}} G(t+u_1, s-u_2)x(s)ds,$$

and

$$((\widetilde{T}(u)\mathscr{A}^{-1})x)(t) = \int_{\mathbb{R}} G(t+u, s+u)x(s)ds.$$

Therefore, if the space E is finite dimensional, we have

$$\|\mathscr{A}^{-1}\| \sim \sup_{t\in\mathbb{R}} \int_{\mathbb{R}} \|G(t,s)\|\, ds,$$

$$\left\|\widetilde{T}(\phi_a)\mathscr{A}^{-1}\right\| \sim \sup_{t\in\mathbb{R}} \int_{\mathbb{R}} \left\| \int_{\mathbb{R}} \phi_a(u)G(t-u, s-u)du \right\| ds,$$

$$\|T(\phi_a)\mathscr{A}^{-1}\| \sim \sup_{t\in\mathbb{R}} \int_{\mathbb{R}} \left\| \int_{\mathbb{R}} \phi_a(u)G(t-u, s)du \right\| ds,$$

$$\|\mathscr{A}^{-1}T(\phi_a)\| \sim \sup_{t\in\mathbb{R}} \int_{\mathbb{R}} \left\| \int_{\mathbb{R}} \phi_a(u)G(t, s+u)du \right\| ds,$$

yielding

$$\|\mathscr{A}^{-1}\|_{\mathcal{W}_v} \sim \int_{\mathbb{R}} \left(\sup_{t\in\mathbb{R}} \int_{\mathbb{R}} \left\| \int_{\mathbb{R}} \phi_a(u)G(t-u, s-u)du \right\| ds \right) v(a)da,$$

$$\|\mathscr{A}^{-1}\|_{\mathfrak{F}_v^2} \sim \sup_{a\in\mathbb{R}} \left(\sup_{t\in\mathbb{R}} \int_{\mathbb{R}} \left\| \int_{\mathbb{R}} \phi_a(u)G(t-u, s-u)du \right\| ds \right) v(a),$$

$$\|\mathscr{A}^{-1}\|_{\mathfrak{F}_w^1} \sim \sup_{a,b\in\mathbb{R}} \left(\sup_{t\in\mathbb{R}} \int_{\mathbb{R}} \left\| \int_{\mathbb{R}} \int_{\mathbb{R}} \phi_a(u_1)\phi_b(u_2)G(t-u_1, s+u_2)du_1du_2 \right\| ds \right) w(a,b),$$

and similarly for $\mathcal{S}_{p,w}$. To describe $\left\|\mathscr{A}^{-1}\right\|_h$, we define $G^{(0)} = G$ and

$$G^{(m)}(t,s) = \frac{d}{du} G^{(m-1)}(t+u, s+u) \bigg|_{u=0}, \quad m \in \mathbb{N}.$$

Then

$$\left\|\mathscr{A}^{-1}\right\|_h \sim \sum_{m=0}^{\infty} \frac{h_m}{m!} \sup_{t\in\mathbb{R}} \int_{\mathbb{R}} \left\|G^{(m)}(t,s)\right\| ds.$$

References

[1] J. M. Almira and U. Luther, *Inverse closedness of approximation algebras*, J. Math. Anal. Appl. **314** (2006), no. 1, 30–44, DOI 10.1016/j.jmaa.2005.03.067. MR2183535

[2] Jose Maria Almira and Uwe Luther, *Generalized approximation spaces and applications*, Math. Nachr. **263/264** (2004), 3–35, DOI 10.1002/mana.200310121. MR2029742

[3] Wolfgang Arendt, Charles J. K. Batty, Matthias Hieber, and Frank Neubrander, *Vector-valued Laplace transforms and Cauchy problems*, 2nd ed., Monographs in Mathematics, vol. 96, Birkhäuser/Springer Basel AG, Basel, 2011. MR2798103

[4] Radu Balan and Ilya Krishtal, *An almost periodic noncommutative Wiener's lemma*, J. Math. Anal. Appl. **370** (2010), no. 2, 339–349, DOI 10.1016/j.jmaa.2010.04.053. MR2651657

[5] A. G. Baskakov, *Methods of abstract harmonic analysis in the theory of perturbations of linear operators* (Russian), Sibirsk. Mat. Zh. **24** (1983), no. 1, 21–39, 191. MR688589

[6] A. G. Baskakov, *The Krylov-Bogolyubov substitution in the theory of perturbations of linear operators* (Russian), Ukrain. Mat. Zh. **36** (1984), no. 5, 606–611. MR764378

[7] A. G. Baskakov, *Garmonicheskiĭ analiz lineĭnykh operatorov* (Russian, with Russian summary), Izdatel′stvo Voronezhskogo Universiteta, Voronezh, 1987. MR1607000

[8] A. G. Baskakov, *Spectral analysis of perturbed non-quasi-analytic and spectral operators* (Russian, with Russian summary), Izv. Ross. Akad. Nauk Ser. Mat. **58** (1994), no. 4, 3–32, DOI 10.1070/IM1995v045n01ABEH001621; English transl., Russian Acad. Sci. Izv. Math. **45** (1994), no. 1, 1–31. MR1307054

[9] A. G. Baskakov, *Semigroups of difference operators in the spectral analysis of linear differential operators* (Russian, with Russian summary), Funktsional. Anal. i Prilozhen. **30** (1996), no. 3, 1–11, 95, DOI 10.1007/BF02509501; English transl., Funct. Anal. Appl. **30** (1996), no. 3, 149–157 (1997). MR1435133

[10] A. G. Baskakov, *Estimates for the elements of inverse matrices, and the spectral analysis of linear operators* (Russian, with Russian summary), Izv. Ross. Akad. Nauk Ser. Mat. **61** (1997), no. 6, 3–26, DOI 10.1070/im1997v061n06ABEH000164; English transl., Izv. Math. **61** (1997), no. 6, 1113–1135. MR1609144

[11] A. G. Baskakov, *Theory of representations of Banach algebras, and abelian groups and semigroups in the spectral analysis of linear operators* (Russian), Sovrem. Mat. Fundam. Napravl. **9** (2004), 3–151, DOI 10.1007/s10958-006-0286-4; English transl., J. Math. Sci. (N.Y.) **137** (2004), no. 4, 4885–5036. MR2123307

[12] A.G. Baskakov, *Linear relations as generators of semigroups of operators* (Russian, with Russian summary), Mat. Zametki **84** (2008), no. 2, 175–192, DOI 10.1134/S0001434608070183; English transl., Math. Notes **84** (2008), no. 1-2, 166–183. MR2475046

[13] A. G. Baskakov, A. V. Derbushev, and A. O. Shcherbakov, *The method of similar operators in the spectral analysis of the nonselfadjoint Dirac operator with nonsmooth potential* (Russian, with Russian summary), Izv. Ross. Akad. Nauk Ser. Mat. **75** (2011), no. 3, 3–28, DOI 10.1070/IM2011v075n03ABEH002540; English transl., Izv. Math. **75** (2011), no. 3, 445–469. MR2847780

[14] A. G. Baskakov and K. A. Sintyaeva, *On the Bohr-Favard inequalities for operators* (Russian, with English and Russian summaries), Izv. Vyssh. Uchebn. Zaved. Mat. **12** (2009), 14–21, DOI 10.3103/S1066369X09120020; English transl., Russian Math. (Iz. VUZ) **53** (2009), no. 12, 11–17. MR2663268

[15] A. G. Baskakov and I. A. Krishtal, *Harmonic analysis of causal operators and their spectral properties* (Russian, with Russian summary), Izv. Ross. Akad. Nauk Ser. Mat. **69** (2005), no. 3, 3–54, DOI 10.1070/IM2005v069n03ABEH000535; English transl., Izv. Math. **69** (2005), no. 3, 439–486. MR2150500

[16] Anatoly G. Baskakov and Ilya A. Krishtal, *Memory estimation of inverse operators*, J. Funct. Anal. **267** (2014), no. 8, 2551–2605, DOI 10.1016/j.jfa.2014.07.025. MR3255468

[17] Anatoly G. Baskakov and Ilya A. Krishtal, *Spectral analysis of abstract parabolic operators in homogeneous function spaces*, Mediterr. J. Math. **13** (2016), no. 5, 2443–2462. MR3554255

[18] Anatoly G. Baskakov, Ilya A. Krishtal, and Elena Yu. Romanova, *Spectral analysis of a differential operator with an involution*, J. Evol. Equ. **17** (2017), no. 2, 669–684, DOI 10.1007/s00028-016-0332-8. MR3665225

[19] I. A. Blatov, *On operator algebras with pseudo-sparse matrices, and their applications* (Russian, with Russian summary), Sibirsk. Mat. Zh. **37** (1996), no. 1, 36–59, i, DOI 10.1007/BF02104758; English transl., Siberian Math. J. **37** (1996), no. 1, 32–52. MR1401079

[20] Marco Cappiello, Todor Gramchev, and Luigi Rodino, *Decay estimates for solutions of nonlocal semilinear equations*, Nagoya Math. J. **218** (2015), 175–198, DOI 10.1215/00277630-2891745. MR3345627

[21] Ju. L. Dalec′kiĭ and M. G. Kreĭn, *Stability of solutions of differential equations in Banach space*, American Mathematical Society, Providence, R.I., 1974. Translated from the Russian by S. Smith; Translations of Mathematical Monographs, Vol. 43. MR0352639

[22] Nelson Dunford and Jacob T. Schwartz, *Linear operators. Part II*, Wiley Classics Library, John Wiley & Sons, Inc., New York, 1988. Spectral theory. Selfadjoint operators in Hilbert space; With the assistance of William G. Bade and Robert G. Bartle; Reprint of the 1963 original; A Wiley-Interscience Publication. MR1009163

[23] Klaus-Jochen Engel and Rainer Nagel, *One-parameter semigroups for linear evolution equations*, Graduate Texts in Mathematics, vol. 194, Springer-Verlag, New York, 2000. With contributions by S. Brendle, M. Campiti, T. Hahn, G. Metafune, G. Nickel, D. Pallara, C. Perazzoli, A. Rhandi, S. Romanelli and R. Schnaubelt. MR1721989

[24] Qiquan Fang, Chang Eon Shin, and Qiyu Sun, *Wiener's lemma for singular integral operators of Bessel potential type*, Monatsh. Math. **173** (2014), no. 1, 35–54, DOI 10.1007/s00605-013-0575-1. MR3148659

[25] I. Gelfand, D. Raikov, and G. Shilov, *Commutative normed rings*, Translated from the Russian, with a supplementary chapter, Chelsea Publishing Co., New York, 1964. MR0205105

[26] Karlheinz Gröchenig, *Weight functions in time-frequency analysis*, Pseudo-differential operators: partial differential equations and time-frequency analysis, Fields Inst. Commun., vol. 52, Amer. Math. Soc., Providence, RI, 2007, pp. 343–366. MR2385335

[27] Karlheinz Gröchenig and Andreas Klotz, *Noncommutative approximation: inverse-closed subalgebras and off-diagonal decay of matrices*, Constr. Approx. **32** (2010), no. 3, 429–466, DOI 10.1007/s00365-010-9101-z. MR2726441

[28] Andreas Klotz, *Spectral invariance of Besov-Bessel subalgebras*, J. Approx. Theory **164** (2012), no. 2, 268–296, DOI 10.1016/j.jat.2011.10.008. MR2864646

[29] Ilya A. Krishtal, *Wiener's lemma and memory localization*, J. Fourier Anal. Appl. **17** (2011), no. 4, 674–690, DOI 10.1007/s00041-010-9152-3. MR2819172

[30] Ilya A. Krishtal, *Wiener's lemma: pictures at an exhibition*, Rev. Un. Mat. Argentina **52** (2011), no. 2, 61–79. MR2952951

[31] V. G. Kurbatov and V. I. Kuznetsova, *Inverse-closedness of the set of integral operators with L_1-continuously varying kernels*, J. Math. Anal. Appl. **436** (2016), no. 1, 322–338, DOI 10.1016/j.jmaa.2015.12.007. MR3440096

[32] Stevan Pilipović, Bojan Prangoski, and Daniel Velinov, *On a solution of the Cauchy problem in the weighted spaces of Beurling ultradistributions*, Rocky Mountain J. Math. **45** (2015), no. 6, 1937–1984, DOI 10.1216/RMJ-2015-45-6-1937. MR3473163

[33] M. A. Shubin, *Pseudodifferential operators and spectral theory*, 2nd ed., Springer-Verlag, Berlin, 2001. Translated from the 1978 Russian original by Stig I. Andersson. MR1852334

[34] Johannes Sjöstrand, *An algebra of pseudodifferential operators*, Math. Res. Lett. **1** (1994), no. 2, 185–192, DOI 10.4310/MRL.1994.v1.n2.a6. MR1266757

[35] Qiyu Sun, *Wiener's lemma for infinite matrices with polynomial off-diagonal decay* (English, with English and French summaries), C. R. Math. Acad. Sci. Paris **340** (2005), no. 8, 567–570, DOI 10.1016/j.crma.2005.03.002. MR2138705

[36] Qiyu Sun, *Wiener's lemma for infinite matrices*, Trans. Amer. Math. Soc. **359** (2007), no. 7, 3099–3123, DOI 10.1090/S0002-9947-07-04303-6. MR2299448

[37] Romain Tessera, *The inclusion of the Schur algebra in $B(\ell^2)$ is not inverse-closed*, Monatsh. Math. **164** (2011), no. 1, 115–118, DOI 10.1007/s00605-010-0216-x. MR2827175

Department of Applied Mathematics and Mechanics, Voronezh State University, Voronezh, 394693, Russia
Email address: anatbaskakov@yandex.ru

Department of Mathematical Sciences, Northern Illinois University, DeKalb, Illinois 60115
Email address: krishtal@math.niu.edu

Contemporary Mathematics
Volume **706**, 2018
http://dx.doi.org/10.1090/conm/706/14216

The Kaczmarz algorithm, row action methods, and statistical learning algorithms

Xuemei Chen

ABSTRACT. The Kaczmarz algorithm is an iterative row action method that typically solves an overdetermined linear system. The randomized Kaczmarz algorithm, which was introduced about a decade ago, revived this simple algorithm and raised a lot of interest in this community. It has come to people's attention that there is overlap between many other iterative methods, statistical learning algorithms and the (randomized) Kaczmarz algorithm. This note brings these methods together and discuss connections and theories, with a focus on the convergence rate of the deterministic Kaczmarz algorithm. Moreover, the convergence under noise for the deterministic case is discussed in Theorem 3.2.

1. Introduction

The original Kaczmarz algorithm [**23**] is an iterative algorithm that solves a consistent and overdetermined system $Ax = b$, where the size of A is $m \times d$. Let a_i be the ith row of A, and b_i be the ith coordinate of b, then the algorithm is as follows:

$$\begin{aligned} &\text{Initialize } x^{(0)} \\ &x^{(k+1)} = x^{(k)} + \alpha_k \frac{b_{i(k)} - \langle a_{i(k)}, x^{(k)}\rangle}{\|a_{i(k)}\|^2} a_{i(k)}, k \geq 0 \qquad (1.1) \\ &\text{with } i(k) = (k \bmod m) + 1. \end{aligned}$$

We define the hyperplane $H_i = \{x : \langle a_i, x\rangle = b_i\}$. At each iteration, $x^{(k+1)}$ is the projection of $x^{(k)}$ onto the convex set $H_{i(k)}$ (if the relaxation parameter $\alpha_k = 1$). The choice of $i(k)$ here allows the algorithm iterate through all the rows sequentially and then start over. Figure 1 shows 8 iterations of the Kaczmarz algorithm for solving a 4×2 system.

The Kaczmarz algorithm is a *row action method* where only one row is used in each iteration [**6**]. Due to its simplicity, the Kaczmarz method has found numerous applications including image reconstruction, distributed computation, signal processing, etc. [**6, 13, 18**]. The Kaczmarz algorithm has also been rediscovered in

2010 *Mathematics Subject Classification.* Primary: 65K15, 65F10, 68Q32; Secondary: 15A60.

Key words and phrases. Kaczmarz, randomized Kaczmarz, stochastic gradient descent, row action method, relaxation method.

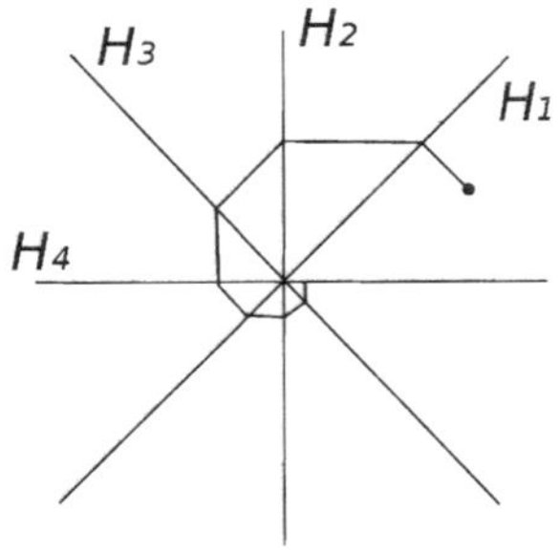

FIGURE 1

the field of image reconstruction and called ART (Algebraic Reconstruction Technique) [**16**]. More work on the (deterministic) Kaczmarz algorithm can be found in [**38**], [**6**], and [**11**].

It is well known that the Kaczmarz algorithm produces monotonically improving approximations as the iteration number increases. However, it can be difficult to quantify the associated rates of convergence, and in some cases, the convergence can be quite slow. To tackle this issue, Strohmer and Vershynon [**37**] introduced a randomized version. For the *randomized Kaczmarz algorithm*, rather than processing each row cyclically, at each iteration, a row of A is randomly selected to perform the projection, as

$$\begin{aligned}&\text{Initialize } x^{(0)}\\ (1.2)\qquad & x^{(k+1)} = x^{(k)} + \frac{b_{i(k)} - \langle a_{i(k)}, x^{(k)}\rangle}{\|a_{i(k)}\|^2} a_{i(k)}, k \geq 0\\ &\text{with } i(k) \text{ picking from } \{1, 2, \cdots, m\} \text{ with probability } \|a_i\|^2/\|A\|_F^2.\end{aligned}$$

Moreover, the relaxation parameter has been set to 1. Throughout this paper, $\|\cdot\|$ indicates the Euclidean norm and $\|\cdot\|_F$ the Frobenius norm of a matrix. Moreover, cond$(\cdot)$ is the condition number of a matrix and $\sigma(\cdot)$ denotes the smallest singular value of a matrix. In contrast to the randomized Kaczmarz algorithm (1.2), we will call algorithm (1.1) the *cyclic Kaczmarz algorithm.*

Strohmer and Vershynin proved that this randomized approach achieves mean squared error with a rate that is quantifiable as

$$(1.3)\qquad \mathbb{E}\|x^* - x^{(k)}\|^2 \leq \left(1 - \frac{\sigma^2(A)}{\|A\|_F^2}\right)^k \|x^* - x^{(0)}\|^2,$$

where x^* is the solution of the system.

Following [**37**], there has been a great amount of work that highlights other favorable properties of the Kaczmarz algorithm. We will only list a few here. The work [**12**] accelerates the convergence of the Kaczmarz algorithm in high dimensions with help of the Johnson-Lindenstrauss Lemma. The work [**9**] discusses the almost sure convergence when the measurement or data is drawn from a more general random distribution. The work in [**33**] and [**10**] extends to processing multiple rows in one iteration. The work in [**41**] deals with inconsistent systems and develops a variation of the randomized Kaczmarz algorithm so that the approximates will converge to the least square solution with a better rate. The paper [**28**] further deals with the underdetermined (A does not have full rank) case (cf. [**39**]). There is also work to apply the Kaczmarz algorithm to compressed sensing [**27**] and to phase retrieval [**40**]. However, there has been relatively less interest in the convergence rate of the cyclic (deterministic) method, which will be a focus of this note.

With this increased interest in the Kaczmarc algorithm, researchers have found many connections with other row action methods, which will be elaborated in Section 2, with some historical notes. Moreover, the ability of reading measurements as it becomes available [**9**] is very reminiscent of online machine learning. As it turns out, the Kaczmarz algorithm (both the cyclic and randomized version) can be viewed as an instance of some gradient methods. This will be discussed in Section 4. This note will first provide a survey of all these connections.

The second contribution of this paper is to analyze the convergence rate of the cyclic Kaczmarz algorithm under noise in Section 3. See Theorem 3.2. Comparison of the current noise-free deterministic Kaczmarz convergence rates are also given, and Theorem 3.2 will recover one of the those results.

2. Connection with projection methods and row action methods

Row action methods are iterative methods such that only one row of the matrix A is utilized in each iteration. Such methods are often applied to large scale and especially sparse systems due to its little computational work per iteration. But this is also a liability as the rate of convergence could be dismally low [**14, 29**]. This liability is the motivation for the randomized Kaczmarz algorithm. In this section we will discuss other row action methods and how they can be reduced to the Kaczmarz algorithm. The randomized versions of these row action methods, to speed up the convergence, are also discussed.

2.1. Projection onto convex sets. The method of alternating projection (MAP) is a broader concept than the row action methods. It is an iterative scheme for finding the best approximation to any given point in a Hilbert space from the intersection of a finite collection of closed subspaces. We let $M_1, M_2, \cdots, M_k$ be k closed subspaces in the Hilbert space X, and let P_{M_i} denote the orthogonal projection onto M_i. Both Von Neuman [**35**] in 1950 and Halperin [**20**] in 1962 proved the following theorem:

$$\lim_{n\to\infty} \| \left(P_{M_k} P_{M_{k-1}} \cdots P_{M_1}\right)^n (x) - P_{\cap_i^k M_i}(x) \| = 0, \tag{2.1}$$

Although no specific convergence rate is given. The cyclic Kaczmarz algorithm (1.1) is a special case here where $M_i = H_i$.

A more general setting was considered later. The projection onto convex sets (POCS) algorithm aims to find a point in the intersection of convex sets Q_i, which is a very common problem in diverse area of mathematics and physical sciences. The POCS algorithm was first introduced by Bregman [**5**] in 1966 with a very general projection function where they discussed application to convex programming. Around the same time, Gubin, Polyak and Raik [**17**] introduced their successive projection algorithm. The algorithm starts with a random guess $x^{(0)}$, and is iterative as

$$x^{(k+1)} = x^{(k)} + \alpha_k (P_{i(k)} x^{(k)} - x^{(k)}),$$

where P_j is the projection onto Q_j, and $i(k)$ indicates the order of which convex sets are selected. For example, we can choose the common cyclic control as $i(k) =$

$(k \bmod m)+1$. It is obvious that the Kaczmarz algorithm is a particular instance of POCS algorithm where each convex set is the affine hyperplane.

Gubin et al. [**17**] showed that the sequence $x^{(k)}$ converges to some point in the intersection when the relaxation parameter $\alpha_k \in (\epsilon_1, 2-\epsilon_2)$. They also noticed that, however, the rate of convergence can be slow in certain cases. The authors discussed the special case for solving a system of linear inequalities, which becomes the relaxation method (See Section 2.2).

Bauschke and Borwein [**2**] wrote a general overview of projection algorithms for the consistent case. Censor and Tom discussed a variation of POCS in [**8**], especially for the inconsistent case. Numerous applications can be found in [**6**], and throughout this paper.

2.2. The relaxation method. The *relaxation method* for linear inequalities, as introduced in 1954 by Agmon [**1**], Motzkin and Schoenberg [**31**], is to find solutions of linear inequalities by orthogonally projecting the current iterate onto chosen halfspace, and therefore is a particular instance of POCS.

Given a linear system of inequalities $H_i = \{x : \langle a_i, x\rangle \le b_i\}_{i=1}^m$, a solution of this system can be found by

$$\begin{aligned} &\text{Initialize } x^{(0)} \\ &c_k = \min\{0, \alpha_k \frac{b_{i(k)} - \langle a_{i(k)}, x^{(k)}\rangle}{\|a_{i(k)}\|^2}\}, \\ &x^{(k+1)} = x^{(k)} + c_k a_{i(k)}, \\ &\text{with } i(k) = (k \bmod m)+1. \end{aligned} \tag{2.2}$$

If we let the relaxation parameter $\alpha_k = 1$, the next iterate is simply projecting the current iterate onto the half space $H_{i(k)} = \{x : \langle a_{i(k)}, x\rangle \le b_{i(k)}\}$: if $x^{(k)}$ has already satisfied the linear constraint, then simply do nothing; otherwise, an orthogonal projection is performed (see Figure 2). The algorithm (2.2) also chooses rows cyclically. Other common implementations include almost cyclic (slightly more general than cyclic) [**24**, Definition 2.3], maximal distance and maximal residual control [**1**].

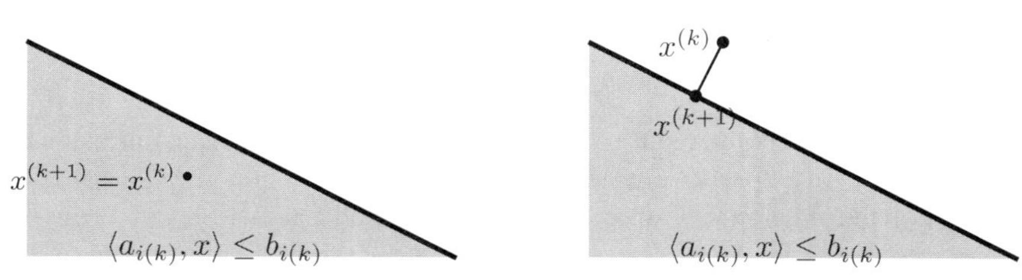

FIGURE 2. The Relaxation Method

In 1984, Mandel [**29**] proved that $\{x^{(k)}\}$ of (2.2) with $\alpha_k = \alpha \in (0,2)$ converges to a point $\hat{x}$ on the boundary of $P = \{x : Ax \le b\}$. This can be viewed as a particular instance of the convergence theorem by Gubin et al. [**17**, Theorem 1] as mentioned in the previous section. The more significant contribution of Mandel is to prove a quantifiable linear convergence rate:

THEOREM 2.1 ([**29**, Theorem 3.1]). *If $P \neq \emptyset$, then the iterates of (2.2) satisfy*

$$d^2(x^{(k+m)}, P) \leq \left(1 - \frac{\alpha(2-\alpha)\mu^2}{1+(m-1)\alpha^2}\right) d^2(x^{(k)}, P),$$

where $d(x, P)$ is the distance of x to P and

$$\mu = \inf_{x \notin P} \frac{\max_i d(x, H_i)}{d(x, P)}. \tag{2.3}$$

The relaxation method is a generalization of the Kaczmarz method in dealing with inequalities. As a result, Mandel applied the convergence result to $Ax = b$ and got the first convergence rate result of the Kaczmarz algorithm [**29**, Corollary 4.3].

Same as the Kaczmarz algorithm, $i(k)$ can be chosen cyclically or randomly. When chosen randomly, Leventhal et al [**25**] proved that the mean square error also converges with a linear rate, and therefore generalized the result of (1.3).

The Perceptron convergence theorem from the theory of machine learning is a rediscovery of the relaxation method; see Minsky and Papert [**30**, p. 248] and Nilsson [**36**]. More connections with machine learning is discussed in Section 4.2.

2.3. The Hildreth's algorithm. The Hildreth's algorithm also solves a system of linear inequalities, but with one more benefit: finding the closest point in the solution set to a given point, i.e., it solves the following problem

$$\begin{aligned} x^* &= \arg\min \|x - x^{(0)}\| \\ &\text{subject to } Ax \leq b. \end{aligned} \tag{2.4}$$

The algorithm is slightly more complicated than the relaxation method. It is defined as

$$\begin{aligned}
&\text{Initialize } x^{(0)}, z^{(0)} = 0 \\
&c_k = \min\{z_{i(k)}^{(k)}, \alpha_k \frac{b_{i(k)} - \langle a_{i(k)}, x^{(k)}\rangle}{\|a_{i(k)}\|^2}\}, \\
&x^{(k+1)} = x^{(k)} + c_k a_{i(k)} \\
&z_i^{(k+1)} = \begin{cases} z_i^{(k)}, & i \neq i(k) \\ z_{i(k)}^{(k)} - c_k, & i = i(k) \end{cases} \\
&\text{with } i(k) = (k \bmod m) + 1.
\end{aligned} \tag{2.5}$$

Figure 3 shows the nice geometric interpretation when $\alpha_k = 1$. If the constraint $\langle a_{i(k)}, x\rangle \leq b_{i(k)}$ is satisfied, we move the last approximate closer to the hyperplane $\{x : \langle a_{i(k)}, x\rangle = b_{i(k)}\}$ (Figure 3 (A)). The definition of c_k guarantees no over projection (staying on the halfspace). This is what's different from the relaxation method. If the constraint is violated, we project the last approximate to the halfspace, just like in the relaxation method (Figure 3 (B)). The vector $z^{(k)}$ only gets updated at ith component, where i is the index of the row/constraint in that iteration. It can be shown that all entries of $z^{(k)}$ are never negative [**24**]. Hildreth's method is a primal-dual optimization method, where $z^{(k)}$ is the sequence of dual iterates.

The algorithm was first presented by Hildreth [**19**] in 1957. Lent and Censor [**24**] studied it extensively in 1980 and supplied a proof of convergence of the

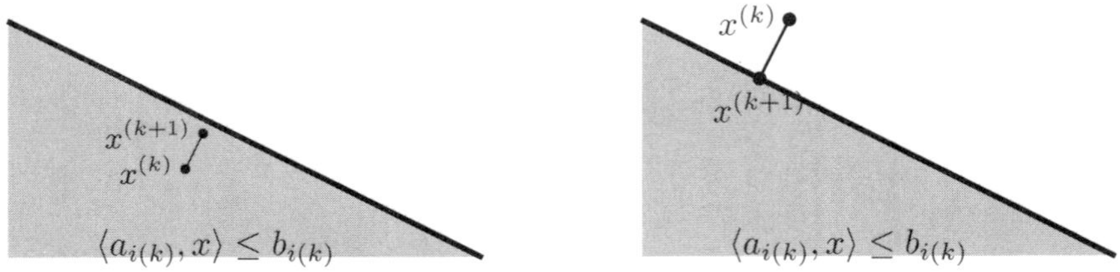

FIGURE 3. Hildreth's Method

Hildreth's algorithm under almost cyclic control. Iusem and De Pierro [**21**] further provided a convergence rate under almost cyclic control in 1990.

THEOREM 2.2 ([**21**, Theorem 1]). *Let x^* and $x^{(k)}$ be given in* (2.4) *and* (2.5), *and the relaxation parameter $\alpha_k = \alpha \in (0, 2)$, then after one cycle,*

$$\|x^{(k+m)} - x^*\|^2 \le \frac{1}{1 + \frac{(2-\alpha)\alpha\mu^2}{1+\alpha^2(m-1)}}\|x^{(k)} - x^*\|^2,$$

where the quantity μ is very similar to the one in (2.3), *but defined slightly differently.*

The Hildreth's algorithm can be once again reduced to the Kaczmarz algorithm since $\langle a_i, x\rangle = b_i$ is equivalent to $\langle a_i, x\rangle \le b_i$ and $\langle -a_i, x\rangle \le -b_i$.

If i is chosen at random from $\{1, 2, \cdots, m\}$ with the discrete uniform distribution $\Pr(i = k) = \|a_i\|^2/\|A\|_F^2$, then the resulted method is called *the randomized Hildreth's algorithm* as proposed in [**22**]. Jamil et al. showed that the randomized version also enjoys a linear convergence rate similar to the randomized Kaczmarz algorithm (1.3) (see [**22**, Theorem 4.5]).

2.4. Schwarz iterative method. The *Schwarz iterative method* (or subspace correction method) is for solving symmetric positive semi-definite linear systems [**39**]. Consider a separable Hilbert space V, let $a(\cdot, \cdot)$ be a continuous symmetric positive sesqui-linear form on V, and let F be a bounded linear functional on V. The method aims to solve the variational problem: Find $u \in V$ such that $a(u, v) = F(v)$. The concept of stable space splittings is used here. Let V_a be the Hilbert space with the scalar product given by the form $a(\cdot, \cdot)$. We will represent V_a by a finite number of Hilbert spaces V_{a_i} with associated scalar products a_i and corresponding linear bounded operators $R_i : V_{a_i} \to V_a$. We define the linear operators $T_i : V_a \to V_{a_i}$ via $a_i(T_i v, v_i) = a(v, R_i v_i)$. The additive Schwarz iteration is given by

$$x^{(k+1)} = x^{(k)} + w_k \sum_{i=1}^{N} R_i T_i(u - x^{(k)}), \tag{2.6}$$

and the sequential Schwarz iteration is

$$x^{(k+1)} = x^{(k)} + w_l R_{i(k)} T_{i(k)}(u - x^{(k)}). \tag{2.7}$$

The interested reader can find details on how to compute $u - x^{(k)}$ in [**39**]. The additive Schwarz iteration could be thought of the generalization of the gradient method

(4.3) and the sequential Schwarz iteration should be compared to the Stochastic gradient descent (4.4) with minibatch size 1.

For a system $Ax = b$, set $V = \text{range}(A^T)$ with $a(x,x) = x^T x$. Let $V_{a_i} = \mathbb{C}, a_i(y_i, y_i) = \|a_i\|^2 |y_i|^2, R_i y_i = y_i a_i$, then (2.7) becomes the Kaczmarz algorithm or the randomized version depending on the ordering of the rows. Osward and Zhou [**39**] analyze the convergence rate of the Schwarz iterative methods, and in return achieves a convergence rate of both versions of the Kaczmarz methods. We will list the cyclic version here in comparison to Theorem 2.1 and Theorem 2.2.

THEOREM 2.3. *if x^* is the only solution of $Ax = b$, and the relaxation parameter α_k is chosen properly, then the iterates of* (1.1) *satisfy*

$$\|x^* - x^{(k+m)}\|^2 \leq \left(1 - \frac{1}{\log(2m)C^2(B)}\right) \|x^* - x^{(k)}\|^2, \tag{2.8}$$

where B is the matrix one obtains after normalizing each row of A.

3. Convergence rate of Kaczmarz algorithm under noise

3.1. Summary of deterministic convergence results. Dai and Schön [**11**] also studied the cyclic Kaczmarz convergence result with $\alpha_k = \alpha$:

THEOREM 3.1 ([**11**, Theorem 1]). *if x^* is the only solution of $Ax = b$, then the iterates of* (1.1) *satisfy*

$$\|x^* - x^{(k+m)}\|^2 \leq \left(1 - \frac{\alpha(2-\alpha)}{(2+\alpha^2 m)\sigma(B)^2}\right) \|x^* - x^{(k)}\|^2, \tag{3.1}$$

where B is the matrix one obtains after normalizing each row of A, and $\sigma(B)$ is the smallest singular value of B.

Proofs of both Theorem 2.1 and Theorem 2.2 can be applied to the cyclic Kaczmarz algorithm, with the setup

$$\begin{aligned} &x^* \text{ is the only solution to } Ax = b \\ &H_i = \{x : \langle a_i, x\rangle = b_i\} \\ &\mu = \inf_{x \neq x^*} \frac{\max_i d(x, H_i)}{d(x, x^*)}. \end{aligned} \tag{3.2}$$

We summarize the results in Table 1 where we assume the rows of A are normalized, i.e. $A = B$. The first column is the linear rates presented in the theorems and the second column is the linear rates with the relaxation parameter optimized. The rates (I) and (II) are quite similar. The reader can refer to [**11**] for a comparison of rates (I), (III) and (IV).

TABLE 1. Comparison of cyclic Kaczmarz convergence rate

	linear rates	Optimal α
Mandel, Theorem 2.1	$1 - \frac{\alpha(2-\alpha)\mu^2}{1+(m-1)\alpha^2}$	$1 - \frac{2\mu^2}{1+\sqrt{4m-3}}$ (I)
Hildreth, Theorem 2.2	$\frac{1}{1+\frac{(2-\alpha)\alpha\mu^2}{1+\alpha^2(m-1)}}$	$\frac{1}{1+\frac{2\mu^2}{1+\sqrt{4m-3}}}$ (II)
Osward, Theorem 2.3	$1 - \frac{1}{\log(2m)C^2(B)}$	$1 - \frac{1}{\log(2m)C^2(B)}$ (III)
Dai, Theorem 3.1	$1 - \frac{\alpha(2-\alpha)}{(2+\alpha^2 m)\sigma(B)^2}$	$1 - \frac{\sqrt{2}}{4m\sigma(B)^2}$ (IV)

3.2. Convergence rate of deterministic Kaczmarz under noise. In this section, we consider the case where the measurements b_i are slightly perturbed. The stability of randomized Kaczmarz algorithm was studied in [**32**], but there is no result on noisy case for the deterministic (cyclic) Kaczmarz algorithm. The following theorem uses proof techniques from [**21**].

THEOREM 3.2. *Let x^* be the solution of $Ax = b$. If the system $Ax = b$ is perturbed as $\langle a_i, x\rangle = b_i + \epsilon_i, i \in [m]$, then the iterates of* (1.1) *with $\alpha_k = 1$ follows*

$$\|x^{(k+m)} - x^*\|^2 \tag{3.3}$$

$$\leq \frac{m}{m+\mu^2}\|x^{(k)} - x^*\|^2 + \frac{m(-\sum_{l=1}^m \frac{\epsilon_l^2}{\|a_l\|^2} + \sum_{l=k}^{k+m-1}\frac{2\epsilon_{i(l)}\|x^{(l)}\|}{\|a_{i(l)}\|}) + \|\sum_{l=1}^m \frac{\epsilon_l}{\|a_l\|^2}a_l\|^2}{m+\mu^2},$$

where $i(l)$ is defined as in (1.1) *and μ as in* (3.2).

PROOF. Without loss of generality, we assume $x^* = 0$.

For an arbitrary k, $x^{(k+1)}$ is the projection of $x^{(k)}$ onto the perturbed hyperplane $H'_{i(k)} = \{x : a^*_{i(k)}x = b_{i(k)} + \epsilon_{i(k)}\}$. Suppose $v^{(k+1)}$ is the projection of $x^{(k)}$ onto $H_{i(k)}$. See the picture below.

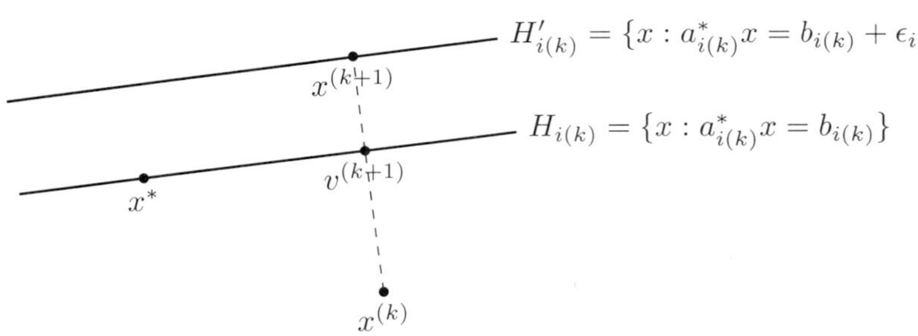

The orthogonality gives

$$\|v^{(k+1)}\|^2 + \|v^{(k+1)} - x^{(k)}\|^2 = \|x^{(k)}\|^2 \tag{3.4}$$

and a simple calculation reaches

$$x^{(k+1)} = v^{(k+1)} + \frac{\epsilon_{i(k)}}{\|a_{i(k)}\|^2}a_{i(k)}. \tag{3.5}$$

(3.4) and (3.5) together implies

$$\begin{aligned}\|x^{(k)}\|^2 - \|x^{(k+1)} - \frac{\epsilon_{i(k)}}{\|a_{i(k)}\|^2}a_{i(k)} - x^{(k)}\|^2 &= \|x^{(k+1)} - \frac{\epsilon_{i(k)}}{\|a_{i(k)}\|^2}a_{i(k)}\|^2 \\ &\geq \left(\|x^{(k+1)}\| - \frac{\epsilon_{i(k)}}{\|a_{i(k)}\|}\right)^2.\end{aligned} \tag{3.6}$$

Let the index in (3.6) run from k to $k+m-1$ and add all these terms up we get

$$\|x^{(k)}\|^2 - \sum_{l=k}^{k+m-1} \|x^{(l+1)} - \frac{\epsilon_{i(l)}}{\|a_{i(l)}\|^2} a_{i(l)} - x^{(l)}\|^2 \geq \|x^{(k+m)}\|^2 + \sum_{l=k}^{k+m-1} \frac{\epsilon_{i(l)}^2}{\|a_{i(l)}\|^2} - \sum_{l=k}^{k+m-1} \frac{2\epsilon_{i(l)}\|x^{(l)}\|}{\|a_{i(l)}\|} \tag{3.7}$$

On the other hand, suppose $H_j = H_{x^{(k+m)}}$, and define

$$r = \max\{l < k+m : i(l) = j\}.$$

It is clear that $k \leq r$.

Let $y^{(k+m)} = P_{H_j}(x^{(k+m)})$, and $y^{(r)} = P_{H_j}(x^{(r)}) = v^{(r+1)}$. By the definition of μ,

$$\begin{aligned}
&\mu\|x^{(k+m)}\| \leq \|x^{(k+m)} - y^{(k+m)}\| \leq \|x^{(k+m)} - y^{(r)}\| \\
&\leq \sum_{l=r+1}^{k+m-1} \|x^{(l+1)} - x^{(l)} - \frac{\epsilon_{i(l)}}{\|a_{i(l)}\|^2} a_{i(l)}\| + \|x^{(r+1)} - y^{(r)} + \sum_{l=r+1}^{k+m-1} \frac{\epsilon_{i(l)}}{\|a_{i(l)}\|^2} a_{i(l)}\| \\
&= \sum_{l=r+1}^{k+m-1} \|x^{(l+1)} - x^{(l)} - \frac{\epsilon_{i(l)}}{\|a_{i(l)}\|^2} a_{i(l)}\| + \|\sum_{l=r}^{k+m-1} \frac{\epsilon_{i(l)}}{\|a_{i(l)}\|^2} a_{i(l)}\| \\
&\leq \sum_{l=k+1}^{k+m-1} \|x^{(l+1)} - x^{(l)} - \frac{\epsilon_{i(l)}}{\|a_{i(l)}\|^2} a_{i(l)}\| + \|\sum_{l=k}^{k+C-1} \frac{\epsilon_{i(l)}}{\|a_{i(l)}\|^2} a_{i(l)}\|.
\end{aligned} \tag{3.8}$$

Square (3.8), we get

$$\mu^2\|x^{(k+m)}\|^2 \leq m \sum_{l=k+1}^{k+m-1} \|x^{(l+1)} - x^{(l)} - \frac{\epsilon_{i(l)}}{\|a_{i(l)}\|^2} a_{i(l)}\|^2 + \|\sum_{l=k}^{k+m-1} \frac{\epsilon_{i(l)}}{\|a_{i(l)}\|^2} a_{i(l)}\|^2. \tag{3.9}$$

Combine (3.7) and (3.9), we have

$$\begin{aligned}
&\mu^2\|x^{(k+m)}\|^2 \\
&\leq m\left[\|x^{(k)}\|^2 - \|x^{(k+m)}\|^2 - \sum_{l=k}^{k+m-1} \frac{\epsilon_{i(l)}^2}{\|a_{i(l)}\|^2} + \sum_{l=k}^{k+m-1} \frac{2\epsilon_{i(l)}\|x^{(l)}\|}{\|a_{i(l)}\|}\right] + \|\sum_{l=k}^{k+m-1} \frac{\epsilon_{i(l)}}{\|a_{i(l)}\|^2} a_{i(l)}\|^2 \\
&= m\left[\|x^{(k)}\|^2 - \|x^{(k+m)}\|^2 - \sum_{l=1}^{m} \frac{\epsilon_l^2}{\|a_l\|^2} + \sum_{l=k}^{k+m-1} \frac{2\epsilon_{i(l)}\|x^{(l)}\|}{\|a_{i(l)}\|}\right] + \|\sum_{l=1}^{m} \frac{\epsilon_l}{\|a_l\|^2} a_l\|^2
\end{aligned}$$

which can be simplified to the desired result. □

The interested reader can generalize this theorem for any relaxation parameter $\alpha_k \in (0, 2)$.

4. Connection with statistical learning methods

The connection between the Kaczmarz algorithm and learning theory is established as early as in 1972 with the Perceptron convergence theorem [**30**]. The randomized Kaczmarz algorithm, by definition, is more connected to some statistical methods as we will see below. Many good work that explores this connection includes Needell et al [**34**] on stochastic gradient descent, and Lin and Zhou [**26**] on least square regression setting. See the rest of this section for more references.

4.1. Incremental gradient method. We consider least square problem of the form

$$\begin{aligned} &\text{minimize } f(x) = \sum_{i=1}^{m} f_i(x) \\ &\text{subject to } x \in \mathbb{R}^d. \end{aligned} \tag{4.1}$$

Least squares problems often arise in contexts of learning theory where we are trying to fit data with a model. In problems where there are many data blocks, and particularly in neural network training problems, one does not wait to process the entire data set before updating; instead, one cycles through the data blocks in sequence and update the estimate of x after each data block is processed. The iteration formula is

$$\begin{aligned} x^{(k+1)} &= x^{(k)} - \alpha_k \nabla f_{i(k)}(x^{(k)}) \\ &\text{with } i(k) = (k \bmod m) + 1. \end{aligned} \tag{4.2}$$

This is called the *incremental gradient method* [**3**]. Notice that the Kaczmarz algorithm (1.1) is a special case of it where $f_i(x) = (\langle a_i, x\rangle - b_i)^2$. Moreover, the incremental gradient method also covers the block Kaczmarz algorithm [**7, 33**] case by letting $f_i(x) = \|A_i x - b_i'\|^2$ where A_i is a submatrx of certain rows of A.

The incremental gradient method has a more general form [**4**], and is similar to the stochastic gradient method, which will be discussed below.

4.2. Stochastic gradient descent. The gradient descent method is an iterative algorithm to find a local minimum of the cost function $F(x)$. With an initial estimate $x^{(0)}$, each iterative step is

$$x^{(k+1)} = x^{(k)} - \alpha_k \nabla F(x^{(k)}). \tag{4.3}$$

A recurring problem in machine learning is that large training sets are necessary for good generalization, but large training sets are also more computationally expensive. In other scenarios like online machine learning, data becomes available in a sequential order and one simply does not have access to the entire training data set at once.

The insight of *stochastic gradient descent* (SGD) is that the cost function is an expectation. The gradient of the expectation may be estimated using a small set of samples. To be specific, let

$$F(x) = \mathbb{E} f_i(x) (= \frac{1}{N} \sum f_i(x) \text{ for example}).$$

We can sample a *minibatch* of data $I \in \{1, 2 \cdots, N\}$ drawn uniformly from the training set. The minibatch size $n = |I|$ is typically chosen to be a relatively small number, ranging from 1 to a few hundred. The SGD update is to approximate the gradient in (4.3) by $\frac{1}{n}\nabla \sum_{i \in I} f_i(x^{(k)})$, as

$$x^{(k+1)} = x^{(k)} - \alpha_k \frac{1}{n} \nabla \sum_{i \in I_k} f_i(x^{(k)})), \tag{4.4}$$

where α_k is called the step size or learning rate. The stochastic gradient method plays an important role in machine learning. Nearly all of deep learning is powered by SGD [**15**].

The randomized Kaczmarz algorithm is a special case of SGD with minibatch size $n = 1$ as mentioned in [**34**]. To reduce to the randomized Kaczmarz case, let $F(x) = \frac{1}{2}\|Ax - b\|^2 = \mathbb{E}_i\left[\frac{1}{2p_i}(\langle a_i, x\rangle - b_i)^2\right]$, where the probability of picking i is $p_i = \frac{\|a_i\|^2}{\|A\|_F^2}$, and $f_i(x) = \frac{1}{2p_i}(\langle a_i, x\rangle - b_i)^2$. Now (4.4) becomes

$$x^{(k+1)} = x^{(k)} - \alpha_k \|A\|_F^2 \frac{\langle a_{i(k)}, x^{(k)}\rangle - b_{i(k)}}{\|a_{i(k)}\|^2} a_{i(k)}, \tag{4.5}$$

which is exactly (1.2) if we let the learning rate $\alpha_k = \frac{1}{\|A\|_F^2}$.

Understanding the randomized Kaczmarz algorithm as SGD allows to obtain improved methods and results for the randomized Kaczmarz method. Some results are available in [**34**].

References

[1] Shmuel Agmon, *The relaxation method for linear inequalities*, Canadian J. Math. **6** (1954), 382–392. MR0062786

[2] Heinz H. Bauschke and Jonathan M. Borwein, *On projection algorithms for solving convex feasibility problems*, SIAM Rev. **38** (1996), no. 3, 367–426, DOI 10.1137/S0036144593251710. MR1409591

[3] Dimitri P. Bertsekas, *A new class of incremental gradient methods for least squares problems*, SIAM J. Optim. **7** (1997), no. 4, 913–926, DOI 10.1137/S1052623495287022. MR1479606

[4] D. Bertsekas. *Incremental gradient, subgradient, and proximal methods for convex optimization: A survey.* Optimization for Machine Learning 2010.1-38 (2011): 3.

[5] L. M. Brègman, *A relaxation method of finding a common point of convex sets and its application to the solution of problems in convex programming* (Russian), Ž. Vyčisl. Mat. i Mat. Fiz. **7** (1967), 620–631. MR0215617

[6] Yair Censor, *Row-action methods for huge and sparse systems and their applications*, SIAM Rev. **23** (1981), no. 4, 444–466, DOI 10.1137/1023097. MR636080

[7] Yair Censor, Paul P. B. Eggermont, and Dan Gordon, *Strong underrelaxation in Kaczmarz's method for inconsistent systems*, Numer. Math. **41** (1983), no. 1, 83–92, DOI 10.1007/BF01396307. MR696552

[8] Yair Censor and Eli Tom, *Convergence of string-averaging projection schemes for inconsistent convex feasibility problems*, Optim. Methods Softw. **18** (2003), no. 5, 543–554, DOI 10.1080/10556780310001610484. MR2015396

[9] Xuemei Chen and Alexander M. Powell, *Almost sure convergence of the Kaczmarz algorithm with random measurements*, J. Fourier Anal. Appl. **18** (2012), no. 6, 1195–1214, DOI 10.1007/s00041-012-9237-2. MR3000980

[10] Xuemei Chen and Alexander M. Powell, *Randomized subspace actions and fusion frames*, Constr. Approx. **43** (2016), no. 1, 103–134, DOI 10.1007/s00365-015-9303-5. MR3439235

[11] L. Dai, and Thomas B. Schön. *On the exponential convergence of the Kaczmarz algorithm.* IEEE Signal Processing Letters 22.10 (2015): 1571-1574.

[12] Yonina C. Eldar and Deanna Needell, *Acceleration of randomized Kaczmarz method via the Johnson-Lindenstrauss lemma*, Numer. Algorithms **58** (2011), no. 2, 163–177, DOI 10.1007/s11075-011-9451-z. MR2835851

[13] H. G. Feichtinger, C. Cenker, M. Mayer, H. Steier, and Thomas Strohmer. *New variants of the POCS method using affine subspaces of finite codimension with applications to irregular sampling.* In Applications in Optical Science and Engineering, pp. 299-310. International Society for Optics and Photonics, 1992.

[14] J.-L. Goffin, *The relaxation method for solving systems of linear inequalities*, Math. Oper. Res. **5** (1980), no. 3, 388–414, DOI 10.1287/moor.5.3.388. MR594854

[15] Ian Goodfellow, Yoshua Bengio, and Aaron Courville, *Deep learning*, Adaptive Computation and Machine Learning, MIT Press, Cambridge, MA, 2016. MR3617773

[16] R. Gordon, R. Bender, and G. T. Herman. *Algebraic reconstruction techniques (ART) for three-dimensional electron microscopy and X-ray photography.* Journal of theoretical Biology 29.3 (1970): 471-481.

[17] L. G. Gurin, B. T. Poljak, and È. V. Raĭk, *Projection methods for finding a common point of convex sets* (Russian), Ž. Vyčisl. Mat. i Mat. Fiz. **7** (1967), 1211–1228. MR0232225

[18] Gabor T. Herman, *Fundamentals of computerized tomography*, 2nd ed., Advances in Pattern Recognition, Springer, Dordrecht, 2009. Image reconstruction from projections. MR2848263

[19] Clifford Hildreth, *A quadratic programming procedure*, Naval Res. Logist. Quart. **4** (1957), 79–85, DOI 10.1002/nav.3800040113. MR0089100

[20] Israel Halperin, *The product of projection operators*, Acta Sci. Math. (Szeged) **23** (1962), 96–99. MR0141978

[21] Alfredo N. Iusem and Alvaro R. De Pierro, *On the convergence properties of Hildreth's quadratic programming algorithm*, Math. Programming **47** (1990), no. 1, (Ser. A), 37–51, DOI 10.1007/BF01580851. MR1054840

[22] Noreen Jamil, Xuemei Chen, and Alexander Cloninger, *Hildreth's algorithm with applications to soft constraints for user interface layout*, J. Comput. Appl. Math. **288** (2015), 193–202, DOI 10.1016/j.cam.2015.04.014. MR3349614

[23] S. Kaczmarz, *Approximate solution of systems of linear equations*, Internat. J. Control **57** (1993), no. 6, 1269–1271, DOI 10.1080/00207179308934446. Translated from the German. MR1220361

[24] Arnold Lent and Yair Censor, *Extensions of Hildreth's row-action method for quadratic programming*, SIAM J. Control Optim. **18** (1980), no. 4, 444–454, DOI 10.1137/0318033. MR579552

[25] D. Leventhal and A. S. Lewis, *Randomized methods for linear constraints: convergence rates and conditioning*, Math. Oper. Res. **35** (2010), no. 3, 641–654, DOI 10.1287/moor.1100.0456. MR2724068

[26] Junhong Lin and Ding-Xuan Zhou, *Learning theory of randomized Kaczmarz algorithm*, J. Mach. Learn. Res. **16** (2015), 3341–3365. MR3450541

[27] D. A. Lorenz, et al. *A sparse Kaczmarz solver and a linearized Bregman method for online compressed sensing.* Image Processing (ICIP), 2014 IEEE International Conference on. IEEE, 2014.

[28] Anna Ma, Deanna Needell, and Aaditya Ramdas, *Convergence properties of the randomized extended Gauss-Seidel and Kaczmarz methods*, SIAM J. Matrix Anal. Appl. **36** (2015), no. 4, 1590–1604, DOI 10.1137/15M1014425. MR3422440

[29] Jan Mandel, *Convergence of the cyclical relaxation method for linear inequalities*, Math. Programming **30** (1984), no. 2, 218–228, DOI 10.1007/BF02591886. MR758005

[30] M. Minsky, and S. Papert. *Perceptron: An introduction to computational geometry.* MIT Press, Cambridge, MA, 1972

[31] T. S. Motzkin and I. J. Schoenberg, *The relaxation method for linear inequalities*, Canadian J. Math. **6** (1954), 393–404. MR0062787

[32] Deanna Needell, *Randomized Kaczmarz solver for noisy linear systems*, BIT **50** (2010), no. 2, 395–403, DOI 10.1007/s10543-010-0265-5. MR2640019

[33] Deanna Needell and Joel A. Tropp, *Paved with good intentions: analysis of a randomized block Kaczmarz method*, Linear Algebra Appl. **441** (2014), 199–221, DOI 10.1016/j.laa.2012.12.022. MR3134343

[34] Deanna Needell, Nathan Srebro, and Rachel Ward, *Stochastic gradient descent, weighted sampling, and the randomized Kaczmarz algorithm*, Math. Program. **155** (2016), no. 1-2, Ser. A, 549–573, DOI 10.1007/s10107-015-0864-7. MR3439812

[35] John von Neumann, *Functional Operators. II. The Geometry of Orthogonal Spaces*, Annals of Mathematics Studies, no. 22, Princeton University Press, Princeton, N. J., 1950. MR0034514

[36] N. J. Nilsson. *Learning machines.* McGraw-Hill, New York, 1965

[37] Thomas Strohmer and Roman Vershynin, *A randomized Kaczmarz algorithm with exponential convergence*, J. Fourier Anal. Appl. **15** (2009), no. 2, 262–278, DOI 10.1007/s00041-008-9030-4. MR2500924

[38] Kunio Tanabe, *Projection method for solving a singular system of linear equations and its applications*, Numer. Math. **17** (1971), 203–214, DOI 10.1007/BF01436376. MR0293824

[39] P. Oswald, and Weiqi Zhou. *Convergence Estimates for Kaczmarz-Type Methods.* preprint (2015).

[40] Ke Wei, *Solving systems of phaseless equations via Kaczmarz methods: a proof of concept study*, Inverse Problems **31** (2015), no. 12, 125008, 23, DOI 10.1088/0266-5611/31/12/125008. MR3540315

[41] Anastasios Zouzias and Nikolaos M. Freris, *Randomized extended Kaczmarz for solving least squares*, SIAM J. Matrix Anal. Appl. **34** (2013), no. 2, 773–793, DOI 10.1137/120889897. MR3069089

Department of Mathematics and Statistics, University of San Francisco, San Francisco, California 94117

Email address: xchen@math.usfca.edu

Contemporary Mathematics
Volume **706**, 2018
http://dx.doi.org/10.1090/conm/706/14205

Lipschitz properties for deep convolutional networks

Radu Balan, Maneesh Singh, and Dongmian Zou

ABSTRACT. In this paper we discuss the stability properties of convolutional neural networks. Convolutional neural networks are widely used in machine learning. In classification they are mainly used as feature extractors. Ideally, we expect similar features when the inputs are from the same class. That is, we hope to see a small change in the feature vector with respect to a deformation on the input signal. This can be established mathematically, and the key step is to derive the Lipschitz properties. Further, we establish that the stability results can be extended for more general networks. We give a formula for computing the Lipschitz bound, and compare it with other methods to show it is closer to the optimal value.

1. Introduction

Recently convolutional neural networks have enjoyed tremendous success in many applications in image and signal processing. According to [**5**], a general convolutional network contains three types of layers: convolution layers, detection layers, and pooling layers. In [**7**], Mallat proposes the scattering network, which is a tree-structured convolutional neural network whose filters in convolution layers are wavelets. Mallat proves that the scattering network satisfies two important properties: (approximately) invariance to translation and stabitity to deformation. However, for those properties to hold, the wavelets must satisfy an admissibility condition. This restricts the adaptability of the theory. The authors in [**11, 12**] use a slightly different setting to relax the conditions. They consider sets of filters that form semi-discrete frames of upper frame bound equal to one. They prove that deformation stability holds for signals that satisfy certain conditions.

In both settings, the deformation stability is a consequence of the Lipschitz property of the network, or feature extractor. The Lipschitz property in itself is important even if we do not consider deformation of the form described in [**7**]. In [**10**], the authors detect some instability of the AlexNet by generating images that are easily recognizable by nude eyes but cause the network to give incorrect classification results. They partially attribute the instability to the large Lipschitz bound of the AlexNet. It is thus desired to have a formula to compute the Lipschitz bound in case the upper frame bound is not one.

2010 *Mathematics Subject Classification.* Primary 62M45.

The first author was partially supported by NSF Grant DMS-1413249 and ARO Grant W911NF-16-1-0008. The third author was partially supported by NSF Grant DMS-1413249.

The lower bound in the frame condition is not used when we analyze the stability properties for scattering networks. In [**12**] the authors conjectured that it has to do with the distinguishability of the two classes for classification. However, certain loss of information should be allowed for classification tasks. A lower frame bound is too strong in this case since it has most to do with injectivity. In this paper, we only consider the semi-discrete Bessel sequence, and discuss a convolutional network of finite depth.

Merging is widely used in convolutional networks. Note that practitioners use a concatenation layer ([**9**]) but that is just a concatenation of vectors and is of no mathematical interest. Nevertheless, aggregation by p-norms and multiplication is frequently used in networks and we still obtain stability to deformation in those cases and the Lipschitz bound increases only by a factor depending on the number of filters to be aggregated.

The organization of this paper is as follows. In Section 2, we introduce the scattering network and state a general Lipschitz property. In Section 3, we discuss the aggregation of filters using p-norms or pointwise multiplication. In Section 4, we use examples of networks to compare different methods for computing the Lipschitz constants.

2. Scattering network

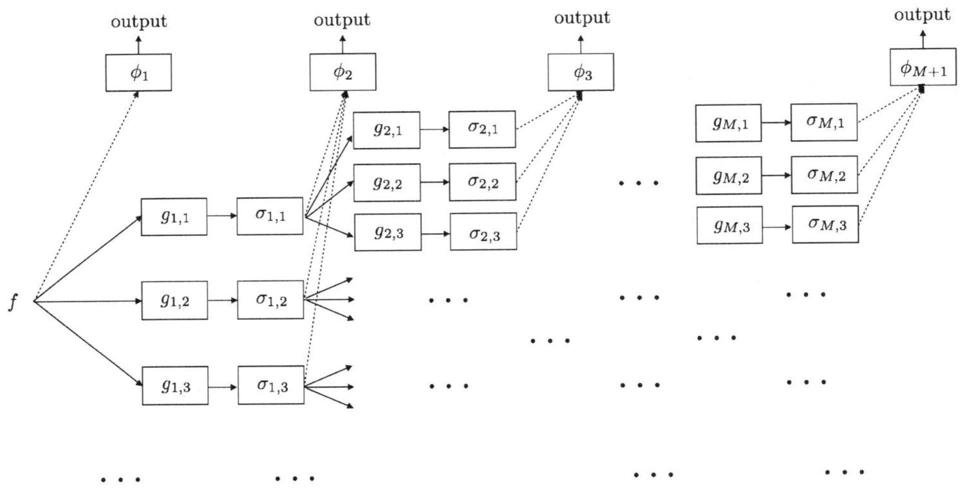

FIGURE 1. Structure of the scattering network of depth M

We first review the theory developed by the authors in [**7, 11, 12**] and give a more general result. Figure 1 shows a typical scattering network. f denotes an input signal (commonly in L^2 or l^2, for our discussion we take $f \in L^2(\mathbb{R}^d)$). $g_{m,l}$'s and ϕ_m's are filters and the corresponding blocks symbolizes the operation of doing convolution with the filter in the block. The blocks marked $\sigma_{m,l}$ illustrate the action of a nonlinear function. This structure clearly shows the three stages of a convolutional neural network: the $g_{m,l}$'s are the convolution stage; the $\sigma_{m,l}$'s are the detection stage; the ϕ_m's are the pooling stage.

The output of the network in Figure 1 is the collection of outputs of each layer. To represent the result clearly, we introduce some notations first.

We call an ordered collection of filters $g_{m,l} \in L^1(\mathbb{R}^d)$ connected in the network starting from $m = 1$ a path, say $q = (g_{1,l_1}, g_{2,l_2}, \cdots, g_{M_q,l_{M_q}})$ (for brevity we also denote it as $q = ((1,l_1),(2,l_2),\cdots,(M_q,l_{M_q}))$, and in this case we denote $|q| = M_q$ to be the number of filters in the collection. We call $|q|$ the length of the path. For $q = \emptyset$, we say that $|q| = 0$. The largest possible $|q|$, say M, is called the depth of the network. For each $m = 1, 2, \cdots, M+1$, there is an output-generating atom $\phi_m \in L^1(\mathbb{R}^d)$, which is usually taken to be a low-pass filter. ϕ_1 generates an output from the original signal f, and ϕ_m generates an output from a filter g_{m-1,l_m} in the $(m-1)$'s layer, for $2 \leq m \leq M+1$. It is clear that a scattering network of finite depth is uniquely determined by ϕ_m's and the collection Q of all paths. We use G_m to denote the set of filters in the m-th layer. For a fixed q with $|q| = m$, we use G^q_{m+1} to denote the set of filters in the $(m+1)$'s layer that are connected with q. Thus G_{m+1} is a disjoint union of G^q_{m+1}'s:

$$G_{m+1} = \dot{\bigcup} G^q_{m+1} \, .$$

$\sigma_{m,l} : \mathbb{C} \to \mathbb{C}$ are Lipschitz continuous functions with Lipschitz bound no greater than 1. That is,

$$\|\sigma_{m,l}(y) - \sigma_{m,l}(\tilde{y})\|_2 \leq \|y - \tilde{y}\|_2$$

for any y, $\tilde{y} \in L^2(\mathbb{R}^d)$. The Lip-1 condition is not restrictive since any other Lipschitz constant can be absorbed by the proceeding $g_{m,l}$ filters.

The scattering propagator $U[q] : L^2(\mathbb{R}^d) \to L^2(\mathbb{R}^d)$ for a path $q = (g_{1,l_1}, g_{2,l_2}, \cdots, g_{M_q,l_{M_q}})$ is defined to be

$$U[q]f := \sigma_{M_q,l_{M_q}}\left(\sigma_{2,l_2}\left(\sigma_{1,l_1}(f * g_{1,l_1}) * g_{2,l_2}\right) * \cdots * g_{M_q,l_{M_q}}\right) . \tag{2.1}$$

If $q = \emptyset$, then by convention we say $U[\emptyset]f := f$.

Given an input $f \in L^2(\mathbb{R}^d)$, the output of the network is the collections $\Phi(f) := \{U[q]f * \phi_{M_q+1}\}_{q \in Q}$. The norm $|||\cdot|||$ is defined by

$$|||\Phi(f)||| := \left(\sum_{q \in Q} \left\|U[q]f * \phi_{M_q+1}\right\|_2^2\right)^{\frac{1}{2}} . \tag{2.2}$$

Given a collection of filters $\{g_i\}_{i \in \mathcal{I}}$ where the index set $\mathcal{I}$ is at most countable and for each i, $g_i \in L^1(\mathbb{R}^d) \cap L^2(\mathbb{R}^d)$, $\{g_i\}_{i \in \mathcal{I}}$ is said to form the atoms of a semi-discrete Bessel sequence if there exists a constant $B > 0$ for which

$$\sum_{i \in \mathcal{I}} \|f * g_i\|_2^2 \leq B \|f\|_2^2$$

for any $f \in L^2$. In this case, $\{g_i\}_{i \in \mathcal{I}}$ is said to form the atoms of a semi-discrete frame if in addition there exists a constant $A > 0$ for which

$$A \|f\|_2^2 \leq \sum_{i \in \mathcal{I}} \|f * g_i\|_2^2 \leq B \|f\|_2^2$$

for any $f \in L^2$.

Conditions (2.1) and (2.5) can be achieved for a larger class of filters. Specifically, we shall introduce a Banach algebra in (3.1), where the Bessel bound is naturally defined.

Throughout this paper, we adapt the definition of Fourier transform of a function f to be

$$\hat{f}(\omega) = \int_{\mathbb{R}^d} f(x)e^{-2\pi i\omega x}dx \ . \tag{2.3}$$

The dilation of f by a factor λ is defined by

$$f_\lambda(x) = \lambda f(\lambda x) \ . \tag{2.4}$$

The first result of this paper compiles and extends previous results obtained in [**7**, **11**, **12**].

THEOREM 2.1 (See also [**7**, **11**, **12**]). *Suppose we have a scattering network of depth M. For each $m = 1, 2, \cdots, M+1$,*

$$B_m = \max_{q:|q|=m-1} \left\| \left(\sum_{g_{m,l}\in G_m^q} |\hat{g}_{m,l}|^2 \right) + \left|\hat{\phi}_m\right|^2 \right\|_\infty < \infty \tag{2.5}$$

with the understanding that $B_{M+1} = \left\|\hat{\phi}_{M+1}\right\|_\infty^2$ (that is, $G_{M+1}^q = \emptyset$). Then the corresponding feature extractor Φ is Lipschitz continuous in the following manner:

$$|||\Phi(f) - \Phi(h)||| \leq \left(\prod_{m=1}^{M+1} \tilde{B}_m \right)^{\frac{1}{2}} \|f-h\|_2 \ ,$$

where

$$\tilde{B}_1 = B_1, \quad \tilde{B}_m = \max\{1, B_m\} \text{ for } m \geq 2 \ . \tag{2.6}$$

PROOF. First we prove a lemma.

LEMMA 2.2. *With the settings in Theorem 2.1, for $0 \leq m \leq M-1$, we have*

$$\begin{aligned} &\sum_{|q|=m+1} \|U[q]f - U[q]h\|_2^2 + \sum_{|q|=m} \|U[q]f * \phi_{m+1} - U[q]h * \phi_{m+1}\|_2^2 \\ &\leq \sum_{|q|=m} B_{m+1} \|U[q]f - U[q]h\|_2^2 \ ; \end{aligned} \tag{2.7}$$

for $m = M$, we have

$$\sum_{|q|=M} \|U[q]f * \phi_{m+1} - U[q]h * \phi_{m+1}\|_2^2 \leq \sum_{|q|=M} B_{M+1} \|U[q]f - U[q]h\|_2^2 \ . \tag{2.8}$$

PROOF OF LEMMA 2.2. Let q be a path with $|q| = m < M$. We go one layer deeper to get

$$
\begin{aligned}
&\sum_{q'\in q\times G^q_{m+1}} \|U[q']f - U[q']h\|_2^2 + \|U[q]f * \phi_{m+1} - U[q]h * \phi_{m+1}\|_2^2 \\
= &\sum_{g_{m+1,l}\in G^q_{m+1}} \|\sigma_{m+1,l}(U[q]f * g_{m+1,l}) - \sigma_{m+1,l}(U[q]h * g_{m+1,l})\|_2^2 + \\
&\|U[q]f * \phi_{m+1} - U[q]h * \phi_{m+1}\|_2^2 \\
\leq &\sum_{g_{m+1,l}\in G^q_{m+1}} \|U[q]f * g_{m+1,l} - U[q]h * g_{m+1,l}\|_2^2 + \\
&\|U[q]f * \phi_{m+1} - U[q]h * \phi_{m+1}\|_2^2 \\
= &\sum_{g_{m+1,l}\in G^q_{m+1}} \|(U[q]f - U[q]h) * g_{m+1,l}\|_2^2 + \|(U[q]f - U[q]h) * \phi_{m+1}\|_2^2 \ .
\end{aligned}
\tag{2.9}
$$

Sum over all q with length m, we have

$$
\begin{aligned}
&\sum_{|q|=m+1} \|U[q]f - U[q]h\|_2^2 + \sum_{|q|=m} \|U[q]f * \phi_{m+1} - U[q]h * \phi_{m+1}\|_2^2 \\
\leq &\sum_{|q|=m} \sum_{g_{m+1,l}\in G_{m+1}} \|(U[q]f - U[q]h) * g_{m+1,l}\|_2^2 + \\
&\sum_{|q|=m} \|U[q]f * \phi_{m+1} - U[q]h * \phi_{m+1}\|_2^2 \\
\leq &\sum_{|q|=m} B_{m+1} \|U[q]f - U[q]h\|_2^2 \ ,
\end{aligned}
\tag{2.10}
$$

which follows the Bessel inequality by the definition of the B_m's. For m = M, directly following the Young's inequality we have (2.8). □

We now continue with the proof of Theorem 2.1. The inqualities (2.7) and (2.8) have two consequences. First, summing over $m = 0, \cdots, M$ we have

$$
\begin{aligned}
&\sum_{m=0}^{M} \sum_{|q|=m} \|U[q]f * \phi_{m+1} - U[q]h * \phi_{m+1}\|_2^2 \\
\leq &\ B_1 \|f - h\|_2^2 + \sum_{m=1}^{M} (B_{m+1} - 1) \sum_{|q|=m} \|U[q]f - U[q]h\|_2^2 \ ;
\end{aligned}
\tag{2.11}
$$

second, we have for each $m = 0, \cdots, M-1$ that

$$
\sum_{|q|=m+1} \|U[q]f - U[q]h\|_2^2 \leq \sum_{|q|=m} B_{m+1} \|U[q]f - U[q]h\|_2^2 \ . \tag{2.12}
$$

Therefore, put (2.11) and (2.12) together, noting that $B_m \leq \tilde{B}_m$ for each m, we have

$$\begin{aligned}
&\sum_{m=0}^{M} \sum_{|q|=m} \|U[q]f * \phi_{m+1} - U[q]h * \phi_{m+1}\|_2^2 \\
&\leq \tilde{B}_1 \|f-h\|_2^2 + \sum_{m=1}^{M} (\tilde{B}_{m+1} - 1) \left(\prod_{m'=1}^{m} \tilde{B}_{m'} \right) \|f-h\|_2^2 \qquad (2.13) \\
&\leq \left(\prod_{m=1}^{M+1} \tilde{B}_m \right) \|f-h\|_2^2 \ .
\end{aligned}$$

We complete the proof by observing that the uppermost object in Inequality (2.13) is nothing but $|||\Phi(f) - \Phi(h)|||^2$.

□

REMARK 2.3. [**11**, **12**] consider the case where each filter in the m-th layer in connected to all the frame vectors from the pre-designed frame for the $(m+1)$-th layer. In practical uses then, a dimension reduction process needs to be done to select a few branches from the numerous tributaries due to such a design manner (see [**1**]). Also, the authors of [**11**, **12**] assume that all the B_m's are less than or equal to one. As can be seen in the above proof, this assumption is not needed.

REMARK 2.4. The infinite-depth case is an immediate extension of the finite-depth case if $\prod \tilde{B}_m < \infty$.

Theorem 2.1, together with Schur's test (for integral operators), lead to the following theorem, which implies the deformation stability of the corresponding network. The proof can be found in [**11**]. We state this result for the completeness of this article.

THEOREM 2.5 ([**11**]). *With the settings in Theorem 2.1, Let H_R be the space of R-band-limited functions defined by*

$$H_R := \{ f \in L^2(\mathbb{R}^d) : supp(\hat{f}) \subset B_R(0) \} \ .$$

Then for all $f \in H_R$, $\omega \in C(\mathbb{R}^d, R)$, $\tau \in C^1(\mathbb{R}^d, R)$ with $\|D\tau\|_\infty \leq (2d)^{-1}$,

$$|||\Phi(f) - \Phi(F_{\tau,\omega} f)||| \leq C(R\|\tau\|_\infty + \|\omega\|_\infty) \|f\|_2 \ ,$$

where $F_{\tau,\omega} f)$ is the deformed version of f defined by

$$F_{\tau,\omega} f(x) := e^{2\pi i \omega(x)} f(x - \tau(x)) \ . \qquad (2.14)$$

Note that the Lipschitz property of $\sigma_{m,l}$'s is not necessary in some cases. For instance, if we use $|\cdot|^2$ in place of all the $\sigma_{m,l}$'s, as illustrated in Figure 2. Then the training process would deal with smooth functions that are not Lipschitz. To guarantee a finite Lipschitz constant for Φ we need to control the L^∞ norm of the input.

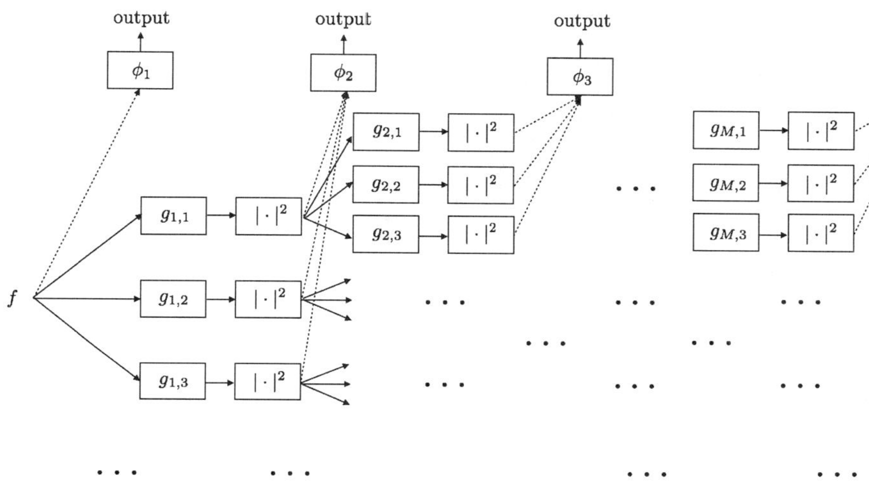

FIGURE 2. Structure of the scattering network of depth M with nonlinearity $|\cdot|^2$

THEOREM 2.6. *Consider the settings in Theorem 2.1, where $\sigma_{m,l}$'s are replaced with $|\cdot|^2$ (see Figure 2). Suppose there is a constant $R > 0$ for which $\|g_{m,l}\|_1 \leq \frac{\min\{1,2\sqrt{R}\}}{2R}$ for all m, l. Then the corresponding feature extractor Φ is Lipschitz continuous on the ball of radius R under infinity norm in the following manner:*

$$|||\Phi(f) - \Phi(h)||| \leq \left(\prod_{m=1}^{M+1} \tilde{B}_m \right)^{\frac{1}{2}} \|f - h\|_2$$

for any f, $h \in L^2(\mathbb{R}^d)$ with $\|f\|_\infty \leq R$, $\|h\|_\infty \leq R$, where $\tilde{B}_m$'s are defined as in (2.6) and (2.5).

REMARK 2.7. In the case of deformation, h is given by $h = F_{\tau,\omega}$ as defined in (2.14). If f satisfies the L^∞ condition $\|f\|_\infty = R$, so does h, since $\|h\|_\infty = \|f\|_\infty$.

PROOF. Notice that $\frac{\min\{1,2\sqrt{R}\}}{2R} = \min\{\frac{1}{\sqrt{R}}, \frac{1}{2R}\}$. Hence $\|g_{m,l}\|_1 \le 1/\sqrt{R}$ and $\|g_{m,l}\|_1 \le 1/2R$. We observe that for any path q with length $|q| = m \ge 1$, say $q = \big((1,l_1),(2,l_2),\cdots,(M_q,l_{M_q})\big)$, and for convenience denote $q_1 = ((1,l_1))$, $q_2 = ((1,l_1),(2,l_2))$, $\cdots$, $q_{M_q-1} = \big((1,l_1),(2,l_2),\cdots,(M_q-1,l_{M_q-1})\big)$, we have

$$
\begin{aligned}
\|U[q]f\|_\infty &= \left\| \left| U[q_{M_q-1}]f * g_{M_q,l_{M_q}} \right|^2 \right\|_\infty \\
&\le \left\|U[q_{M_q-1}]f\right\|_\infty^2 \left\|g_{M_q,l_{M_q}}\right\|_1^2 \\
&\le \left\|U[q_{M_q-2}]f\right\|_\infty^4 \left\|g_{M_q-1,l_{M_q-1}}\right\|_1^4 \left\|g_{M_q,l_{M_q}}\right\|_1^2 \\
&\le \cdots \\
&\le \|U[q_1]f\|_\infty^{2^{M_q-1}} \prod_{j=2}^{M_q} \|g_{j,l_j}\|_1^{2^{M_q-j+1}} \\
&\le \|f\|_\infty^{2^{M_q}} \prod_{j=1}^{M_q} \|g_{j,l_j}\|_1^{2^{M_q-j+1}} \\
&\le R^{2^{M_q}} \prod_{j=1}^{M_q} \left(\frac{1}{\sqrt{R}}\right)^{2^{M_q-j+1}} \\
&= R^{2^{M_q}} \cdot \left(\frac{1}{\sqrt{R}}\right)^{(2^{M_q}-1)} \\
&= R\,.
\end{aligned}
$$

With this, let q be a path of length $|q| = m < M$, we have for each l that

$$
\begin{aligned}
&\left\| |U[q]f * g_{m+1,l}|^2 - |U[q]h * g_{m+1,l}|^2 \right\|_2^2 \\
= \ & \|(|U[q]f * g_{m+1,l}| + |U[q]h * g_{m+1,l}|)\,(|U[q]f * g_{m+1,l}| - |U[q]h * g_{m+1,l}|)\|_2^2 \\
\le \ & \||U[q]f * g_{m+1,l}| + |U[q]h * g_{m+1,l}|\|_1^2 \, \||U[q]f * g_{m+1,l}| - |U[q]h * g_{m+1,l}|\|_2^2 \\
\le \ & (\|U[q]f\|_\infty + \|U[q]h\|_\infty)^2 \, \|g_{m+1,l}\|_1^2 \, \||U[q]f * g_{m+1,l}| - |U[q]h * g_{m+1,l}|\|_2^2 \\
\le \ & (R+R)^2(1/2R)^2 \, \||U[q]f * g_{m+1,l}| - |U[q]h * g_{m+1,l}|\|_2^2 \\
= \ & \||U[q]f * g_{m+1,l}| - |U[q]h * g_{m+1,l}|\|_2^2 \\
\le \ & \|U[q]f * g_{m+1,l} - U[q]h * g_{m+1,l}\|_2^2 \ .
\end{aligned}
$$

Therefore,

$$
\begin{aligned}
&\sum_{q' \in q \times G^q_{m+1}} \|U[q']f - U[q']h\|_2^2 + \|U[q]f * \phi_{m+1} - U[q]h * \phi_{m+1}\|_2^2 \\
\le \ & \sum_{g_{m+1,l} \in G^q_{m+1}} \|U[q]f * g_{m+1,l} - U[q]h * g_{m+1,l}\|_2^2 + \\
& \qquad \|U[q]f * \phi_{m+1} - U[q]h * \phi_{m+1}\|_2^2 \\
= \ & \sum_{g_{m+1,l} \in G^q_{m+1}} \|(U[q]f - U[q]h) * g_{m+1,l}\|_2^2 + \|(U[q]f - U[q]h) * \phi_{m+1}\|_2^2 \ .
\end{aligned}
$$

Then by exactly the same inequality as (2.10), for $0 \leq m \leq M-1$,

$$\sum_{|q|=m+1} \|U[q]f - U[q]h\|_2^2 + \sum_{|q|=m} \|U[q]f * \phi_{m+1} - U[q]h * \phi_{m+1}\|_2^2$$
$$\leq \sum_{|q|=m} B_{m+1} \|U[q]f - U[q]h\|_2^2 \ ;$$

and for $m = M$,

$$\sum_{|q|=M} \|U[q]f * \phi_{m+1} - U[q]h * \phi_{m+1}\|_2^2 \leq \sum_{|q|=M} B_{M+1} \|U[q]f - U[q]h\|_2^2 \ .$$

The rest of the proof is a minimal modification to that of Theorem 2.1. It is obvious that $\|f\|_\infty = \|F_{\tau,\omega}\|_\infty$. □

In most applications, the L^∞-norm of the input is well bounded. For instance, normalized grayscale images have pixel valued between 0 and 1. Even if it is not the case, we can pre-filter the input by widely used sigmoid functions, such as tanh. For instance, in the above case of $|\cdot|^2$, we can use the structure as follows.

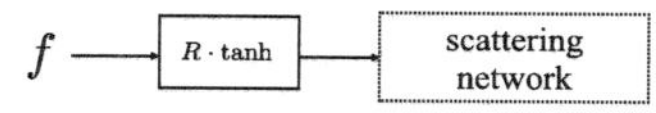

FIGURE 3. Restrict $\|f\|_\infty$ at the first layer using $R \cdot \tanh$

3. Filter aggregation

3.1. Aggregation by taking norm across filters. We use filter aggregation to model the pooling stage after convolution. In deep learning there are two widely used pooling operation, max pooling and average pooling. Max pooling is the operation of extracting local maximum of the signal, and can be modeled by an L^∞-norm aggregation of copies of shifted and dilated signals. Average pooling is the operation of taking local average of the signal, and can be modeled by a L^1-norm aggregation of copies of shifted and dilated signals. When those pooling operations exist, it is still desired that the feature extractor is stable. We analyze this type of aggregation in detail as follows.

We consider filter aggregation by taking pointwise p-norms of the inputs. That is, suppose the inputs of the aggregation are $y_1, y_2, \cdots, y_L$ from L different filters, the output is given by $(\sum_{l=1}^{L} |y_l|^p)^{1/p}$ for some p with $1 \leq p \leq \infty$. Note that $y_1, y_2, \cdots, y_L$ are all L^2 functions and thus the output is also a L^2 function. A typical structure is illustrated in Figure 4. Recall all the nonlinearities $\sigma_{m,l}$'s are assumed to be pointwise Lipschitz functions, with Lipschitz bound less than or equal to one.

Note that we do not necessarily aggregate filters in the same layer. For instance, in Figure 4, $f * g_{1,1}$, $f * g_{1,2}$ are aggregated with f. Nevertheless, for the purpose of analysis it suffices to consider the case where the filters to be aggregated are in the same layer of the network. To see this, note that the equivalence relation in Figure 5. We can coin a block which does not change the input (think of a δ-function if we want to make the block "convolutional"). Since a δ-function is not in $L^1(\mathbb{R}^d)$,

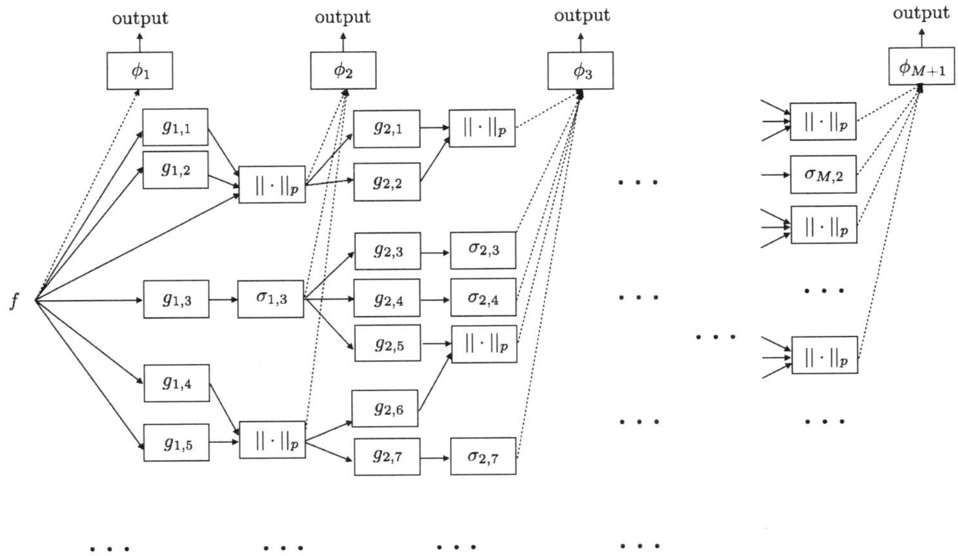

FIGURE 4. A typical structure of the scattering network with pointwise p-norms

if we want to apply the theory we have to consider a larger space where the filters stay. In this case, it is natural to consider the Banach algebra

$$\mathcal{B} = \left\{ f \in \mathcal{S}'(\mathbb{R}^d), \left\| \hat{f} \right\|_\infty < \infty \right\} . \tag{3.1}$$

Without loss of generality, we can consider only networks in which the aggregation only takes inputs from the same layer.

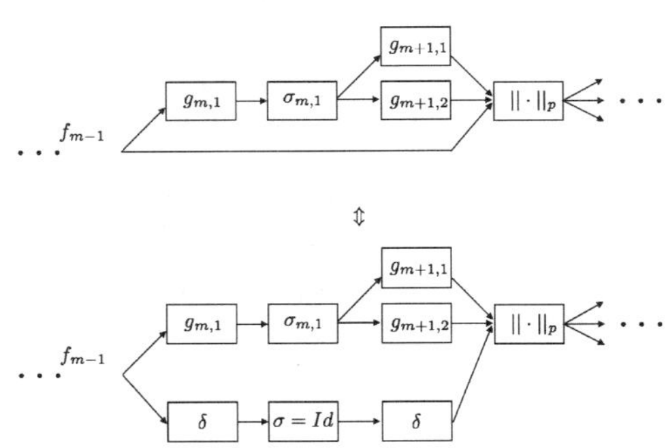

FIGURE 5. Equivalence for aggregating from different layers

Our purpose is to derive inequalities similar to (2.7) and (2.8). We define a path q to be a sequence of filters in the same manner as in Section 2. Note that by aggregating the filters we no longer have a scattering structure but a general convolutional network. That is, we might have two different filters in the m-th layer that flows into the same filter in the $(m+1)$-th layer. Although a scattering network with aggregation by the p-norm is still uniquely determined by the collection Q of its paths, the notation $U[q]$ is meaningless since it does not take into account the aggregation. The output in this case may not depend on a single path.

Note that for each $m = 1, \cdots, M$, the m-th layer of filters is followed by blocks of $\|\cdot\|_p$'s and nonlinearity $\sigma_{m,l}$'s. Let μ_m be the total number of the blocks in the m-th layer. Also take $\mu_0 = 1$. Further, we denote the blocks to be $K_{m,1}, \cdots, K_{m,\mu_m}$. For a block K and a filter g, we denote $g \leftrightarrow K$ if they are connected in the network. For a block $K_{m,\lambda}$, $1 \leq m \leq M$, $1 \leq \lambda \leq \mu_m$, we denote $G^{\text{in}}_{m,\lambda}$ to be the collection of filters in the m-th layer that are connected to $K_{m,\lambda}$ ("in" implies the filters "flow into" the block), and denote $G^{\text{out}}_{m+1,\lambda}$ to be the collection of filters in the $(m+1)$-th layer connected to $K_{m,\lambda}$. Then for each $m = 1, \cdots, M$, $G_m = \dot{\bigcup}_{1 \leq \lambda \leq \mu_m} G^{\text{in}}_{m,\lambda}$; also, for each $m = 1, \cdots, M-1$, $G_m = \dot{\bigcup}_{1 \leq \lambda' \leq \mu_{m+1}} G^{\text{out}}_{m,\lambda'}$.

We define the scattering propagator $\{U^1_m, \cdots, U^{\mu_m}_m\}_{m=0}^M$ recursively as follows. Define $U^1_0 f := f$. Suppose $\{U^1_m, \cdots, U^{\mu_m}_m\}$ has been defined for some $m < M$, then for each $\lambda = 1, \cdots, \mu_{m+1}$, we define

$$U^\lambda_{m+1} f := \left(\sum_{g_{m+1,l_{m+1}} \leftrightarrow K_{m+1,\lambda}} \left| U^{\lambda'}_m f * g_{m+1,l_{m+1}} \right|^p \right)^{\frac{1}{p}}, \tag{3.2}$$

where λ' satisfies $g_{m+1,l_{m+1}} \leftrightarrow K_{m,\lambda'}$, which is unique by the structure of the network. Now the output $\Phi(f) := \{U^\lambda_m * \phi_{m+1}\}_{0 \leq m \leq M, 1 \leq \lambda \leq m}$ is naturally defined.

To proceed we first prove the following lemma.

LEMMA 3.1. *Let $\{g_{m,l}\}_{l=1}^L$ be the filters to be aggregated using p-norm with $1 \leq p \leq \infty$, then we have the following: suppose $\{f_{m-1,l}\}_{l=1}^L$ and $\{h_{m-1,l}\}_{l=1}^L$ are two sets of inputs to those filters and f_m and h_m are the outputs respectively, then*

$$\|f_m - h_m\|_2^2 \leq \max(1, L^{2/p-1}) \sum_{l=1}^L \|(f_{m-1,l} - h_{m-1,l}) * g_{m,l}\|_2^2 . \tag{3.3}$$

PROOF. For $1 \leq p \leq \infty$, applying $\left| \|v_1\|_p - \|v_2\|_p \right| \leq \|v_1 - v_2\|_p$ and $\|v_1\|_p \leq \max(1, L^{1/p-1/2}) \|v_1\|_2$ for any vectors v_1, v_2 of length L, we have

$$\begin{aligned}
\|f_m - h_m\|_2^2 &= \left\| \left(\sum_{l=1}^L |f_{m-1,l} * g_{m,l}|^p \right)^{1/p} - \left(\sum_{l=1}^L |h_{m-1,l} * g_{m,l}|^p \right)^{1/p} \right\|_2^2 \\
&\leq \left\| \left(\sum_{l=1}^L |(f_{m-1,l} - h_{m-1,l}) * g_{m,l}|^p \right)^{1/p} \right\|_2^2 \\
&\leq \left\| \max(1, L^{1/p-1/2}) \left(\sum_{l=1}^L |(f_{m-1,l} - h_{m-1,l}) * g_{m,l}|^2 \right)^{1/2} \right\|_2^2 \\
&= \max(1, L^{2/p-1}) \int \sum_{l=1}^L |(f_{m-1,l} - h_{m-1,l}) * g_{m,l}|^2 \\
&= \max(1, L^{2/p-1}) \sum_{l=1}^L \|(f_{m-1,l} - h_{m-1,l}) * g_{m,l}\|_2^2 .
\end{aligned}$$

□

With Lemma 3.1 we can compute for any $m = 0, \cdots, M$ that

$$\sum_{\lambda=1}^{\mu_{m+1}} \left\| U_{m+1}^{\lambda} f - U_{m+1}^{\lambda} h \right\|_2^2$$
$$\leq \sum_{\lambda'=1}^{\mu_m} \sum_{l: g_{m+1,l} \in G^{\text{out}}_{m,\lambda'}} \max\left(1, \left|G^{\text{in}}_{m+1,\lambda}\right|^{2/p-1}\right) \left\| U_m^{\lambda'} f * g_{m+1,l} - U_m^{\lambda'} h * g_{m+1,l} \right\|_2^2 ,$$

where for each m, l, $G^{\text{in}}_{m+1,\lambda}$ is the unique class of filters that contains $g_{m+1,l}$. We can then proceed similar to Inequality (2.9) with minor changes. We get the following result on the Lipschitz properties for Φ.

THEOREM 3.2. *Suppose we have a scattering network of depth M including only p-norm aggregations. For $m = 1, 2, \cdots, M+1$, set*

$$B_m = \max_{1 \leq \lambda' \leq \mu_{m-1}} \left\| \sum_{l: g_{m,l} \in G^{out}_{m,\lambda'}} \max\left(1, \left|G^{in}_{m+1,\lambda}\right|^{2/p-1}\right) |\hat{g}_{m,l}|^2 + \left|\hat{\phi}_m\right|^2 \right\|_\infty < \infty$$

(with the understanding that $B_{M+1} = \left\|\hat{\phi}_{M+1}\right\|_\infty^2$, that is, $G^{out}_{M+1,\lambda'} = \emptyset$ for any $1 \leq \lambda' \leq \mu_M$), where for each m, l, $G^{in}_{m,\lambda}$ is the unique class of filters that contains $g_{m,l}$ and $G^{in}_{m,\lambda}$ denotes its cardinal. Then the corresponding feature extractor Φ is Lipschitz continuous in the following manner:

$$|||\Phi(f) - \Phi(h)||| \leq \left(\prod_{m=1}^{M+1} \tilde{B}_m\right)^{\frac{1}{2}} \|f - h\|_2 \ , \ \forall f, h \in L^2(\mathbb{R}^d) \ ,$$

where $\tilde{B}_m$'s are defined as in (2.6) and (2.5).

3.2. Aggregation by pointwise multiplication. In convolutional networks that includes time sequences, it is often useful to take the pointwise product of two intermediate outputs. For instance, in the Long Short-Term Memory (LSTM) networks introduced in [**3**,**8**], multiplication is used when we have two branches and want to use one branch for information extraction and the other for controlling, or so called "gating". A typical structure is illustrated in Figure 6. The multiplication brings two outputs into one.

Similar to the previous section, we consider multiplication blocks (if a filter is not followed by a multiplication block, such as $g_{1,3}$ in Figure 6, we still consider a block after $|\cdot|$), $J_{m,\lambda}$, $1 \leq m \leq M$, $1 \leq \lambda \leq \mu_m$. We define $G^{\text{in}}_{m,\lambda}$ and $G^{\text{out}}_{m+1,\lambda}$ to be the filters in the m-th and the $(m+1)$-th layer that are connected to $J_{m,\lambda}$, respectively. Note that $\left|G^{\text{in}}_{m,\lambda}\right| \in \{1, 2\}$. The scattering propagator U_m^{λ}'s and output generating operator Φ are defined similarly. The Lipschitz property is given by the following Theorem.

THEOREM 3.3. *Suppose we have a scattering network of depth M involving only pointwise multiplication blocks. For $m = 1, 2, \cdots, M+1$,*

$$B_m = \max_{1 \leq \lambda \leq \mu_m} \left\| \sum_{g_{m,l} \in G^{out}_{m,\lambda}} \left|G^{in}_{m,\lambda}\right| |\hat{g}_{m,l}|^2 + \left|\hat{\phi}_m\right|^2 \right\|_\infty < \infty$$

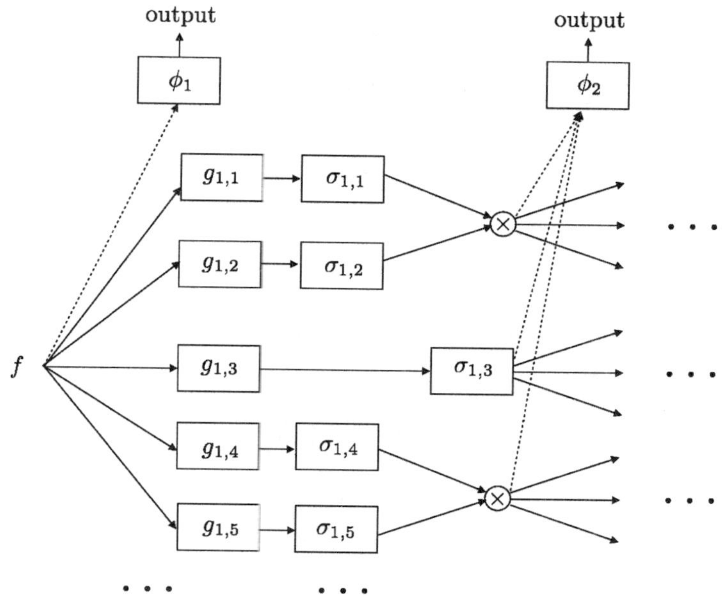

FIGURE 6. A typical structure of a scattering network with multiplication

(with the understanding that $B_{M+1} = \left\|\hat{\phi}_{M+1}\right\|_\infty^2$*, that is,* $G^{out}_{M+1,\lambda} = \emptyset$ *for all* $1 \le \lambda \le M$*), where for each* m*,* l*,* $G^{in}_{m,\lambda}$ *is the unique class of filters that contains* $g_{m,l}$*. Suppose* $\|g_{m,l}\|_1 \le 1$ *for all* m*,* l*. Then the corresponding feature extractor* Φ *is Lipschitz continuous on the ball of radius* 1 *under infinity norm in the following manner:*

$$|||\Phi(f) - \Phi(h)||| \le \left(\prod_{m=1}^{M+1} \tilde{B}_m\right)^{\frac{1}{2}} \|f - h\|_2 \ ,$$

for any f*,* $h \in L^2(\mathbb{R}^d)$ *with* $\|f\|_\infty \le 1$*,* $\|h\|_\infty \le 1$*, where* $\tilde{B}_m$*'s are defined as in (2.6) and (2.5).*

This follows by minimal modification in the proof of Theorem 2.1 once we prove the following two lemmas. Lemma 3.4 implies that the infinite norm of the inputs to each layer have the same bound. Lemma 3.5 gives a similar inequality to (2.9).

LEMMA 3.4. *(1) Let* $g_{m,1}$ *and* $g_{m,2}$ *be the two filters to be aggregated using multiplication with* $\|g_{m,j}\|_1 \le 1$ *for* $j = 1, 2$*. We have the following: suppose* $f_{m-1,1}$ *and* $f_{m-1,2}$ *are the inputs to the filters respectively with* $\|f_{m-1,j}\|_\infty \le 1$ *for* $j = 1, 2$*, then the output* f_m *satisfies* $\|f_m\|_\infty \le 1$*;*
(2) Let g_m *be a filter not to be aggregated with* $\|g_m\|_1 \le 1$*, then suppose* f_{m-1} *is the input to the filter with* $\|f_{m-1}\|_\infty \le 1$*, we have the output* f_m *satisfies* $\|f_m\|_\infty \le 1$.

PROOF. (2) directly follows from Young's Inequality. For (1), we have

$$\begin{aligned}\|f_m\|_\infty &= \|\sigma_{m,1}(f_{m-1,1} * g_{m,1}) \cdot \sigma_{m,2}(f_{m-1,2} * g_{m,2})\|_\infty \\ &\leq \|f_{m-1,1} * g_{m,1}\|_\infty \|f_{m-1,2} * g_{m,2}\|_\infty \\ &\leq \|f_{m-1,1}\|_\infty \|f_{m-1,2}\|_\infty \|g_{m,1}\|_1 \|g_{m,2}\|_1 \\ &\leq 1 .\end{aligned}$$

□

LEMMA 3.5. *Let $g_{m,1}, g_{m,2}$ be the two filters to be aggregated using a multiplication block with $\|g_{m,j}\|_1 \leq 1$ for $j = 1, 2$. We have the following: suppose $\{f_{m-1,j}\}_{j=1}^2$ and $\{h_{m-1,j}\}_{j=1}^2$ are two sets of inputs to those filters with infinite norm bounded by 1, and f_m and h_m are the outputs respectively, then*

$$\|f_m - h_m\|_2^2 \leq 2\,\|(f_{m-1,1} - h_{m-1,1}) * g_{m,1}\|_2^2 + 2\,\|(f_{m-1,2} - h_{m-1,2}) * g_{m,2}\|_2^2 \ .$$

PROOF.

$$\begin{aligned}&\|f_m - h_m\|_2^2 \\ &= \|\sigma_{m,1}(f_{m-1,1} * g_{m,1})\sigma_{m,2}(f_{m-1,2} * g_{m,2}) - \\ &\qquad \sigma_{m,1}(h_{m-1,1} * g_{m,1})\sigma_{m,2}(h_{m-1,2} * g_{m,2})\|_2^2 \\ &= \|\sigma_{m,1}(f_{m-1,1} * g_{m,1})\sigma_{m,2}(f_{m-1,2} * g_{m,2}) - \\ &\qquad \sigma_{m,1}(f_{m-1,1} * g_{m,1})\sigma_{m,2}(h_{m-1,2} * g_{m,2}) + \\ &\qquad \sigma_{m,1}(f_{m-1,1} * g_{m,1})\sigma_{m,2}(h_{m-1,2} * g_{m,2}) - \\ &\qquad \sigma_{m,1}(h_{m-1,1} * g_{m,1})\sigma_{m,2}(h_{m-1,2} * g_{m,2})\|_2^2 \\ &\leq 2\|\sigma_{m,1}(f_{m-1,1} * g_{m,1})\sigma_{m,2}(f_{m-1,2} * g_{m,2}) - \\ &\qquad \sigma_{m,1}(f_{m-1,1} * g_{m,1})\sigma_{m,2}(h_{m-1,2} * g_{m,2})\|_2^2 + \\ &\qquad 2\|\sigma_{m,1}(f_{m-1,1} * g_{m,1})\sigma_{m,2}(h_{m-1,2} * g_{m,2}) - \\ &\qquad \sigma_{m,1}(h_{m-1,1} * g_{m,1})\sigma_{m,2}(h_{m-1,2} * g_{m,2})\|_2^2 \\ &\leq 2\|\sigma_{m,1}(f_{m-1,1} * g_{m,1})\|_\infty^2 \|\sigma_{m,2}(f_{m-1,2} * g_{m,2}) - \\ &\qquad \sigma_{m,2}(h_{m-1,2} * g_{m,2})\|_2^2 + \\ &\qquad 2\,\|\sigma_{m,2}(h_{m-1,2} * g_{m,2})\|_\infty^2\,\|\sigma_{m,1}(f_{m-1,1} * g_{m,1}) - \\ &\qquad \sigma_{m,1}(h_{m-1,1} * g_{m,1})\|_2^2 \\ &\leq 2\,\|f_{m-1,1}\|_\infty^2\,\|g_{m,1}\|_1^2\,\|(f_{m-1,2} - h_{m-1,2}) * g_{m,2}\|_2^2 + \\ &\qquad 2\,\|h_{m-1,2}\|_\infty^2\,\|g_{m,2}\|_1^2\,\|(f_{m-1,1} - h_{m-1,1}) * g_{m,1}\|_2^2 \\ &\leq 2\,\|(f_{m-1,1} - h_{m-1,1}) * g_{m,1}\|_2^2 + \\ &\qquad 2\,\|(f_{m-1,2} - h_{m-1,2}) * g_{m,2}\|_2^2 \ .\end{aligned}$$

□

For a general $f \in L^2(\mathbb{R}^d)$, as discussed in the end of Section 2, we can first let it go through a sigmoid-like function, then go through the scattering network.

3.3. Mixed aggregations. The two types of aggregation blocks can be mixed together in the same networks (which is the common case in applications). The precise statement of the Lipschitz property becomes a little cumbersome to state in full generality. However, L^2-norm estimates can be combined using Theorem 2.1, 2.6, 3.2 and 3.3. This is illustrated in the next section.

4. Examples of estimating the Lipschitz constant

We use three different approaches to estimate the Lipschitz constant. The first is by propagating backward from the outputs, regardless of what we have done above. The second is by directly applying what we have discussed above. The third is by deriving a lower bound, either because of the specifies of the network (the first example), or by numerical simulating (the second example).

4.1. A standard scattering network. We first give an example of a standard scattering networks of three layers. The structure is as Figure 2.1 in [**7**]. We consider the 1D case and the wavelet given by the Haar wavelets

$$\phi(t) = \begin{cases} 1, & \text{if } 0 \leq t < 1 \\ 0, & \text{otherwise} \end{cases} \qquad \text{and} \qquad \psi(t) = \begin{cases} 1, & \text{if } 0 \leq t < 1/2 \\ -1, & \text{if } 1/2 \leq t < 1 \\ 0, & \text{otherwise} \end{cases} .$$

In this section, the sinc function is defined as $\operatorname{sinc}(x) = \sin(\pi x)/(\pi x)$ if $x \neq 0$ and 0 if $x = 0$.

We first look at real input functions. In this case the Haar wavelets ϕ and ψ readily satisfies Equation (2.7) in [**7**]. We take $J = 3$ in our example and consider all possible three-layer paths for $j = 0, -1, -2$. We have three branches from each node. Therefore we have outputs from $1 + 3 + 3^2 + 3^3 = 40$ nodes.

To convert the settings to our notations in this paper, we have a three-layer convolutional network (as in Section 2) for which the filters are given by $g_{1,l_1}, l_1 \in \{1, 2, 3\}$, $g_{2,l_2}, l_2 \in \{1, \cdots, 9\}$ and $g_{3,l_3}, l_3 \in \{1, \cdots, 27\}$, where

$$g_{m,l} = \begin{cases} \psi, & \text{if} \quad \operatorname{mod}(l, 3) = 1; \\ \psi_{2^{-1}}, & \text{if} \quad \operatorname{mod}(l, 3) = 2; \\ \psi_{2^{-2}}, & \text{if} \quad \operatorname{mod}(l, 3) = 0. \end{cases}$$

$q = ((1, l_1), (2, l_2), (3, l_3))$ is a path if and only if $l_2 \in \{3l_1 - k, k = 1, 2, 3\}$ and $l_3 \in \{3l_2 - k, k = 1, 2, 3\}$. $q = ((1, l_1), (2, l_2))$ is a path if and only if $l_2 \in \{3l_1 - k, k = 1, 2, 3\}$. The set of all paths is

$$\begin{aligned} Q = \{&\emptyset, \{(1,1)\}, \{(1,2)\}, \{(1,3)\}, \{(1,1),(2,1)\}, \{(1,1),(2,2)\}, \{(1,1),(2,3)\}, \\ &\{(1,2),(2,4)\}, \{(1,2),(2,5)\}, \{(1,2),(2,6)\}, \{(1,3),(2,7)\}, \{(1,3),(2,8)\}, \\ &\{(1,3),(2,9)\} \; \cup \; \{(1,l_1),(2,l_2),(3,l_3), 1 \leq l_1 \leq 3, \\ &\quad l_2 \in \{3l_1 - k, k = 1,2,3\}, l_3 \in \{3l_2 - k, k = 1,2,3\}\} \;. \end{aligned}$$

Also, for the output generation, $\phi_1 = \phi_2 = \phi_3 = \phi_4 = 2^{-J}\phi(2^{-J}\cdot)$. An illustration of the network is as in Figure 7.

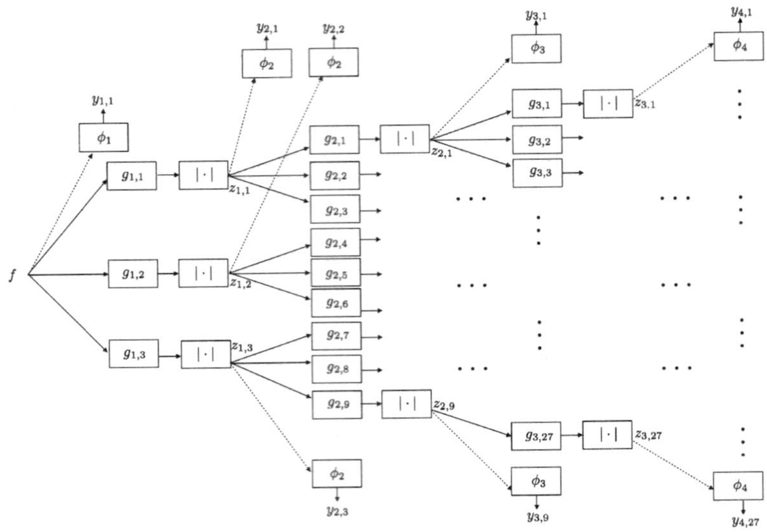

FIGURE 7. The scattering network in the example

The list of sets of filters G_m^q and G_m are

$$\begin{aligned}
G_1^{\emptyset} &= \{g_{1,1}, g_{1,2}, g_{1,3}\}\ ; \\
G_2^{(1,1)} &= \{g_{2,1}, g_{2,2}, g_{2,3}\}\ ; \\
G_2^{(1,2)} &= \{g_{2,4}, g_{2,5}, g_{2,6}\}\ ; \\
G_2^{(1,3)} &= \{g_{2,7}, g_{2,8}, g_{2,9}\}\ ; \\
G_3^{((1,1),(2,1))} &= \{g_{3,1}, g_{3,2}, g_{3,3}\}\ ; \\
&\cdots \\
G_3^{((1,3),(2,9))} &= \{g_{3,25}, g_{3,26}, g_{3,27}\}\ ;
\end{aligned}$$

and

$$\begin{aligned}
G_1 &= \{g_{1,1}, g_{1,2}, g_{1,3}\}\ ; \\
G_2 &= \{g_{2,1}, \cdots, g_{2,9}\}\ ; \\
G_3 &= \{g_{3,1}, \cdots, g_{3,27}\}\ .
\end{aligned}$$

The first approach. We use backpropagation and the chain rule. Note that $\psi_{2^j}(t) = 2^j\psi(2^j t)$ and thus $\|\psi\|_1 = \|\psi_{2^j}\|_1 = 1$. Therefore $\|g_{m,l}\|_1 = 1$ for all m, l. Similarly, $\|\phi_j\|_1 = 1$ for all j. Let y's denote the outputs and z's denote the intermediate values, as marked in Figure 7. Note that each y is associated with a unique path. Consider two inputs f and $\tilde{f}$, and $r \geq 1$. Take a path

$q = ((1, l_1), (2, l_2), (3, l_3))$ we have

$$\begin{aligned}
\|y_{4,l_3} - \tilde{y}_{4,l_3}\|_r &= \|(z_{3,l_3} - \tilde{z}_{3,l_3}) * \phi_4\|_r \le \|z_{3,l_3} - \tilde{z}_{3,l_3}\|_r \|\phi_4\|_1 = \|z_{3,l_3} - \tilde{z}_{3,l_3}\|_r \ ; \\
\|z_{3,l_3} - \tilde{z}_{3,l_3}\|_r &= \||z_{2,l_2} * g_{3,l_3}| - |\tilde{z}_{2,l_2} * g_{3,l_3}|\|_r \le \\
&\quad \|z_{2,l_2} - \tilde{z}_{2,l_2}\|_r \|g_{3,l_3}\|_1 = \|z_{2,l_2} - \tilde{z}_{2,l_2}\|_r \ ; \\
\|z_{2,l_2} - \tilde{z}_{2,l_2}\|_r &= \||z_{1,l_1} * g_{2,l_2}| - |\tilde{z}_{1,l_1} * g_{2,l_2}|\|_r \le \\
&\quad \|z_{1,l_1} - \tilde{z}_{1,l_1}\|_r \|g_{2,l_2}\|_1 = \|z_{1,l_1} - \tilde{z}_{1,l_1}\|_r \ ; \\
\|z_{1,l_1} - \tilde{z}_{1,l_3}\|_r &= \left\| |f * g_{1,l_1}| - \left|\tilde{f} * g_{1,l_1}\right| \right\|_r \le \left\|f - \tilde{f}\right\|_r \|g_{1,l_1}\|_1 = \left\|f - \tilde{f}\right\|_r .
\end{aligned}$$

and similarly for all output y_{m,l_m}'s. Therefore, we have

$$|||\Phi(f) - \Phi(\tilde{f})|||^2 = \sum_{m,l_m} \|y_{m,l_m} - \tilde{y}_{m,l_m}\|_2^2 \le 40 \left\|f - \tilde{f}\right\|_2^2 .$$

The second approach. According to the result from multi-resolution analysis, we have $\left|\hat{\phi}_{2^{-J}}(\omega)\right| + \sum_{j=-2}^{0} \left|\hat{\psi}_{2^j}(\omega)\right|^2 \le 1$ (plotted in Figure 8), we have $\tilde{B}_1 = \tilde{B}_2 = \tilde{B}_3 = \tilde{B}_4 = 1$. Indeed, we can compute that

$$\begin{aligned}
\left|\hat{\phi}_{2^{-J}}(\omega)\right| + \sum_{j=-2}^{0} \left|\hat{\psi}_{2^j}(\omega)\right|^2 &= \operatorname{sinc}^2(8\omega) + \operatorname{sinc}^2(\omega/2)\sin^2(\pi\omega/2) + \\
&\quad \operatorname{sinc}^2(\omega)\sin^2(\pi\omega) + \operatorname{sinc}^2(2\omega)\sin^2(2\pi\omega) .
\end{aligned}$$

Thus in this way, according to our discussion in Section 2, we have $|||\Phi(f) - \Phi(\tilde{f})|||^2 \le \left\|f - \tilde{f}\right\|_2^2$.

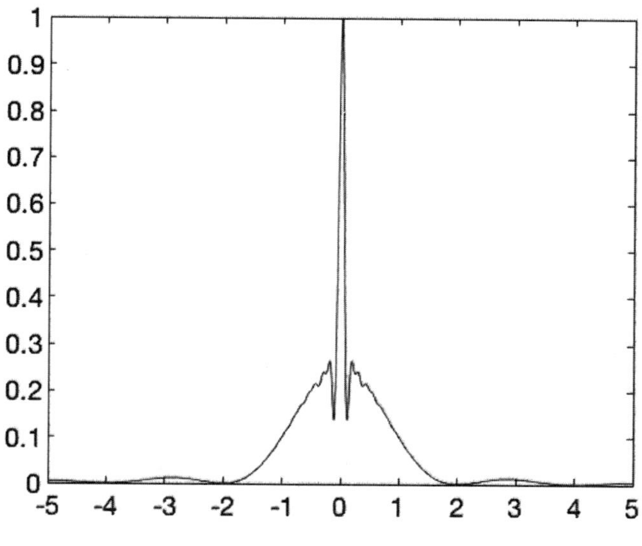

FIGURE 8. Plot of $\left|\hat{\phi}_{2^{-J}}(\omega)\right| + \sum_{j=-2}^{0} \left|\hat{\psi}_{2^j}(\omega)\right|^2$

The third approach. A lower bound is derived by considering only the output $y_{1,1}$ from the input layer. Obviously

$$|||\Phi(f) - \Phi(\tilde{f})|||^2 \geq \left\|(f - \tilde{f}) * \phi_1\right\|_1^2 .$$

Thus

$$\sup_{f \neq \tilde{f}} \frac{|||\Phi(f) - \Phi(\tilde{f})|||^2}{\left\|f - \tilde{f}\right\|_2^2} \geq \sup_{f \neq \tilde{f}} \frac{\left\|(f - \tilde{f}) * \phi_1\right\|_1^2}{\left\|f - \tilde{f}\right\|_2^2} = \left\|\hat{\phi}_1\right\|_\infty^2 = 1 .$$

Therefore, 1 is the exact Lipschitz bound (and Lipschitz constant) in our example.

4.2. A general 3-layer network. We now give an example of how to compute the Lipschitz constant as in Figure 9. In Figure 9 f is the input, y's are the outputs and z's are the intermediate values within the network. We assume that $p \geq 2$. We also assume that the nonlinearities σ_1, σ_2 are Lipschitz continuous with Lipschitz constant 1 and their L^∞ norms are bounded above by 1.

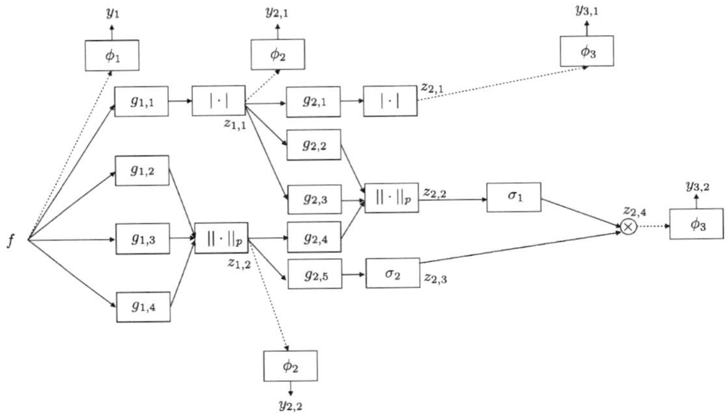

FIGURE 9. An example for computing the Lipschitz constant

Again we use three approaches to estimate the Lipschitz constant.

The first approach. In this approach we do not analyze the network by layers, but directly look at the outputs. We make use of the following rules: (1) backpropagation using the product rule and the chain rule; (2) each p-norm block is a multi-input-single-output nonlinear system with Lipschitz constant 1 for each channel.

Take two signals f and $\tilde{f}$. We use $\tilde{y}$'s and $\tilde{z}$'s to denote the outputs and intermediate values corresponding to $\tilde{f}$. Starting from the leftmost channels, we have for the first layer that

$$|y_1 - \tilde{y}_1| = \left|(f - \tilde{f}) * \phi_1\right| ,$$

and thus for any $1 \leq r \leq \infty$,

$$\|y_1 - \tilde{y}_1\|_r \leq \left\|f - \tilde{f}\right\|_r \|\phi_1\|_1 . \tag{4.1}$$

For the second layer we have

$$|y_{2,1} - \tilde{y}_{2,1}| = |(z_{1,1} - \tilde{z}_{1,1}) * \phi_{2,2}| \ ,$$

and thus

$$\|y_{2,1} - \tilde{y}_{2,1}\|_r \leq \|z_{1,1} - \tilde{z}_{1,1}\|_r \|\phi_2\|_1 \ .$$

With

$$\|z_{1,1} - \tilde{z}_{1,1}\|_r \leq \left\|f - \tilde{f}\right\|_r \|g_{1,1}\|_1 \ ,$$

we have

$$\|y_{2,1} - \tilde{y}_{2,1}\|_r \leq \left\|f - \tilde{f}\right\|_r \|g_{1,1}\|_1 \|\phi_2\|_1 \ . \tag{4.2}$$

Similarly,

$$\|y_{2,2} - \tilde{y}_{2,2}\|_r \leq \|z_{1,2} - \tilde{z}_{1,2}\|_r \|\phi_2\|_1 \ ,$$

and with

$$\begin{aligned} |z_{1,2} - \tilde{z}_{1,2}| &= \Big| \left(|f * g_{1,2}|^p + |f * g_{1,3}|^p + |f * g_{1,4}|^p\right)^{1/p} - \\ &\qquad \left(\left|\tilde{f} * g_{1,2}\right|^p + \left|\tilde{f} * g_{1,3}\right|^p + \left|\tilde{f} * g_{1,4}\right|^p\right)^{1/p} \Big| \\ &\leq \left(\left|(f - \tilde{f}) * g_{1,2}\right|^p + \left|(f - \tilde{f}) * g_{1,3}\right|^p + \left|(f - \tilde{f}) * g_{1,4}\right|^p\right)^{1/p} \\ &\leq \left|(f - \tilde{f}) * g_{1,2}\right| + \left|(f - \tilde{f}) * g_{1,3}\right| + \left|(f - \tilde{f}) * g_{1,4}\right| \end{aligned}$$

we have

$$\|z_{1,2} - \tilde{z}_{1,2}\|_r \leq \left\|f - \tilde{f}\right\| \left(\|g_{1,2}\|_1 + \|g_{1,3}\|_1 + \|g_{1,4}\|_1\right) \ .$$

Therefore

$$\|y_{2,2} - \tilde{y}_{2,2}\|_r \leq \left\|f - \tilde{f}\right\| \left(\|g_{1,2}\|_1 + \|g_{1,3}\|_1 + \|g_{1,4}\|_1\right) \|\phi_2\|_1 \ . \tag{4.3}$$

For the third layer we have

$$\|y_{3,1} - \tilde{y}_{3,1}\|_r \leq \|z_{2,1} - \tilde{z}_{2,1}\|_r \|\phi_3\|_1 \ .$$

With

$$\|z_{2,1} - \tilde{z}_{2,1}\|_r \leq \|z_{1,1} - \tilde{z}_{1,1}\|_r \|g_{2,1}\|_1 \ ,$$

we have

$$\|y_{3,1} - \tilde{y}_{3,1}\|_r \leq \left\|f - \tilde{f}\right\|_r \|g_{1,1}\|_1 \|g_{2,1}\|_1 \|\phi_3\|_1 \ . \tag{4.4}$$

Also,

$$\begin{aligned} |z_{2,2} - \tilde{z}_{2,2}| &= \Big| \left(|z_{1,1} * g_{2,2}|^p + |z_{1,1} * g_{2,3}|^p + |z_{1,2} * g_{2,4}|^p\right)^{1/p} - \\ &\qquad \left(|\tilde{z}_{1,1} * g_{2,2}|^p + |\tilde{z}_{1,1} * g_{2,3}|^p + |\tilde{z}_{1,2} * g_{2,4}|^p\right)^{1/p} \Big| \\ &\leq \left(|(z_{1,1} - \tilde{z}_{1,1}) * g_{2,2}|^p + |(z_{1,1} - \tilde{z}_{1,1}) * g_{2,3}|^p + \right. \\ &\qquad \left. |(z_{1,2} - \tilde{z}_{1,2}) * g_{2,4}|^p\right)^{1/p} \\ &\leq |(z_{1,1} - \tilde{z}_{1,1}) * g_{2,2}| + |(z_{1,1} - \tilde{z}_{1,1}) * g_{2,3}| + |(z_{1,2} - \tilde{z}_{1,2}) * g_{2,4}| \ , \end{aligned}$$

which gives

$$\|z_{2,2} - \tilde{z}_{2,2}\|_r \leq \|z_{1,1} - \tilde{z}_{1,1}\|_r \left(\|g_{2,2}\|_1 + \|g_{2,3}\|_1\right) + \|z_{1,2} - \tilde{z}_{1,2}\|_r \|g_{2,4}\|_1 \ .$$

A more obvious relation is

$$\|z_{2,3} - \tilde{z}_{2,3}\|_r \leq \|z_{1,2} - \tilde{z}_{1,2}\|_r \|g_{2,5}\|_1 \ .$$

Under conditions in Theorem 3.3, we have

$$\begin{aligned}\|z_{2,4}-\tilde{z}_{2,4}\|_r &= \|z_{2,3}z_{2,2}-\tilde{z}_{2,3}\tilde{z}_{2,2}\|_r\\ &= \|z_{2,3}z_{2,2}-\tilde{z}_{2,3}z_{2,2}+\tilde{z}_{2,3}z_{2,2}-\tilde{z}_{2,3}\tilde{z}_{2,2}\|_r\\ &\le \|z_{2,3}-\tilde{z}_{2,3}\|_r\|z_{2,2}\|_\infty+\|\tilde{z}_{2,3}\|_\infty\|z_{2,2}-\tilde{z}_{2,2}\|_r\\ &\le \|z_{2,2}-\tilde{z}_{2,2}\|_r+\|z_{2,3}-\tilde{z}_{2,3}\|_r\ ,\end{aligned}$$

and consequently we have

(4.5)

$$\begin{aligned}\|y_{3,2}-\tilde{y}_{3,2}\|_r &\le \|z_{2,4}-\tilde{z}_{2,4}\|_r\|\phi_3\|_1\\ &\le (\|z_{2,2}-\tilde{z}_{2,2}\|_r+\|z_{2,3}-\tilde{z}_{2,3}\|_r)\|\phi_3\|_1\\ &\le \|z_{1,1}-\tilde{z}_{1,1}\|_r(\|g_{2,2}\|_1+\|g_{2,3}\|_1)\|\phi_3\|_1+\\ &\qquad \|z_{1,2}-\tilde{z}_{1,2}\|_r(\|g_{2,4}\|_1+\|g_{2,5}\|_1)\|\phi_3\|_1\\ &\le \left\|f-\tilde{f}\right\|_r\Big(\|g_{1,1}\|_1(\|g_{2,2}\|_1+\|g_{2,3}\|_1)+\\ &\qquad (\|g_{1,2}\|_1+\|g_{1,3}\|_1+\|g_{1,4}\|_1)(\|g_{2,4}\|_1+\|g_{2,5}\|_1)\Big)\|\phi_3\|_1\ .\end{aligned}$$

Collecting (4.1)-(4.5) we have

$$\begin{aligned}&\sum_{m,l}\|y_{m,l}-\tilde{y}_{m,l}\|_r\\ \le\ & \left\|f-\tilde{f}\right\|_r\Bigg(\|\phi_1\|_1+\|g_{1,1}\|_1\|\phi_2\|_1+\\ &(\|g_{1,2}\|_1+\|g_{1,3}\|_1+\|g_{1,4}\|_1)\|\phi_2\|_1+\\ &\|g_{1,1}\|_1\|g_{2,1}\|_1\|\phi_3\|_1+\Big(\|g_{1,1}\|_1(\|g_{2,2}\|_1+\|g_{2,3}\|_1)+\\ &(\|g_{1,2}\|_1+\|g_{1,3}\|_1+\|g_{1,4}\|_1)(\|g_{2,4}\|_1+\|g_{2,5}\|_1)\Big)\|\phi_3\|_1\Bigg)\\ =\ & \left\|f-\tilde{f}\right\|_r\Bigg(\|\phi_1\|_1+(\|g_{1,1}\|_1+\|g_{1,2}\|_1+\|g_{1,3}\|_1+\|g_{1,4}\|_1)\\ &\|\phi_2\|_1+\Big(\|g_{1,1}\|_1(\|g_{2,1}\|_1+\|g_{2,2}\|_1+\|g_{2,3}\|_1)+\\ &(\|g_{1,2}\|_1+\|g_{1,3}\|_1+\|g_{1,4}\|_1)(\|g_{2,4}\|_1+\|g_{2,5}\|_1)\Big)\|\phi_3\|_1\Bigg)\ .\end{aligned}$$

On the other hand we also have

$$\begin{aligned}&|||\Phi(f)-\Phi(\tilde{f})|||^2\\ =\ & \sum_{m,l}\|y_{m,l}-\tilde{y}_{m,l}\|_2^2\\ \le\ & \left\|f-\tilde{f}\right\|_2^2\Bigg(\|\phi_1\|_1^2+\|g_{1,1}\|_1^2\|\phi_2\|_1^2+\\ &(\|g_{1,2}\|_1+\|g_{1,3}\|_1+\|g_{1,4}\|_1)^2\|\phi_2\|_1^2+\\ &\|g_{1,1}\|_1^2\|g_{2,1}\|_1^2\|\phi_3\|_1^2+\Big(\|g_{1,1}\|_1(\|g_{2,2}\|_1+\|g_{2,3}\|_1)+\\ &(\|g_{1,2}\|_1+\|g_{1,3}\|_1+\|g_{1,4}\|_1)(\|g_{2,4}\|_1+\|g_{2,5}\|_1)\Big)^2\|\phi_3\|_1^2\Bigg)\ .\end{aligned} \tag{4.6}$$

The second approach. To apply our formula, we first add δ's and form a network as in Figure 10. We have a three-layer network and since $p \geq 2$, we obtain

$$\begin{aligned}
\tilde{B}_1 &= \left\| |\hat{g}_{1,1}|^2 + |\hat{g}_{1,2}|^2 + |\hat{g}_{1,3}|^2 + |\hat{g}_{1,4}|^2 + \left|\hat{\phi}_1\right|^2 \right\|_\infty ; \\
\tilde{B}_2 &= \max\left\{1, \left\| |\hat{g}_{2,1}|^2 + |\hat{g}_{2,2}|^2 + |\hat{g}_{2,3}|^2 + \left|\hat{\phi}_2\right|^2 \right\|_\infty , \left\| |\hat{g}_{2,4}|^2 + |\hat{g}_{2,5}|^2 + \left|\hat{\phi}_2\right|^2 \right\|_\infty \right\} ; \\
\tilde{B}_3 &= \max\left\{2, \left\|\hat{\phi}_3\right\|_\infty^2 \right\} ; \\
\tilde{B}_4 &= \max\left\{1, \left\|\hat{\phi}_3\right\|_\infty^2 \right\} .
\end{aligned}$$

Then the Lipschitz constant is given by $(\tilde{B}_1\tilde{B}_2\tilde{B}_3\tilde{B}_4)^{1/2}$, that is,

$$|||\Phi(f) - \Phi(\tilde{f})|||^2 \leq (\tilde{B}_1\tilde{B}_2\tilde{B}_3\tilde{B}_4) \left\| f - \tilde{f} \right\|_2^2 . \tag{4.7}$$

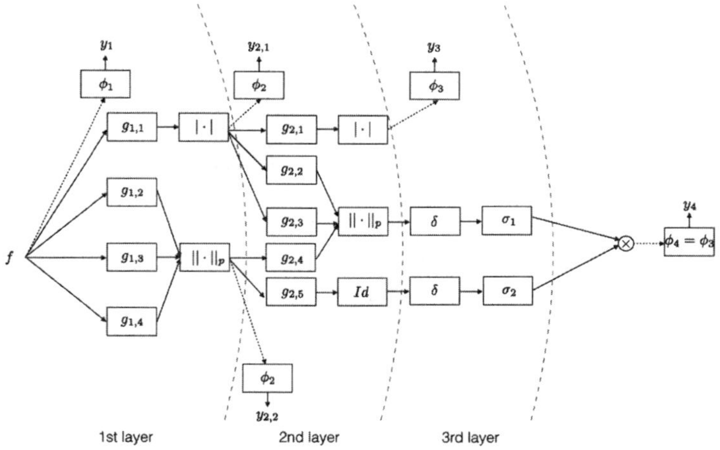

FIGURE 10. An equivalent reformulation of the same network as in Figure 9

The third approach. In general (4.7) provides a more optimal bound than (4.6) because the latter does not consider the intrinsic relations of the filters that are grouped together in the same layer. The actual Lipschitz bound can depend on the actual design of filters, not only on the Bessel bounds. We do a numerical experiment in which the Fourier transform of the filters in the same layer are the (smoothed) characteristic functions supported disjointly in the frequency domain.

Define $F(\omega) = \exp(4\omega^2/(4\omega^2-1)) \cdot \chi_{(-1/2,0)}(\omega)$, and $G(\omega) = F(-\omega)$. The fourier transform of the filters are defined to be

$$\begin{aligned}
\hat{\phi}_1(\omega) &= F(\omega+1) + \chi_{(-1,1)}(\omega) + G(\omega-1)\\
\hat{g}_{1,1}(\omega) &= F(\omega+3) + \chi_{(-3,-2)}(\omega) + G(\omega+2) + F(\omega-2) + \chi_{(2,3)}(\omega) + G(\omega-3)\\
\hat{g}_{1,2}(\omega) &= F(\omega+5) + \chi_{(-5,-4)}(\omega) + G(\omega+4) + F(\omega-4) + \chi_{(4,5)}(\omega) + G(\omega-5)\\
\hat{g}_{1,3}(\omega) &= F(\omega+7) + \chi_{(-7,-6)}(\omega) + G(\omega+6) + F(\omega-6) + \chi_{(6,7)}(\omega) + G(\omega-7)\\
\hat{g}_{1,4}(\omega) &= F(\omega+9) + \chi_{(-9,-8)}(\omega) + G(\omega+8) + F(\omega-8) + \chi_{(8,9)}(\omega) + G(\omega-9)\\
\hat{\phi}_2(\omega) &= F(\omega+2) + \chi_{(-2,2)}(\omega) + G(\omega-2)\\
\hat{g}_{2,1}(\omega) &= F(\omega+4) + \chi_{(-4,-3)}(\omega) + G(\omega+3) + F(\omega-3) + \chi_{(3,4)}(\omega) + G(\omega-4)\\
\hat{g}_{2,2}(\omega) &= F(\omega+6) + \chi_{(-6,-5)}(\omega) + G(\omega+5) + F(\omega-5) + \chi_{(5,6)}(\omega) + G(\omega-6)\\
\hat{g}_{2,3}(\omega) &= F(\omega+8) + \chi_{(-8,-7)}(\omega) + G(\omega+7) + F(\omega-7) + \chi_{(7,8)}(\omega) + G(\omega-8)\\
\hat{g}_{2,4}(\omega) &= F(\omega+5) + \chi_{(-5,-3)}(\omega) + G(\omega+3) + F(\omega-3) + \chi_{(3,5)}(\omega) + G(\omega-5)\\
\hat{g}_{2,5}(\omega) &= F(\omega+8) + \chi_{(-8,-6)}(\omega) + G(\omega+6) + F(\omega-6) + \chi_{(6,8)}(\omega) + G(\omega-8)\\
\hat{\phi}_3(\omega) &= F(\omega+9) + \chi_{(-9,9)}(\omega) + G(\omega-9)
\end{aligned}$$

Then each function is in $C_C^\infty(\hat{\mathbb{R}})$.

We numerically compute the L^1 norms of the inverse transform of the above functions using IFFT and numerical integration with stepsize 0.025: $\|\phi_1\|_1 = 1.8265$, $\|g_{1,1}\|_1 = 2.0781$, $\|g_{1,2}\|_1 = 2.0808$, $\|g_{1,3}\|_1 = 2.0518$, $\|g_{1,4}\|_1 = 2.0720$, $\|\phi_2\|_1 = 2.0572$, $\|g_{2,1}\|_1 = 2.0784$, $\|g_{2,2}\|_1 = 2.0734$, $\|g_{2,3}\|_1 = 2.0889$, $\|g_{2,4}\|_1 = 2.2390$, $\|g_{2,5}\|_1 = 2.3175$, $\|\phi_3\|_1 = 2.6378$. Then the constant on the right-hand side of Inequality (4.6) is 966.26, and by taking the square root we get the Lipschitz bound computed using the first approach is equal to $\Gamma_1 = 31.08$.

It is no effort to conclude that in the second approach, $\tilde{B}_1 = \tilde{B}_2 = \tilde{B}_4 = 1$ and $\tilde{B}_3 = 2$. Therefore the Lipschitz bound computed using the second approach is $\Gamma_2 = \sqrt{2} = 1.4142$. Note that in this example the conditions in Lemma 3.5 is satisfied.

The experiment suggests that the Lipschitz bound associated with our setting of filters is $\Gamma_3 = 1.1937$. We numerically compute the output of the network and record the largest ratio $|||\Phi(f) - \Phi(\tilde{f})|||/||f - \tilde{f}||_2$ over one million iterations. Numerically, we consider the range $[-20, 20]$ for both the time domain and the frequency domain and take stepsize to be 0.025. For each iteration we generate two randomly signals on $[-20, 20]$ with stepsize 1 and then upsample to the same scale with stepsize 0.025.

We conclude that the naïve first approach may lead to a much larger Lipschitz bound for analysis, and the second approach gives a more reasonable estimation.

References

[1] J. Bruna and S. Mallat, *Invariant scattering convolution networks*, IEEE Transactions on Pattern Analysis and Machine Intelligence **35** (2013), no. 8, 1872–1886.

[2] Joan Bruna, Soumith Chintala, Yann LeCun, Serkan Piantino, Arthur Szlam, and Mark Tygert, *A theoretical argument for complex-valued convolutional networks*, CoRR **abs/1503.03438** (2015).

[3] Sepp Hochreiter and Jürgen Schmidhuber, *Long short-term memory*, Neural Comput. **9** (1997), no. 8, 1735–1780.

[4] Ian Goodfellow, Yoshua Bengio, and Aaron Courville, *Deep learning*, Adaptive Computation and Machine Learning, MIT Press, Cambridge, MA, 2016. MR3617773

[5] Yann Lecun, Yoshua Bengio, and Geoffrey Hinton, *Deep learning*, Nature **521** (2015), no. 7553, 436–444.

[6] Roi Livni, Shai Shalev-shwartz, and Ohad Shamir, *On the computational efficiency of training neural networks*, Advances in Neural Information Processing Systems 27 (Z. Ghahramani, M. Welling, C. Cortes, N.d. Lawrence, and K.q. Weinberger, eds.), Curran Associates, Inc., 2014, pp. 855–863.

[7] Stéphane Mallat, *Group invariant scattering*, Comm. Pure Appl. Math. **65** (2012), no. 10, 1331–1398, DOI 10.1002/cpa.21413. MR2957703

[8] Tara N. Sainath, Oriol Vinyals, Andrew W. Senior, and Hasim Sak, *Convolutional, long short-term memory, fully connected deep neural networks*, 2015 IEEE International Conference on Acoustics, Speech and Signal Processing, ICASSP 2015, South Brisbane, Queensland, Australia, April 19-24, 2015, 2015, pp. 4580–4584.

[9] Christian Szegedy, Wei Liu, Yangqing Jia, Pierre Sermanet, Scott Reed, Dragomir Anguelov, Dumitru Erhan, Vincent Vanhoucke, and Andrew Rabinovich, *Going deeper with convolutions*, CVPR 2015, 2015.

[10] Christian Szegedy, Wojciech Zaremba, Ilya Sutskever, Joan Bruna, Dumitru Erhan, Ian J. Goodfellow, and Rob Fergus, *Intriguing properties of neural networks*, CoRR **abs/1312.6199** (2013).

[11] Thomas Wiatowski and Helmut Bölcskei, *Deep convolutional neural networks based on semi-discrete frames*, Proc. of IEEE International Symposium on Information Theory (ISIT), June 2015, pp. 1212–1216.

[12] Thomas Wiatowski and Helmut Bölcskei, *A mathematical theory of deep convolutional neural networks for feature extraction*, IEEE Transactions on Information Theory (2015).

Department of Mathematics and Center for Scientific Computation and Mathematical Modeling, University of Maryland, College Park, Maryland 20742
Email address: `rvbalan@math.umd.edu`

Verisk Analytics, 545 Washington Boulevard Jersey City, New Jersey 07310
Email address: `Maneesh.Singh@verisk.com`

Department of Mathematics and Center for Scientific Computation and Mathematical Modeling, University of Maryland, College Park, Maryland 20742
Email address: `zou@math.umd.edu`

Contemporary Mathematics
Volume **706**, 2018
http://dx.doi.org/10.1090/conm/706/14207

Invertibility of graph translation and support of Laplacian Fiedler vectors

Matthew Begué and Kasso A. Okoudjou

Abstract. The graph Laplacian operator is widely studied in spectral graph theory largely due to its importance in modern data analysis. Recently, the Fourier transform and other time-frequency operators have been defined on graphs using Laplacian eigenvalues and eigenvectors. We extend these results and prove that the translation operator to the i'th node is invertible if and only if all eigenvectors are nonzero on the i'th node. Because of this dependency on the support of eigenvectors we study the characteristic set of Laplacian eigenvectors. We prove that the Fiedler vector of a planar graph cannot vanish on large neighborhoods and then explicitly construct a family of non-planar graphs that do exhibit this property.

1. Introduction

1.1. Preliminaries. Techniques and methods from spectral graph theory and applied and computational harmonic analysis are increasingly being used to analyze, process and make predictions on the huge data sets being generated by the technological advances of the last few decades, e.g., see [**6**, **11**, **18**, **22**]. At the same time these tasks on large data sets and networks require new mathematical technologies which are leading to the golden age of Mathematical Engineering [**4**, **8**].

As a result, theories like the vertex-frequency analysis have emerged in an effort to investigate data from both a computational harmonic analysis and spectral graph theoretical point of views [**23**]. In particular, analogues of fundamental concepts and tools such as time-frequency analysis [**24**], wavelets [**15**], sampling theory [**1**], are being developed in the graph context. Much of this is done via spectral properties of the graph Laplacian, more specifically through the choice of appropriate eigenbases of the graph Laplacian. However, and to the best of our knowledge, a qualitative analysis of the effect of this choice on the resulting theory has not been undertaken. In this paper we initiate such qualitative analysis, focusing on the effect of the choice of eigenbasis for the graph Laplacian, on the graph translation operator defined in [**24**].

2010 *Mathematics Subject Classification.* Primary 94A12, 42C15; Secondary 65F35, 90C22.
Key words and phrases. Signal processing on graphs, time-vertex analysis, generalized translation and modulation, spectral graph theory, Fiedler vectors.
The second named author was partially supported by a grant from the Simons Foundation (#319197 to Kasso Okoudjou), and ARO grant W911NF1610008.

Throughout this paper we shall consider finite unweighted and undirected graphs. To be specific, a graph is defined by the pair (V, E) where V denotes the set of vertices and E denotes the set of edges. When the vertex and edge set (V, E) are clear, we will simply denote the graph by G. We assume that the cardinality of V is N. Each element in the edge set E is denoted as an ordered pair (x, y) where $x, y \in V$. If $(x, y) \in E$, we will often write $x \sim y$ to indicate that vertex x is connected to y. In such case, we say that y is a neighbor of x.

A graph is *undirected* if the edge set is symmetric, i.e., $(x, y) \in E$ if and only if $(y, x) \in E$. In the sequel, we only consider undirected graphs. The graph is *simple* if there are no loops or multiple edges. Additionally, we assume that graphs have only zero or one edge between any two pair of vertices, i.e., we do not allow multiple-edges between vertices.

The *degree* of vertex $x \in V$ in an undirected graph equals the number of edges emminating from (equivalently, to) x and is denoted d_x. A graph is called *regular* if every vertex has the same degree; it is called k-regular when that degree equals $k \in \mathbb{N}$. We refer to [**5, 7, 28**] for more background on graphs.

We will consider functions on graphs that take on real (or complex) values on the vertices of the graph. Since $V = \{x_i\}_{i=1}^N$ is finite, it is often useful, especially when doing numerical computations, to represent $f : V \to \mathbb{R}$ (or $\mathbb{C}$) as a vector of length N whose i'th component equals $f(x_i)$.

A *path* (of length m), denoted p, is defined to be a sequence of adjacent edges, $p = \{(p_{j-1}, p_j)\}_{j=1}^m$. We say that the path p connects p_0 to p_m. A path is said to be simple if no edge is repeated in it. A graph is *connected* if for any two distinct vertices $x, y \in V$, there exists some path connecting x and y.

We make the following definition of a ball on a graph that is motivated by the definition of a closed ball on a metric space.

DEFINITION 1.1. Given any $x \in V$ and any integer $r \geq 1$, we define the *ball of radius r centered at x*,

$$B_r(x) = \{y \in V : d(x, y) \leq r\},$$

where $d(x, y)$ is the length of a shortest path from x to y in G.

1.2. The graph Laplacian. Given a finite graph $G(V, E)$, the *adjacency matrix* is the $N \times N$ matrix, A, defined by

$$A(i, j) = \begin{cases} 1, & \text{if } x_i \sim x_j \\ 0, & \text{otherwise.} \end{cases}$$

The *degree matrix* is the $N \times N$ diagonal matrix D whose entries equal the degrees d_{x_i}, i.e.,

$$D(i, j) = \begin{cases} d_{x_i}, & \text{if } i = j \\ 0, & \text{otherwise.} \end{cases}$$

The main differential operator that we shall study is L, the Laplacian (Laplace's operator, or graph Laplacian). The pointwise formulation of the Laplacian applied to a function $f : V \to \mathbb{C}$ is given by

$$Lf(x) = \sum_{y \sim x} f(x) - f(y). \tag{1.1}$$

The graph Laplacian, L, can be conveniently represented as a matrix, which, by an abuse of notations, we shall also denote by L. It follows from (1.1) that the

$(i,j)^{th}$ entry of L is given

$$L(i,j) = \begin{cases} d_{x_i} & \text{if } i = j \\ -1 & \text{if } x_i \sim x_j \\ 0 & \text{otherwise,} \end{cases} \tag{1.2}$$

or, equivalently, $L = D - A$. Matrix L is called the unnormalized Laplacian to distinguish it from the normalized Laplacian, $\mathcal{L} = D^{-1/2}LD^{-1/2} = I - D^{-1/2}AD^{-1/2}$, used in some of the literature on graphs, e.g., [**7**]. However, we shall work exclusively with the unnormalized Laplacian and shall henceforth just refer to it as the Laplacian. See [**20**] for a well-developed survey of algebraic, analytic, and applied results on graph Laplacians.

It is not difficult to see that

$$\langle Lf, f\rangle = \sum_{x\sim y} |f(x) - f(y)|^2$$

for any $f \in \mathbb{C}^N$. Consequently, L is a positive semidefinite matrix whose eigenvalues, $\{\lambda_k\}_{k=0}^{N-1} \subset [0,\infty)$. Furthermore, if the graph $G(V,E)$ is connected the spectrum of the Laplacian L is given by

$$0 = \lambda_0 < \lambda_1 \leq \cdots \leq \lambda_{N-1}.$$

Throughout, we shall denote by Φ the set of orthonormal eigenvectors $\{\varphi_k\}_{k=0}^{N-1}$. We abuse notations, and view Φ as a $N \times N$ orthogonal matrix whose $(k-1)$th column is the vector φ_k. Note that Φ is not unique, but for the theory that follows, we assume that one has fixed an eigenbasis and hence the matrix Φ is assumed to be fixed.

In fact, the following result completely characterizes the relationship between eigenvalues of a graph's Laplacian and connectedness properties of the graph.

THEOREM 1.2 ([**7**]). *If the graph G is connected then $\lambda_0 = 0$ and $\lambda_i > 0$ for all $1 \leq k \leq N-1$. In this case $\varphi_0 \equiv 1/\sqrt{N}$. More generally, if the graph G has m connected components, then $\lambda_0 = \lambda_1 = \cdots = \lambda_{m-1} = 0$ and $\lambda_k > 0$ for all $k = m, ..., N-1$. The indicator function on each connected component (properly renormalized), forms an orthonormal eigenbasis for the m-dimensional eigenspace associated to eigenvalue 0.*

As seen from Theorem 1.2, the first nonzero eigenvalue of L is directly related to whether or not the graph is connected. In fact, λ_1 is known as the *algebraic connectivity* of the graph, see [**13**], and is widely studied. Its corresponding eigenvector, φ_1, is known as the *Fiedler vector* [**13, 14**] and will be discussed more in-depth in Section 3. If λ_1 has multiplicity 1, then the corresponding Fiedler vector is unique up to a sign. The Fiedler vector is used extensively in dimensionality reduction techniques [**4, 8, 9**], data clustering [**21**], image segmentation [**12**], and graph drawing [**25**]. Finally, we observe that the highest λ_1 can be is N, which happens only for the complete graph in which case the spectrum is $\{0, N, ..., N\}$.

1.3. Outline of the paper. The rest of the paper is organized as follows. In Section 2 we review the theory of vertex-frequency analysis on graphs introduced in [**24**]. We primarily focus on the graph translation operator since it has substantial differences to the classical Euclidean analogue of translation. In general, the translation operator is not invertible in the graph setting. We prove when the graph translation operator acts as a semigroup and completely characterize conditions in which the operator is invertible and derive its inverse.

In Section 3 we investigate characteristic sets (sets of zeros) of eigenvectors of the Laplacian because it is directly related to the theory of translation developed in Section 2. In particular, we focus on the support of the Fiedler vector of the graph. We prove in Section 3.1 that planar graphs cannot have large neighborhoods of vertices on which the Fiedler vector vanishes. We then introduce a family of (non-planar) graphs, called barren graphs, that have arbitrarily large neighborhoods on which the Fiedler vector does vanish in Section 3.2. In Section 3.3 we prove results about the algebraic connectivity and Fiedler vector of a graph formed by adding multiple graphs.

2. Translation operator on graphs

The notions of graph Fourier transform, vertex-frequency analysis, convolution, translation, and modulation operators were recently introduced in [**24**]. In this section, we focus on the translation operator and investigate certain of its properties including, semi-group (Theorem 2.2), invertibility and isometry (Theorem 2.5).

Analogously to the classical Fourier transform on the real line which expressed a function f in terms of the eigenfunctions of the Laplace operator, we define the *graph Fourier transform*, $\hat{f}$, of a functions $f : V \to \mathbb{C}$ as the expansion of f in terms of the eigenfunctions of the graph Laplacian.

DEFINITION 2.1. Given the graph, G, and its Laplacian, L, with spectrum $\sigma(L) = \{\lambda_k\}_{k=0}^{N-1}$ and eigenvectors $\{\varphi_k\}_{k=0}^{N-1}$, the *graph Fourier transform* of $f : V \to \mathbb{C}$ is given by

$$\hat{f}(\lambda_k) = \langle f, \varphi_k \rangle = \sum_{n=1}^{N} f(n)\bar{\varphi}_k(n). \tag{2.1}$$

Notice that the graph Fourier transform is only defined on values of $\sigma(L)$. In particular, one should interpret the notation $\hat{f}(\lambda_k)$ to designate the inner product of f with the k'th eigenfunction of L. However to emphasize the interplay between the vertex and spectral domains, we shall abuse the notation as defined here.

The *graph inverse Fourier transform* is then given by

$$f(n) = \sum_{k=0}^{N-1} \hat{f}(\lambda_k)\varphi_k(n). \tag{2.2}$$

It immediately follows from the above definition that Parseval's equality holds in this setting as well. Indeed, for any $f, g : V \to \mathbb{C}$ we have $\langle f, g \rangle = \langle \hat{f}, \hat{g} \rangle$. Consequently,

$$\|f\|_{\ell^2}^2 = \sum_{n=1}^{N} |f(n)|^2 = \sum_{l=0}^{N-1} |\hat{f}(\lambda_\ell)|^2 = \left\|\hat{f}\right\|_{\ell^2}^2.$$

Recall that the convolution of two signals $f, g \in L^2(\mathbb{R})$ can be defined via the Fourier transform as $\widehat{(f * g)}(\xi) = \hat{f}(\xi)\hat{g}(\xi)$. Using this approach, and by taking the inverse graph Fourier transform, (2.2), we can define convolution in the graph domain. For $f, g : V \to \mathbb{R}$, we define the *graph convolution* of f and g as

$$f * g(n) = \sum_{l=0}^{N-1} \hat{f}(\lambda_\ell)\hat{g}(\lambda_\ell)\varphi_\ell(n). \tag{2.3}$$

Many of the classical time-frequency properties of the convolution including commutativity, distributivity, and associativity hold for the graph convolution, see [**24**, Proposition 1], and all follow directly from the definition of graph convolution (2.3).

For any $k = 0, 1, ..., N-1$ the *graph modulation operator* $M_k : \mathbb{R}^N \to \mathbb{R}^N$ is defined as

$$(M_k f)(n) = \sqrt{N} f(n)\varphi_k(n). \tag{2.4}$$

Notice that since $\varphi_0 \equiv \frac{1}{\sqrt{N}}$ then M_0 is the identity operator.

An important remark is that in the classical case, modulation in the time domain represents translation in the frequency domain, i.e., $\widehat{M_\xi f}(\omega) = \hat{f}(\omega - \xi)$. The graph modulation does not exhibit this property due to the discrete nature of the spectral domain. However, it is worthy to notice the special case if $\hat{g}(\lambda_\ell) = \delta_0(\lambda_\ell)$, i.e., g is a constant function, then

$$\widehat{M_k g}(\lambda_\ell) = \sum_{n=1}^{N} \bar{\varphi}_\ell(n)(M_k g)(n) = \sum_{n=1}^{N} \bar{\varphi}_\ell(n)\sqrt{N}\varphi_k(n)\frac{1}{\sqrt{N}} = \delta_\ell(k).$$

Consequently, if g is the constant unit function, $M_k g = \varphi_k$.

Formally, the translation of a function defined on $\mathbb{R}$ is given by

$$(T_u f)(t) = f(t-u) = (f * \delta_u)(t).$$

Motivated by this example, for any $f : V \to \mathbb{R}$ we can define the *graph translation operator*, $T_i : \mathbb{R}^N \to \mathbb{R}^N$ via the graph convolution of the Dirac delta function centered at the i'th vertex:

$$(T_i f)(n) = \sqrt{N}(f * \delta_i)(n) = \sqrt{N} \sum_{k=0}^{N-1} \hat{f}(\lambda_k)\bar{\varphi}_k(i)\varphi_k(n). \tag{2.5}$$

We can express $T_i f$ in matrix notation as follows:

$$T_i f = \sqrt{N} \begin{pmatrix} \bar{\varphi}_0(i)\varphi_0(1) & \bar{\varphi}_1(i)\varphi_1(1) & \cdots & \bar{\varphi}_{N-1}(i)\varphi_{N-1}(1) \\ \vdots & \vdots & \cdots & \vdots \\ \bar{\varphi}_0(i)\varphi_0(N) & \bar{\varphi}_1(i)\varphi_1(N) & \cdots & \bar{\varphi}_{N-1}(i)\varphi_{N-1}(N) \end{pmatrix} \begin{pmatrix} \hat{f}(\lambda_0) \\ \vdots \\ \hat{f}(\lambda_{N-1}) \end{pmatrix}. \tag{2.6}$$

Graph translation exhibits commutative properties, i.e., $T_i T_j f = T_j T_i f$, and distributive properties under the convolution, i.e., $T_i(f*g) = (T_i f)*g = f*(T_i g)$, see [**24**, Corollary 1]. Also, using the definitions of graph convolution, it is elementary to show that for any $i, n \in \{1, ..., N\}$ and for any function $g : V \to \mathbb{C}$ we have

$$T_i g(n) = \overline{T_n \bar{g}(i)}.$$

Observe that if we choose real-valued eigenfunction in the definition of the graph Fourier transform, then we simply have

$$T_i g(n) = T_n g(i).$$

These results can be further generalized in the following theorem.

THEOREM 2.2. *Assume G is a graph whose Laplacian has real-valued eigenvectors $\{\varphi_k\}_{k=0}^{N-1}$. Let α be a multi-index, i.e. $\alpha = (\alpha_1, \alpha_2, ..., \alpha_K)$ where $\alpha_j \in \{1, ..., N\}$ for $1 \leq j \leq K$ and let $\alpha_0 \in \{1, ..., N\}$. We let T_α denote the composition $T_{\alpha_K} \circ \cdots T_{\alpha_2} \circ T_{\alpha_1}$. Then for any $f : V \to \mathbb{R}$, we have $T_\alpha f(\alpha_0) = T_\beta f(\beta_0)$ where $\beta = (\beta_1, ..., \beta_K)$ and $(\beta_0, \beta_1, \beta_2, ..., \beta_K)$ is any permutation of $(\alpha_0, \alpha_1, ..., \alpha_K)$.*

PROOF. There exists a bijection between the collection of all possible $T_\alpha f(\alpha_0)$ for $|\alpha| = K$, $1 \leq \alpha_0 \leq N$, and the space of $(K+1)$-tuples with values in $\{1, ..., N\}$. That is, the map that sends $T_\alpha f(\alpha_0)$ to $(\alpha_0, \alpha_1, ..., \alpha_K)$ is a bijection. This enables us to define an equivalence relation on the space $\{1, ..., N\}^{K+1}$. We write $(a_0, ..., a_K) \cong (b_0, ..., b_K)$ if and only if $T_{a_K} \circ \cdots \circ T_{a_1} f(a_0) = T_{b_K} \circ \cdots \circ T_{b_1} f(b_0)$.

By the commutavity of the graph translation operators, $(a_0, a_1 ..., a_K) \cong \sigma_1(a_0, a_1 ..., a_K) = (a_1, a_0 ..., a_K)$, i.e., σ_1 is the permutation $(1, 2)$. In general, we write σ_i to denote the permutation $(i, i{+}1)$. Similarly, $(a_0, a_1 ..., a_K) \cong \sigma_i(a_0, a_1 ..., a_K)$ for any $i = 2, 3, ..., K-1$.

We now have that any permutation σ_i for $i = 1, ..., K-1$ preserves equivalency. This collection of $K-1$ transpositions allow for any permutation, which proves the corollary. □

The graph translation operators are distributive with the convolution and the operators commute among themselves. However, the niceties end here; other properties of translation on the real line do not carry over to the graph setting. For example, we do not have the collection of graph translation operators forming a group, i.e., $T_i T_j \neq T_{i+j}$. In fact, we cannot even assert that the translation operators form a semigroup, i.e. $T_i T_j = T_{i \bullet j}$ for some semigroup operator $\bullet : \{1, ..., N\} \times \{1, ..., N\} \to \{1, ..., N\}$. The following theorem characterizes graphs which do exhibit a semigroup structure of the translation operators.

THEOREM 2.3. *Consider the graph, $G(V, E)$, with real-valued (resp. complex-valued) eigenvector matrix $\Phi = [\varphi_0 \cdots \varphi_{N-1}]$. Graph translation on G is a semigroup, i.e. $T_i T_j = T_{i \bullet j}$ for some semigroup operator $\bullet : \{1, ..., N\} \times \{1, ..., N\} \to \{1, ..., N\}$, only if $\Phi = (1/\sqrt{N})H$, where H is a real-valued (resp. complex-valued) Hadamard matrix.*

PROOF. *i.* We first show that graph translation on G is a semigroup, i.e. $T_i T_j = T_{i \bullet j}$ for some semigroup operator $\bullet : \{1, ..., N\} \times \{1, ..., N\} \to \{1, ..., N\}$, if and only if $\sqrt{N}\varphi_k(i)\varphi_k(j) = \varphi_k(i \bullet j)$ for all $l = 0, ..., N-1$. By the definition of graph translation, we have,

$$T_i T_j f(n) = N \sum_{k=0}^{N-1} \hat{f}(\lambda_k) \bar{\varphi}_k(j) \bar{\varphi}_k(i) \varphi_k(n)$$

and

$$T_\ell f(n) = \sqrt{N} \sum_{k=0}^{N-1} \hat{f}(\lambda_k) \bar{\varphi}_k(\ell) \varphi_k(n).$$

Therefore, $T_i T_j f = T_{i \bullet j} f$ will hold for any function $f : V \to \mathbb{R}$ if and only if $\sqrt{N}\varphi_k(i)\varphi_k(j) = \varphi_k(i \bullet j)$ for every $k \in \{0, ..., N-1\}$.

ii. We show next that $\sqrt{N}\varphi_k(i)\varphi_k(j) = \varphi_k(i \bullet j)$ for all $k = 0, ..., N-1$ only if the eigenvectors are constant amplitude, namely $1/\sqrt{N}$ by the orthonormality of the eigenvectors. Assume $\sqrt{N}\varphi_k(i)\varphi_k(j) = \varphi_k(i \bullet j)$, which, in particular, implies $\sqrt{N}\varphi_k(i)\varphi_k(i) = \sqrt{N}\varphi_k(i)^2 = \varphi_k(i \bullet i)$.

Suppose that $|\varphi_k(a_1)| < 1/\sqrt{N}$ for some $a_1 \in \{1, ..., N\}$ and for some $k \in \{0, ..., N-1\}$. Then $\sqrt{N}\varphi_k(a_1)^2 < |\varphi_k(a_1)|$ and so $a_1 \bullet a_1 = a_2$ for some $a_2 \in \{1, ..., N\} \setminus \{a_1\}$. Then since $|\varphi_k(a_2)| = \sqrt{N}\varphi_k(a_1)^2 < |\varphi_k(a_1)| < 1/\sqrt{N}$ we can repeat the same argument to assert $a_2 \bullet a_2 = a_3$ for some $a_3 \in \{1, ..., N\} \setminus \{a_1, a_2\}$. This procedure can be repeated producing an infinite number of unique indices $\{a_i\}$ on a graph, G, which contradicts the graph having only $N < \infty$ nodes. A similar argument gives a contradiction if $|\varphi_k(i)| > 1/\sqrt{N}$ for any l, i. Therefore, the graph translation operators form a semigroup only if $|\varphi_k(n)| = 1/\sqrt{N}$ for all $k = 0, 1, ..., N-1$ and $n = 1, ..., N$. Since Φ is an orthogonal matrix, i.e. $\Phi\Phi^* = \Phi^*\Phi = I$, then $\Phi = (1/\sqrt{N})H$, where H is a Hadamard matrix and Φ^* is the complex conjugate of matrix Φ. □

REMARK 2.4. (a) If we relax the constraint that Φ must be real-valued, we can obtain graphs with constant-amplitude eigenfunctions that allow the translation operators to form a (semi)group. For the cycle graph on N nodes, C_N, one can choose Φ equal to the discrete Fourier transform (DFT) matrix, where $\Phi_{nm} = e^{-2\pi i(n-1)(m-1)/N}$. Under this construction, we have $T_i T_j = T_{i+j \pmod{N}}$.

(b) It is shown in [**3**, Theorem 5] that if $\Phi = (1/\sqrt{N})H$ for Hadamard H, then the spectrum of the Laplacian, $\sigma(L)$, must consist entirely of even integers. The authors of [**10**] explore graphs with integer spectrum but do not address the case of a spectrum of only even integers.

(c) The converse to Theorem 2.3 does not necessarily hold. That is, if the eigenvector matrix $\Phi = 1/\sqrt{N}H$, for a renormalized Hadamard matrix H, then the translation operators on G need not form a semigroup. For example, consider the real Hadamard matrix, H, of order 12 given by

$$H = \begin{bmatrix} 1 & 1 & 1 & 1 & 1 & 1 & 1 & 1 & 1 & 1 & 1 & 1 \\ 1 & -1 & 1 & -1 & 1 & 1 & 1 & -1 & -1 & -1 & 1 & -1 \\ 1 & -1 & -1 & 1 & -1 & 1 & 1 & 1 & -1 & -1 & -1 & 1 \\ 1 & 1 & -1 & -1 & 1 & -1 & 1 & 1 & 1 & -1 & -1 & -1 \\ 1 & -1 & 1 & -1 & -1 & 1 & -1 & 1 & 1 & 1 & -1 & -1 \\ 1 & -1 & -1 & 1 & -1 & -1 & 1 & -1 & 1 & 1 & 1 & -1 \\ 1 & -1 & -1 & -1 & 1 & -1 & -1 & 1 & -1 & 1 & 1 & 1 \\ 1 & 1 & -1 & -1 & -1 & 1 & -1 & -1 & 1 & -1 & 1 & 1 \\ 1 & 1 & 1 & -1 & -1 & -1 & 1 & -1 & -1 & 1 & -1 & 1 \\ 1 & 1 & 1 & 1 & -1 & -1 & -1 & 1 & -1 & -1 & 1 & -1 \\ 1 & -1 & 1 & 1 & 1 & -1 & -1 & -1 & 1 & -1 & -1 & 1 \\ 1 & 1 & -1 & 1 & 1 & 1 & -1 & -1 & -1 & 1 & -1 & -1 \end{bmatrix}.$$

Then the second and third columns multiplied componentwise equals the vector

$$[1, -1, 1, -1, -1, 1, 1, -1, 1, 1, -1, -1]^\top,$$

which does not equal any of the columns of H.

(d) What kinds of graphs have a Hadamard eigenvector matrix? The authors of [**3**] prove that if N is a multiple of 4 for which a Hadamard matrix exists, then the complete graph on N vertices, K_N, is one such graph.

Unlike in the classical case in $\mathbb{R}^d$, graph translation is not an isometric operation, i.e., $\|T_i f\|_2 \neq \|f\|_2$. However, [**24**, Lemma 1] provides the following estimates:

$$|\hat{f}(0)| \leq \|T_i f\|_2 \leq \sqrt{N} \max_{k\in\{0,1,\dots,N-1\}} |\varphi_\ell(i)| \, \|f\|_2 \leq \sqrt{N} \max_{k\in\{0,1,\dots,N-1\}} \|\varphi_\ell\|_\infty \, \|f\|_2 \tag{2.7}$$

Furthermore, unlike the Euclidean notion of translation, graph translation need not be invertible. Theorem 2.5 characterizes all graphs for which the operator T_i is not invertible. Additionally, Hadamard matrices appear again in characterizing when graph translation does act as a unitary operator.

THEOREM 2.5. *The graph translation operator T_i fails to be invertible if and only if there exists some $k = 1, \dots, N-1$ for which $\varphi_k(i) = 0$. In particular, the nullspace of T_i has a basis equal to those eigenvectors that vanish on the i'th vertex.*

Additionally, T_i is unitary if and only if $|\varphi_k(i)| = 1/\sqrt{N}$ for all $k = 0, 1, \dots, N-1$ and all graph translation operators are unitary if and only if $\sqrt{N}\Phi$ is a Hadamard matrix.

PROOF. By (2.6), the operator T_i can be written as the matrix

$$\begin{aligned} T_i &= \sqrt{N} \begin{pmatrix} \bar{\varphi}_0(i)\varphi_0(1) & \bar{\varphi}_1(i)\varphi_1(1) & \cdots & \bar{\varphi}_{N-1}(i)\varphi_{N-1}(1) \\ \vdots & \vdots & \cdots & \vdots \\ \bar{\varphi}_0(i)\varphi_0(N) & \bar{\varphi}_1(i)\varphi_1(N) & \cdots & \bar{\varphi}_{N-1}(i)\varphi_{N-1}(N) \end{pmatrix} \Phi^* \\ &=: \sqrt{N} A_i \Phi^* \end{aligned} \tag{2.8}$$

We can compute the rank of $T_i^* T_i = N\Phi A_i^* A_i \Phi^*$. Since Φ is an $N \times N$ matrix of full rank, we can express the rank of T_i solely in terms of the matrix A_i, i.e.,

$$\text{rank}(T_i) = \text{rank}(T_i^* T_i) = \text{rank}(\Phi A_i^* A_i \Phi^*) = \text{rank}(A_i^* A_i).$$

We can explicitly compute for any indices $n, m \in \{1, \dots, N\}$,

$$\begin{aligned} (A_i^* A_i)(n,m) = \sum_{k=1}^{N} \overline{A_i(k,n)} A_i(k,m) &= \sum_{k=0}^{N-1} \bar{\varphi}_n(k)\varphi_n(i)\varphi_m(k)\bar{\varphi}_m(i) \\ &= \varphi_n(i)\bar{\varphi}_m(i) \sum_{k=0}^{N-1} \bar{\varphi}_n(k)\varphi_m(k) \\ &= \varphi_n(i)\bar{\varphi}_m(i)\delta_n(m). \end{aligned}$$

Hence, $A_i^* A_i$ is a diagonal matrix with diagonal entries $(A_i^* A_i)(n,n) = |\varphi_n(i)|^2$. Therefore,

$$T_i^* T_i(n,n) = N \sum_{k=0}^{N-1} |\varphi_n(i)|^2 \varphi_k \otimes \varphi_k^*. \tag{2.9}$$

So $T_i T_i^* = \frac{N}{N} \sum_{k=0}^{N-1} \varphi_k \otimes \varphi_n^* = I$ if and only if $|\varphi_n(i)| = 1/\sqrt{N}$ for all $n = 0, 1, \dots, N-1$. Additionally, this proves that $\text{rank}(T_i) = |\{k : \varphi_k(i) \neq 0\}|$ and hence T_i is invertible if and only if $\varphi_k(i) \neq 0$ for all k.

Suppose now that $\varphi_{k_j}(i) = 0$ for $\{k_j\}_{j=1}^{K} \subseteq \{1, ..., N-1\}$. Hence, $\text{rank}(T_i) = N - K$. Then for each $j \in \{1, ..., K\}$ and any $n \in \{1, ..., N\}$ we have

$$T_i\varphi_{k_j}(n) = \sqrt{N}\sum_{k=0}^{N-1} \hat{\varphi}_{k_j}(\lambda_k)\bar{\varphi}_k(i)\varphi_k(n) = \sqrt{N}\bar{\varphi}_{k_j}(i)\varphi_{k_j}(n) = 0.$$

Therefore, φ_{k_j} is in the null space of T_i for every $j = 1, ..., K$. Thus $\{\varphi_{k_j}\}_{j=1}^{K}$ is a collection of K orthogonal unit-norm vectors in the null space which has dimension $N - \text{rank}(T_i) = K$, hence they form an orthonormal basis for the null space of T_i which proves the claim about the null space of T_i.

Finally, if $\sqrt{N}|\varphi_n(i)| = 1$ for all $n, i = 1, ..., N$ then $\sqrt{N}\Phi$ is Hadamard which concludes proof.

□

COROLLARY 2.6. If $\varphi_k(i) \neq 0$ for all $k = 1, ..., N-1$, then the graph translation operator T_i is invertible and its inverse is given by

$$T_i^{-1}=\frac{1}{\sqrt{N}}\Phi\begin{pmatrix} \bar{\varphi}_0(1)\bar{\varphi}_0(i)^{-1} & \bar{\varphi}_0(2)\bar{\varphi}_0(i)^{-1} & \cdots & \bar{\varphi}_0(N)\bar{\varphi}_0(i)^{-1} \\ \vdots & \vdots & \cdots & \vdots \\ \bar{\varphi}_{N-1}(1)\bar{\varphi}_{N-1}(i)^{-1} & \bar{\varphi}_{N-1}(2)\bar{\varphi}_{N-1}(i)^{-1} & \cdots & \bar{\varphi}_{N-1}(N)\bar{\varphi}_{N-1}(i)^{-1} \end{pmatrix}.$$

PROOF. We shall first prove that the inverse to the matrix A_i given in (2.8) is given by

$$A_i^{-1}=\begin{pmatrix} \bar{\varphi}_0(1)\bar{\varphi}_0(i)^{-1} & \bar{\varphi}_0(2)\bar{\varphi}_0(i)^{-1} & \cdots & \bar{\varphi}_0(N)\bar{\varphi}_0(i)^{-1} \\ \vdots & \vdots & \cdots & \vdots \\ \bar{\varphi}_{N-1}(1)\bar{\varphi}_{N-1}(i)^{-1} & \bar{\varphi}_{N-1}(2)\bar{\varphi}_{N-1}(i)^{-1} & \cdots & \bar{\varphi}_{N-1}(N)\bar{\varphi}_{N-1}(i)^{-1} \end{pmatrix}.$$

We can then compute

$$\begin{aligned} A_iA_i^{-1}(n,m) &= \sum_{k=1}^{N} A_i(n,k)A_i^{-1}(k,m) = \sum_{k=0}^{N-1} \bar{\varphi}_k(i)\varphi_k(n)\bar{\varphi}_k(m)\bar{\varphi}_k(i)^{-1} \\ &= \sum_{k=0}^{N-1} \varphi_k(n)\bar{\varphi}_k(m) = \delta_n(m), \end{aligned}$$

and similarly

$$\begin{aligned} A_i^{-1}A_i(n,m) &= \sum_{k=1}^{N} A_i^{-1}(n,k)A_i(k,m) = \sum_{k=1}^{N} \bar{\varphi}_{n-1}(k)\bar{\varphi}_{n-1}(i)^{-1}\varphi_{m-1}(k)\bar{\varphi}_{m-1}(i) \\ &= \bar{\varphi}_{n-1}(i)^{-1}\bar{\varphi}_{m-1}(i)\sum_{k=1}^{N} \bar{\varphi}_{n-1}(k)\varphi_{m-1}(k) = \bar{\varphi}_{n-1}(i)^{-1}\bar{\varphi}_{m-1}(i)\delta_n(m), \end{aligned}$$

which proves $A_i^{-1}A_i = A_iA_i^{-1} = I_N$.

Thus we can verify by the orthonormality of Φ that

$$T_iT_i^{-1} = A_i\Phi^*\Phi A_i^{-1} = I_N = \Phi A_i^{-1}A_i\Phi^* = T_i^{-1}T_i.$$

□

Since the invertibility of the graph translation operators depends entirely on when and where eigenvectors vanish, Section 3 is devoted to studying the support of graph eigenvectors.

REMARK 2.7. The results of Theorem 2.5 and its corollary are not applicable solely to the graph translation operators. They can be generalized to a broader class of operators on graphs, in particular, operators that act as Fourier operators. We say that an operator A that acts on a function f is a Fourier multiplier if the Fourier transform of Af can be written as the product

$$\widehat{Af}(\xi) = \hat{a}(\xi)\hat{f}(\xi)$$

for some function a defined in the spectral domain.

Indeed graph translation is defined as a Fourier multiplier since it is defined as

$$\widehat{T_i f}(\lambda_k) = \varphi_k(i)\hat{f}(\lambda_k).$$

Hence, Theorem 2.5 and Corollary 2.6 can be generalized to Fourier multipliers in the following way

COROLLARY 2.8. Let A be a Fourier multiplier whose action on $f : V \to \mathbb{C}$ is defined in the spectral domain $\widehat{Af}(\lambda_k) = \hat{a}(\lambda_k)\hat{f}(\lambda_k)$. Then A is invertible if and only if $\hat{a}(\lambda_k) \neq 0$ for all $\lambda = 0, 1, ..., N-1$. Furthermore, its inverse A^{-1} will be given by the Fourier multiplier

$$\widehat{A^{-1}f}(\lambda_k) = \hat{a}(\lambda_k)^{-1}\hat{f}(\lambda_k).$$

3. Support of Laplacian Fiedler vectors on graphs

This section proves results about the support of Laplacian eigenvectors on graphs. In particular, we characterize and describe the set on which eigenvectors vanish. The Fiedler vector, φ_1, has unique properties that enable us to prove our main result, Theorem 3.9, that planar graphs cannot have large regions on which φ_1 vanishes. We then construct a family of (non-planar) graphs, called the barren graphs, and prove in Theorem 3.12 that their Fiedler vectors do vanish on large regions. As seen in Theorem 2.5, the support of eigenvectors will influence the behavior of the graph translation operators defined in the last section.

3.1. The characteristic set of the Fiedler vector. Let φ_1 denote a Fiedler vector for L on G. We can decompose the vertex set, V, into three disjoint subsets, $V = V_+ \cup V_- \cup V_0$, where $V_+ = \{x \in V : \varphi_1(x) > 0\}$, $V_- = \{x \in V : \varphi_1(x) < 0\}$, and $V_0 = \{x \in V : \varphi_1(x) = 0\}$. The set V_0, the set of vertices on which the Fiedler vector vanishes, is referred to in literature as the *characteristic set* of the graph [**2**]. This vertex decomposition is not a unique property to the graph G; any graph can allow multiple such decompositions of the vertex set V. In the case that the algebraic connectivity has higher multiplicities, i.e., $\lambda_1 = \lambda_2 = \cdots = \lambda_m$ for some $2 \leq m \leq N-1$, then each φ_s is a Fiedler vector for $1 \leq s \leq m$. Futhermore, any linear combination of $\{\varphi_s\}_{s=1}^m$ will also be a Fiedler vector and yield a different vertex decomposition. Even in the case when the algebraic connectivity of G is simple, then $-\varphi_1$ is also a Fiedler vector for G. In this case, V_+ and V_- can be interchanged but the set V_0 is unique to G.

We wish to describe and characterize the sets V_+, V_-, and V_0 for graphs. Fiedler proved in [**14**] that the subgraph induced on the vertices $\{v \in V : \varphi_1(v) \geq 0\} = V_+ \cup V_0$ forms a connected subgraph of G. Similarly, $V_- \cup V_0$ form a connect subgraph of G. Recently, it was proved in [**26**] that we can relax the statement and show that the subgraphs on V_+ and V_- are connected subgraphs of G.

The following result guarantees that V_+ and V_- are always close in terms of the shortest path graph distance.

LEMMA 3.1. *Let $G(V, E)$ with Fiedler vector φ_1 inducing the partition of vertices $V = V_+ \cup V_- \cup V_0$. Then $d(V_+, V_-) \leq 2$.*

PROOF. First consider the case in which $V_0 = \emptyset$. In this case, there necessarily exists an edge $e = (x, y)$ with $x \in V_+$ and $y \in V_-$ and hence $d(V_+, V_-) = 1$.

Now consider the case in which $V_0 \neq \emptyset$. Since G is connected we are guaranteed the existence of some $x \in V_0$ and some $y \sim x$ with either $y \in V_+$ or $y \in V_-$. Since $x \in V_0$, we have

$$0 = \lambda_1 \varphi_1(x) = L\varphi_1(x) = \sum_{z \sim x} \varphi_1(z) - \varphi_1(x) = \sum_{z \sim x} \varphi_1(z). \tag{3.1}$$

We have established that at least one neighbor of x, namely y, satisfied $\varphi_1(y) \neq 0$. Then by (3.1), there must exist at least one other neighbor of x, call it y', such that $\varphi_1(y')$ has the opposite sign of $\varphi_1(y)$. Hence we have now constructed a path, namely (y, x, y') connecting V_+ and V_- and the lemma is proved. □

Many graphs exhibit the property that eigenvectors φ_k for large values of k are highly localized and vanish on large regions of the graph; see [**16**] for an experimental excursion on this phenomenon. It is perhaps a misconception that eigenvectors corresponding to small eigenvalues, or in particular, the Fiedler vector of graphs have full support. Indeed the Fiedler vector of the Minnesota graph never achieves value zero. On the other hand the Fiedler vector of the the graph approximations to the Sierpeinski gasket SG_n can vanish but only along the small number of vertices symmetrically in the center of the graph.

It was shown in [**2**], that the cardinality of V_0 can be arbitrarily large. Figure 1 shows a family of graphs that yield sets V_0 with arbitrarily large cardinality. The family is a path graph P_N on an odd number of vertices, except the middle vertex and its edges are duplicated an arbitrarily large number of times. As evident from Figure 1, the set V_0 is not connected; in fact, no vertex in V_0 is connected to any other vertex of V_0.

For the sake of thoroughness we introduce a family of graphs also with arbitrarily large V_0 but that is also connected. We call this family of graphs the *generalized ladder graphs*, denoted $\mathrm{Ladder}(n, m)$. The standard ladder graphs, $\mathrm{Ladder}(n, 2)$, is simply the graph Cartesian product, see [**17**], of the path graph of length n, P_n, and the path graph of length 1, P_1. The graph $\mathrm{Ladder}(n, 2)$ resembles a ladder with n rungs. The generalized ladder graphs, $\mathrm{Ladder}(n, m)$, are ladders with n rungs and each rung contains m vertices. Provided that the number of rungs, n, is odd, then V_0 will be the middle rung and will clearly be connected. This gives $|V_0| = m$. Figure 2 shows a generalized ladder graph and its Fiedler vector.

The generalized ladder graph provides an example of a graph with a connected characteristic set. Observe in Figure 2 however, that each vertex in V_0 is connected to at most two vertices. We then pose the question as to whether or not there exist graphs for which a vertex in V_0 has three or more neighbors all contained in V_0. It is simpler to state this property using the definition of a graph ball, given in Definition 1.1.

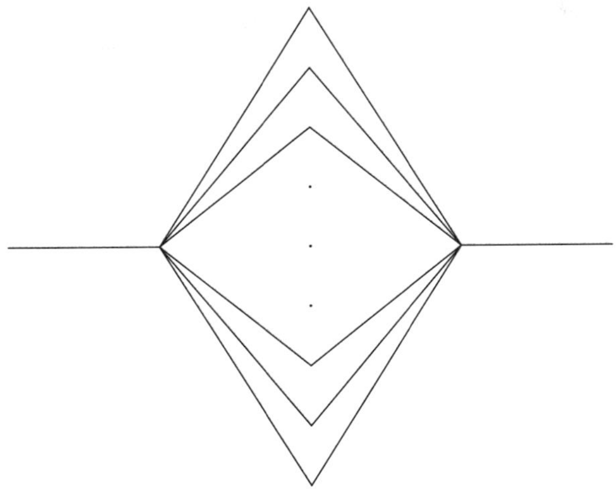

FIGURE 1. A graph with arbitrarily large set $V_0(\varphi)$

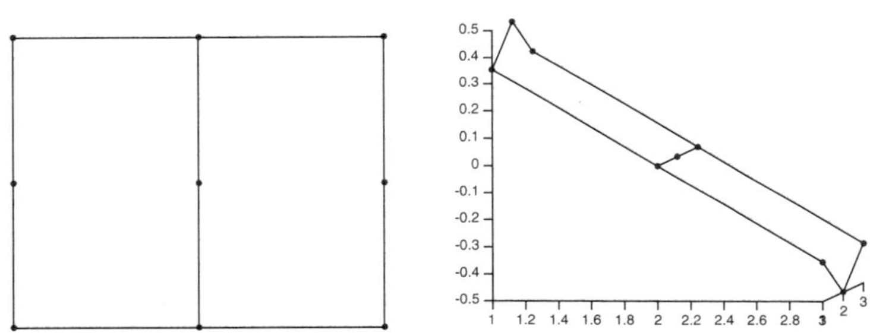

FIGURE 2. Left: The generalized ladder graph, Ladder(3, 3). Right: A 3-dimensional plot of the Fideler vector φ_1 on Ladder(3, 3). The set V_0 consists of the three vertices making the middle rung of the ladder and contains one ball of three vertices.

The following proposition shows that the Fiedler vector cannot be constant-valued on any balls within V_+ and V_-.

PROPOSITION 3.2. Let φ_1 be the Fiedler vector for the Laplacian on graph G and suppose $B_r(x) \subseteq V_+$ or $B_r(x) \subseteq V_-$ for $r \geq 1$. Then φ_1 cannot be constant-valued on $B_r(x)$.

PROOF. It suffices to prove the claim for $r = 1$. Without loss of generality, assume $B_1(x) \subseteq V_+$ and suppose that φ_1 is constant on $B_1(x)$. Then

$$L\varphi_1(x) = \sum_{y \sim x} (\varphi_1(x) - \varphi_1(y)) = 0,$$

since $y \sim x$ implies $y \in B_1(x)$ and φ_1 is constant on that ball. However, $L\varphi_1(x) = \lambda_1\varphi_1(x) > 0$ since $\lambda_1 > 0$ and $\varphi_1(x) > 0$ on V_+. This is a contradiction and the proof is complete. □

The result of Proposition 3.2 can be formulated in terms on *any* non-constant eigenvector of the Laplacian, not just a Fiedler vector.

COROLLARY 3.3. Any non-constant eigenvector of the Laplacian, φ_k, associated with eigenvalue $\lambda_k > 0$ cannot be constant on any ball contained in the positive vertices $\{i \in V : \varphi_k(i) > 0\}$ or negative vertices $\{i \in V : \varphi_k(i) < 0\}$ associated to that eigenvector.

PROOF. Suppose there existed a ball $B_1(x) \subseteq \{i \in V : \varphi_k(i) > 0\}$ on which φ_k was constant. Then just as in the previous proof we could calculate

$$L\varphi_k(x) = \sum_{y \sim x}(\varphi_k(x) - \varphi_k(y)) = 0,$$

which contradicts $L\varphi_k(x) = \lambda_k\varphi_k(x) > 0$. □

Let us remark that Proposition 3.2 and its corollary also hold for the Laplacian of a weighted graph. See [**7**] for the definition and theory of the Laplacian of a weighted graph.

We wish to extend Proposition 3.2 to the set V_0. However, as seen in generalized ladder graphs, Ladder(n, m) for n odd and $m > 2$, for which V_0 contains a ball of radius 1. This ball, however, contains 3 vertices (the center vertex and its two neighbors). The next goal is to characterize graphs whose characteristic set V_0 contains a ball of radius 1 containing at least four vertices. We prove that this is impossible for planar graphs.

DEFINITION 3.4. A *planar graph* is a graph whose vertices and edges can be embedded in $\mathbb{R}^2$ with edges intersecting only at vertices.

In 1930, Kazimierz Kuratowski characterized all planar graphs in terms of subdivisions.

DEFINITION 3.5. A *subdivision* of a graph $G(V, E)$, also referred to as an *expansion*, is the graph $H(\tilde{V}, \tilde{E})$ where the vertex set is the original vertex set with an added vertex, w, and the edge set replaces an edge (u, v) with the two edges (u, w) and (w, v). That is, $\tilde{V} = V \cup \{w\}$ and $\tilde{E} = E \setminus \{(u, v)\} \cup \{(u, w), (w, v)\}$.

THEOREM 3.6 (Kuratowski's Theorem, [**19**]). *A finite graph, G, is planar if and only if it does not contain a subgraph that is a subdivision of K_5 or $K_{3,3}$, where K_5 is the complete graph on 5 vertices and $K_{3,3}$ is the complete bipartite graph on six vertices (also known as the utility graph), see Figure* 3.

A equivalent formulation of Kuratowski's Theorem can be stated in terms of graph minors.

DEFINITION 3.7. Given an undirected graph $G(V, E)$, consider edge $e = (u, v) \in E$. *Contracting* the edge e entails deleting edge e and identifying u and v as the same vertex. The resulting graph $H(\tilde{V}, \tilde{E})$ has one fewer edge and vertex as G.

An undirected graph is called a *minor* of G if it can be formed by contracting edges of G, deleting edges, and deleting isolated vertices (vertices not contained in any edges).

THEOREM 3.8 (Wagner's Theorem, [**27**]). *A finite graph is planar if and only if it does not have K_5 or $K_{3,3}$ as a minor.*

Because of the importance of K_5 and $K_{3,3}$ in identifying non-planar graphs, there are referred to as *forbidden minors.*

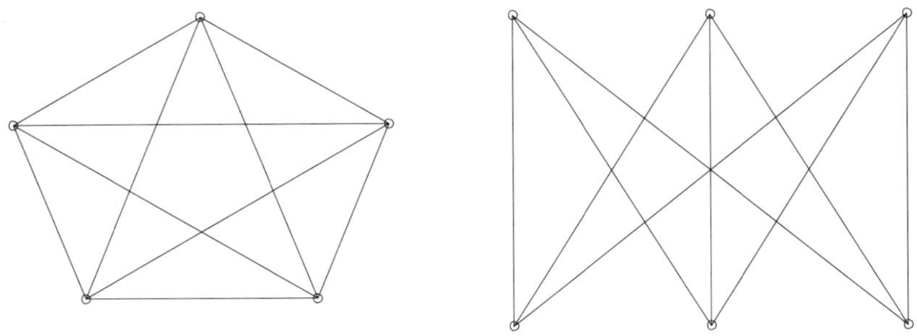

FIGURE 3. The forbidden minors. Left: The complete graph on five vertices. Right: The complete bipartite graph on six vertices.

One of the main results in this section shows that planar graphs cannot have large balls contained in the characteristic set V_0.

THEOREM 3.9. *Let $G(V,E)$ be a planar graph with Fiedler vector φ_1. Then the zero set of φ_1 contains no balls of radius $r = 1$ with more than three vertices.*

PROOF. Suppose that V_0 contains a ball, $B_1(x)$, centered at vertex $x \in V_0$ and comprised of at least four vertices. Without loss of generality, we can assume that the connected component of V_0 containing x equals $B_1(x)$. If not, then contract edges so that the connected component of V_0 containing x equals a ball of radius 1. Since $|B_1(x)| \geq 4$, then we have $d_x \geq 3$ and let $\{y_i\}_{i=1}^{d_x}$ denote the neighbors of x. Then as constructed, $B_1(x) = \{x, y_1, y_2, ..., y_{d_x}\}$.

By Lemma 3.1, for $i = 1, 2, 3$, each vertex y_i has at least one neighbor in V_+ and at least one in V_-; pick one neighbor from each set and denote them p_i and n_i, respectively. It is proved in [**26**] that V_+ and V_- are connected subgraphs of G. Therefore, there is a path of edges that connect p_1, p_2, and p_3 (if $p_1 = p_2 = p_3$, then this path is empty). We create a minor of G by contracting the path connecting p_1, p_2, and p_3 to create one vertex $p \in V_+$. Similarly, since V_- is connected, we can contract the path connecting n_1, n_2, and n_3, to create one vertex $n \in V_-$.

Consider the subgraph of the now minorized version of G consisting of vertices $\{x, p, n, y_1, y_2, y_3\}$. This subgraph is $K_{3,3}$, the complete bipartite graph on six vertices since the vertices $\{x, p, n\}$ are all connected to $\{y_1, y_2, y_3\}$. Thus by Wagner's Theorem, G is not a planar graph, which is a contradiction. This completes the proof. □

The result of Theorem 3.9 does not hold for general graphs. We construct a family of (nonplanar) graphs for which V_0 contains a ball with a large number of vertices. Since the set of vertices for which the Fiedler vector vanishes is large, we call this family of graphs the *barren graphs*. The barren graph with $|V| = N + 7$ and $|V_0| = N + 1$ is denoted Barr(N).

3.2. Construction of the barren graph, Barr(N). The barren graph will be constructed as a sum of smaller graphs.

DEFINITION 3.10. Let $G_1(V, E_1)$ and $G_2(V, E_2)$ be two graphs. The *sum* of graphs G_1 and G_2, denoted $G_1 + G_2$, is the graph $G(V, E)$ where $E = E_1 \cup E_2$.

The barren graph Barr(N) is defined as follows

DEFINITION 3.11. Let $K(V_i, V_j)$ denote the bipartite complete graph between vertex sets V_i and V_j, that is, the graph with vertex set $V = V_i \cup V_j$ and edge set $E = \{(x, y) : x \in V_i, y \in V_j\}$. For $N \geq 3$ the barren graph, Barr(N), is a graph with $N+7$ vertices. Let $\{V_i\}_{i=1}^6$ denote distinct vertex sets with given cardinalities $\{|V_i|\}_{i=1}^6 = \{N, 1, 2, 2, 1, 1\}$. Then the barren graph is the following graph sum of the 5 complete bipartite graphs

$$\mathrm{Barr}(N) = K(V_1, V_2) + K(V_1, V_3) + K(V_1, V_4) + K(V_3, V_5) + K(V_4, V_6).$$

As constructed, Barr(N) itself is bipartite; all edges connect the sets $V_2 \cup V_3 \cup V_4$ to $V_1 \cup V_5 \cup V_6$. Figure 4 shows two examples of barren graphs.

We shall show that for any N, the Fiedler vector for Barr(N) vanishes on $V_1 \cup V_2$ which has cardinality $N + 1$. Hence, the Fiedler vector for Barr(N) has support on exactly six vertices for any $N \geq 3$. In order to prove this, we explicitly derive the entire spectrum and all eigenvectors of the Laplacian.

THEOREM 3.12. *The barren graph, Barr(N), has the spectrum given in Table 1. In particular, the Fiedler vector of Barr(N) vanishes on vertices $V_1 \cup V_2$ and hence $|\mathrm{supp}(\varphi_1)| = 6$ for any N.*

λ_k	value	eigenvector
λ_0	0	constant function
λ_1	$\frac{1}{2}\left(N + 3 - \sqrt{N^2 - 2N + 9}\right)$	Figure 5
λ_2	y_1	Figure 6
$\lambda_3 = \cdots = \lambda_{N+1}$	5	ON basis on V_1
λ_{N+2}	y_2	Figure 6
$\lambda_{N+3} = \lambda_{N+4}$	$N + 1$	Figure 7
λ_{N+5}	$\frac{1}{2}\left(N + 3 + \sqrt{N^2 - 2N + 9}\right)$	Figure 5
λ_{N+6}	y_3	Figure 6

TABLE 1. The spectrum of the barren graph, Barr(N). The values y_1, y_2, y_3 are the roots to the cubic polynomial (3.2).

PROOF. Firstly, the graph Barr(N) is connected and so we have $\lambda_0 = 0$ with $\varphi_0 \equiv (N + 7)^{-1/2}$. All other eigenvalues must be positive.

We will next show that the structure and support of the function shown in Figure 5 is an eigenvector for two eigenvalues of Barr(N). One can check upon inspection that the shown function φ is orthogonal to the constant function. Then, if the function shown in Figure 5, call it φ, is an eigenvector, then the eigenvalue equation, $Lx = \lambda x$ is satisfied at each vertex. It is easy to verify that $L\varphi(x) = 0$ for each $x \in V_1 \cup V_2$. For $x \in V_5$ or $x \in V_6$ the eigenvalue equation becomes $L\varphi(x) = 2(b - a) = \lambda b$. For any $x \in V_3$ or $x \in V_4$, the eigenvalue equation gives

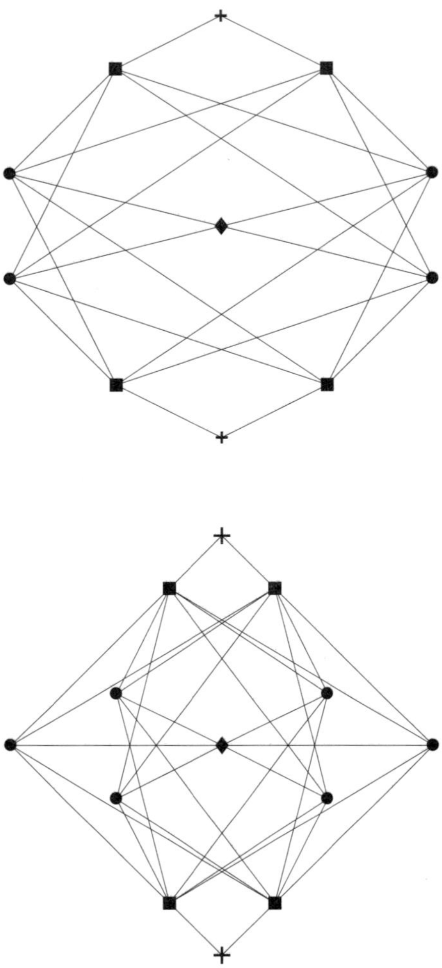

FIGURE 4. The barren graph Barr(4) (top) and Barr(6) (bottom). The set V_1 is denoted with N dots; the vertex set V_2 is the diamond vertex in the center; the sets V_3 and V_4 are denoted with squares; the sets V_5 and V_6 are denoted with crosses.

$L\varphi(x) = Na + (a - b) = \lambda a$. Finally, we also impose that the condition that the eigenvectors are normalized so that $\|\varphi\| = 1$. Therefore, the function φ shown in Figure 5 is an eigenvector of L if and only if the following system of equations has a nontrivial solution:

$$\left\{ \begin{array}{rcl} 4a^2 + 2b^2 & = & 1 \\ 2(b-a) & = & \lambda b \\ Na + (a-b) & = & \lambda a. \end{array} \right.$$

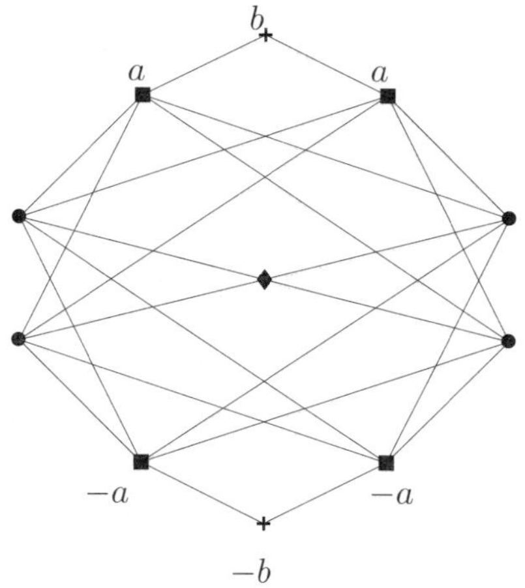

FIGURE 5. Support and function values for the eigenvectors associated with eigenvalues λ_1 and λ_{N+5}.

The first equation is not linear, but we can still solve this system by hand with substitution to obtain the following two solutions:

$$\begin{cases} a &= \frac{1}{2}\sqrt{\frac{N^2-2N+9\mp(N-1)\sqrt{N^2-2N+9}}{2(N^2-2N+9)}} \\ b &= \frac{1}{2}\sqrt{\frac{N^2-2N+9\pm(N-1)\sqrt{N^2-2N+9}}{N^2-2N+9}} \\ \lambda &= \frac{1}{2}\left(N+3\pm\sqrt{N^2-2N+9}\right). \end{cases}$$

This gives two orthogonal eigenvectors and their eigenvalues.

Consider now the vector shown in Figure 6 with full support, yet only taking on four distinct values. Similar to the previous example, we obtain a system of

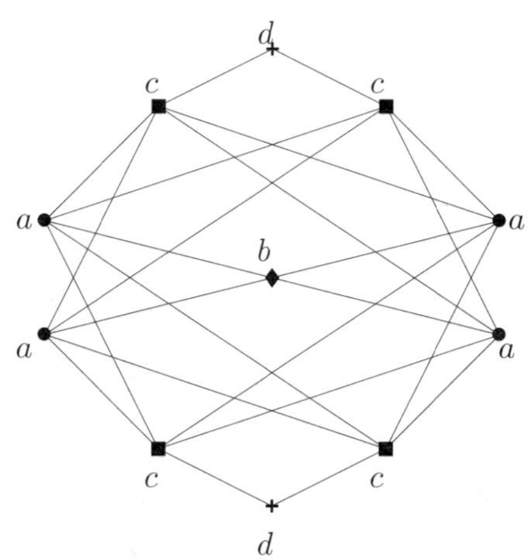

FIGURE 6. Support and function values for the eigenvectors associated with eigenvalues λ_2, λ_{N+2}, and λ_{N+6}.

equations by imposing the conditions $\|\varphi\| = 1$, $\langle \varphi, 1\rangle = 0$, and from writing out the eigenvalue equations at each vertex class from V_1, V_2, V_3 and V_5 which gives:

$$\begin{cases} Na^2 + b^2 + 4c^2 + 2d^2 &= 1 & (\|\varphi\|^2 = 1) \\ Na + b + 4c + 2d &= 0 & (\varphi \perp 1) \\ 4(a-c) + (a-b) &= \lambda a & (L\varphi(x) = \lambda\varphi(x) : x \in V_1) \\ N(b-1) &= \lambda b & (L\varphi(x) = \lambda\varphi(x) : x \in V_2) \\ (c-d) + N(c-a) &= \lambda c & (L\varphi(x) = \lambda\varphi(x) : x \in V_3 \cup V_4) \\ 2(d-c) &= \lambda d & (L\varphi(x) = \lambda\varphi(x) : x \in V_5 \cup V_6) \end{cases}.$$

Again, this system cannot be solved with linear methods. However, by tedious substitutions we can reduce the system (assuming each of the variables a, b, c, d, λ are nonzero) to solving for the roots of the following cubic polynomial in λ:

$$\lambda^3 + (-2N - 8)\lambda^2 + (N^2 + 10N + 15)\lambda + (-2N^2 - 14N) = 0 \tag{3.2}$$

The cubic polynomial $x^3 + c_2x^2 + c_1x + c_0 = 0$ has three distinct real roots if its discriminant, $\Delta = 18c_0c_1c_2 - 4c_2^3c_0 + c_2^2c_1^2 - 4c_1^3 - 27c_0^2$, is positive. The discriminant of (3.2) is positive for all $N > 0$ and hence we let $y_1 < y_2 < y_3$ denote the three positive roots which make up λ_2, λ_{N+2}, and λ_{N+6}, respectively. By substituting back into the system of equations, one can obtain values for a, b, c, d for each of the $\lambda = y_1, y_2, y_3$.

The roots y_1, y_2, y_3 monotonically increase in N. A simple calculation shows that $y_1 = 2$ for $N = 3$ and $y_1 > 2$ for $N > 3$. Hence $\lambda_1 < \lambda_2 = y_1$ for all N. Also observe that $y_2 < 5$ for $N < 5$, so the ordering of the eigenvalues in Table 1 can vary but their values are accurate.

One can verify that the three eigenvectors obtained from Figure 6 are linearly independent and orthogonal to each eigenvector derived so far.

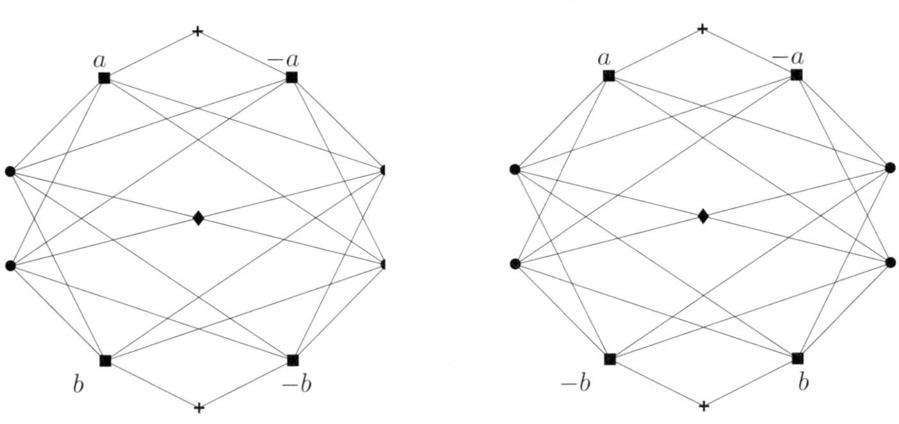

FIGURE 7. Support and function values for the eigenvectors associated with eigenvalues λ_{N+3} and λ_{N+4}.

Consider now the two functions shown in Figure 7. The eigenvalue equation gives $L\varphi(x) = 0$ for every except for those $x \in V_3 \cup V_4$ in which case we have $L\varphi(x) = (N+1)\varphi(x)$. The two functions shown in Figure 7 are orthogonal and linearly independent to each other and every eigenvector derived thus far and hence $N+1$ is an eigenvalue of Barr(N) with multiplicity two.

Finally, we will construct eigenfunctions that are supported only on the N vertices in V_1. Observe that if a function, f, is supported on V_1 then for any $x \in V_1$, the eigenvalue equation gives $Lf(x) = 5f(x)$ since x neighbors five vertices on which f vanishes. Therefore, Barr(N) has eigenvalue 5. To construct the corresponding eigenbasis, one can choose any orthonormal basis for the subspace of the N-dimensional vector space that is orthogonal to the constant vector. Any basis for this $(N-1)$-dimensional vector space will give an eigenbasis for on V_1. Finally one can verify by inspection that these $N-1$ eigenvectors are orthogonal and linearly independent to each eigenvector derived in this proof.

As such, we have now constructed an orthonormal, linearly independent eigenbasis for Barr(N) corresponding to the eigenvalues given in Table 1 □

As a remark, observe the behavior of the spectrum of Barr(N) as $N \to \infty$. For every natural number N, $\lambda_1 < 2$ and $\lim_{N\to\infty} \lambda_1 = 2$. Using a symbolic solver, one can prove that $\lim_{N\to\infty} \lambda_2 = \lim_{N\to\infty} y_1 = 2$ as well. Furthermore, the other two roots of the polynomial (3.2) tend to infinity as $N \to \infty$. Therefore, as $N \to \infty$, Barr(N) has spectrum approaching 0 (with multiplicity 1), 2 (with multiplicity 2), 5 (with multiplicity $N-2$), and the rest of the eigenvalues tending to ∞.

3.3. Characteristic vertices and special graph sums. In this subsection, we prove results about eigenvectors and their characteristic vertices for graph sums as defined in Definition 3.10. We borrow the following notation from [**26**] for clarity.

DEFINITION 3.13. For any function f, we make the following definitions,

$$i_0(f) = \{i \in V : f(i) = 0\},$$

$$i_+(f) = \{i \in V : f(i) > 0\},$$

$$i_-(f) = \{i \in V : f(i) < 0\}.$$

Observe that the set V_0 (resp. V_+ and V_-) from Section 3.1 is equal to $i_0(\varphi_1)$ (resp. $i_+(\varphi_1)$ and $i_-(\varphi_1)$).

THEOREM 3.14. *Consider $n \geq 2$ connected graphs, $\{G_j(V_j, E_j)\}_{j=1}^n$ and suppose that all n graph Laplacians, L_j, share a common eigenvalue $\lambda > 0$ with corresponding eigenvectors $\varphi_{(j)}$. Each graph's vertex set, V_j, assumes a decomposition $V_j = i_+(\varphi_{(j)}) \cup i_-(\varphi_{(j)}) \cup i_0(\varphi_{(j)})$ and suppose that $i_0(\varphi_{(j)}) \neq \emptyset$ for all j. Consider the graph $G(V,E) = G(\cup_{j=1}^n V_j, \cup_{j=1}^n E_j \cup E_0)$ where the edge set $E_0 = \{(x_i, y_i)\}_{i=1}^K$ for $x_i \in i_0(\varphi_j)$, $y_i \in i_0(\varphi_\ell)$, and $j \neq \ell$ is nonempty. Define φ on G by $\varphi(x) = \varphi_{(j)}(x)$ for $x \in V_j$. Then, λ is an eigenvalue of G and φ is a corresponding eigenvector.*

Furthermore, if we add the assumption that the common eigenvalue $\lambda > 0$ is the algebraic connectivity, i.e., the lowest nonzero eigenvalue of the graphs G_j, then λ is an eigenvalue of $G(V,E) = G(\cup_{j=1}^n V_j, \cup_{j=1}^n E_j \cup E_0)$ but not the smallest positive eigenvalue. Hence, $\varphi(x)$ is an eigenvector of G but not its Fiedler vector.

PROOF. We will verify that $L\varphi(x) = \lambda\varphi(x)$ for every $x \in V$. Every $x \in V$ lies in exactly one V_j and every edge connecting to x must be in either E_j or E_0. Suppose x contains no edges from E_0. Then $L\varphi(x) = L_j\varphi_{(j)}(x) = \lambda\varphi_{(j)}(x) = \lambda\varphi(x)$.

Suppose instead that x does contain at least one edge from E_0. Then by construction of the set E_0, we have $\varphi(x) = 0$ and $\varphi(y) = 0$ for all $(x, y) \in E_0$. This allows us to compute

$$\begin{aligned} L\varphi(x) = \sum_{y\sim x} (\varphi(x) - \varphi(y)) &= \sum_{\substack{y\sim x \\ (x,y)\in E_j}} (\varphi(x) - \varphi(y)) + \sum_{\substack{y\sim x \\ (x,y)\in E_0}} (\varphi(x) - \varphi(y)) \\ &= L_j\varphi_{(j)}(x) + 0 = \lambda\varphi_{(j)}(x) = \lambda\varphi(x). \end{aligned}$$

Hence for any vertex in V, the vector φ satisfies the eigenvalue equation and the first part of the proof is complete.

For the second claim of the theorem, let $G_1(V_1, E_2)$ and $G_2(V_2, E_2)$ have equal algebraic connectivities and Fiedler vectors $\varphi_{(1)}$ and $\varphi_{(2)}$, respectively. We can decompose each vertex set into it's positive, negative, and zero sets, i.e., $V_j = i_+(\varphi_{(j)}) \cup i_-(\varphi_{(j)}) \cup i_0(\varphi_{(j)})$. Furthermore, $i_+(\varphi_{(j)})$ and $i_-(\varphi_{(j)})$ are connected subgraphs of G_j.

Now consider the larger graph $G(V, E)$. The function $\varphi(x) := \varphi_{(j)}(x)$ for $x \in V_j$ is an eigenfunction of G by the first part of the theorem. However, now, the sets $i_+(\varphi)$ and $i_-(\varphi)$ are disconnected. Indeed, let $x \in i_+(\varphi_{(1)})$ and $y \in i_+(\varphi_{(2)})$. Then any path connecting x and y must contain an edge in E_0 since all E_0 contains all edges connecting G_1 to G_2. And hence any path connect x to y will contain at least two vertices in $i_0(\varphi)$.

Then by **[26]** since $i_+(\varphi)$ and $i_-(\varphi)$ are both disconnected, then φ cannot be the Fiedler vector of G and λ is not the smallest nonzero eigenvalue. □

We can prove a stronger statement in the specific case where the graphs share algebraic connectivity, λ_1.

We can state a generalization of Theorem 3.14 for eigenvectors supported on subgraphs of G.

THEOREM 3.15. *Consider the graph $G(V, E)$. Let $S \subseteq V$ and let $H(S, E_S)$ be the resulting subgraph defined by just the vertices of S. Suppose that $\varphi^{(S)}$ is an eigenvector of L_S, the Laplacian of subgraph H, with corresponding eigenvalue λ. If $E(S, V \setminus S) = E(i_0(\varphi^{(S)}), V \setminus S)$, that is, if all edges connecting graph H to its complement have a vertex in the zero-set of $\varphi^{(S)}$, then λ is an eigenvalue of G with eigenvector*

$$\varphi(x) = \begin{cases} \varphi^{(S)}(x) & x \in S \\ 0 & x \notin S. \end{cases}$$

PROOF. The proof is similar to the proof of Theorem 3.14 in that we will simply verify that $L\varphi(x) = \lambda\varphi(x)$ at every point $x \in V$. For any x in the interior of S, then $L\varphi(x) = L_S\varphi^{(S)}(x) = \lambda\varphi^{(S)}(x) = \lambda\varphi(x)$. For any x in the interior of $(V \setminus S)$, then $L\varphi(x) = 0$ since φ vanishes at x and all of its neighbors. For $x \in \delta(S)$ (recall $\delta(S) = \{x \in S : (x, y) \in E$ and $y \in V \setminus S\}$), we have

$$\begin{aligned} L\varphi(x) = \sum_{y\sim x} (\varphi(x) - \varphi(y)) &= \sum_{\substack{y\sim x \\ y\in S}} (\varphi(x) - \varphi(y)) + \sum_{\substack{y\sim x \\ y\notin S}} (\varphi(x) - \varphi(y)) \\ &= L_S\varphi^{(S)}(x) + (0 - 0) = \lambda\varphi^{(S)}(x) = \lambda\varphi(x), \end{aligned}$$

where the term $(0 - 0)$ arises from the fact that $\varphi(y) = 0$ since $y \notin S$ and since $(x, y) \in E$ then by assumption $x \in i_0(\varphi^{(S)})$ and hence $\varphi(x) = 0$. The same logic shows that $L\varphi(x) = 0$ for $x \in \delta(V \setminus S)$. Hence, we have shown that $L\varphi(x) = \lambda\varphi(x)$ for every possible vertex $x \in V$ and the proof is complete. □

Theorem 3.15 is interesting because it allows us to obtain eigenvalues and eigenvectors of graphs by inspecting for certain subgraphs. Furthermore since the eigenvector is supported on the subgraph, it is sparse and has a large nodal set.

EXAMPLE 3.16. Consider the star graph $S_N(V, E)$ which is complete bipartite graph between N vertices in one class (V_A) and 1 vertex in the other (V_B). Let S be the subgraph formed by any two vertices in V_A and the one vertex in V_B. Then the resulting subgraph on S is the path graph on 3 vertices, P_3. It is known that P_3 has Fiedler vector $\varphi^{(S)} = (\sqrt{2}, 0, -\sqrt{2})$ and eigenvalue $\lambda = 1$. Then by Theorem 3.15, the star graph S_N has eigenvalue $\lambda = 1$ with eigenvector supported on two vertices. In fact, S_N contains exactly $\binom{N}{2}$ path subgraphs all of which contain the center vertex and have $\varphi^{(S)}$ as an eigenvector. However, only $N-1$ of them will be linearly independent. This method of recognizing subgraphs explains why S_N has eigenvalue 1 with multiplicity $N-1$ and we have identified a set of basis vectors for that eigenspace.

References

[1] Aamir Anis, Akshay Gadde, and Antonio Ortega, *Towards a sampling theorem for signals on arbitrary graphs*, IEEE International Conference on Acoustics, Speech and Signal Processing (ICASSP), 2014.

[2] R. B. Bapat and Sukanta Pati, *Algebraic connectivity and the characteristic set of a graph*, Linear and Multilinear Algebra **45** (1998), no. 2-3, 247–273, DOI 10.1080/03081089808818590. MR1671627

[3] S. Barik, S. Fallat, and S. Kirkland, *On Hadamard diagonalizable graphs*, Linear Algebra Appl. **435** (2011), no. 8, 1885–1902, DOI 10.1016/j.laa.2011.03.033. MR2810634

[4] Mikhail Belkin and Partha Niyogi, *Laplacian eigenmaps for dimensionality reduction and data representation*, Neural computation **15** (2003), no. 6, 1373–1396.

[5] J. A. Bondy and U. S. R. Murty, *Graph theory*, Graduate Texts in Mathematics, vol. 244, Springer, New York, 2008. MR2368647

[6] Maria Cameron and Eric Vanden-Eijnden, *Flows in complex networks: theory, algorithms, and application to Lennard-Jones cluster rearrangement*, J. Stat. Phys. **156** (2014), no. 3, 427–454, DOI 10.1007/s10955-014-0997-8. MR3217531

[7] Fan R. K. Chung, *Spectral graph theory*, CBMS Regional Conference Series in Mathematics, vol. 92, Published for the Conference Board of the Mathematical Sciences, Washington, DC; by the American Mathematical Society, Providence, RI, 1997. MR1421568

[8] Ronald R. Coifman and Stéphane Lafon, *Diffusion maps*, Appl. Comput. Harmon. Anal. **21** (2006), no. 1, 5–30, DOI 10.1016/j.acha.2006.04.006. MR2238665

[9] Wojciech Czaja and Martin Ehler, *Schroedinger eigenmaps for the analysis of bio-medical data*, IEEE Trans Pattern Anal Mach Intell. **35** (2013), no. 5, 1274–80.

[10] Shaun M. Fallat, Stephen J. Kirkland, Jason J. Molitierno, and M. Neumann, *On graphs whose Laplacian matrices have distinct integer eigenvalues*, J. Graph Theory **50** (2005), no. 2, 162–174, DOI 10.1002/jgt.20102. MR2165035

[11] Pedro F Felzenszwalb and Daniel P Huttenlocher, *Efficient graph-based image segmentation*, International journal of computer vision **59** (2004), no. 2, 167–181.

[12] Pedro F. Felzenszwalb and Daniel P. Huttenlocher, *Efficient graph-based image segmentation*, International Journal of Computer Vision **59** (2004), no. 2, 167–181.

[13] Miroslav Fiedler, *Algebraic connectivity of graphs*, Czechoslovak Math. J. **23(98)** (1973), 298–305. MR0318007

[14] Miroslav Fiedler, *A property of eigenvectors of nonnegative symmetric matrices and its application to graph theory*, Czechoslovak Mathematical Journal **25** (1975), no. 4, 619–633.

[15] David K. Hammond, Pierre Vandergheynst, and Rémi Gribonval, *Wavelets on graphs via spectral graph theory*, Appl. Comput. Harmon. Anal. **30** (2011), no. 2, 129–150, DOI 10.1016/j.acha.2010.04.005. MR2754772
[16] Steven M. Heilman and Robert S. Strichartz, *Localized eigenfunctions: here you see them, there you don't*, Notices Amer. Math. Soc. **57** (2010), no. 5, 624–629. MR2664041
[17] Wilfried Imrich, Sandi Klavžar, and Douglas F. Rall, *Topics in graph theory*, A K Peters, Ltd., Wellesley, MA, 2008. Graphs and their Cartesian product. MR2468851
[18] David Kempe, Jon Kleinberg, and Éva Tardos, *Maximizing the spread of influence through a social network*, Proceedings of the ninth ACM SIGKDD international conference on Knowledge discovery and data mining, ACM, 2003, pp. 137–146.
[19] Casimir Kuratowski, *Sur le problème des courbes gauches en topologie*, Fundamenta mathematicae **1** (1930), no. 15, 271–283.
[20] Russell Merris, *Laplacian matrices of graphs: a survey*, Linear Algebra Appl. **197/198** (1994), 143–176, DOI 10.1016/0024-3795(94)90486-3. Second Conference of the International Linear Algebra Society (ILAS) (Lisbon, 1992). MR1275613
[21] M. E. J. Newman, *The structure and function of complex networks*, SIAM Rev. **45** (2003), no. 2, 167–256, DOI 10.1137/S003614450342480. MR2010377
[22] Lawrence Page, Sergey Brin, Rajeev Motwani, and Terry Winograd, *The pagerank citation ranking: Bringing order to the web.*, Tech. report, Stanford InfoLab, 1999.
[23] David I. Shuman, Sunil K. Narang, Pascal Frossard, Antonio Ortega, and Pierre Vandergheynst, *The emerging field of signal processing on graphs: Extending high-dimensional data analysis to networks and other irregular domains*, IEEE Signal Processing Magazine **30** (2013), no. 3, 83–98.
[24] David I. Shuman, Benjamin Ricaud, and Pierre Vandergheynst, *Vertex-frequency analysis on graphs*, Appl. Comput. Harmon. Anal. **40** (2016), no. 2, 260–291, DOI 10.1016/j.acha.2015.02.005. MR3440174
[25] Daniel A. Spielman, *Spectral graph theory*, Lecture Notes, Yale University (2009), 740–776.
[26] John C. Urschel and Ludmil T. Zikatanov, *Spectral bisection of graphs and connectedness*, Linear Algebra Appl. **449** (2014), 1–16, DOI 10.1016/j.laa.2014.02.007. MR3191855
[27] K. Wagner, *Über eine Eigenschaft der ebenen Komplexe* (German), Math. Ann. **114** (1937), no. 1, 570–590, DOI 10.1007/BF01594196. MR1513158
[28] Douglas B. West, *Introduction to graph theory*, Prentice Hall, Inc., Upper Saddle River, NJ, 1996. MR1367739

Department of Mathematics, University of Maryland, College Park, Maryland, 20742
Email address: begue@math.umd.edu

Department of Mathematics, University of Maryland, College Park, Maryland, 20742
Email address: kasso@math.umd.edu

Contemporary Mathematics
Volume **706**, 2018
http://dx.doi.org/10.1090/conm/706/14213

Weighted convolution inequalities and Beurling density

Jean-Pierre Gabardo

ABSTRACT. The upper and lower Beurling density of a positive Borel measure on $\mathbb{R}^d$ are related to the value of the best constants in certain convolution inequalities satisfied by that measure. We consider here extensions of these type of results where the best constants in certain weighted convolution inequalities yield information on "weighted" versions of the upper and lower Beurling density of a measure. The weight involved in characterizing the classical upper and lower Beurling density of a positive Borel measure using convolution inequalities is simply the constant weight $w = 1$. In the weighted version of these convolution inequalities considered in this paper, we assume that the weight w is moderate.

1. Introduction

In problems dealing with (weighted or unweighted) frames or Bessel sequences of exponentials for certain Hilbert spaces of functions or distributions whose norm is defined by a weighted L^2-norm in the Fourier domain (such as the standard Sobolev spaces $H^s(\mathbb{R}^d)$, $s \in \mathbb{R}$, for example) convolution inequalities appear naturally as necessary conditions for the frame or Bessel inequalities to hold. It is therefore of interest to determine under which conditions such convolution inequalities might hold and it turn out that the best constants appearing in some of these inequalities are related to the upper or lower Beurling density of the measure involved. Some results in that direction in the unweighted case were obtained by the author in [**2**]. This approach can also be used to define the notion of upper and lower Beurling density for measures on locally compact groups as shown in [**3**]. The main goal of this paper will be to generalize these results to the weighted case. We will start by giving an overview of the unweighted case, but before doing this, we need to introduce some notations and definitions. If $\xi \in \mathbb{R}^d$ and A, B are subsets of $\mathbb{R}^d$, we define the sets $\xi + A$ and $A + B$ as

$$\xi + A = \{\xi + \lambda,\ \lambda \in A\}, \quad A + B = \{\eta + \lambda,\ \eta \in A,\ \lambda \in B\}.$$

We will denote by $|E|$ the Lebesgue measure of a measurable subset E and by χ_E its characteristic function. We let $I_R = [-R/2, R/2]^d$ be the d-dimensional box of side length $R > 0$ centered at the origin. When we talk about a positive Borel measure on $\mathbb{R}^d$, we always assume implicitly that that such measure is finite when restricted to any compact subset of $\mathbb{R}^d$. If $f \geq 0$ is Borel measurable and μ is a

2010 *Mathematics Subject Classification.* Primary 42C15; Secondary 44A35.
The author was supported by an NSERC grant.

positive Borel measure on $\mathbb{R}^d$, we define their convolution $\mu * f$ to be the measure defined by

$$(\mu * f)(E) = \int_{\mathbb{R}^d\times\mathbb{R}^d} \chi_E(x+y)\, f(x)\, dx\, d\mu(y), \quad E\subset \mathbb{R}^d,\ E \text{ Borel}.$$

It is quite clear from this definition that $\mu * f$ is absolutely continuous with respect to the Lebesgue measure. Note also that this definition can be extended to any Lebesgue measurable function $f\geq 0$ although we will always assume f to be Borel measurable in this paper for simplicity and also since f is in fact continuous or the characteristic function of a Borel set in most applications. Using the Radon-Nikodym theorem, there exists thus a Lebesgue measurable function H with values in $[0,\infty]$ such that

$$(\mu * f)(E) = \int_E H(x)\, dx, \quad E\subset \mathbb{R}^d,\ E \text{ Borel}.$$

In the following we will identify $\mu * f$ with the function H, so when we write an inequality between $\mu * f$ and a certain function G, we have in mind the same inequality between H and G.

The notion of Beurling density plays an central role in many areas of modern Fourier analysis (see [**1, 4, 6, 8**–**13, 15**]) and, in particular, in the study of sampling and interpolation in spaces of band-limited functions as illustrated in the classical pioneering work of H. Landau ([**14**]). If μ is a positive Borel measure on $\mathbb{R}^d$, the quantities

$$\text{(1.1)}\quad \mathcal{D}^+(\mu) = \limsup_{R\to\infty}\sup_{\xi\in\mathbb{R}^d}\frac{\mu\left(\xi+I_R\right)}{R^d} \quad\text{and}\quad \mathcal{D}^-(\mu) = \liminf_{R\to\infty}\inf_{\xi\in\mathbb{R}^d}\frac{\mu\left(\xi+I_R\right)}{R^d}$$

are called the *upper and lower Beurling density* of the measure μ, respectively. If both these densities are equal and finite, we say that the Beurling density of the measure μ exists and we define it to be the quantity $\mathcal{D}(\mu) := \mathcal{D}^+(\mu) = \mathcal{D}^-(\mu)$. If $\Lambda\subset\mathbb{R}^d$ is a discrete set, the corresponding Beurling densities $\mathcal{D}^+(\Lambda)$, $\mathcal{D}^-(\Lambda)$ and $\mathcal{D}(\Lambda)$ are defined as the Beurling densities of the measure $\mu = \sum_{\lambda\in\Lambda}\delta_\lambda$, where δ_λ is the Dirac mass at λ. A positive Borel measure μ on $\mathbb{R}^d$ is called *translation-bounded* if there exists a constant $C\geq 0$ such that

$$\mu\left(\xi+[0,1]^d\right)\leq C, \quad \xi\in\mathbb{R}^d.$$

This is equivalent to the condition $\mathcal{D}^+(\mu)<\infty$ or to the existence of a Borel measurable function with $f\geq 0$ and with $\int_{\mathbb{R}^d} f(\xi)\, d\xi = 1$ such that, for some constant $C\geq 0$, $\mu * f\leq C$ pointwise, as shown in [**2**]. In fact, the relationship between Beurling densities and convolution inequalities can be made very precise in the following sense. Consider the collection

$$\mathcal{P}(\mathbb{R}^d) = \{f :\ f \text{ Borel measurable},\ f\geq 0 \text{ and } \int_{\mathbb{R}^d} f(x)\, dx = 1\}.$$

and, for a fixed positive Borel measure μ on $\mathbb{R}^d$, define the number

$$\mathcal{C}^+(\mu) = \inf\left\{C\geq 0,\ \mu * f\leq C \text{ pointwise for some } f\in\mathcal{P}(\mathbb{R}^d)\right\},$$

with the convention that $\mathcal{C}^+(\mu)=\infty$ if the set where the infimum above is taken happens to be empty. Similarly, let

$$\mathcal{C}^-(\mu) = \sup\left\{D\geq 0,\ \mu * f\geq D \text{ pointwise for some } f\in\mathcal{P}(\mathbb{R}^d)\right\}.$$

THEOREM 1.1 ([2]). *Let μ be a positive Borel measure on $\mathbb{R}^d$. Then,*

(a) $\mathcal{C}^+(\mu) = \mathcal{D}^+(\mu)$.

(b) $\mathcal{C}^-(\mu) \geq \mathcal{D}^-(\mu)$ *and if we assume in addition that μ is translation-bounded, then we have the equality $\mathcal{C}^-(\mu) = \mathcal{D}^-(\mu)$.*

This result was proved in [2] but we will give a somewhat easier proof of it in Section 3. The proof given here has also the advantage that it can be generalized to certain moderate weights as explained below. Theorem 1.1 can be used to estimate the upper and lower Beurling density of a positive measure satisfying certain convolution inequalities as the following corollary shows.

COROLLARY 1.2 ([2]). *Let μ be a positive Borel measure on $\mathbb{R}^d$ and $h \geq 0$ be Borel measurable and integrable on $\mathbb{R}^d$.*

(a) *If there exists a constant $C > 0$ such that the inequality $\mu * h \leq C$ holds pointwise on $\mathbb{R}^d$, then $\mathcal{D}^+(\mu) \int_{\mathbb{R}^d} h(x)\,dx \leq C$.*

(b) *If μ is translation-bounded and there exists a constant $C > 0$ such that the inequality $\mu * h \geq C$ holds pointwise on $\mathbb{R}^d$, then we have $\mathcal{D}^-(\mu) \leq \mathcal{D}^+(\mu) < \infty$ and $\mathcal{D}^-(\mu) \int_{\mathbb{R}^d} h(x)\,dx \geq C$.*

In the following, our goal will be to extend the previous results to situations involving a measure μ and a weight w satisfying an inequality of the form

$$\mu * F \leq C\,(w * F) \quad \text{or, of the form } \mu * F \geq C\,(w * F), \text{ pointwise on } \mathbb{R}^d,$$

where F is a positive function which is "small" in the sense that it is integrable with respect to a certain weight related to w and defined below. We will consider measurable weights $w > 0$ on $\mathbb{R}^d$ with the property that the translation operators $T_\alpha : L^2_w(\mathbb{R}^d) \to L^2_w(\mathbb{R}^d), f \mapsto f(\cdot - \alpha)$, where $\alpha \in \mathbb{R}^d$, are all continuous. Here $L^2_w(\mathbb{R}^d)$ denotes the Hilbert space of square-integrable functions with respect to the weight w with norm defined by

$$\|f\|^2_{L^2_w} = \int_{\mathbb{R}^d} |f(\xi)|^2\, w(\xi)\, d\xi, \quad f \in L^2_w(\mathbb{R}^d).$$

The continuity of the translation operators is easily seen to be equivalent to the existence of a positive function v on $\mathbb{R}^d$ such that the inequality

$$w(\xi + \tau) \leq w(\xi)\, v(\tau), \quad \xi, \tau \in \mathbb{R}^d. \tag{1.2}$$

holds. In fact, for a fixed τ, the inequality (1.2) only needs to hold for almost every $\xi \in \mathbb{R}^d$, but we will assume, for simplicity, that w is continuous on $\mathbb{R}^d$ which implies in particular, that (1.2) holds for all $\xi \in \mathbb{R}^d$. This is not an important restriction as any weight satisfying (1.2) is equivalent to a smooth one as Lemma 2.1 shows. (See also the comment before the lemma). A continuous weight w satisfying (1.2) will be called *moderate*. Furthermore, one can take $v(\tau) = \|T_{-\tau}\|$, where the previous norm is the usual operator norm of an operator acting on $L^2_w(\mathbb{R}^d)$. With this choice for v, we have the properties

$$v(0) = 1, \quad v(\xi + \tau) \leq v(\xi)\, v(\tau), \quad \xi, \tau \in \mathbb{R}^d, \tag{1.3}$$

which we will always assume to hold. A weight v satisfying the inequality in (1.3) is called *submultiplicative*. A typical example of a weight w satisfying (1.2) is the weight

$$w(\xi) = (1 + |\xi|^2)^s, \quad \xi \in \mathbb{R}^d,$$

for any $s \in \mathbb{R}$ which is used in the definition of the standard Sobolev $H^s(\mathbb{R}^d)$. Using Peetre's inequality, it is easily seen that the corresponding v satisfies

$$v(\xi) \sim (1 + |\xi|^2)^{|s|},$$

where $w_1 \sim w_2$ means that $A\,w_1 \leq w_2 \leq B\,w_1$ pointwise for two positive constants A and B. We refer the reader to Gröchenig's paper [**5**] for more examples of weights satisfying (1.2) as well as an extensive overview of their properties and applications in various areas of harmonic analysis. (See also [**7**].)

In order to extend both Theorem 1.1 and Corollary 1.2 to the weighted case, we need to introduce a "weighted" version of the notions of Beurling densities as defined in (1.1). In order to simplify the notation, we first let $w(E) = \int_E w(\lambda)\,d\lambda$ for any Borel set $E \subset \mathbb{R}^d$. Perhaps the most natural choice would consist in defining these weighted densities as

$$(1.4)\quad \mathcal{D}^+_w(\mu) = \limsup_{R\to\infty} \sup_{\xi\in\mathbb{R}^d} \frac{\mu(\xi + I_R)}{w(\xi + I_R)}, \quad \text{and} \quad \mathcal{D}^-_w(\mu) = \liminf_{R\to\infty} \inf_{\xi\in\mathbb{R}^d} \frac{\mu(\xi + I_R)}{w(\xi + I_R)}.$$

These definitions reduce to the classical ones in the case where $w = 1$. It turns out that, beside the requirement that the weight be moderate, two additional technical conditions are needed in order to be able to fully extend Theorem 1.1 to the weighted case using the previous definition of weighted densities. One of them is related to the annular decay condition and puts a restriction on the behaviour of the weight $w(\xi)$ as $|\xi| \to \infty$. More explicitly, it requires that, for some $r_0 > 0$, we have

$$\lim_{N\to\infty} \sup_{\xi\in\mathbb{R}^d} \frac{w(\xi + (I_{N+r_0} \setminus I_N))}{w(\xi + I_N)} = 0.$$

A similar condition was used in a recent paper by Führ et al. ([**1**]) to investigate analogues of the notion of densities in abstract metric spaces. This condition fails to be satisfied for the moderate weight $w(\xi) = e^{\xi}$ on $\mathbb{R}$ for example or for integrable weights. The other one requires that $v \geq 1$ and is automatically satisfied if the weight is even, for example. We will consider this situation in Section 3 and show, in particular, that both conditions holds for the weights $w(\xi) = (1 + |\xi|^2)^s$ in one dimension when they are not integrable i.e. if $s \geq -1/2$.

For weights w for which at least one of the two conditions above fails, it is not possible using our technique to extend the result of Theorem 1.1 with the versions of weighted Beurling densities defined above. However the results of Corollary 1.2 can be generalize in a certain sense to the general weighted case (involving a moderate weight w) if different definitions of weighted densities are used. Indeed, we will show in section 2 that the existence of certain weighted convolution inequalities satisfied by the measure μ and the weight w are "controlled" by the upper and lower Beurling densities of the measure $w^{-1}\,\mu$, i.e. the quantities $\mathcal{D}^+(w^{-1}\,\mu)$ and $\mathcal{D}^-(w^{-1}\,\mu)$ (see Theorem 2.2). These last densities appear to play a role more fundamental than that of the weighted densities defined in (1.4) as they do not require any special assumption on the weight w besides the property of being moderate. In particular, we show in Section 4, that for the one-dimensional class of exponential weights the Beurling densities are $\mathcal{D}^+(w^{-1}\,\mu)$ and $\mathcal{D}^-(w^{-1}\,\mu)$ are the correct one to consider when dealing with these types of weighted convolution inequalities.

2. Weighted convolution inequalities and Beurling densities of the measure $w^{-1}\mu$

The following lemma will be a useful tool. It shows, in particular, that if w is a weight satisfying (1.2) and (1.3), then $w \sim \tilde{w}$, where $\tilde{w}$ is smooth and satisfies the same two conditions. Indeed, it suffices to take $\tilde{w} = w * F$, where $F \geq 0$ and $F \in C_0^\infty(\mathbb{R}^d) \setminus \{0\}$ and apply the lemma.

LEMMA 2.1. *Suppose that w and v satisfy both (1.2) and (1.3) and let $F \geq 0$ be a measurable function on $\mathbb{R}^d$ satisfying*

$$\int_{\mathbb{R}^d} F(\xi)\, v(-\xi)\, d\xi < \infty.$$

We have then the inequalities

$$w(\xi) \left(\int_{\mathbb{R}^d} F(\gamma)\, v^{-1}(\gamma)\, d\gamma \right) \leq (w * F)(\xi) \leq w(\xi) \left(\int_{\mathbb{R}^d} F(\gamma)\, v(-\gamma)\, d\gamma \right), \tag{2.1}$$

for $\xi \in \mathbb{R}^d$. Furthermore, we have the inequality

$$(w * F)(\xi + \tau) \leq (w * F)(\xi)\, v(\tau), \quad \xi, \tau \in \mathbb{R}^d. \tag{2.2}$$

PROOF. Note first that, since $1 = v(0) \leq v(\gamma)\, v(-\gamma)$, we have

$$\int_{\mathbb{R}^d} F(\gamma)\, v^{-1}(\gamma)\, d\gamma \leq \int_{\mathbb{R}^d} F(\gamma)\, v(-\gamma)\, d\gamma < \infty,$$

Hence,

$$\begin{aligned}(w * F)(\xi) &= \int_{\mathbb{R}^d} F(\xi - \gamma)\, w(\gamma)\, d\gamma \leq \int_{\mathbb{R}^d} F(\xi - \gamma)\, v(\gamma - \xi)\, w(\xi)\, d\gamma \\ &= w(\xi) \int_{\mathbb{R}^d} F(\gamma)\, v(-\gamma)\, d\gamma,\end{aligned}$$

and

$$\begin{aligned}(w * F)(\xi) &= \int_{\mathbb{R}^d} F(\xi - \gamma)\, w(\gamma)\, d\gamma \geq \int_{\mathbb{R}^d} F(\xi - \gamma)\, v^{-1}(\xi - \gamma)\, w(\xi)\, d\gamma \\ &= w(\xi) \int_{\mathbb{R}^d} F(\gamma)\, v^{-1}(\gamma)\, d\gamma,\end{aligned}$$

which proves the inequalities in (2.1). We have also

$$\begin{aligned}(w * F)(\xi + \tau) &= \int_{\mathbb{R}^d} F(\gamma)\, w(\xi + \tau - \gamma)\, d\gamma \leq \int_{\mathbb{R}^d} F(\gamma)\, w(\xi - \gamma)\, v(\tau)\, d\gamma \\ &= (w * F)(\xi)\, v(\tau),\end{aligned}$$

proving (2.2). □

The following result generalizes Corollary 1.2 to the weighted case.

THEOREM 2.2. *Let μ be a positive Borel measure on $\mathbb{R}^d$ and let w and v be weights on $\mathbb{R}^d$ satisfying both (1.2) and (1.3). Let $F \geq 0$ be a Borel measurable function defined on $\mathbb{R}^d$ such that $0 < \int_{\mathbb{R}^d} F(\gamma)\, v(-\gamma)\, d\gamma < \infty$ and suppose that, for some constant $B > 0$, we have the inequality*

$$\mu * F \leq B\,(w * F) \quad \textit{pointwise on } \mathbb{R}^d. \tag{2.3}$$

Then, we have the inequality

$$\mathcal{D}^+(w^{-1}\mu) \le B\, \frac{\int_{\mathbb{R}^d} F(\gamma)\, v(-\gamma)\, d\gamma}{\int_{\mathbb{R}^d} F(\gamma)\, v^{-1}(\gamma)\, d\gamma}. \tag{2.4}$$

Furthermore if, for some positive constants A, B, we have

$$A\,(w * F) \le \mu * F \le B\,(w * F) \quad \textit{pointwise on } \mathbb{R}^d, \tag{2.5}$$

then,

$$A\, \frac{\int_{\mathbb{R}^d} F(\gamma)\, v^{-1}(\gamma)\, d\gamma}{\int_{\mathbb{R}^d} F(\gamma)\, v(-\gamma)\, d\gamma} \le \mathcal{D}^-(w^{-1}\mu) \le \mathcal{D}^+(w^{-1}\mu) \le B\, \frac{\int_{\mathbb{R}^d} F(\gamma)\, v(-\gamma)\, d\gamma}{\int_{\mathbb{R}^d} F(\gamma)\, v^{-1}(\gamma)\, d\gamma}. \tag{2.6}$$

PROOF. We have

$$(w * F)(\xi) = \int_{\mathbb{R}^d} F(\gamma)\, w(\xi - \gamma)\, d\gamma \le w(\xi) \int_{\mathbb{R}^d} F(\gamma)\, v(-\gamma)\, d\gamma, \quad \xi \in \mathbb{R}^d.$$

Furthermore, since $w(\xi) \le w(\gamma)\, v(\xi - \gamma)$, we have also

$$\begin{aligned}(\mu * F)(\xi) &= \int_{\mathbb{R}^d} F(\xi - \gamma)\, d\mu(\gamma) = \int_{\mathbb{R}^d} F(\xi - \gamma)\, w(\gamma)\, w^{-1}(\gamma)\, d\mu(\gamma) \\ &\ge w(\xi) \int_{\mathbb{R}^d} \frac{F(\xi - \gamma)}{v(\xi - \gamma)}\, w^{-1}(\gamma)\, d\mu(\gamma).\end{aligned}$$

Using our assumption (2.3), it follows thus that

$$(F\, v^{-1}) * (w^{-1}\mu) \le B \int_{\mathbb{R}^d} F(\gamma)\, v(-\gamma)\, d\gamma. \tag{2.7}$$

Since we have $1 = v(0) \le v(\gamma)\, v(-\gamma)$ for $\gamma \in \mathbb{R}^d$, we have also

$$\int_{\mathbb{R}^d} F(\gamma)\, v^{-1}(\gamma)\, d\gamma \le \int_{\mathbb{R}^d} F(\gamma)\, v(-\gamma)\, d\gamma < \infty. \tag{2.8}$$

The inequality (2.4) follows now immediately from (2.7) and (2.8) using part (a) of Corollary 1.2.

If, in addition, we have also the inequality $A\,(w * F) \le \mu * F$, we just need to prove the first inequality in (2.6) since the second one is obvious and we just proved the third one. Note first that

$$(w * F)(\xi) = \int_{\mathbb{R}^d} F(\gamma)\, w(\xi - \gamma)\, d\gamma \ge w(\xi) \int_{\mathbb{R}^d} F(\gamma)\, v^{-1}(\gamma)\, d\gamma, \quad \xi \in \mathbb{R}^d$$

and

$$\begin{aligned}(\mu * F)(\xi) &= \int_{\mathbb{R}^d} F(\xi - \gamma)\, d\mu(\gamma) = \int_{\mathbb{R}^d} F(\xi - \gamma)\, w(\gamma)\, w^{-1}(\gamma)\, d\mu(\gamma) \\ &\le w(\xi) \int_{\mathbb{R}^d} F(\xi - \gamma)\, v(\gamma - \xi)\, w^{-1}(\gamma)\, d\mu(\gamma), \quad \xi \in \mathbb{R}^d.\end{aligned}$$

Letting $\check{v}(\xi) = v(-\xi)$ and using the first inequality in (2.5), we deduce that

$$(F\, \check{v}) * (w^{-1}\mu) \ge A \int_{\mathbb{R}^d} F(\gamma)\, v^{-1}(\gamma)\, d\gamma.$$

Since $\mathcal{D}^+(w^{-1}\mu) < \infty$ from the first part of the proof, we deduce the first inequality in (2.6) from the convolution inequality above using part (b) of Corollary 1.2. This proves our claim.

□

As consequences of Theorem 2.2, we have the following characterizations.

THEOREM 2.3. *Let μ be a positive Borel measure on $\mathbb{R}^d$ and let w and v be weights on $\mathbb{R}^d$ satisfying both (1.2) and (1.3). Then, the following are equivalent.*

(a) $\mathcal{D}^+(w^{-1}\mu) < \infty$.

(b) *There exists $r_0 > 0$ such that* $\sup_{\xi\in\mathbb{R}^d} \frac{\mu(\xi+I_{r_0})}{w(\xi+I_{r_0})} < \infty$.

(c) *We have* $\sup_{\xi\in\mathbb{R}^d} \frac{\mu(\xi+I_r)}{w(\xi+I_r)} < \infty$ *for all $r > 0$ large enough.*

(d) *There exists a Borel measurable function $F \geq 0$ on $\mathbb{R}^d$ satisfying*

$$0 < \int_{\mathbb{R}^d} F(\xi)\, v(-\xi)\, d\xi < \infty$$

and a constant $B > 0$ such that $\mu * F \leq B\,(w * F)$ pointwise on $\mathbb{R}^d$.

Moreover, the following statements are also equivalent.

(a′) $0 < \mathcal{D}^-(w^{-1}\mu) \leq \mathcal{D}^+(w^{-1}\mu) < \infty$.

(b′) *There exists $r_0 > 0$ such that* $0 < \inf_{\xi\in\mathbb{R}^d} \frac{\mu(\xi+I_{r_0})}{w(\xi+I_{r_0})} \leq \sup_{\xi\in\mathbb{R}^d} \frac{\mu(\xi+I_{r_0})}{w(\xi+I_{r_0})} < \infty$.

(c′) *We have* $0 < \inf_{\xi\in\mathbb{R}^d} \frac{\mu(\xi+I_r)}{w(\xi+I_r)} \leq \sup_{\xi\in\mathbb{R}^d} \frac{\mu(\xi+I_r)}{w(\xi+I_r)} < \infty$, *for all $r > 0$ large enough.*

(d′) *There exists a Borel measurable function $F \geq 0$ on $\mathbb{R}^d$ satisfying*

$$0 < \int_{\mathbb{R}^d} F(\xi)\, v(-\xi)\, d\xi < \infty$$

and constants $A, B > 0$ such that $A\,(w * F) \leq \mu * F \leq B\,(w * F)$ pointwise on $\mathbb{R}^d$.

PROOF. If (a) holds, we can find a constant $B > 0$ such that

$$\sup_{\xi\in\mathbb{R}^d} \frac{\int_{\xi+I_r} w^{-1}(\lambda)\, d\mu(\lambda)}{r^d} \leq B$$

if $r > 0$ is large enough. Fix such an r. If $\lambda, \lambda' \in \xi + I_r$, we have

$$w(\lambda) \leq A\, w(\xi), \quad w(\xi) \leq A\, w(\lambda'),$$

where $A = \sup_{\eta\in I_r} v(\eta)$. Hence,

$$\frac{\int_{\xi+I_r} w^{-1}(\lambda)\, d\mu(\lambda)}{r^d} \geq A^{-1} \frac{\int_{\xi+I_r} 1\, d\mu(\lambda)}{w(\xi) r^d} \geq A^{-2} \frac{\int_{\xi+I_r} 1\, d\mu(\lambda)}{\int_{\xi+I_r} w(\lambda)\, d\lambda},$$

which shows that

$$\sup_{\xi\in\mathbb{R}^d} \frac{\mu\,(\xi + I_r)}{w\,(\xi + I_r)} \leq A^2\, B < \infty$$

and thus that (c) holds. Clearly, (c) implies (b) and, if (b) holds, then so does (d) by simply defining $F = \chi_{I_r}$ where $r > 0$ is chosen large enough. Finally, the implication (d) $\Longrightarrow$ (a) follows from the first statement in Theorem 2.2. The proof of the equivalence of (a′), (b′) and (c′) is similar and left to the interested reader. □

We should point out that the condition $\int_{\mathbb{R}^d} F(\lambda)\, v(-\lambda)\, d\lambda < \infty$ in statements (c) and (c′) is needed for the equivalence with the other respective conditions in the previous theorem. For example, in one dimension, if

$$d\mu = \frac{1}{(1+\xi^2)^2}\, d\xi, \quad w(\xi) = \frac{1}{(1+\xi^2)^3} \quad \text{and} \quad F(\xi) = \xi^2,$$

we have

$$(w * F)(\xi) \le (\mu * F)(\xi) = \frac{\pi}{2}\,(\xi^2+1) \le 4\,(w * F)(\xi) = \frac{\pi}{2}\,(3\,\xi^2+1), \quad \xi \in \mathbb{R},$$

but the other statements clearly fails for these measures. The integrability condition fails for F since $v(\xi) \sim (1+\xi^2)^3$.

We will need the following lemma.

LEMMA 2.4. *Let μ be a positive Borel measure on $\mathbb{R}^d$ and let w and v be weights on $\mathbb{R}^d$ satisfy both (1.2) and (1.3). Suppose that there exist constants C_1, C_2 with $0 \le C_1, C_2 < \infty$ and Borel measurable functions F_1, F_2 with $F_i \ge 0$ and*

$$\int_{\mathbb{R}^d} F_i(\tau)\, v(-\tau)\, d\tau < \infty,$$

$i = 1, 2$, such that

$$\mu * F_1 \le C_1\,(w * F_1) \quad \textit{and} \quad \mu * F_2 \ge C_2\,(w * F_2) \textit{ pointwise on } \mathbb{R}^d.$$

Then, $F := F_1 * F_2$ satisfies

$$\int_{\mathbb{R}^d} F(\tau)\, v(-\tau)\, d\tau < \infty,$$

and the inequalities

$$C_2\,(w * F) \le \mu * F \le C_1\,(w * F) \textit{ pointwise on } \mathbb{R}^d.$$

PROOF. Using Tonelli's theorem and the submultiplicativity inequality (1.3), we have

$$\begin{aligned}
\int_{\mathbb{R}^d} F(\lambda)\, v(-\lambda)\, d\lambda &= \int_{\mathbb{R}^d} \left(\int_{\mathbb{R}^d} F_1(\lambda-\tau)\, F_2(\tau)\, d\tau \right) v(-\lambda+\tau-\tau)\, d\lambda \\
&\le \int_{\mathbb{R}^d} \left(\int_{\mathbb{R}^d} F_1(\lambda-\tau)\, v(-\lambda+\tau)\, d\lambda \right) F_2(\tau)\, v(-\tau)\, d\tau \\
&= \int_{\mathbb{R}^d} F_1(\lambda)\, v(-\lambda)\, d\lambda \int_{\mathbb{R}^d} F_2(\tau)\, v(-\tau)\, d\tau < \infty.
\end{aligned}$$

Moreover,

$$\mu * F = (\mu * F_1) * F_2 \le C_1\,(w * F_1) * F_2 = C_1\,(w * F)$$

and

$$\mu * F = (\mu * F_2) * F_1 \ge C_2\,(w * F_2) * F_1 = C_2\,(w * F),$$

which proves the lemma. □

The following result shows how to construct functions F satisfying the weighted convolution inequality (c) and (c′) in the previous theorem starting from a function satisfying an unweighted convolution inequality.

PROPOSITION 2.5. *Let μ be a positive Borel measure on $\mathbb{R}^d$ and let w and v be weights on $\mathbb{R}^d$ satisfying both (1.2) and (1.3).*

(a) *Suppose that* $\mathcal{D}^+(w^{-1}\mu) < \infty$ *and that* $G \geq 0$ *is a Borel measurable function on* $\mathbb{R}^d$ *with* $0 < \int_{\mathbb{R}^d} G(\xi)\,d\xi < \infty$ *such that* $(w^{-1}\mu) * G \leq C$ *pointwise on* $\mathbb{R}^d$, *where* $C \geq 0$ *is a constant. Define*

$$F(\xi) = G(\xi)\,v^{-1}(-\xi), \quad \xi \in \mathbb{R}^d.$$

Then, F *satisfies*

$$F(\xi) \geq 0, \quad 0 < \int_{\mathbb{R}^d} F(\xi)\,v(-\xi)\,d\xi < \infty \quad \text{and} \quad \mu * F \leq B\,(w * F) \tag{2.9}$$

pointwise on $\mathbb{R}^d$, *where* $B = C\left(\int_{\mathbb{R}^d} F(\xi)\,v^{-1}(\xi)\,d\xi\right)^{-1}$.

(b) *Suppose that* $0 < \mathcal{D}^-(w^{-1}\mu) \leq \mathcal{D}^+(w^{-1}\mu) < \infty$ *and that* $G \geq 0$ *is a Borel measurable function on* $\mathbb{R}^d$ *with* $0 < \int_{\mathbb{R}^d} G(\xi)\,v(\xi)\,v(-\xi)d\xi < \infty$ *such that the inequalities* $C_1 \leq (w^{-1}\mu) * G \leq C_2$ *hold pointwise on* $\mathbb{R}^d$, *where* $C_1, C_2 > 0$ *are constants. Define* $F = F_1 * F_2$ *where*

$$F_1(\xi) = G(\xi)\,v(\xi) \quad \text{and} \quad F_2(\xi) = G(\xi)\,v^{-1}(-\xi), \quad \xi \in \mathbb{R}^d.$$

Then, F *satisfies*

$$F(\xi) \geq 0, \quad 0 < \int_{\mathbb{R}^d} F(\xi)\,v(-\xi)\,d\xi < \infty \quad \text{and} \quad A\,(w * F) \leq \mu * F \leq B\,(w * F) \tag{2.10}$$

pointwise on $\mathbb{R}^d$ *with*

$$A = C_1\left(\int_{\mathbb{R}^d} F_1(\xi)\,v(-\xi)\,d\xi\right)^{-1} \quad \text{and} \quad B = C_2\left(\int_{\mathbb{R}^d} F_2(\xi)\,v^{-1}(\xi)\,d\xi\right)^{-1}.$$

PROOF. The first two inequalities in (2.9) are obvious. The third one follows from the following computation.

$$\begin{aligned}
(\mu * F)(\xi) &= \int_{\mathbb{R}^d} F(\xi - \gamma)\,d\mu(\gamma) = \int_{\mathbb{R}^d} F(\xi - \gamma)\,w(\gamma)\,w^{-1}(\gamma)\,d\mu(\gamma) \\
&= \int_{\mathbb{R}^d} G(\xi - \gamma)\,v^{-1}(\gamma - \xi)\,w(\gamma)\,w^{-1}(\gamma)\,d\mu(\gamma) \\
&\leq \int_{\mathbb{R}^d} G(\xi - \gamma)\,w(\xi)\,w^{-1}(\gamma)\,d\mu(\gamma) \\
&= w(\xi)\,\left((w^{-1}\mu) * G\right)(\xi) \leq C\,w(\xi) \leq B\,(w * F)(\xi), \quad \xi \in \mathbb{R}^d,
\end{aligned}$$

with $B = C\left(\int_{\mathbb{R}^d} F(\xi)\,v^{-1}(\xi)\,d\xi\right)^{-1}$ using the first inequality in (2.1). If, the assumptions in (b) hold, it is clear that $F \geq 0$. We have also

$$\int_{\mathbb{R}^d} F_1(\xi)\,v(-\xi)\,d\xi = \int_{\mathbb{R}^d} G(\xi)\,v(\xi)\,v(-\xi)\,d\xi < \infty$$

and

$$\int_{\mathbb{R}^d} F_2(\xi)\,v(-\xi)\,d\xi = \int_{\mathbb{R}^d} G(\xi)\,d\xi < \infty,$$

where the last inequality follows from the fact that $1 \leq v(\xi)\,v(-\xi)$ for $\xi \in \mathbb{R}^d$. By (a), we have the pointwise inequality $\mu * F_2 \leq B\,(w * F_2)$, where

$$B = C_2\left(\int_{\mathbb{R}^d} F_2(\xi)\,v^{-1}(\xi)\,d\xi\right)^{-1}.$$

We have also

$$\begin{aligned}(\mu * F_1)(\xi) &= \int_{\mathbb{R}^d} F_1(\xi-\gamma)\,d\mu(\gamma) = \int_{\mathbb{R}^d} F_1(\xi-\gamma)\,w(\gamma)\,w^{-1}(\gamma)\,d\mu(\gamma)\\ &= \int_{\mathbb{R}^d} G(\xi-\gamma)\,v(\xi-\gamma)\,w(\gamma)\,w^{-1}(\gamma)\,d\mu(\gamma)\\ &\geq \int_{\mathbb{R}^d} G(\xi-\gamma)\,w(\xi)\,w^{-1}(\gamma)\,d\mu(\gamma)\\ &= w(\xi)\,\left((w^{-1}\mu)*G\right)(\xi) \geq C_1\,w(\xi) \geq A\,(w*F_1)(\xi), \quad \xi\in\mathbb{R}^d,\end{aligned}$$

where $A = C_1\left(\int_{\mathbb{R}^d} F_1(\xi)\,v(-\xi)\,d\xi\right)^{-1}$ using the second inequality in (2.1). The statement (b) follows then immediately from Lemma 2.4. This proves our claim. □

3. Best constants in weighted convolution inequalities

If w and v are weights on $\mathbb{R}^d$ satisfiying both (1.2) and (1.3), we denote by $\mathcal{P}_w(\mathbb{R}^d)$ the set of Borel measurable functions $F\geq 0$ on $\mathbb{R}^d$ such that

$$\int_{\mathbb{R}^d} F(\lambda)\,v(-\lambda)\,d\lambda = 1.$$

As before, we let $w(E)=\int_E w(\lambda)\,d\lambda$ for any Borel set $E\subset\mathbb{R}^d$. If μ is a (locally finite) positive Borel measure on $\mathbb{R}^d$, we then define the following quantity

$$\mathcal{C}_w^+(\mu) = \inf\left\{C\geq 0,\ \mu*F\leq C\,(w*F),\ \text{for some } F\in\mathcal{P}_w(\mathbb{R}^d)\right\},$$

with the convention that $\mathcal{C}_w^+(\mu)=\infty$ if no constant $C>0$ exists such that the inequality $\mu*F\leq C\,(w*F)$ holds for some $F\in\mathcal{P}_w(\mathbb{R}^d)$. We also let

$$\mathcal{C}_w^-(\mu) = \sup\left\{C\geq 0,\ \mu*F\geq C\,(w*F),\ \text{for some } F\in\mathcal{P}_w(\mathbb{R}^d)\right\}.$$

Recall also the definition of upper and lower weighted densities:

$$\mathcal{D}_w^+(\mu) = \limsup_{R\to\infty}\,\sup_{\xi\in\mathbb{R}^d}\frac{\mu(\xi+I_R)}{w(\xi+I_R)}, \quad \text{and} \quad \mathcal{D}_w^-(\mu) = \liminf_{R\to\infty}\,\inf_{\xi\in\mathbb{R}^d}\frac{\mu(\xi+I_R)}{w(\xi+I_R)}.$$

By considering, for large values of R, functions of the form $F=c\,\chi_{I_R}$ where $c>0$ is chosen so that $F\in\mathcal{P}_w(\mathbb{R}^d)$, it is clear that we have the inequalities

$$\mathcal{C}_w^+(\mu)\leq\mathcal{D}_w^+(\mu) \quad\text{and}\quad \mathcal{C}_w^-(\mu)\geq\mathcal{D}_w^-(\mu).$$

(see the proof of Theorem 3.3 for more details.) We will need the following result.

PROPOSITION 3.1. *Let μ be a positive Borel measure on $\mathbb{R}^d$ and let w and v satisfy both (1.2) and (1.3) on $\mathbb{R}^d$. Then, the following are equivalent.*

(a) $\mathcal{D}_w^+(\mu)<\infty$.
(b) $\mathcal{D}^+(w^{-1}\mu)<\infty$.

PROOF. If (a) holds, then we have

$$\sup_{\xi\in\mathbb{R}^d}\frac{\mu(\xi+I_r)}{w(\xi+I_r)}<\infty$$

if $r>0$ is large enough. Hence (b) holds, by the equivalence of (a) and (b) in Theorem 2.3.

If (b) holds, we can write $d\mu = w\,d\rho$ where ρ is a translation-bounded measure. Given $r > 1$, let N be the unique positive integer such that $N-1 < r \le N$. If $\xi \in \mathbb{R}^d$, we can find finitely many $\xi_j \in \mathbb{R}^d$ such that

$$\xi + I_r = \cup_j\,(\xi_j + I_{r/N})$$

and with $|\xi_j + I_{r/N} \cap \xi_{j'} + I_{r/N}| = 0$ for $j \neq j'$. Let $C := \sup_{\tau\in\mathbb{R}^d} \rho(\tau + I_1)$ and let $D = \sup_{\tau\in I_1}\, v(\tau)$. We have then

$$\begin{aligned}
\mu(\xi + I_r) &= \mu(\cup_j\,(\xi_j + I_{r/N})) \le \sum_j \mu(\xi_j + I_{r/N}) = \sum_j \int_{\xi_j + I_{r/N}} w(\lambda)\,d\rho(\lambda)\\
&\le \sum_j \rho(\xi_j + I_{r/N}) \sup_{\tau\in\xi_j + I_{r/N}} w(\tau) \le C\,D \sum_j w(\xi_j)\\
&= C\,D\,\frac{N^d}{r^d} \sum_j \int_{\xi_j + I_{r/N}} w(\xi_j)\,d\lambda \le C\,D^2\,2^d \sum_j \int_{\xi_j + I_{r/N}} w(\lambda)\,d\lambda\\
&= A\,w(\xi + I_r),
\end{aligned}$$

where the constant $A > 0$ does not depend on r. This shows that (a) holds.

□

LEMMA 3.2. *Let μ be a positive Borel measure on $\mathbb{R}^d$ and let $w > 0$ be a moderate weight defined on $\mathbb{R}^d$. Assume also that for some $r_0 > 0$, we have*

$$\lim_{N\to\infty} \sup_{\xi\in\mathbb{R}^d} \frac{w(\xi + (I_{N+r_0} \setminus I_N))}{w(\xi + I_N)} = 0. \tag{3.1}$$

Then,

(a) *For any $r > 0$, we have*

$$\lim_{N\to\infty} \sup_{\xi\in\mathbb{R}^d} \frac{w(\xi + (I_{N+r} \setminus I_N))}{w(\xi + I_N)} = 0.$$

(b) *If $\mathcal{D}_w^+(\mu) < \infty$, we have, for any $r > 0$, that*

$$\lim_{N\to\infty} \sup_{\xi\in\mathbb{R}^d} \frac{\mu(\xi + (I_{N+r} \setminus I_N))}{w(\xi + I_N)} = 0$$

PROOF. It is enough to prove (a) for r of the form $r = k\,r_0$ where k is an integer with $k \ge 1$, which we proceed to do by induction on k. If $k = 1$, the assertion is true by hypothesis. Suppose that it is true for k. We have then

$$\begin{aligned}
\frac{w(\xi + (I_{N+(k+1)\,r_0} \setminus I_N))}{w(\xi + I_N)} &= \frac{w(\xi + (I_{N+r_0+k\,r_0} \setminus I_{N+r_0}))}{w(\xi + I_N)} + \frac{w(\xi + (I_{N+r_0} \setminus I_N))}{w(\xi + I_N)}\\
&:= A_N(\xi) + B_N(\xi).
\end{aligned}$$

We have $B_N(\xi) \to 0$ uniformly in ξ as $N \to \infty$ by assumption and, since

$$\begin{aligned}
A_N(\xi) &= \frac{w(\xi + (I_{N+r_0-k\,r_0} \setminus I_{N+r_0}))}{w(\xi + I_{N+r_0})}\,\frac{w(\xi + I_{N+r_0})}{w(\xi + I_N)}\\
&= \frac{w(\xi + (I_{N+r_0+k\,r_0} \setminus I_{N+r_0}))}{w(\xi + I_{N+r_0})} \left(1 + \frac{w(\xi + (I_{N+r_0} \setminus I_N))}{w(\xi + I_N)}\right),
\end{aligned}$$

it follows that $A_N(\xi) \to 0$ uniformly in ξ as well as $N \to \infty$, showing that our assertion is true for $k+1$, which proves (a).

To prove the statement in (b), we can assume that $r > 1$ and we note that, by Theorem 2.3 and Proposition 3.1, the condition $\mathcal{D}_w^+(\mu) < \infty$ is equivalent to the existence of a constant $B > 0$ such that $\mu(\xi + I_r) \leq B\, w(\xi + I_r)$ for all $\xi \in \mathbb{R}^d$. Let us first assume that $N = Mr$ where $M > 0$ is an integer. Note that we have the inclusion $\xi + (I_{(M+1)r} \setminus I_{Mr}) \subset \cup_{j=1}^J \xi_j + I_r$ for some $\xi_j \in \mathbb{R}^d$, where the sets $\xi_j + I_r$, $j = 1, \ldots, J$ are pairwise disjoint (up to sets of zero Lebesgue measure) and $\left|\left(\cup_{j=1}^J \xi_j + I_r\right) \setminus \left(\xi + (I_{(M+1)r} \setminus I_{Mr})\right)\right| = 0$. We have thus

$$\begin{aligned}\mu\left(\xi + (I_{(M+1)r} \setminus I_{Mr})\right) &\leq \mu\left(\cup_{j=1}^J \xi_j + I_r\right) \leq \sum_{j=1}^J \mu\left(\xi_j + I_r\right) \leq B \sum_{j=1}^J w\left(\xi_j + I_r\right) \\ &= B\, w\left(\xi + (I_{(M+1)r} \setminus I_{Mr})\right)\end{aligned}$$

In general, if $N > 0$, let $M > 0$ be the unique integer such that $M\,r \leq N < (M+1)\,r$. We have then

$$\begin{aligned}\mu\left(\xi + (I_{N+r} \setminus I_N)\right) &\leq \mu\left(\xi + (I_{(M+2)r} \setminus I_{Mr})\right) \\ &= \mu\left(\xi + (I_{(M+2)r} \setminus I_{(M+1)r})\right) + \mu\left(\xi + (I_{(M+1)r} \setminus I_{Mr})\right) \\ &\leq B\left\{w\left(\xi + (I_{(M+2)r} \setminus I_{(M+1)r})\right) + w\left(\xi + (I_{(M+1)r} \setminus I_{Mr})\right)\right\} \\ &= B\, w\left(\xi + (I_{(M+2)r} \setminus I_{Mr})\right).\end{aligned}$$

Hence, we have the inequality

$$\sup_{\xi \in \mathbb{R}^d} \frac{\mu(\xi + (I_{N+r} \setminus I_N))}{w(\xi + I_N)} \leq B \sup_{\xi \in \mathbb{R}^d} \frac{w(\xi + (I_{(M+2)r} \setminus I_{Mr}))}{w(\xi + I_{Mr})}$$

which, with the help of part (a), yields the statement in (b). □

We can now state the main result in this section. Note that the assumptions of the theorem below are easily seen to be satisfied by the constant weight $w = 1$. Hence, this result generalizes Theorem 1.1 and provides also a more direct proof of it than the one given in [**2**].

THEOREM 3.3. *Let μ be a positive Borel measure on $\mathbb{R}^d$ and let w and v satisfy (1.2) and (1.3). Assume also that for some (and thus for every) $r > 0$, we have*

$$\lim_{N \to \infty} \sup_{\xi \in \mathbb{R}^d} \frac{w(\xi + (I_{N+r} \setminus I_R))}{w(\xi + I_N)} = 0 \tag{3.2}$$

and that $v(\gamma) \geq 1$. Then,

(a) $\mathcal{C}_w^+(\mu) = \mathcal{D}_w^+(\mu)$.
(b) $\mathcal{C}_w^-(\mu) \geq \mathcal{D}_w^-(\mu)$ *and if, in addition, $\mathcal{D}_w^+(\mu) < \infty$, then we have the equality* $\mathcal{C}_w^-(\mu) = \mathcal{D}_w^-(\mu)$.

PROOF. We first prove the equality in (a). By Theorem 2.3 and Proposition 3.1, we have $\mathcal{D}_w^+(\mu) = \infty$ if and only if $\mathcal{C}_w^+(\mu) = \infty$, so we can assume that both $\mathcal{D}_w^+(\mu)$ and $\mathcal{C}_w^+(\mu)$ are finite. Given $\epsilon > 0$, we can find $R > 0$ such that

$$\frac{\mu(\xi + I_R)}{w(\xi + I_R)} \leq \mathcal{D}_w^+(\mu) + \epsilon, \quad \xi \in \mathbb{R}^d. \tag{3.3}$$

Letting $F = c\,\chi_{I_R}$, where $c = \left(\int_{I_R} v(-\lambda)\, d\lambda\right)^{-1}$, we have $F \in \mathcal{P}_w(\mathbb{R}^d)$, and, using both (3.3) and the definition of $\mathcal{C}_w^+(\mu)$, we have that

$$(\mu * F)(\xi) = c\,\mu(\xi + I_R) \leq c\,(\mathcal{D}_w^+(\mu) + \epsilon)\, w(\xi + I_R) = (\mathcal{D}_w^+(\mu) + \epsilon)\,(w * F)(\xi), \quad \xi \in \mathbb{R}^d,$$

showing that $\mathcal{C}_w^+(\mu) \leq \mathcal{D}_w^+(\mu)$, since $\epsilon > 0$ is arbitrary.

Suppose now that $C > 0$ is a constant such that $\mu * F \leq C(w * F)$ pointwise for some function $F \in \mathcal{P}_w(\mathbb{R}^d)$. Note that $F \in L^1(\mathbb{R}^d)$ since $v \geq 1$. Fix ϵ with $0 < \epsilon < 1$ and choose $r > 0$ large enough so that

$$\int_{I_r} F(\gamma)\,d\gamma \geq (1-\epsilon)\int_{\mathbb{R}^d} F(\gamma)\,d\gamma \quad \text{and} \quad \int_{I_r} F(\gamma)\,v(-\gamma)\,d\gamma \geq 1-\epsilon.$$

Note that

$$F * \chi_{I_{N+r}}(\xi) \leq \int_{\mathbb{R}^d} F(\gamma)\,d\gamma, \quad \xi \in \mathbb{R}^d,$$

and

$$F * \chi_{I_{N+r}}(\xi) \geq (1-\epsilon)\int_{\mathbb{R}^d} F(\gamma)\,d\gamma, \quad \xi \in I_N.$$

We have also

$$\mu * (F * \chi_{I_{N+r}}) \leq C\left(w * (F * \chi_{I_{N+r}})\right) \quad \text{on } \mathbb{R}^d.$$

It follows that

$$\begin{aligned}
&(1-\epsilon)\left(\int_{\mathbb{R}^d} F(\gamma)\,d\gamma\right)\mu(\xi - I_N) \leq \int_{\xi+I_N} \left(F * \chi_{I_{N+r}}\right)(\xi-\lambda)\,d\mu(\lambda)\\
&\leq \int_{\mathbb{R}^d} \left(F * \chi_{I_{N+r}}\right)(\xi-\lambda)\,d\mu(\lambda) \leq C\int_{\mathbb{R}^d} \left(F * \chi_{I_{N+r}}\right)(\xi-\lambda)\,w(\lambda)\,d\lambda\\
&= C\left\{\int_{\xi+I_N} \left(F * \chi_{I_{N+r}}\right)(\xi-\lambda)\,w(\lambda)\,d\lambda\right.\\
&\left.+\int_{\mathbb{R}^d\setminus(\xi+I_N)} \left(F * \chi_{I_{N+r}}\right)(\xi-\lambda)\,w(\lambda)\,d\lambda\right\}\\
&\leq C\left\{\left(\int_{\mathbb{R}^d} F(\gamma)\,d\gamma\right)w(\xi+I_N) + \int_{\mathbb{R}^d\setminus(\xi+I_N)} \left(F * \chi_{I_{N+r}}\right)(\xi-\lambda)\,w(\lambda)\,d\lambda\right\}
\end{aligned} \tag{3.4}$$

We have

$$\begin{aligned}
&\int_{\mathbb{R}^d\setminus(\xi+I_N)} \left(F * \chi_{I_{N+r}}\right)(\xi-\lambda)\,w(\lambda)\,d\lambda\\
&= \int_{\mathbb{R}^d\setminus(\xi+I_N)} \left(\int_{\mathbb{R}^d} F(\tau)\,\chi_{I_{N+r}}(\xi-\lambda-\tau)\,d\tau\right)w(\lambda)\,d\lambda\\
&= \int_{\mathbb{R}^d} F(\tau)\left(\int_{\mathbb{R}^d\setminus(\xi+I_N)} \chi_{I_{N+r}}(\xi-\lambda-\tau)\,w(\lambda)\,d\lambda\right)d\tau\\
&= \int_{\mathbb{R}^d} F(\tau)\,w\left((\xi-\tau+I_{N+r})\setminus(\xi+I_N)\right)d\tau\\
&= \int_{I_r} F(\tau)\,w\left((\xi-\tau+I_{N+r})\setminus(\xi+I_N)\right)d\tau\\
&+\int_{\mathbb{R}^d\setminus I_r} F(\tau)\,w\left((\xi-\tau+I_{N+r})\setminus(\xi+I_N)\right)d\tau\\
&:= A_{N,r}(\xi) + B_{N,r}(\xi).
\end{aligned}$$

Note that if $\tau \in I_r$, we have the inclusion $\xi - \tau + I_{N+r} \subset \xi + I_{N+2r}$. Hence,

$$A_{N,r}(\xi) \leq \int_{I_r} F(\tau)\, w\left((\xi + I_{N+2r}) \setminus (\xi + I_N)\right) d\tau$$

and

$$\sup_{\xi \in \mathbb{R}^d} \frac{A_{N,r}(\xi)}{w(\xi + I_N)} \leq \int_{\mathbb{R}^d} F(\tau)\, d\tau \left(\sup_{\xi \in \mathbb{R}^d} \frac{w\left((\xi + I_{N+2r}) \setminus (\xi + I_N)\right)}{w(\xi + I_N)} \right) \to 0,$$

as $N \to \infty$, using our assumption (3.2). Furthermore,

$$\begin{aligned} B_{N,r}(\xi) &= \int_{\mathbb{R}^d \setminus I_r} F(\tau)\, w\left((\xi - \tau + I_{N+r}) \setminus (\xi + I_N)\right) d\tau \\ &\leq \int_{\mathbb{R}^d \setminus I_r} F(\tau)\, w\left(\xi - \tau + I_{N+r}\right) d\tau \end{aligned}$$

and

$$\begin{aligned} w\left(\xi - \tau + I_{N+r}\right) &= \int_{\xi - \tau + I_{N+r}} w(\lambda)\, d\lambda = \int_{\xi + I_{N+r}} w(\lambda - \tau)\, d\lambda \\ &\leq v(-\tau) \int_{\xi + I_{N+r}} w(\lambda)\, d\lambda = v(-\tau)\, w\left(\xi + I_{N+r}\right), \end{aligned}$$

showing that

$$B_{N,r}(\xi) \leq \left(\int_{\mathbb{R}^d \setminus I_r} F(\tau)\, v(-\tau)\, d\tau \right) w\left(\xi + I_{N+r}\right) \leq \epsilon\, w\left(\xi + I_{N+r}\right).$$

Therefore

$$\sup_{\xi \in \mathbb{R}^d} \frac{B_{N,r}(\xi)}{w(\xi + I_N)} \leq \epsilon \left\{ \frac{w\left(\xi + I_N\right)}{w(\xi + I_N)} + \frac{w\left((\xi + I_{N+r}) \setminus (\xi + I_N)\right)}{w(\xi + I_N)} \right\}$$

which implies, using (3.2) again, that

$$\limsup_{N \to \infty} \sup_{\xi \in \mathbb{R}^d} \frac{B_{N,r}(\xi)}{w(\xi + I_N)} \leq \epsilon.$$

Using (3.4), it follows that

$$(1 - \epsilon) \frac{\mu(\xi + I_N)}{w(\xi + I_N)} \leq C + C \left\{ \frac{A_{N,r}(\xi)}{w(\xi + I_N)} + \frac{B_{N,r}(\xi)}{w(\xi + I_N)} \right\} \left(\int_{\mathbb{R}^d} F(\gamma)\, d\gamma \right)^{-1}$$

and we obtain that

$$(1 - \epsilon) \limsup_{N \to \infty} \sup_{\xi \in \mathbb{R}^d} \frac{\mu(\xi + I_N)}{w(\xi + I_N)} = (1 - \epsilon)\, \mathcal{D}_w^+(\mu) \leq C + \epsilon \left(\int_{\mathbb{R}^d} F(\gamma)\, d\gamma \right)^{-1}$$

Since $\epsilon > 0$ is arbitrary, we obtain, using the definition of $\mathcal{C}_w^+(\mu)$, the inequality $\mathcal{D}_w^+(\mu) \leq \mathcal{C}_w^+(\mu)$. This, together with the reverse inequality proved earlier, shows that $\mathcal{D}_w^+(\mu) = \mathcal{C}_w^+(\mu)$ as claimed in (a). We now prove the statements in (b). To prove the inequality $\mathcal{C}_w^-(\mu) \geq \mathcal{D}_w^-(\mu)$, we can assume that $\mathcal{D}_w^-(\mu) > 0$. Suppose first that $\mathcal{D}_w^-(\mu) < \infty$. Let ϵ with $0 < \epsilon < \mathcal{D}_w^-(\mu)$ and choose $R > 0$ such that

$$\frac{\mu(\xi + I_R)}{w(\xi + I_R)} \geq \mathcal{D}_w^-(\mu) - \epsilon, \quad \xi \in \mathbb{R}^d. \tag{3.5}$$

Letting $F = c\,\chi_{I_R}$, where $c = \left(\int_{I_R} v(-\lambda)\,d\lambda\right)^{-1}$, we have $F \in \mathcal{P}_w(\mathbb{R}^d)$, and, using (3.5), we have that

$$(\mu*F)(\xi) = c\,\mu(\xi+I_R) \geq c\,(\mathcal{D}_w^-(\mu)-\epsilon)\,w(\xi+I_R) = (\mathcal{D}_w^-(\mu)-\epsilon)\,(w*F)(\xi), \quad \xi \in \mathbb{R}^d,$$

showing, using the definition of $\mathcal{C}_w^-(\mu)$, that $\mathcal{C}_w^-(\mu) \geq \mathcal{D}_w^-(\mu)$, since $\epsilon > 0$ is arbitrary. If, on the other hand, $\mathcal{D}_w^-(\mu) = \infty$, we can find, for any $N > 0$, a number $R > 0$ such that

$$\frac{\mu(\xi + I_R)}{w(\xi + I_R)} \geq N, \quad \xi \in \mathbb{R}^d, \tag{3.6}$$

and a computation similar to the one above shows that $\mathcal{C}_w^-(\mu) = \infty$. We have thus the inequality $\mathcal{C}_w^-(\mu) \geq \mathcal{D}_w^-(\mu)$. Conversely, if $\mathcal{D}_w^+(\mu) < \infty$, we will show that $\mathcal{C}_w^-(\mu) \leq \mathcal{D}_w^-(\mu)$ and, in doing so, we can clearly assume that $\mathcal{C}_w^-(\mu) > 0$. Using Lemma 2.4, we can assume the existence of a function $F \in \mathcal{P}_w(\mathbb{R}^d)$ and of two positive constants C and D, with C arbitrarily close to $\mathcal{C}_w^-(\mu)$, such that

$$C\,(w*F) \leq \mu * F \leq D\,(w*F).$$

Note that $F \in L^1(\mathbb{R}^d)$ since $v \geq 1$. Fix ϵ with $0 < \epsilon < 1$ and choose $r > 0$ large enough so that

$$\int_{I_r} F(\gamma)\,d\gamma \geq (1-\epsilon)\int_{\mathbb{R}^d} F(\gamma)\,d\gamma \quad \text{and} \quad \int_{I_r} F(\gamma)\,v(-\gamma)\,d\gamma \geq 1-\epsilon.$$

As in the first part of the proof, we have

$$F * \chi_{I_{N+r}}(\xi) \leq \int_{\mathbb{R}^d} F(\gamma)\,d\gamma, \quad \xi \in \mathbb{R}^d,$$

and

$$F * \chi_{I_{N+r}}(\xi) \geq (1-\epsilon)\int_{\mathbb{R}^d} F(\gamma)\,d\gamma, \quad \xi \in I_N.$$

We have also

$$C(w*(F*\chi_{I_{N+r}})) \leq (\mu*(F*\chi_{I_{N+r}})) \quad \text{on } \mathbb{R}^d.$$

It follows that

$$\begin{aligned}
&C\,(1-\epsilon)\left(\int_{\mathbb{R}^d} F(\gamma)\,d\gamma\right) w(\xi+I_N) \leq C\int_{\xi+I_N} (F*\chi_{I_{N+r}})\,(\xi-\lambda)\,w(\lambda)\,d\lambda\\
&\leq C\int_{\mathbb{R}^d} (F*\chi_{I_{N+r}})\,(\xi-\lambda)\,w(\lambda)\,d\lambda \leq \int_{\mathbb{R}^d} (F*\chi_{I_{N+r}})\,(\xi-\lambda)\,d\mu(\lambda)\\
&= \left\{\int_{\xi+I_N} (F*\chi_{I_{N+r}})\,(\xi-\lambda)\,d\mu(\lambda) + \int_{\mathbb{R}^d\setminus(\xi+I_N)} (F*\chi_{I_{N+r}})\,(\xi-\lambda)\,d\mu(\lambda)\right\}\\
&\leq \left\{\left(\int_{\mathbb{R}^d} F(\gamma)\,d\gamma\right)\mu(\xi+I_N) + \int_{\mathbb{R}^d\setminus(\xi+I_N)} (F*\chi_{I_{N+r}})\,(\xi-\lambda)\,d\mu(\lambda)\right\}
\end{aligned} \tag{3.7}$$

We have

$$\begin{aligned}
&\int_{\mathbb{R}^d\setminus(\xi+I_N)} \left(F*\chi_{I_{N+r}}\right)(\xi-\lambda)\,d\mu(\lambda)\\
&= \int_{\mathbb{R}^d\setminus(\xi+I_N)} \left(\int_{\mathbb{R}^d} F(\tau)\,\chi_{I_{N+r}}(\xi-\lambda-\tau)\,d\tau\right) d\mu(\lambda)\\
&= \int_{\mathbb{R}^d} F(\tau)\left(\int_{\mathbb{R}^d\setminus(\xi+I_N)} \chi_{I_{N+r}}(\xi-\lambda-\tau)\,d\mu(\lambda)\right) d\tau\\
&= \int_{\mathbb{R}^d} F(\tau)\,\mu\left((\xi-\tau+I_{N+r})\setminus(\xi+I_N)\right) d\tau.
\end{aligned}$$

Hence,

$$\begin{aligned}
&\int_{\mathbb{R}^d\setminus(\xi+I_N)} \left(F*\chi_{I_{N+r}}\right)(\xi-\lambda)\,d\mu(\lambda)\\
&\le \int_{I_r} F(\tau)\,\mu\left((\xi-\tau+I_{N+r})\setminus(\xi+I_N)\right) d\tau\\
&+ \int_{\mathbb{R}^d\setminus I_r} F(\tau)\,\mu\left((\xi-\tau+I_{N+r})\setminus(\xi+I_N)\right) d\tau\\
&:= C_{N,r}(\xi)+D_{N,r}(\xi).
\end{aligned}$$

Note that if $\tau\in I_r$ we have the inclusion $\xi-\tau+I_{N+r}\subset\xi+I_{N+2r}$. Therefore,

$$C_{N,r}(\xi)\le\int_{I_r} F(\tau)\,\mu\left((\xi+I_{N+2r})\setminus(\xi+I_N)\right) d\tau$$

and

$$\sup_{\xi\in\mathbb{R}^d}\frac{C_{N,r}(\xi)}{w(\xi+I_N)}\le\int_{\mathbb{R}^d}F(\tau)\,d\tau\left(\sup_{\xi\in\mathbb{R}^d}\frac{\mu\left((\xi+I_{N+2r})\setminus(\xi+I_N)\right)}{w(\xi+I_N)}\right)\to 0,$$

as $N\to\infty$, using part (b) of Lemma 3.2. We have also

$$D_{N,r}(\xi)\le\int_{\mathbb{R}^d\setminus I_r} F(\tau)\,\mu\left(\xi-\tau+I_{N+r}\right) d\tau$$

and, for N large enough,

$$\mu\left(\xi+I_{N+r}\right)\le\left(\mathcal{D}_w^+(\mu)+1\right)w\left(\xi+I_{N+r}\right),\quad \xi\in\mathbb{R}^d,$$

which implies that, for $\xi,\tau\in\mathbb{R}^d$,

$$\begin{aligned}
\mu\left(\xi-\tau+I_{N+r}\right)&\le\left(\mathcal{D}_w^+(\mu)+1\right)w\left(\xi-\tau+I_{N+r}\right)\\
&\le\left(\mathcal{D}_w^+(\mu)+1\right)v(-\tau)\,w\left(\xi+I_{N+r}\right).
\end{aligned}$$

Hence,

$$\begin{aligned}
D_{N,r}(\xi)&\le\left(\mathcal{D}_w^+(\mu)+1\right)\left(\int_{\mathbb{R}^d\setminus I_r}F(\tau)\,v(-\tau)\,d\tau\right)w\left(\xi+I_{N+r}\right)\\
&\le\epsilon\left(\mathcal{D}_w^+(\mu)+1\right)w\left(\xi+I_{N+r}\right),
\end{aligned}$$

and

$$\sup_{\xi\in\mathbb{R}^d} \frac{D_{N,r}(\xi)}{w(\xi+I_N)} \le \epsilon\,(\mathcal{D}_w^+(\mu)+1) \sup_{\xi\in\mathbb{R}^d} \frac{w\,(\xi+I_{N+r})}{w(\xi+I_N)}$$
$$= \epsilon\,(\mathcal{D}_w^+(\mu)+1) \sup_{\xi\in\mathbb{R}^d} \left\{1+\frac{w\,((\xi+I_{N+r})\setminus(\xi+I_N))}{w(\xi+I_N)}\right\}.$$

Using (3.7), we deduce that, for N large enough,

$$(1-\epsilon)\,C \le \frac{\mu(\xi+I_N)}{w(\xi+I_N)} + \epsilon\,(\mathcal{D}_w^+(\mu)+2)\left(\int_{\mathbb{R}^d} F(\gamma)\,d\gamma\right)^{-1}, \quad \xi\in\mathbb{R}^d,$$

which implies, since $\epsilon>0$ is arbitrary that

$$\mathcal{D}_w^-(\mu) = \liminf_{N\to\infty} \inf_{\xi\in\mathbb{R}^d} \frac{\mu(\xi+I_N)}{w(\xi+I_N)} \ge C.$$

The definition of $\mathcal{C}_w^-(\mu)$ shows then that $\mathcal{D}_w^-(\mu)\ge\mathcal{C}_w^-(\mu)$ as claimed. Hence, we have $\mathcal{D}_w^-(\mu)=\mathcal{C}_w^-(\mu)$ which concludes the proof. □

Note that the condition $v\ge 1$ is automatically satisfied if the weight w is even, i.e. if $w(-\xi)=w(\xi)$. Indeed, in that case we have $v(-\xi)=v(\xi)$ as well and the submultiplicativity property of v shows that

$$1=v(0)\le v(-\xi)\,v(\xi)=v(\xi)^2.$$

The next result show that weights w which are integrable always fail to satisfy the condition (3.2).

PROPOSITION 3.4. *Let w be a moderate weight defined on $\mathbb{R}^d$ and assume that*

$$\int_{\mathbb{R}^d} w(\xi)\,d\xi<\infty.$$

Then, w fails to satisfy (3.2).

PROOF. If $N>0$, define $\xi_N=(N/2,N/2,\ldots,N/2)\in\mathbb{R}^d$. We have then the inclusion

$$(-r/2,0)^d \subset \xi_N+(I_{N+r}\setminus I_N)$$

and thus

$$w(\xi_N+(I_{N+r}\setminus I_N)) \ge w((-r/2,0)^d)>0.$$

We have also

$$w(\xi_N+I_N)\le w(\mathbb{R}^d)=\int_{\mathbb{R}^d} w(\xi)\,d\xi<\infty.$$

Hence,

$$\sup_{\xi\in\mathbb{R}^d} \frac{w(\xi+(I_{N+r}\setminus I_N))}{w(\xi+I_N)} \ge \frac{w(\xi_N+(I_{N+r}\setminus I_N))}{w(\xi_N+I_N)} \ge c$$

where

$$c=\frac{w((-r/2,0)^d)}{\int_{\mathbb{R}^d} w(\xi)\,d\xi}>0.$$

This shows that (3.2) fails. □

Our next result shows that for the one-dimensional moderate weights $w(\xi)=(1+|\xi|^2)^s$, where $s\in\mathbb{R}$, which appear in the definition of the classical Sobolev spaces H^s, the conditions of the theorem above are satisfied as long as w is not integrable on $\mathbb{R}^d$.

PROPOSITION 3.5. *If $s \in \mathbb{R}$, let*

$$w(\xi) = (1 + |\xi|^2)^s, \quad \xi \in \mathbb{R}.$$

Then, given $r > 0$, we have

$$\lim_{N\to\infty} \sup_{\xi\in\mathbb{R}} \frac{w(\xi + (I_{N+r} \setminus I_N))}{w(\xi + I_N)} = 0 \tag{3.8}$$

if and only if $s \geq -1/2$.

PROOF. Note that

$$2^{-|s|}\,(1 + |\xi|)^{2s} \leq (1 + |\xi|^2)^s \leq 2^{|s|}\,(1 + |\xi|)^{2s}, \quad \xi \in \mathbb{R},$$

so it is enough to prove the claim for $w_1(\xi) := (1 + |\xi|)^{2s}$ instead of $w(\xi)$. We can also assume that $s \neq 0$ since the case $s = 0$ is obvious. By Proposition 3.4, (3.8) fails if $s < -1/2$ as w is integrable in that case. Hence we need to show that (3.8) holds if $s \geq -1/2$. We first consider the case $s > -1/2$. If $\xi \in \mathbb{R}$, we have

$$\begin{aligned}
&w_1(\xi + I_N) = \int_{\xi-N/2}^{\xi+N/2} (1 + |\lambda|)^{2s}\, d\lambda \\
&= \chi_{(N/2,\infty)}(\xi) \int_{\xi-N/2}^{\xi+N/2} (1 + \lambda)^{2s}\, d\lambda + \chi_{(-\infty,-N/2)}(\xi) \int_{\xi-N/2}^{\xi+N/2} (1 - \lambda)^{2s}\, d\lambda \\
&+ \chi_{[-N/2,N/2]}(\xi) \left\{ \int_{\xi-N/2}^{0} (1 - \lambda)^{2s}\, d\lambda + \int_{0}^{\xi+N/2} (1 + \lambda)^{2s}\, d\lambda \right\} \\
&= \chi_{(N/2,\infty)}(|\xi|) \int_{|\xi|-N/2}^{|\xi|+N/2} (1 + \lambda)^{2s}\, d\lambda \\
&+ \chi_{[0,N/2]}(|\xi|) \left\{ \int_{0}^{-|\xi|+N/2} (1 + \lambda)^{2s}\, d\lambda + \int_{0}^{|\xi|+N/2} (1 + \lambda)^{2s}\, d\lambda \right\} \\
&= \chi_{(N/2,\infty)}(|\xi|)\, \frac{1}{2s+1}\, \left\{ (1 + |\xi| + N/2)^{2s+1} - (1 + |\xi| - N/2)^{2s+1} \right\} \\
&+ \chi_{[0,N/2]}(|\xi|)\, \frac{1}{2s+1}\, \left\{ (1 - |\xi| + N/2)^{2s+1} + (1 + |\xi| + N/2)^{2s+1} - 2 \right\}.
\end{aligned}$$

Hence,

$$\begin{aligned}
&(2s + 1)\, w_1(\xi + (I_{N+r} \setminus I_N)) \\
&= \chi_{(N/2+r/2,\infty)}(|\xi|)\, \left\{ (1 + |\xi| + N/2 + r/2)^{2s+1} - (1 + |\xi| - N/2 - r/2)^{2s+1} \right\} \\
&+ \chi_{[0,N/2+r/2]}(|\xi|)\, \left\{ (1 - |\xi| + N/2 + r/2)^{2s+1} + (1 + |\xi| + N/2 + r/2)^{2s+1} - 2 \right\} \\
&- \chi_{(N/2,\infty)}(|\xi|)\, \left\{ (1 + |\xi| + N/2)^{2s+1} - (1 + |\xi| - N/2)^{2s+1} \right\} \\
&- \chi_{[0,N/2]}(|\xi|)\, \left\{ (1 - |\xi| + N/2)^{2s+1} + (1 + |\xi| + N/2)^{2s+1} - 2 \right\}
\end{aligned}$$

If $|\xi| > N/2 + r/2$, we have

$$(2s+1)\, w_1(\xi + (I_{N+r} \setminus I_N)) = \{(1+|\xi|+N/2+r/2)^{2s+1} - (1+|\xi|+N/2)^{2s+1} \\ +(1+|\xi|-N/2)^{2s+1} - (1+|\xi|-N/2-r/2)^{2s+1}\}$$

and

$$\frac{w_1(\xi + (I_{N+r} \setminus I_N))}{w_1(\xi + I_N)} = \frac{(1+|\xi|+N/2+r/2)^{2s+1} - (1+|\xi|+N/2)^{2s+1}}{(1+|\xi|+N/2)^{2s+1} - (1+|\xi|-N/2)^{2s+1}} \\ + \frac{(1+|\xi|-N/2)^{2s+1} - (1+|\xi|-N/2-r/2)^{2s+1}}{(1+|\xi|+N/2)^{2s+1} - (1+|\xi|-N/2)^{2s+1}}$$

We will make use of the fact that, for $t \geq -a$ and $a > b$, the function $t \mapsto (t+a)^{2s+1} - (t+b)^{2s+1}$ is increasing if $s > 0$ and decreasing if $-1/2 < s < 0$. We consider first the case where $N/2 + r/2 < |\xi| \leq 3N/2$. If $s > 0$, we have then

$$\frac{w_1(\xi + (I_{N+r} \setminus I_N))}{w_1(\xi + I_N)} \leq \frac{(1+2N+r/2)^{2s+1} - (1+2N)^{2s+1}}{(1+N+r/2)^{2s+1} - (1+r/2)^{2s+1}} \\ + \frac{(1+N)^{2s+1} - (1+N-r/2)^{2s+1}}{(1+N+r/2)^{2s+1} - (1+r/2)^{2s+1}} \\ = \frac{(2\,N)^{2s+1}\left[(1+\frac{1+r/2}{2N})^{2s+1} - (1+\frac{1}{2N})^{2s+1}\right]}{N^{2s+1}\left[(1+\frac{1+r/2}{N})^{2s+1} - (\frac{1+r/2}{N})^{2s+1}\right]} \\ + \frac{N^{2s+1}\left[(1+\frac{1}{N})^{2s+1} - (1+\frac{1-r/2}{N})^{2s+1}\right]}{N^{2s+1}\left[(1+\frac{1+r/2}{N})^{2s+1} - (\frac{1+r/2}{N})^{2s+1}\right]} \to 0, \quad N \to \infty.$$

For $-1/2 < s < 0$, we have instead,

$$\frac{w_1(\xi + (I_{N+r} \setminus I_N))}{w_1(\xi + I_N)} \leq \frac{(1+N+r)^{2s+1} - (1+N+r/2)^{2s+1}}{(1+2N)^{2s+1} - (1+N)^{2s+1}} \\ + \frac{(1+r/2)^{2s+1} - 1}{(1+2N)^{2s+1} - (1+N)^{2s+1}} \\ = \frac{N^{2s+1}\left[(1+\frac{1+r}{N})^{2s+1} - (1-\frac{1+r/2}{N})^{2s+1}\right] + (1+r/2)^{2s+1} - 1}{N^{2s+1}\left[(2+\frac{1}{N})^{2s+1} - (1+\frac{1}{N})^{2s+1}\right]} \to 0, \quad N \to \infty.$$

If $3N/2 \leq |\xi|$ and $s > 0$, we use the mean-value theorem to obtain the inequalities

$$(3.9)\quad (2s+1)\,(t_2-t_1)\,t_1^{2s} \leq |t_1^{2s+1} - t_2^{2s+1}| \leq (2s+1)\,(t_2-t_1)\,t_2^{2s}, \quad \text{if} \quad 0 \leq t_1 < t_2.$$

Hence,

$$\begin{aligned}\frac{w_1(\xi+(I_{N+r}\setminus I_N))}{w_1(\xi+I_N)} &\le \frac{\left[(1+|\xi|+N/2+r/2)^{2s}+(1+|\xi|-N/2)^{2s}\right](r/2)}{(1+|\xi|-N/2)^{2s}\,N}\\ &= \frac{r}{2N}\left(\left(1+\frac{N+r/2}{(1+|\xi|-N/2)}\right)^{2s}+1\right)\\ &\le \frac{r}{2N}\left\{\left(1+\frac{N+r/2}{1+N}\right)^{2s}+1\right\}\to 0,\quad N\to\infty.\end{aligned}$$

Similarly, if $3N/2\le|\xi|$ and $-1/2<s<0$, we use the mean-value theorem to obtain the inequalities that

$$(3.10)\quad (2s+1)\,(t_2-t_1)\,t_2^{2s}\le|t_1^{2s+1}-t_2^{2s+1}|\le(2s+1)\,(t_2-t_1)\,t_1^{2s},\quad \text{if}\quad 0\le t_1<t_2.$$

Hence,

$$\begin{aligned}\frac{w_1(\xi+(I_{N+r}\setminus I_N))}{w_1(\xi+I_N)} &\le \frac{\left[(1+|\xi|+N/2)^{2s}+(1+|\xi|-N/2-r/2)^{2s}\right](r/2)}{(1+|\xi|+N/2)^{2s}\,N}\\ &= \frac{r}{2N}\left(\left(1-\frac{N+r/2}{(1+|\xi|+N/2)}\right)^{2s}+1\right)\\ &\le \frac{r}{2N}\left\{\left(1-\frac{N+r/2}{1+2N}\right)^{2s}+1\right\}\to 0,\quad N\to\infty.\end{aligned}$$

It follows that

$$\sup_{|\xi|>N/2+r/2}\frac{w_1(\xi+(I_{N+r}\setminus I_N))}{w_1(\xi+I_N)}\to 0,\quad N\to\infty.$$

If $N/2<|\xi|\le N/2+r/2$ and $s>0$, we have, using (3.9),

$$\begin{aligned}&(2s+1)\,w_1(\xi+(I_{N+r}\setminus I_N))\\ &=\left\{(1-|\xi|+N/2+r/2)^{2s+1}+(1+|\xi|+N/2+r/2)^{2s+1}-2\right.\\ &\quad\left.-(1+|\xi|+N/2)^{2s+1}+(1+|\xi|-N/2)^{2s+1}\right\}\\ &\le\left\{(1+r/2)^{2s+1}+(1+|\xi|+N/2+r/2)^{2s+1}\right.\\ &\quad\left.-(1+|\xi|+N/2)^{2s+1}+(1+r/2)^{2s+1}\right\}\\ &\le\left\{2\,(1+r/2)^{2s+1}+(2s+1)\,(1+|\xi|+N/2+r/2)^{2s}\,(r/2)\right\}\\ &\le\left\{2\,(1+r/2)^{2s+1}+(2s+1)\,(1+N+r)^{2s}\,(r/2)\right\}\end{aligned}$$

and

$$\begin{aligned}(2s+1)\,w_1(\xi+I_N)&=\left\{(1+|\xi|+N/2)^{2s+1}-(1+|\xi|-N/2)^{2s+1}\right\}\\ &\ge\left\{(1+N)^{2s+1}-(1+r/2)^{2s+1}\right\}.\end{aligned}$$

Hence,

$$\frac{w_1(\xi+(I_{N+r}\setminus I_N))}{w_1(\xi+I_N)}\le\frac{2\,(1+r/2)^{2s+1}+(2s+1)\,(1+N+r)^{2s}\,(r/2)}{(1+N)^{2s+1}-(1+r/2)^{2s+1}}\to 0,$$

as $N \to \infty$. On the other hand, if $N/2 < |\xi| \le N/2 + r/2$ and $-1/2 < s < 0$, we have, using (3.10),

$$\begin{aligned}
&(2s+1)\, w_1(\xi + (I_{N+r} \setminus I_N)) \\
&\le \big\{(1+r/2)^{2s+1} + (1+|\xi| + N/2 + r/2)^{2s+1} \\
&\quad -(1+|\xi|+N/2)^{2s+1} + (1+r/2)^{2s+1}\big\} \\
&\le \big\{2\,(1+r/2)^{2s+1} + (2s+1)\,(1+|\xi|+N/2)^{2s}\,(r/2)\big\} \\
&\le \big\{2\,(1+r/2)^{2s+1} + (2s+1)\,(1+N)^{2s}\,(r/2)\big\}
\end{aligned}$$

and

$$\begin{aligned}
(2s+1)\, w_1(\xi + I_N) &= \big\{(1+|\xi|+N/2)^{2s+1} - (1+|\xi|-N/2)^{2s+1}\big\} \\
&\ge \big\{(1+N)^{2s+1} - (1+r/2)^{2s+1}\big\}.
\end{aligned}$$

Hence,

$$\frac{w_1(\xi + (I_{N+r} \setminus I_N))}{w_1(\xi + I_N)} \le \frac{2\,(1+r/2)^{2s+1} + (2s+1)\,(1+N)^{2s}\,(r/2)}{(1+N)^{2s+1} - (1+r/2)^{2s+1}} \to 0,$$

as $N \to \infty$. This shows that

$$\sup_{N/2 < |\xi| \le N/2 + r/2} \frac{w_1(\xi + (I_{N+r} \setminus I_N))}{w_1(\xi + I_N)} \to 0, \quad N \to \infty.$$

Finally, if $|\xi| \le N/2$, we have, using (3.9) and (3.10) again, that

$$\begin{aligned}
&(2s+1)\, w_1(\xi + (I_{N+r} \setminus I_N)) \\
&= \big\{(1-|\xi|+N/2+r/2)^{2s+1} - (1-|\xi|+N/2)^{2s+1} \\
&\quad +(1+|\xi|+N/2+r/2)^{2s+1} - (1+|\xi|+N/2)^{2s+1}\big\} \\
&\le \begin{cases} (2s+1)(r/2)\,\big[(1-|\xi|+N/2+r/2)^{2s} + (1+|\xi|+N/2+r/2)^{2s}\big], & s > 0, \\ (2s+1)(r/2)\,\big[(1-|\xi|+N/2)^{2s} + (1+|\xi|+N/2)^{2s}\big], & -1/2 < s < 0, \end{cases} \\
&\le \begin{cases} (2s+1)(r/2)\,\big[(1+N/2+r/2)^{2s} + (1+N+r/2)^{2s}\big], & s > 0, \\ (2s+1)(r/2)\,\big[1 + (1+N/2)^{2s}\big], & -1/2 < s < 0, \end{cases} \\
&= O(N^{2s})
\end{aligned}$$

and

$$\begin{aligned}
(2s+1)\, w_1(\xi + I_N) &= \big\{(1-|\xi|+N/2)^{2s+1} + (1+|\xi|+N/2)^{2s+1} - 2\big\} \\
&\ge \big\{(1+N/2)^{2s+1} - 1\big\}.
\end{aligned}$$

This shows that

$$\sup_{|\xi| \le N/2} \frac{w_1(\xi + (I_{N+r} \setminus I_N))}{w_1(\xi + I_N)} \to 0, \quad N \to \infty.$$

We now consider the case $s = -\frac{1}{2}$. We can assume that $0 < r \leq 1$. We have

$$\begin{aligned}
w_1(\xi + I_N) &= \chi_{(N/2,\infty)}(|\xi|) \int_{|\xi|-N/2}^{|\xi|+N/2} (1+\lambda)^{-1}\, d\lambda \\
&\quad + \chi_{[0,N/2]}(|\xi|) \left\{ \int_0^{-|\xi|+N/2} (1+\lambda)^{-1}\, d\lambda + \int_0^{|\xi|+N/2} (1+\lambda)^{-1}\, d\lambda \right\} \\
&= \chi_{(N/2,\infty)}(|\xi|) \log\left(\frac{1+|\xi|+N/2}{1+|\xi|-N/2}\right) \\
&\quad + \chi_{[0,N/2]}(|\xi|) \log\left((1+N/2)^2 - |\xi|^2\right).
\end{aligned}$$

Thus if $r > 0$, we have

$$\begin{aligned}
w_1(\xi + (I_{N+r} \setminus I_N)) = &= \chi_{(N/2+r/2,\infty)}(|\xi|) \log\left(\frac{1+|\xi|+N/2+r/2}{1+|\xi|-N/2-r/2}\right) \\
&\quad + \chi_{[0,N/2+r/2]}(|\xi|) \log\left((1+N/2+r/2)^2 - |\xi|^2\right) \\
&\quad - \chi_{(N/2,\infty)}(|\xi|) \log\left(\frac{1+|\xi|+N/2}{1+|\xi|-N/2}\right) \\
&\quad - \chi_{[0,N/2]}(|\xi|) \log\left((1+N/2)^2 - |\xi|^2\right).
\end{aligned}$$

Consider first the case $|\xi| > N/2 + r/2$. We have then

$$\begin{aligned}
w_1(\xi + (I_{N+r} \setminus I_N)) &= \log\left(\frac{1+|\xi|+N/2+r/2}{1+|\xi|-N/2-r/2}\right) - \log\left(\frac{1+|\xi|+N/2}{1+|\xi|-N/2}\right) \\
= \log &\left(\frac{(1+|\xi|+N/2)\left(1+\frac{r/2}{1+|\xi|+N/2}\right)}{(1+|\xi|-N/2)\left(1-\frac{r/2}{1+|\xi|-N/2}\right)}\right) - \log\left(\frac{1+|\xi|+N/2}{1+|\xi|-N/2}\right) \\
= \log &\left(1+\frac{r/2}{1+|\xi|+N/2}\right) - \log\left(1-\frac{r/2}{1+|\xi|-N/2}\right)
\end{aligned}$$

Note that we have the inequalities

$$\frac{x}{2} \leq \log(1+x) \leq x \quad \text{and} \quad x \leq -\log(1-x) \leq 2\,x, \quad 0 \leq x \leq 1/2.$$

Since $0 \leq r \leq 1$, we have

$$w_1(\xi + (I_{N+r} \setminus I_N)) \leq \frac{r}{1+|\xi|+N/2} + \frac{r}{1+|\xi|-N/2}$$

and

$$w_1(\xi + I_N) = \log\left(\frac{1+|\xi|+N/2}{1+|\xi|-N/2}\right) = \log\left(1+\frac{N}{1+|\xi|-N/2}\right).$$

If $|\xi| \geq \frac{5\,N}{2}$, we have $\frac{N}{1+|\xi|-N/2} \leq 1/2$ and

$$\frac{w_1(\xi + (I_{N+r} \setminus I_N))}{w_1(\xi + I_N)} \leq 2\,\frac{\frac{r}{1+|\xi|+N/2} + \frac{r}{1+|\xi|-N/2}}{\frac{N}{1+|\xi|-N/2}} \leq \frac{4\,r}{N}.$$

If $N/2 + r/2 \le |\xi| \le 5N/2$, we have

$$\frac{r/2}{1+|\xi|+N/2} \le \frac{r/2}{1+|\xi|-N/2} \le \frac{r/2}{1+r/2} = \frac{1}{(2/r)+1} \le \frac{1}{2}$$

since $0 < r \le 1$ and thus

$$\begin{aligned}\frac{w_1(\xi+(I_{N+r}\setminus I_N))}{w_1(\xi+I_N)} &= \frac{\log\left(1+\frac{r/2}{1+|\xi|+N/2}\right)-\log\left(1-\frac{r/2}{1+|\xi|-N/2}\right)}{\log\left(1+\frac{N}{1+|\xi|-N/2}\right)}\\ &\le \frac{\frac{r/2}{1+|\xi|+N/2}+2\,\frac{r/2}{1+|\xi|-N/2}}{\log\left(1+\frac{N}{1+|\xi|-N/2}\right)}.\\ &\le \frac{\frac{3\,r}{2N}\,\frac{N}{1+|\xi|-N/2}}{\log\left(1+\frac{N}{1+|\xi|-N/2}\right)}.\end{aligned}$$

Note that the function $f(x) = \frac{x}{\log(1+x)}$ is increasing on $(0,\infty)$. Indeed,

$$f'(x) = \frac{\log(1+x)-\frac{x}{1+x}}{\log^2(1+x)} > 0, \quad x > 0,$$

since

$$\log(1+x) = \int_0^x \frac{1}{1+t}\,dt > \int_0^x \frac{1}{1+x}\,dt = \frac{x}{1+x}, \quad x > 0.$$

Hence,

$$\frac{w_1(\xi+(I_{N+r}\setminus I_N))}{w_1(\xi+I_N)} \le \frac{\frac{3\,r}{2N}\,\frac{N}{1+2N}}{\log\left(1+\frac{N}{1+2N}\right)} \to 0, \quad N\to\infty.$$

This shows that

$$\sup_{|\xi|\ge N/2+r/2} \frac{w_1(\xi+(I_{N+r}\setminus I_N))}{w_1(\xi+I_N)} \to 0, \quad N\to\infty.$$

If $N/2 < |\xi| \le N/2 + r/2$, we have

$$\begin{aligned}w_1(\xi+(I_{N+r}\setminus I_N)) &= \log\left((1+N/2+r/2)^2-|\xi|^2\right)-\log\left(\frac{1+|\xi|+N/2}{1+|\xi|-N/2}\right)\\ &\le \log((1+r/2)^2+r\,(1+N/2))+\log(1+r/2)-\log(1+N)\\ &= \log\left(\frac{(1+r/2)^2+r\,(1+N/2))\,(1+r/2)}{(1+N)}\right) \le C, \quad N\to\infty,\end{aligned}$$

and

$$w_1(\xi+I_N) = \log\left(\frac{1+|\xi|+N/2}{1+|\xi|-N/2}\right) \ge \log\left(\frac{1+N}{1+r/2}\right)$$

from which it follows directly that

$$\sup_{N/2\le|\xi|\le N/2+r/2} \frac{w_1(\xi+(I_{N+r}\setminus I_N))}{w_1(\xi+I_N)} \to 0, \quad N\to\infty.$$

Finally, if $|\xi| \le N/2$,

$$w_1(\xi+(I_{N+r}\setminus I_N)) = \log\left((1+N/2+r/2)^2-|\xi|^2\right)-\log\left((1+N/2)^2-|\xi|^2\right)$$

and

$$\frac{w_1(\xi+(I_{N+r}\setminus I_N))}{w_1(\xi+I_N)}=\frac{\log\left((1+N/2+r/2)^2-|\xi|^2\right)-\log\left((1+N/2)^2-|\xi|^2\right)}{\log\left((1+N/2)^2-|\xi|^2\right)}.$$

Note that, if $a>0$, the function $g(x)=\frac{\log(x+a)-\log x}{\log x}$ is decreasing for $x>1$ since

$$g'(x)=\frac{\frac{1}{x+a}\log x-\frac{1}{x}\log(x+a)}{\log^2 x}<0,\quad x>1.$$

This shows that

$$\begin{aligned}&\sup_{|\xi|\le R/2}\frac{w_1(\xi+(I_{N+r}\setminus I_N))}{w_1(\xi+I_N)}\\&\le\frac{\log\left((1+N/2+r/2)^2-(N/2)^2\right)-\log\left((1+N/2)^2-(N/2)^2\right)}{\log\left((1+N/2)^2-(N/2)^2\right)}\\&=\frac{\log\left(\frac{(1+r/2)^2+N(1+r/2)}{1+N}\right)}{\log(1+N)}\to 0,\quad N\to\infty.\end{aligned}$$

We have thus

$$\sup_{\xi\in\mathbb{R}^d}\frac{w_1(\xi+(I_{N+r}\setminus I_N))}{w_1(\xi+I_N)}\to 0,\quad N\to\infty,$$

for $s=-1/2$ and this proves our claim. □

4. Exponential weights

In this last section, we consider the class of exponential weights in one dimension. These are the weights $w(\xi)=e^{r\xi}$, where $r\in\mathbb{R}\setminus\{0\}$, for which the associated sub-multiplicative weight is $v(\xi)=w(\xi)=e^{r\xi}$. These are moderate weights but they do not satisfy the condition (3.2) of Theorem 3.3 as can be easily checked. Suppose that there exists a function $F(\gamma)$ with

$$\int_{\mathbb{R}}F(\lambda)\,v(-\lambda)\,d\lambda=\int_{\mathbb{R}}F(\lambda)\,e^{-r\lambda}\,d\lambda=1 \tag{4.1}$$

such that

$$(\mu*F)(\xi)\le C\,(w*F)(\xi)=C\int_{\mathbb{R}}F(\lambda)\,e^{r(\xi-\lambda)}\,d\lambda=C\,e^{r\xi},\quad \xi\in\mathbb{R}, \tag{4.2}$$

for some positive Borel measure μ on $\mathbb{R}$. Letting $G(\lambda)=F(\lambda)\,e^{-r\lambda}$, we have

$$\int_{\mathbb{R}}G(\lambda)\,d\lambda=1 \tag{4.3}$$

and

$$\begin{aligned}(\mu*F)(\xi)&=\int_{\mathbb{R}}G(\xi-\lambda)\,e^{r(\xi-\lambda)}\,d\mu(\lambda)=e^{r\xi}\int_{\mathbb{R}}G(\xi-\lambda)\,e^{-r\lambda}\,d\mu(\lambda)\\&=e^{r\xi}\left(G*(e^{-r\lambda}\,\mu)\right)(\xi).\end{aligned}$$

It follows that the existence of F satisfying (4.1) and (4.2) is equivalent to the existence of G satisfying (4.3) and

$$\left((e^{-r\lambda}\,\mu)*G\right)(\xi)\le C,\quad \xi\in\mathbb{R}, \tag{4.4}$$

By Theorem 1.1 , this is equivalent to $\mathcal{D}^+(e^{-r\lambda}\,\mu) \leq C$, i.e. to

$$\limsup_{N\to\infty}\sup_{\xi\in\mathbb{R}} \frac{\int_{\xi+I_N} e^{-r\lambda}\,d\mu(\lambda)}{N} \leq C.$$

For these particular weights, we have thus $\mathcal{C}_w^+(\mu) = \mathcal{D}^+(w^{-1}\,\mu)$ and one can show, in a similar way, that if $\mathcal{D}^+(w^{-1}\,\mu) < \infty$, we have $\mathcal{C}_w^-(\mu) = \mathcal{D}^-(w^{-1}\,\mu)$. We now show that the constants $\mathcal{C}_w^+(\mu)$ and $\mathcal{D}_w^+(\mu)$ might be different for the exponential weight $w(\xi) = e^{\xi}$. Of course we always have the inequality

$$\mathcal{C}_w^+(\mu) \leq \mathcal{D}_w^+(\mu).$$

To get a strict inequality, consider the set

$$E = \cup_{n\geq 1}\,[2^n - 1, 2^n]$$

and let $d\mu(\xi) = e^{\xi}\,\chi_E(\xi)\,d\xi$. We have

$$\mathcal{C}_w^+(\mu) = \mathcal{D}^+(w^{-1}\,\mu) = \mathcal{D}^+(\chi_E) = 0.$$

Furthermore, for $n, N \geq 1$

$$\mu(2^n - N/2 + I_N) \geq \int_{[2^n-1,2^n]} e^{\xi}\,d\xi = e^{2^n} - e^{2^n-1} = e^{2^n}\,(1 - e^{-1}),$$

while

$$w(2^n - N/2 + I_N) = \int_{[2^n-N,2^n]} e^{\xi}\,d\xi = e^{2^n} - e^{2^n-N} = e^{2^n}\,(1 - e^{-N}).$$

This shows that

$$\sup_{\xi\in\mathbb{R}} \frac{\mu(\xi + I_N)}{w(\xi + I_N)} \geq \frac{1 - e^{-1}}{1 - e^{-N}}$$

and thus

$$\mathcal{D}_w^+(\mu) = \limsup_{N\to\infty}\sup_{\xi\in\mathbb{R}} \frac{\mu(\xi + I_N)}{w(\xi + I_N)} \geq 1 - e^{-1} > 0.$$

Acknowledgements

The author would like to thank the referees for their valuable comments which helped to improve the manuscript.

References

[1] H. Führ, K. Gröchenig, A. Haimi, A. Klotz, J.-L. Romero, Density of sampling and interpolation in reproducing kernel Hilbert spaces, preprint.

[2] J.-P. Gabardo, *Convolution inequalities for positive Borel measures on $\mathbb{R}^d$ and Beurling density*, Excursions in harmonic analysis. Volume 2, Appl. Numer. Harmon. Anal., Birkhäuser/Springer, New York, 2013, pp. 23–47, DOI 10.1007/978-0-8176-8379-5_3. MR3050313

[3] J.-P. Gabardo, *Convolution inequalities in locally compact groups and unitary systems*, Numer. Funct. Anal. Optim. **33** (2012), no. 7-9, 1005–1030, DOI 10.1080/01630563.2012.682142. MR2966142

[4] J.-P. Gabardo and C.-K. Lai, *Frames of multi-windowed exponentials on subsets of $\mathbb{R}^d$*, Appl. Comput. Harmon. Anal. **36** (2014), no. 3, 461–472, DOI 10.1016/j.acha.2013.08.004. MR3175088

[5] K. Gröchenig, *Weight functions in time-frequency analysis*, Pseudo-differential operators: partial differential equations and time-frequency analysis, Fields Inst. Commun., vol. 52, Amer. Math. Soc., Providence, RI, 2007, pp. 343–366. MR2385335

[6] K. Gröchenig and H. Razafinjatovo, *On Landau's necessary density conditions for sampling and interpolation of band-limited functions*, J. London Math. Soc. (2) **54** (1996), no. 3, 557–565, DOI 10.1112/jlms/54.3.557. MR1413898

[7] C. Heil, An introduction to weighted Wiener amalgams. In: Wavelets and their Applications (Chennai, January 2002), M. Krishna, R. Radha and S. Thangavelu, eds., Allied Publishers, New Delhi (2003), pp. 183–216.

[8] C. Heil, *The density theorem and the homogeneous approximation property for Gabor frames*, Representations, wavelets, and frames, Appl. Numer. Harmon. Anal., Birkhäuser Boston, Boston, MA, 2008, pp. 71–102, DOI 10.1007/978-0-8176-4683-7_5. MR2459314

[9] C. Heil, *History and evolution of the density theorem for Gabor frames*, J. Fourier Anal. Appl. **13** (2007), no. 2, 113–166, DOI 10.1007/s00041-006-6073-2. MR2313431

[10] M. N. Kolountzakis, *The study of translational tiling with Fourier analysis*, Fourier analysis and convexity, Appl. Numer. Harmon. Anal., Birkhäuser Boston, Boston, MA, 2004, pp. 131–187. MR2087242

[11] M. N. Kolountzakis and J. C. Lagarias, *Structure of tilings of the line by a function*, Duke Math. J. **82** (1996), no. 3, 653–678, DOI 10.1215/S0012-7094-96-08227-7. MR1387688

[12] G. Kutyniok, *Beurling density and shift-invariant weighted irregular Gabor systems*, Sampl. Theory Signal Image Process. **5** (2006), no. 2, 163–181. MR2233970

[13] G. Kutyniok, *Affine density in wavelet analysis*, Lecture Notes in Mathematics, vol. 1914, Springer, Berlin, 2007. MR2340835

[14] H. J. Landau, *Necessary density conditions for sampling and interpolation of certain entire functions*, Acta Math. **117** (1967), 37–52, DOI 10.1007/BF02395039. MR0222554

[15] S. Nitzan and A. Olevskii, *Revisiting Landau's density theorems for Paley-Wiener spaces* (English, with English and French summaries), C. R. Math. Acad. Sci. Paris **350** (2012), no. 9-10, 509–512, DOI 10.1016/j.crma.2012.05.003. MR2929058

Department of Mathematics and Statistics, McMaster University, 1280, Main St. W., Hamilton, Ontario, Canada

Email address: gabardo@mcmaster.ca

Contemporary Mathematics
Volume **706**, 2018
http://dx.doi.org/10.1090/conm/706/14215

p-Riesz bases in quasi shift invariant spaces

Laura De Carli and Pierluigi Vellucci

ABSTRACT. Let $1 \leq p < \infty$ and let $\psi \in L^p(\mathbb{R}^d)$. We study $p-$Riesz bases of quasi shift invariant spaces $V^p(\psi; Y)$.

1. Introduction

Let $1 \leq p < \infty$ and let $\psi \in L^p(\mathbb{R}^d)$. We consider the shift invariant space $V^p(\psi) = \overline{\text{Span}\{\tau_k\psi\}_{k\in\mathbb{Z}^d}}$, where $\tau_s f(x) = f(x+s)$ is the translation and "bar" denotes the closure in $L^p(\mathbb{R}^d)$. Shift-invariant spaces appear naturally in signal theory and in other branches of applied sciences. In [**2**] [**13**] and in the recent preprint [**14**] *quasi-shift invariant spaces* of functions are considered. Given $X = \{x_j\}_{j\in\mathbb{Z}^d}$, a countable and discrete[1] subset of $\mathbb{R}^d$ and a function $\psi \in L^p(\mathbb{R}^d)$, we let

$$V^p(\psi;\, X) = \overline{\text{Span}\{\tau_{x_j}\psi\}}. \tag{1.1}$$

Thus, $V^p(\psi) = V^p(\psi; \mathbb{Z}^d)$. Quasi-shift invariant spaces are also called *Spline-type spaces* in [**9**], [**10**] [**11**], [**23**].

Following [**2**], [**7**], we say that the translates $\{\tau_{x_j}\psi\}_{j\in\mathbb{Z}^d}$ form a p-Riesz basis in $V^p(\psi;\, X)$ if there exist constants $A,\ B > 0$ such that, for every finite set of coefficients $\vec{d} = \{d_j\} \subset \mathbb{C}$,

$$A\|\vec{d}\|_{\ell^p} \leq \|\sum_j d_j\tau_{x_j}\psi\|_p \leq B\|\vec{d}\|_{\ell^p}. \tag{1.2}$$

Here and throughout the paper, we have let $\|f\|_p = \left(\int_{\mathbb{R}^d}|f(x)|^p dx\right)^{\frac{1}{p}}$ and $\|\vec{c}\|_{\ell^p} = (\sum_j |c_j|^p)^{\frac{1}{p}}$. If $x = (x_1, ..., x_d),\ y = (y_1, ..., y_d) \in \mathbb{R}^d$, we will often let $x\cdot y = x_1y_1 + ... + x_dy_d$ and we will write $|x|_p$ instead of $||x||_{\ell^p}$. We will let $|x|_\infty = \sup_{1\leq j\leq d}|x_j|$.

If (1.2) holds, then it is possible to prove that

$$^2V^p(\psi; X) = \{f = \sum_{k\in\mathbb{Z}^d} d_k\tau_{x_k}\psi(x),\ \vec{d} \in \ell^p\ \} \tag{1.3}$$

and the sequence $\{d_k\}_{k\in\mathbb{Z}^d}$ is uniquely determined by f.

$p-$Riesz bases allow a stable reconstruction of functions in $V^p(\psi; X)$; when $X = \mathbb{Z}^d$ and $\mathcal{B} = \{\tau_j\psi\}_{j\in\mathbb{Z}^d}$ is a $p-$Riesz basis of $V^p(\psi)$, the coefficient d_j in (1.3)

2010 *Mathematics Subject Classification.* Primary 46A35, 41A04.

[1] A countable set $X \subset \mathbb{R}^d$ is *discrete* if for every $x_j \in X$ there exists $\delta_j > 0$ such that $|x_j - x_k|_2 > \delta_j$ for every $k \neq j$.

[2] A proof of this ientity was kindly provided to us by K. Hamm.

can be expressed in an unique way in terms of the functions in the dual basis of $\mathcal{B}$. See [**30**], [**5**] and also [**1**] for explicit reconstruction formulas.

When ψ has compact support, it is known (see e.g. [**1**, Prop. 1.1], [**16**], [**24**]) that $\mathcal{B}$ is a $p-$Riesz basis in $V^p(\psi)$ if and only if $\sum_{m\in\mathbb{Z}^d}|\hat{\psi}(y+m)|^2\neq 0$ for every $y\in[-\frac{1}{2},\frac{1}{2})^d$ and every $m\in\mathbb{Z}^d$. See also Lemma 5 in Section 2.

We have denoted with $\hat{\psi}(y)=\int_{\mathbb{R}^d}e^{2\pi ix\cdot y}f(x)dx$ the Fourier transform of ψ. The proof of the aforementioned result relies on the lattice structure of $\mathbb{Z}^d$ and on standard Fourier analysis technique and does not easily generalize to other sets of translations.

Let $\psi\in L^p(\mathbb{R}^d)$, $1\leq p<\infty$, and let $X=\{x_j\}_{j\in\mathbb{Z}^d}$ be a discrete set of $\mathbb{R}^d$. It is natural to consider the following problem:

Problem 1. *Let $\mathcal{B}_X=\{\tau_{x_j}\psi\}_{j\in\mathbb{Z}^d}$ be a $p-$Riesz basis for $V^p(\psi;X)$; can we find $\delta>0$ so that, for every $Y=\{y_j\}_{j\in\mathbb{Z}^d}\subset\mathbb{R}^d$ with $\sup_j|y_j-x_j|_2<\delta$, the set $\mathcal{B}_Y=\{\tau_{y_j}\psi\}_{j\in\mathbb{Z}^d}$ is a $p-$Riesz basis for $V^p(\psi;X)$?*

This problem cannot be solved if ψ has compact support. For example, let $\psi(x)=\mathrm{rect}(x)$ be the characteristic function of the interval $[-\frac{1}{2},\frac{1}{2})$ and let $X=\mathbb{Z}$; let $Y=\{y_n\}_{n\in\mathbb{Z}}$ be such that $y_0=\delta>0$ and $y_n=n$ when $n\neq 0$. All functions in $V^p(\mathrm{rect};Y)$ vanish in the interval $[-\frac{1}{2},-\frac{1}{2}+\delta]$ and so $V^p(\mathrm{rect};Y)\neq V^p(\mathrm{rect})$.

We prove in Section 3 that Problem 1 can be solved when $p=2$ and ψ is *band-limited*, i.e., when the Fourier transform of ψ has compact support. See also Section 5 for more remarks and comments on problem 1

We are concerned with the following problem:

Problem 2. *With the notation of Problem 1: let $\mathcal{B}_X$ be a $p-$Riesz basis for $V^p(\psi;X)$ and let $Y=\{y_n\}_{n\in\mathbb{Z}^d}$ that satisfies $\sup_n|y_n-x_n|_2<\delta$; is $\mathcal{B}_Y$ a $p-$Riesz basis for $V^p(\psi;Y)$ whenever δ is sufficiently small?*

It is proved in [**10**] that Problem 2 has always solution when X is *relatively separated*, i.e., when $X=X_1\cup\ldots\cup X_k$, with $X_j=\{x_{j,n}\}_{n\in\mathbb{Z}^d}$ and $\inf_{n\neq m}|x_{j,n}-x_{j,m}|_2>0$.

In Section 2 we prove the following theorem.

THEOREM 1. *Suppose that that ψ is in the Sobolev space $W^{1,p}(\mathbb{R}^d)$, with $1<p<\infty$, and that $\{\tau_{x_j}\psi\}_{j\in\mathbb{Z}^d}$ is a $p-$Riesz basis of $V^p(\psi;X)$. For every $j\in\mathbb{Z}^d$ there exists $\delta_j>0$ such that $\{\tau_{y_j}\psi\}_{j\in\mathbb{Z}^d}$ is a $p-$Riesz basis of $V^p(\psi;Y)$ whenever $|x_j-y_j|_2<\delta_j$.*

We recall that $W^{1,p}(D)$ is the space of $L^p(D)$ functions whose partial distributional derivatives are also in $L^p(D)$ and that $W_0^{1,p}(D)$ is the closure of $C_0^\infty(D)$ in $W^{1,p}(D)$.

When X is not relatively separated the δ_j's in Theorem 1 may not have a positive lower bound, but we can still solve Problem 2 in the cases considered in Theorems 2 and 3 below.

THEOREM 2. *Assume that $\psi\in L^1(\mathbb{R}^d)\cap L^2(\mathbb{R}^d)$ satisfies*

$$0<c=\sum_{k\in\mathbb{Z}^d}\inf_{x\in[0,1)^d}|\hat{\psi}(x+k)|^2<\sum_{k\in\mathbb{Z}^d}\sup_{x\in[0,1)^d}|\hat{\psi}(x+k)|^2=C<\infty. \tag{1.4}$$

Then, Problem 2 can be solved when $p = 2$ *and* $\{e^{2\pi i x_n \cdot x}\}_{n\in\mathbb{Z}^d}$ *is a Riesz basis in* $L^2([0,1)^d)$.

We recall that the *amalgam space* $W(L^\infty,\, \ell^q)$ is the set of measurable functions $f : \mathbb{R}^d \to \mathbb{C}$ for which $||f||_{W(L^\infty,\,\ell^q)} = \left(\sum_{n\in\mathbb{Z}^d} \sup_{x\in[0,1)^d} |f(x+n)|_2^q\right)^{\frac{1}{q}} < \infty$. The amalgam space $W(L^r,\, \ell^q)$ can be defined in a similar manner.

The assumption (1.4) implies that $\hat{\psi}$ is in the *amalgam space* $W(L^\infty,\, \ell^2)$.

From Theorem 2 follows that if ψ satisfies (1.4), then $\{\tau_{y_k}\psi\}_{k\in\mathbb{Z}^d}$ is a $2-$Riesz basis of $V^2(\psi, Y)$ whenever $|k - y_k|_\infty < \frac{1}{4}$. See the remark after the proof of Theorem 2 in Section 4.

Exponential Riesz bases in $L^2(0,1)$ are completely understood and classified [**20**]. To the best of our knowledge, no such characterization exists for exponential bases on $L^2((0,1)^d)$ when $d > 1$.

For our next theorem we consider ψ in the Sobolev space $W_0^{1,p}(\mathbb{R}^d)$; we denote with $\partial_j\psi = \frac{\partial\psi}{\partial x_j}$ the partial derivative (in distribution sense) of ψ and we let $\nabla\psi = (\partial_1\psi,\, ...,\, \partial_d\psi)$ be the gradient of ψ. Let $Y = \{y_k\}_{k\in\mathbb{Z}^d}$ and

$$L = \sup_{k\in\mathbb{Z}^d} |y_k - k|_2 < \infty.$$

We prove the following

THEOREM 3. *Let* $D = (a_1,\, b_1) \times ... \times (a_d,\, b_d)$ *be a bounded rectangle in* $\mathbb{R}^d$. *Let* $\psi \in W_0^{1,p}(D)$, *with* $1 \le p < \infty$, *and let* $\{\tau_k\psi\}_{k\in\mathbb{Z}^d}$ *be a* $p-$*Riesz basis of* $V^p(\psi)$ *with frame constants* $0 < A \le B < \infty$. *If*

$$C = L\sum_{j=1}^{d}(1 + [b_j - a_j + L])^{p-1}||\partial_j\psi||_p^p < A, \tag{1.5}$$

the set $\{\tau_{y_k}\psi\}_{k\in\mathbb{Z}^d}$ *is a* $p-$*Riesz basis of* $V^p(\psi;\, Y)$ *with constants* $B+C$ *and* $A-C$.

The proofs of Theorems 2 and 3 are in Section 3.

Our Theorem 3 can be compared to [**10**, Theorem 3.5]. In that theorem it is $|\nabla(\psi)|$ is in the amalgam space $W(L^\infty,\, \ell^1)$, and $\inf_{x\in[0,1)^d}\sum_{k\in\mathbb{Z}^d}|\nabla\psi(x+k)|_2 > 0$; it is proved that $\{\tau_{y_k}\psi\}_{k\in\mathbb{Z}^d}$ is a Riesz basis of $V^2(\psi;\, Y)$ if $C' = L^2(1+2L)^{2d}||\nabla\psi||^2_{W(L^\infty,\,\ell^1)} < A$. Generalizations to functions for which $|\nabla(\psi)|$ is in the amalgam space $W(L^q,\, \ell^1)$, with $q > d$ are also possible (see Remark 3.2 in [**10**]).

Our Theorem 3 reduces to [**10**, Theorem 3.5] when $p > d$ and ψ has compact support. For example, when ψ has support in $[0,1)^d$, the norm in $W(L^p, \ell^1)$ reduces to $||\nabla\psi||_p$. The constant C in Theorem 3 may be smaller than C' defined above when the support of ψ is small.

Theorem 3 does not apply when $\psi = \text{rect}$ or when ψ is a step function; For $J \ge 1$, we let $\mathcal{S}_J = \left\{s(t) = \sum_{|j|\le J} s_j\, \text{rect}(t-j),\ s_j \in \mathbb{R}\right\}$. We let $p' = \frac{p}{p-1}$ be the dual exponent of p. The following theorem is proved in Section 4.2.

THEOREM 4. *Assume that* $g \in \mathcal{S}_J$ *and that* $\{\tau_k g\}_{k\in\mathbb{Z}}$ *is a* $p-$*Riesz basis for* $V^p(g)$, *with frame constants* A *and* B. *If*

$$2^p LJ\, ||g||_{p'}^p < A\,,$$

the sequence $\{\tau_{y_k} g\}_{k\in\mathbb{Z}}$ *is a Riesz basis for* $V^p(g;Y)$.

2. Preliminaries

2.1. Notation. We denote with $\langle f, g\rangle = \int_{\mathbb{R}^d} f(x)\bar{g}(x)dx$ and $\|f\|_2 = \sqrt{\langle f, f\rangle}$ the standard inner product and norm in $L^2(\mathbb{R}^d)$. For a given $p \in \mathbb{R}^d$ and $\delta > 0$, we let $B(p,\delta) = \{x \in \mathbb{R}^d : |x-p|_2 < \delta\}$.

We let $\text{rect}(x) = \chi_{[-\frac{1}{2},\frac{1}{2})}(x)$ be the characteristic function of the interval $[-\frac{1}{2},\frac{1}{2})$ and $\beta^s = \text{rect}^{(s+1)}(x) = \text{rect} * ... * \text{rect}(x)$ be the $s+1$−times iterated convolution of rect. The function $\beta^s(x)$, a piecewise polynomial function of degree s, is a *B-spline* of order s. See [**27**], where the B-splines were first introduced, and [**21**], [**32**] and the references cited there.

2.2. $p-$ **Riesz bases.** Recall that a Schauder basis in a separable Banach space V is a linearly independent set $\mathcal{B} = \{v_j\}_{j\in\mathbb{Z}}$ such that: $\overline{span(\mathcal{B})} = V$, and there exists a sequence of bounded linear functions $f_j : X \to \mathbb{C}$ (the *functional coefficients of the basis*) such that $x = \sum_j f_j(x)v_j$ for every $x \in V$.

Following [**6**], [**7**] and [**2**], we say that $\mathcal{B}$ is a $p-$Riesz basis of V, with $1 \le p < \infty$, if $\overline{\text{Span}(\mathcal{B})} = V$, if every series $\sum_n a_n v_n$ converges in V when $\vec{a} = (a_n)_{n\in\mathbb{Z}} \in \ell^p$ and if there exist constants $A,\ B > 0$ such that, for every finite sequence of coefficients $\{d_j\}_{j\in\mathbb{Z}} \subset \mathbb{C}$, we have

$$A\|\vec{d}\|_{\ell^p} \le \|\sum_j d_j v_j\|_p \le B\|\vec{d}\|_{\ell^p}.$$

Every $p-$Riesz basis is a Schauder basis. As mentioned in the introduction, when $V = V^p(\psi)$ and ψ has compact support, the functional coefficients of the basis $\{\tau_k\psi\}_{k\in\mathbb{Z}}$ of $V^p(\psi)$ can be written in terms of the dual functions of the basis.

The following results are well known (see e.g. [**1**, Prop. 1.1], [**16**], [**24**]).

LEMMA 5. *a) Let $\psi \in L^p_0(\mathbb{R}^d)$. The set $\mathcal{B} = \{\tau_k\psi\}_{k\in\mathbb{Z}^d}$ is a $p-$Riesz basis in $V^p(\psi)$ if and only if*

$$\sum_{m\in\mathbb{Z}^d} |\hat{\psi}(y+m)|^2 \neq 0 \quad \textit{for every } y \in [-\tfrac{1}{2},\tfrac{1}{2})^d. \tag{2.1}$$

b) If $\psi \in W(L^\infty,\ell^1)$ is continuous and if $\mathcal{B}$ is Riesz basis in $V^2(\psi)$, then $\mathcal{B}$ is a p-Riesz basis in $V^p(\psi)$ for every $1 \le p < \infty$.

PROOF. For the convenience of the reader we prove that if $\psi \in L^2_0(\mathbb{R}^d)$, $\mathcal{B}$ is a Riesz basis of $V^2(\psi)$ with constants $0 < A \le B < \infty$ if and only if the following inequality holds for every $y \in Q = [-\frac{1}{2},\frac{1}{2})^d$.

$$A \le \sum_{m\in\mathbb{Z}^d} |\hat{\psi}(y+m)|^2 \le B. \tag{2.2}$$

We can verify (using e.g. the Poisson summation formula) that the function in (2.1) is continuous in $\overline{Q}$, and so (2.2) is equivalent to (2.1).

Let $\{c_k\}_{k\in\mathbb{Z}^d} \subset \mathbb{C}$ be a finite set of coefficients such that $\sum_k |c_k|^2 = 1$. The Fourier transform of $f = \sum_{k\in\mathbb{Z}^d} c_k\tau_k\psi$ is

$$\hat{f}(y) = \hat{\psi}(y)\sum_{k\in\mathbb{Z}^d} c_k e^{2\pi i y\cdot k} = \hat{\psi}(y)M(y).$$

and by Plancherel's theorem

$$\|f\|_2^2 = \|\hat{f}\|_2^2 = \sum_{m\in\mathbb{Z}^d} \int_{m+[-\frac{1}{2},\frac{1}{2})^d} |\hat{f}(y)|^2 dy = \sum_{m\in\mathbb{Z}^d} \int_Q |\hat{\psi}(y+m)|^2 |M(y)|^2 dy$$
$$= \int_Q |M(y)|^2 \sum_{m\in\mathbb{Z}^d} |\hat{\psi}(y+m)|^2 dy.$$

Let $g = \sum_{m\in\mathbb{Z}^d} |\hat{\psi}(y+m)|^2$; if (2.2) holds, from $||f||_2^2 = \int_Q |M(y)|^2 g(y) dy$ and $\int_Q |M(y)|^2 dy = \sum_k |c_k|^2 = 1$ follows that $A \le ||f||_2^2 \le B$.

Conversely, from $A\sum_k |c_k|^2 \le \|f\|_2^2 \le B\sum_k |c_k|^2$ and the above considerations, follows that

$$A\|M^2\|_{L^1(Q)} \le \int_Q |M(y)|^2 g(y) dy \le B\|M^2\|_{L^1(Q)}. \tag{2.3}$$

Every non-negative $h \in L^1(Q)$ can be written as $h = |M|^2$, with $M \in L^2(Q)$. The dual of $L^1(Q)$ is $L^\infty(Q)$ and so $\|g\|_{L^\infty(Q)} = \sup_{\|h\|_{L^1(Q)}=1} \int_Q f(y)g(y)dy$. From (2.3) follows that $A \le \|g\|_{L^\infty(Q)} \le B$ as required.

□

We will use the following Paley-Wiener type result.

LEMMA 6. *Let X, $Y \subset \mathbb{R}^d$ be countable and discrete. Suppose that $\{\tau_{x_j}\psi\}_{j\in\mathbb{Z}^d}$ is a $p-$Riesz basis of $V^p(\psi; X)$ with constants $A \le B$. If the inequality*

$$\left\|\sum_j a_j(\tau_{x_j}\psi - \tau_{y_j}\psi)\right\|_p^p \le C\sum_n |a_n|^p$$

holds for all finite sequences $\{a_n\}_{n\in\mathbb{Z}^d} \in \mathbb{C}$ with a constant $C < A$, the sequence $\{\tau_{y_j}\psi\}_{j\in\mathbb{Z}^d}$ is a $p-$Riesz basis of $V^p(\psi; Y)$ with constants $B + C$ and $A - C$.

PROOF. Assume that $\sum_n |a_n|^p = 1$; we have:

$$\left\|\sum_j a_j\tau_{y_j}\psi\right\|_p \le \left\|\sum_j a_j(\tau_{x_j}\psi - \tau_{y_j}\psi)\right\|_p + \left\|\sum_j a_j\tau_{x_j}\psi\right\|_p \le C + B$$

and

$$\left\|\sum_j a_j\tau_{y_j}\psi\right\|_p \ge \left\|\sum_j a_j\tau_{x_j}\psi\right\|_p - \left\|\sum_j a_j(\tau_{x_j}\psi - \tau_{y_j}\psi)\right\|_p \ge A - C.$$

□

PROOF OF THEOREM 1 . Assume $p \in (1, \infty)$ and $\sum_j |a_j|^p = 1$. Let $p' = \frac{p}{p-1}$ is the dual exponent of p and let $\{\delta_j\}_{j\in\mathbb{Z}^d}$ be a sequence of positive constants such that $\sum_j |\delta_j|^{p'} < \infty$, We recall that, when $1 < p < \infty$, a function $f \in L^p(\mathbb{R}^d)$ is in the Sobolev space $W^{1,p}(\mathbb{R}^d)$ if and only if there is a constant $c > 0$ that depends on f but not on δ, such that

$$\omega_p(\delta, f) = \sup_{|t|<\delta} ||\tau_t f - f||_p \le c\delta \tag{2.4}$$

for every $\delta > 0$. Furthermore, one can choose $c = \|\nabla f\|_p$. See e.g. Prop. 9.3 in [**4**]. By (2.4) and Hölder' s inequality,

$$\left\|\sum_j a_j(\tau_{x_j}\psi - \tau_{y_j}\psi)\right\| \le \sum_j |a_j|\, \|\tau_{x_j}\psi - \tau_{y_j}\psi\|_p \le \sum_j |a_j|\, \delta_j$$

$$\le c\left(\sum_j |a_j|^p\right)^{\frac{1}{p}} \left(\sum_j |\delta_j|^{p'}\right)^{\frac{1}{p'}} = c\left(\sum_j |\delta_j|^{p'}\right)^{\frac{1}{p'}}.$$

We can chose the δ_j so small that $c\left(\sum_j |\delta_j|^{p'}\right)^{\frac{1}{p'}} < A$ and use Lemma 6 to complete the proof. □

3. Problem 1 ($p = 2$)

In this section we prove that Problem 1 can be solved when $p = 2$ and $\hat{\psi}$ has compact support.

THEOREM 7. *Let $\psi \in L^2(\mathbb{R}^d)$. Assume that $\hat{\psi}$ has compact support and that there exist constants $c,\ C > 0$ such that*

$$c \le \inf_{x\in\mathbb{R}^d} |\hat{\psi}(x)| \le \sup_{x\in\mathbb{R}^d} |\hat{\psi}(x)| \le C.$$

Let $\{\tau_{x_j}\psi\}_{j\in\mathbb{Z}^d}$ be a Riesz basis in $V^2(\psi, X)$. There exists $\delta > 0$ such that if $Y = \{y_j\}_{j\in\mathbb{Z}^d} \subset \mathbb{R}^d$ satisfies $\sup_j |x_j - y_j|_2 < \delta$, then also $\{\tau_{x_j}\psi\}_{j\in\mathbb{Z}^d}$ is a Riesz basis of $V^2(\psi)$.

PROOF. Let $D = \operatorname{supp}(\hat{\psi})$. When $p = 2$, Plancherel theorem implies that the set $\{\tau_{x_j}\psi\}_{j\in\mathbb{Z}^d}$ is a Riesz basis in $V^2(\psi)$ if and only if the set $\{e^{2\pi i x_j\cdot x}\}_{j\in\mathbb{Z}^d}$ is a Riesz basis on $L^2(\mathbb{R}^d,\ \hat{\psi}\, dx)$. Our assumptions on $\hat{\psi}$ imply that the norm on $L^2(\mathbb{R}^d,\ \hat{\psi}\, dx)$ is equivalent to the norm on $L^2(D)$ and that $\{e^{2\pi i x_j\cdot x}\}_{j\in\mathbb{Z}^d}$ is an exponential Riesz basis on $L^2(D)$. Exponential Riesz bases on bounded domains of $\mathbb{R}^d$ are stable under small perturbations (see [**22**] and also Section 2.3 in [**18**]); we can find $\delta > 0$ such that, if $Y = \{y_j\}_{j\in\mathbb{Z}^d} \subset \mathbb{R}^d$ satisfies $\sup_j |x_j - y_j| < \delta$, then also $\{e^{2\pi i y_j\cdot x}\}_{j\in\mathbb{Z}^d}$ is a Riesz basis on $L^2(D)$ and hence also in $L^2(\mathbb{R}^d,\ \hat{\psi}\, dx)$. □

Example. Let $d = 1$ and let $\psi = \operatorname{sinc}(x) = \frac{\sin(\pi x)}{\pi x}$; the Fourier transform of $\tau_k\psi(x) = \operatorname{sinc}(x - k)$ is $e^{2\pi i k x}\operatorname{rect}(x) = e^{2\pi i k x}\chi_{[-\frac{1}{2},\frac{1}{2})}(x)$, and so $V^2(\psi)$ is isometrically isomorphic to $\overline{\operatorname{Span}\{e^{2\pi i j x}\operatorname{rect}(x)\}_{j\in\mathbb{Z}^d}} = L^2(-\frac{1}{2}, \frac{1}{2})$. By Kadec's theorem ([**17**], [**33**]) if $Y = \{y_n\}_{n\in\mathbb{Z}^d} \subset \mathbb{R}$ is such that $\sup_n |y_n - n| \le \delta < \frac{1}{4}$, the set $\{e^{2\pi i y_n x}\}_{n\in\mathbb{Z}^d}$ is still a Riesz basis of $L^2(-\frac{1}{2}, \frac{1}{2})$ and therefore, the set $\{\operatorname{sinc}(x - y_n)\}_{n\in\mathbb{Z}^d}$ is a Riesz basis for $V^2(\operatorname{sinc})$. Thus, $V^2(\operatorname{sinc}; Y) = V^2(\operatorname{sinc})$.

Things are not so clear when $p \ne 2$. For example, the trigonometric system $\mathcal{B} = \{e^{2\pi i n x}\}_{n\in\mathbb{Z}^d}$ is a Schauder basis in $L^p(-\frac{1}{2}, \frac{1}{2})$ when $1 < p < \infty$, but it is not a $p-$Riesz basis and the previous example cannot be generalized in an obvious way. Stability results for the Schauder basis $\mathcal{B}$ in $L^p(-\frac{1}{2}, \frac{1}{2})$ are proved in [**25**] and in [**26**].

4. Problem 2

In this section we prove Theorems 2 and 3. Let $X = \{x_n\}_{n\in\mathbb{Z}^d}$ and $\mathcal{B} = \{e^{2\pi i x\cdot x_n}\}_{n\in\mathbb{Z}^d}$. We first prove the following

LEMMA 8. *Let $\psi \in L^2(\mathbb{R}^d)\cap L^1(\mathbb{R}^d)$ be as in (1.4); if $\mathcal{B}$ is a Riesz basis in $L^2([0,1)^d)$ with constants A_1 and B_1 then $\{\tau_{x_n}\psi\}$ is a Riesz basis of $V^2(\psi, X)\}$ with constants $A = A_1 c$ and $B = B_1 C$.*

PROOF. For $k \in \mathbb{Z}^d$, we let $c_k = \inf_{x\in(0,1]^d} |\hat{\psi}(x+k)|^2$ and $C_k = \sup_{x\in(0,1]^d} |\hat{\psi}(x+k)|^2$. Let $\{d_j\}$ be a finite set of complex coefficient such that $\sum_j |d_j|^2 = 1$. Since $\mathcal{B}$ is a Riesz basis in $L^2((0,1]^d)$, for every given $k\in\mathbb{Z}^d$ we have that

$$A_1 \le \left\|\sum_n d_n e^{-2\pi i x_n\cdot k}\, e^{2\pi i x_n\cdot y}\right\|^2_{L^2((0,1]^d)} \le B_1.$$

From this inequality follows at cnce that

$$c_k A_1 \le \left\|\sum_n d_n e^{-2\pi i x_n\cdot k} e^{2\pi i x_n\cdot y}\hat{\psi}(.-k)\right\|^2_{L^2((0,1]^d)} \le C_k B_1.$$

With $c = \sum_{k\in\mathbb{Z}^d} c_k$ and $C = \sum_{k\in\mathbb{Z}^d} C_k = ||\psi||^2_{W(L^\infty,\ell^2)}$, we have

$$A_1 c \le \sum_{k\in\mathbb{Z}^d}\left\|\sum_n d_n e^{2\pi i x_n\cdot(.-k)}\hat{\psi}(.-k)\right\|^2_{L^2((0,1]^d)} \le B_1 C.$$

In view of $\sum_{k\in\mathbb{Z}^d} \|g(.-k)\|_{L^2((0,1]^d)} = ||g||_2$, we obtain

$$A_1 c \le \left\|\sum_n d_n e^{2\pi i x_n\cdot y}\hat{\psi}\right\|^2_2 \le B_1 C.$$

By Plancherel's theorem, the latter is equivalent to $A_1 c \le \|\sum_n d_n \tau_{x_n}\psi\|_2 \le B_1 C$ and so $\{\tau_{x_k}\psi\}_{k\in\mathbb{Z}^d}$ is a Riesz basis of $V^2(\psi, X)$, as required. □

PROOF OF THEOREM 2. Let $\mathcal{B} = \{e^{2\pi i x\cdot x_n}\}_{n\in\mathbb{Z}^d}$ be a Riesz basis in $L^2([0,1)^d)$; it is proved in [**22**] (see also Section 2.3 in [**18**]) that we can find $\delta > 0$ such that, if $Y = \{y_j\}_{j\in\mathbb{Z}^d} \subset \mathbb{R}^d$ satisfies $\sup_j |x_j - y_j|_2 < \delta$, then also $\{e^{2\pi i y_j\cdot x}\}_{j\in\mathbb{Z}^d}$ is a Riesz basis in $L^2([0,1)^d)$. By Lemma 8, $\{\tau_{y_n}\psi\}$ is a Riesz basis of $V^2(\psi, Y)$. □

Remark. When $Y = \{y_k\}_{k\in\mathbb{Z}^d}$ is such that $\sup_{k\in\mathbb{Z}^d} |k - y_k|_\infty < \frac{1}{4}$, by the multi-dimensional generalization of Kadec's theorem proved in [**31**] we have that $\{e^{2\pi i y_j\cdot x}\}_{j\in\mathbb{Z}^d}$ is a Riesz basis in $L^2([0,1)^d)$ and by Lemma 8, $\{\tau_{y_n}\psi\}_{n\in\mathbb{Z}^d}$ is a Riesz basis of $V^2(\psi, Y)\}$.

4.1. Proof of Theorem 3. In order to prove Theorem 3 we need some preliminary result: first, we prove the following

LEMMA 9. *Let $(a,b) \subset \mathbb{R}$, with $a < b < \infty$, and let $1 \le p < \infty$. Let $\psi \in L^p_0(a,b)$. For every finite set of coefficients $\{\alpha_j\} \subset \mathbb{C}$, we have that*

$$\left\|\sum_k \alpha_k \tau_k \psi\right\|^p_p \le \|\psi\|^p_p([b-a]+1)^{p-1}\sum_k |\alpha_k|^p$$

where [ ] denotes the integer part.

PROOF. For simplicity we let $a = 0$. When $b \le 1$ the supports of the $\tau_k\psi$'s are disjoint and so $\|f\|_p^p = \|\sum_k \alpha_k \tau_k \psi\|_p^p = \|\psi\|_p^p \sum_k |\alpha_k|^p$. When $b > 1$ the supports of the $\tau_k\psi$ overlap, and there are at most $[b]+1$ of such supports that intersect at each point. By the elementary inequality $(x_1 + \cdots + x_m)^p \le m^{p-1}(x_1^p + \cdots + x_m^p)$ (which is valid when the x_j are non-negative) we have that

$$|f(t)|^p = |\sum_k a_k \tau_k \psi(t)|^p \le ([b]+1)^{p-1} \sum_k |a_k|^p |\tau_k \psi(t)|^p$$

and so $\|f\|_p^p \le ([b]+1)^{p-1} \|\psi\|_p^p \sum_k |\alpha_k|^p$ as required. □

Let $Y = \{y_k\}_{k\in\mathbb{Z}^d}$ be a discrete subset of $\mathbb{R}^d$. Assume that $L = \sup_{k\in\mathbb{Z}^d} |y_k - k|_2 < \infty$. We prove the following

LEMMA 10. *Let $D = \prod_{j=1}^d [a_j, b_j]$ and let $\psi \in W_0^{1,p}(D)$. Then, for every finite set of coefficients $\{\alpha_j\} \subset \mathbb{C}$ such that $\sum_k |\alpha_k|^p = 1$, we have that*

$$\left\| \sum_k \alpha_k (\tau_k \psi - \tau_{y_k} \psi) \right\|_p^p \le L \sum_{j=1}^d (1 + [b_j - a_j + L])^{p-1} \|\partial_j \psi\|_p^p. \tag{4.1}$$

PROOF. When $d = 1$ and $D = (a, b)$, we prove that

$$\left\| \sum_k \alpha_k (\tau_k \psi - \tau_{y_k} \psi) \right\|_p^p \le L(1 + [b - a + L])^{p-1} \|\psi'\|_p^p \tag{4.2}$$

where $\psi'(t)$ denotes the distributional derivative of ψ. Assume first that $y_k > k$. Observing that $\psi(t + y_k) - \psi(t + k) = \int_k^{y_k} \psi'(t + x)dx$ and that $|k - y_k| \le L$, we have that

$$\left\| \sum_k \alpha_k (\tau_k \psi - \tau_{y_k} \psi) \right\|_p^p = \left\| \sum_k \alpha_k \int_{t+k}^{t+y_k} \psi'(x)dx \right\|_p^p$$
$$\le \left\| \sum_k |\alpha_k| \int_{t+k}^{t+k+L} |\psi'(x)|dx \right\|_p^p = \left\| \sum_k |\alpha_k| \tau_k g \right\|_p^p$$

where we have let $g(t) = \int_t^{t+L} |\psi'(x)|dx$. It is easy to verify that $g(t)$ is supported in the interval $[a - L, b]$. Indeed, ψ' is supported in $[a, b]$ and so $g(t) \equiv 0$ whenever $t + [0, L] \cap [a, b] = \emptyset$. Thus, $g(t) \equiv 0$ when $t + L < a$ or $t > b$, or: $g(t) \equiv 0$ when $t \in \mathbb{R} - [a - L, b]$, as required.

By Lemma 9

$$\left\| \sum_k \alpha_k (\tau_k \psi - \tau_{y_k} \psi) \right\|_p^p \le \left\| \sum_k |\alpha_k| \tau_k g \right\|_p^p \le (1 + [b - a + L])^{p-1} \|g\|_p^p. \tag{4.3}$$

We apply a change of variables and Minkowsky's integral inequality; we gather

$$\|g\|_p = \left\| \int_t^{t+L} |\psi'(x)|dx \right\|_p = \left\| \int_0^L |\psi'(x + t)|dx \right\|_p$$
$$\le L\|\psi'\|_p$$

which together with the inequality (4.3) concludes the proof of (4.2). When $y_k < k$ the proof if similar, but the function $g(t)$ defined above should be replaced by $g(t) = \int_t^{t-L} |\psi'(x)|dx$, a function supported in the interval $[a, b + L]$.

When $d = 2$ we can let $y_k = (y_{k,1}, y_{k,2})$ and $k = (k_1, k_2)$ and write

$$\left\| \sum_k \alpha_k (\tau_k \psi - \tau_{y_k} \psi) \right\|_p$$

$$\leq \left\| \sum_k \alpha_k (\tau_{(k_1,k_2)} \psi - \tau_{(y_{k,1},k_2)} \psi) \right\|_p + \left\| \sum_k \alpha_k (\tau_{(y_{k,1},k_2)} \psi - \tau_{(y_{k,1},y_{k,2})} \psi) \right\|_p$$

$$= \left\| \sum_k \alpha_k (\tau_{k_1} \psi_1 - \tau_{y_{k,1}} \psi_1) \right\|_p + \left\| \sum_k \alpha_k (\tau_{k_2} \psi_2 - \tau_{y_{k,2}} \psi_2) \right\|_p$$

where we have let $\psi_1 = \tau_{(0,k_2)} \psi$ and $\psi_2 = \tau_{(y_{k_1},0)} \psi$. The inequality (4.2), applied to ψ_1 and ψ_2, yields (4.1). The case $d > 2$ is similar.

□

PROOF OF THEOREM 3. Follows from Lemmas 6 and 10. □

4.2. rect **and step functions.** Since Sobolev spaces $W^{1,p}(\mathbb{R})$ do not contain discontinuous functions, we cannot apply Theorem 3 when ψ is a step function.

Let ψ = rect; it is immediate to verify that, for every $1 \leq p < \infty$, the set $\{\tau_j \operatorname{rect}\}_{j \in \mathbb{Z}}$ is a p–Riesz basis of $V^p(\operatorname{rect})$ with frame constants $A = B = 1$. Throughout this section we let $Y = \{y_k\}_{k \in \mathbb{Z}} \subset \mathbb{R}$, with $L = \sup_{k \in \mathbb{Z}^d} |y_k - k|$ and we assume $1 \leq p < \infty$.

Lemma 11 below is an easy generalization of Lemma 10 in [**8**].

LEMMA 11. *Assume* $0 \leq L < 1$. *For every finite set of coefficients* $\{\alpha_k\}_{n \in \mathbb{Z}^d} \subset \mathbb{C}$ *we have that*

$$\left\| \sum_k \alpha_k (\operatorname{rect}(t-k) - \operatorname{rect}(t-y_k)) \right\|_p^p < 2^p L \sum_k |\alpha_k|^p. \tag{4.4}$$

PROOF. Assume $\sum_k |\alpha_k|^p = 1$. Let

$$f(t) = \sum_k \alpha_k (\operatorname{rect}(t-k) - \operatorname{rect}(t-y_k)) = \sum_k \alpha_k \chi_{I_k}, \tag{4.5}$$

where I_j denotes the support of $\operatorname{rect}(t-j) - \operatorname{rect}(t-y_j)$. When $y_j \neq j$, I_j is union of two intervals that we denote with I_j^+ and I_j^-. When $y_j > j$, we let

$$I_j^- = (j - \frac{1}{2},\, y_j - \frac{1}{2}), \quad I_j^+ = (j + \frac{1}{2},\, y_j + \frac{1}{2}).$$

We use (improperly) the same notation to denote I_j^+ and I_j^- also when $y_j < j$.

Since we have assumed $|y_h - h| \leq L < 1$, for every given interval $J = I_h^{\pm}$ there is at most another interval $I_k^{\pm}$ that overlap with J; thus, for every $t \in \mathbb{R}$, the sum in (4.5) has at most 2 terms. By the elementary inequality $(x_1 + \cdots + x_m)^p \leq m^{p-1}(x_1^p + \cdots + x_m^p)$ we have that $|f(t)|^p \leq 2^{p-1} \sum_k |\alpha_k|^p \chi_{I_k}(t)$, and $\|f\|_p^p \leq 2^{p-1} \sup_k |I_k| = 2^{p-1}(2L) = 2^p L$ and the proof of the Lemma is concluded. □

Lemma 11 and Lemma 6 yield the following

THEOREM 12. *With the notation of Lemma 11, the set* $\{\tau_{y_k} \operatorname{rect}\}_{k \in \mathbb{Z}}$ *is a* p*–Riesz basis in* $V^p(\operatorname{rect}; Y)$ *if* $2^p L < 1$.

COROLLARY 13. *Let $\psi_0 \in L^1(\mathbb{R})$ and let $\psi = \mathrm{rect} * \psi_0$. Suppose that $\{\tau_k \psi\}_{k\in\mathbb{Z}}$ is a $p-$Riesz basis for $V^p(\psi)$. For every finite set of coefficients $\{\alpha_k\}_{n\in\mathbb{Z}} \subset \mathbb{C}$ with $\sum_k |\alpha_k|^p = 1$, we have that*

$$\left\| \sum_k \alpha_k(\psi(t-k) - \psi(t-y_k)) \right\|_p^p < 2^p L \|\psi_0\|_1^p$$

and the set $\{\psi(t-y_k)\}_{k\in\mathbb{Z}}$ is a $p-$Riesz basis for for $V^p(\psi; Y)$ whenever

$$2^p L \|\psi_0\|_1^p < A. \tag{4.6}$$

Remark. If $\hat{\psi}_0(x) \neq 0$ for every $x \in \mathbb{R}$, then the set $\{\tau_k \psi\}_{k\in\mathbb{Z}}$ is a $p-$Riesz basis for $V^p(\psi)$. Indeed, $\sum_{m\in\mathbb{Z}} |\widehat{\mathrm{rect}}(y+m)|^2 = \sum_{m\in\mathbb{Z}} |\operatorname{sinc}(x+k)|^2 \neq 0$ whenever $x \in [-\frac{1}{2}, \frac{1}{2})$ and so also $\sum_{m\in\mathbb{Z}} |\hat{\psi}(x+k)|^2 = \sum_{m\in\mathbb{Z}} |\hat{\psi}_0(x+k)\widehat{\mathrm{rect}}(x+k)|^2 \neq 0$; by Lemma 5 the set $\{\tau_k \psi\}_{k\in\mathbb{Z}}$ is a $p-$Riesz basis for $V^p(\psi)$.

PROOF OF COROLLARY 13. Let

$$F(t) = \sum_k \alpha_k \left(\psi(t-k) - \psi(t-y_k)\right), \quad f(y) = \sum_{k\in\mathbb{Z}} \alpha_k \left(\mathrm{rect}(y-k) - \mathrm{rect}(y-x_k)\right)$$

and we show that $\|F\|_p^p \leq 2^p L \|\psi_0\|_1^p$. We gather

$$\begin{aligned} F(t) &= \int_{-\infty}^{\infty} \psi_0(t-y) \sum_{k\in\mathbb{Z}} \alpha_k \left(\mathrm{rect}(y-k) - \mathrm{rect}(y-x_k)\right) dy \\ &= \psi_0 * f(t). \end{aligned}$$

Thus, by Young's inequality and Lemma 11,

$$\|F\|_p^p \leq \|\psi_0\|_1^p \|f\|_p^p \leq 2^p L \|\psi_0\|_1^p$$

and the proof of the corollary is concluded. □

Let $\beta_m(x) = \mathrm{rect}^{(m+1)}$ be the B-spline of order $m > 1$. We recall that β_m is supported in the interval $[-\frac{m+1}{2}, \frac{m+1}{2}]$ and $\beta_m(x) \in W^{1,p}(\mathbb{R})$ whenever $m \geq 1$. It is easy to verify by induction on m that $\|\beta^m\|_p \leq 1$ and $\|\beta'_m\|_p \leq 2$. It is known that $\{\tau_k \beta_m\}_{k\in\mathbb{Z}}$ is a Riesz basis of $V^2(\beta_m)$ whose Riesz constants $A(m)$ and $B(m)$ are explicitly evaluated in [**19**]. See also [**28**]. By the observations after Lemma 5, $\{\tau_k \beta_m\}_{k\in\mathbb{Z}}$ is a $p-$Riesz basis of $V^p(\beta_m)$ with constants $A_p(m) > 0$ and $B_p(m) < \infty$.

We prove the following

COROLLARY 14. *Assume that $L < 2^{-p} A_p(m)$. Then, the set $\{\tau_{y_k} \beta_m\}_{k\in\mathbb{Z}}$ is a $p-$Riesz basis of $V^p(\psi, Y)$.*

PROOF. We apply Corollary 13 with $\psi_0 = \beta^{m-1}$. □

Remark. We could have used Theorem 3 to prove Corollary 14, but we would have obtained a lower upper bound for L (namely, $L < \frac{A_p(m)}{2(2+m)^{p-1}}$).

PROOF OF THEOREM 4. Let $g(t) = \sum_{|j|\le J} s_j \operatorname{rect}(t-j)$. Let $\{\alpha_k\}_{n\in\mathbb{Z}^d} \subset \mathbb{C}$ be a finite set of coefficients such that $\sum_k |\alpha_k|^p = 1$. Let

$$f(t) = \sum_k \alpha_k \left(g\left(t-k\right) - g\left(t-x_k\right)\right).$$

As in previous theorems, we find conditions on L for which $\|f\|_p^p < A$. We have

$$\begin{aligned} f(t) &= \sum_{|j|\le J} s_j \sum_k \alpha_k \left(\operatorname{rect}(t-j-k) - \operatorname{rect}(t-j-x_k)\right) \\ &= \sum_{|j|\le J} s_j f_j(t). \end{aligned}$$

By Minkowski and Hölder inequalities, and noting that $\sum_{|j|\le J} |s_j|^q = \|g\|_q^q$, it follows that

$$\begin{aligned} \|f\|_p \le \sum_{|j|\le J} |s_j| \|f_j\|_p &\le \left(\sum_{|j|\le J} |s_j|^{p'}\right)^{\frac{1}{p'}} \left(\sum_{|j|\le J} \|f_j\|_p^p\right)^{\frac{1}{p}} \\ &= \|g\|_{p'} \left(\sum_{|j|\le J} \|f_j\|_p^p\right)^{\frac{1}{p}}, \end{aligned} \tag{4.7}$$

With the change of variables $t - j = t'$ in the integral below, we obtain

$$\begin{aligned} \int_{\mathbb{R}} |f_j(t)|^p dt &= \int_{\mathbb{R}} \left|\sum_k \alpha_k \left(\operatorname{rect}(t-j-k) - \operatorname{rect}(t-j-x_k)\right)\right|^p dt \\ &= \int_{\mathbb{R}} \left|\sum_k \alpha_k \left(\operatorname{rect}(t'-k) - \operatorname{rect}(t'-x_k)\right)\right|^p dt' \\ &= \left\|\sum_k \alpha_k (\operatorname{rect}(t-k) - \operatorname{rect}(t-y_k)\right\|_p^p. \end{aligned}$$

From Lemma 11, follows that the integral above is $\le 2^p L$. We gather: $\|f\|_p^p \le 2^p LJ \|g\|_{p'}^p$. By assumption $2^p LJ \|g\|_{p'}^p < A$, and by Lemma 6 Theorem 4 follows. □

5. Remarks and open problems

We have discussed Problem 1 when $p = 2$ and the Fourier transform of the window function ψ has compact support. When ψ is not band-limited, Plancherel's theorem implies that the set $\{\tau_{x_j}\psi\}_{j\in\mathbb{Z}^d}$ is a Riesz basis in $V^2(\psi, X)$ if and only if the set $\mathcal{V} = \{e^{2\pi i x_j\cdot x}|\hat{\psi}|\}_{j\in\mathbb{Z}^d}$ is a Riesz sequence in $L^2(\mathbb{R}^d)$, and hence a Riesz basis in $V = \overline{\operatorname{Span}(\mathcal{V})}$). By a theorem of Krein-Milman-Rutman (see e.g. [**33**, Theorem 11]) for every $j \in \mathbb{Z}^d$ there exists $\epsilon_j > 0$ such that every set of functions $\{g_j\}_{j\in\mathbb{Z}^d} \subset V$ is a Riesz basis of V whenever $||g_j - e^{\pi i x_j\cdot x}|\hat{\psi}|||_2 < \epsilon_j$. We can find $\delta_j > 0$ such that $||(e^{\pi i x_j\cdot x} - e^{\pi i y_j\cdot x})\hat{\psi}||_2 < \epsilon_j$ whenever $|x_j - y_j|_2 < \delta_j$, but we do not know whether the δ_j's have a lower bound or not.

For functions ψ in $L^p(\mathbb{R}^d)$ for every $p \in [1, \infty)$ it would be interesting to prove conditions that would ensure that a q-Riesz basis in $V^q(\psi, X)$ for some q is automatically a p-Riesz basis in $V^p(\psi, X)$ for all p. Lemma 5 (b) shows that, for certain ψ, if the set $\{\tau_k\psi\}_{k\in\mathbb{Z}^d}$ is a 2-Riesz basis of $V^2(\psi)$, it is also a p-Riesz in $V^p(\psi)$ but the method of proof of this result does not generalize well to other sets of translations. Results in [**3**] and [**29**] may help generalize Lemma 5.

It would also be interesting to define and investigate p-Riesz bases in quasi-shift invariant spaces $V^p(\psi, X)$ when $0 < p < 1$. Wavelet in L^p with $0 < p < 1$ have been considered in [**12**]. We feel that the results contained in Section 3 of [**12**] may help the understanding of $V^p(\psi, X)$ when $0 < p < 1$.

Acknowledgments

We are grateful to the anonymous referee of this paper for her/his thorough reading of our manuscript and for providing suggestions that have improved the quality of our work.

We also wish to thank K. Hamm for providing a proof of the identity (1.3) for $p \neq 2$.

References

[1] A. Aldroubi, Q. Sun, *Connection between p−frames and p−Riesz bases in locally finite SIS of $L^p(\mathbb{R})$* Proceedings of SPIE - The International Society for Optical Engineering, February 1970

[2] A. Aldroubi, Q. Sun, and W.-S. Tang, *p-frames and shift invariant subspaces of L^p*, J. Fourier Anal. Appl. **7** (2001), no. 1, 1–21, DOI 10.1007/s00041-001-0001-2. MR1812992

[3] A. Aldroubi, A. Baskakov, and I. Krishtal, *Slanted matrices, Banach frames, and sampling*, J. Funct. Anal. **255** (2008), no. 7, 1667–1691, DOI 10.1016/j.jfa.2008.06.024. MR2442078

[4] H. Brezis, *Functional analysis, Sobolev spaces and partial differential equations*, Universitext, Springer, New York, 2011. MR2759829

[5] A. Ben-Artzi and A. Ron, *On the integer translates of a compactly supported function: dual bases and linear projectors*, SIAM J. Math. Anal. **21** (1990), no. 6, 1550–1562, DOI 10.1137/0521085. MR1075591

[6] P. Casazza, O. Christensen, and D. T. Stoeva, *Frame expansions in separable Banach spaces*, J. Math. Anal. Appl. **307** (2005), no. 2, 710–723, DOI 10.1016/j.jmaa.2005.02.015. MR2142455

[7] O. Christensen and D. T. Stoeva, *p-frames in separable Banach spaces*, Adv. Comput. Math. **18** (2003), no. 2-4, 117–126, DOI 10.1023/A:1021364413257. Frames. MR1968115

[8] L. De Carli, P. Vellucci, *Stability theorems for the n-order hold model*, Linear Algebra Appl. 536C (2018) pp. 186–200, DOI 10.1016/j.laa.2017.09.020.

[9] H. G. Feichtinger, *Spline-type spaces in Gabor analysis*, Wavelet analysis (Hong Kong, 2001), Ser. Anal., vol. 1, World Sci. Publ., River Edge, NJ, 2002, pp. 100–122. MR1941607

[10] H. G. Feichtinger, U. Molter, and J. L. Romero, *Perturbation techniques in irregular spline-type spaces*, Int. J. Wavelets Multiresolut. Inf. Process. **6** (2008), no. 2, 249–277, DOI 10.1142/S0219691308002331. MR2450410

[11] H. G. Feichtinger and D. M. Onchiş, *Constructive reconstruction from irregular sampling in multi-window spline-type spaces*, Progress in analysis and its applications, World Sci. Publ., Hackensack, NJ, 2010, pp. 257–265, DOI 10.1142/9789814313179_0034. MR2766970

[12] G. Garrigós, R. Hochmuth, and A. Tabacco, *Wavelet characterizations for anisotropic Besov spaces with $0 < p < 1$*, Proc. Edinb. Math. Soc. (2) **47** (2004), no. 3, 573–595, DOI 10.1017/S001309150300107X. MR2096621

[13] K. Gröchenig and J. Stöckler, *Gabor frames and totally positive functions*, Duke Math. J. **162** (2013), no. 6, 1003–1031, DOI 10.1215/00127094-2141944. MR3053565

[14] K. Hamm, J. Ledford, *On the structure and interpolation properties of quasi-invariant shift spaces*, arxiv:1703.01533 (2017)

[15] C. Heil, *A basis theory primer*, Expanded edition, Applied and Numerical Harmonic Analysis, Birkhäuser/Springer, New York, 2011. MR2744776

[16] R. Q. Jia and C. A. Micchelli, *Using the refinement equations for the construction of pre-wavelets. II. Powers of two*, Curves and surfaces (Chamonix-Mont-Blanc, 1990), Academic Press, Boston, MA, 1991, pp. 209–246. MR1123739

[17] M. Ĭ. Kadec′, *The exact value of the Paley-Wiener constant* (Russian), Dokl. Akad. Nauk SSSR **155** (1964), 1253–1254. MR0162088

[18] G. Kozma and S. Nitzan, *Combining Riesz bases*, Invent. Math. **199** (2015), no. 1, 267–285, DOI 10.1007/s00222-014-0522-3. MR3294962

[19] E. V. Mishchenko, *Determination of the Riesz bounds for the spline basis using trigonometric polynomials* (Russian, with Russian summary), Sibirsk. Mat. Zh. **51** (2010), no. 4, 829–837, DOI 10.1007/s11202-010-0067-7; English transl., Sib. Math. J. **51** (2010), no. 4, 660–666. MR2732301

[20] B. Pavlov, *Basicity of an exponential system and Muckenhoupt's condition*, Soviet Math. Dokl. 20 (1979) 655–659.

[21] H. Prautzsch, W. Boehm, and M. Paluszny, *Bézier and B-spline techniques*, Mathematics and Visualization, Springer-Verlag, Berlin, 2002. MR1993341

[22] R. E. A. C. Paley and N. Wiener, *Fourier transforms in the complex domain*, American Mathematical Society Colloquium Publications, vol. 19, American Mathematical Society, Providence, RI, 1987. Reprint of the 1934 original. MR1451142

[23] J. L. Romero, *Explicit localization estimates for spline-type spaces*, Sampl. Theory Signal Image Process. **8** (2009), no. 3, 249–259. MR2590916

[24] A. Ron, *A necessary and sufficient condition for the linear independence of the integer translates of a compactly supported distribution*, Constr. Approx. **5** (1989), no. 3, 297–308, DOI 10.1007/BF01889611. MR996932

[25] A. M. Sedletskii, Izv. Vyssh. Uchebn. Zaved., Mat., No. 7, 85–91 (1973).

[26] A. M. Sedletskiĭ, *Equivalence of a trigonometric system and its perturbations in* $L^p(-\pi,\pi)$ (Russian, with Russian summary), Dokl. Akad. Nauk **469** (2016), no. 6, 662–665. MR3585795

[27] I. J. Schoenberg, *Cardinal interpolation and spline functions*, J. Approximation Theory **2** (1969), 167–206. MR0257616

[28] A. A. Selvan and R. Radha, *Sampling and reconstruction in shift invariant spaces of B-spline functions*, Acta Appl. Math. **145** (2016), 175–192, DOI 10.1007/s10440-016-0053-6. MR3543232

[29] C. E. Shin and Q. Sun, *Stability of localized operators*, J. Funct. Anal. **256** (2009), no. 8, 2417–2439, DOI 10.1016/j.jfa.2008.09.011. MR2502521

[30] Q. Sun, *Stability of the shifts of global supported distributions*, J. Math. Anal. Appl. **261** (2001), no. 1, 113–125, DOI 10.1006/jmaa.2001.7479. MR1850960

[31] W. Sun and X. Zhou, *On the stability of multivariate trigonometric systems*, J. Math. Anal. Appl. **235** (1999), no. 1, 159–167, DOI 10.1006/jmaa.1999.6386. MR1758675

[32] M. Unser, A. Aldroubi, M. Eden, *B-spline signal processing. I. Theory*, Signal Processing, IEEE Transactions on 41 (2), 821–833.

[33] R. M. Young, *An introduction to nonharmonic Fourier series*, 1st ed., Academic Press, Inc., San Diego, CA, 2001. MR1836633

Department of Mathematics, Florida International University, Modesto Maidique Campus, Miami, Florida 33199
Email address: `decarlil@fiu.edu`

Dipartimento di Matematica, Univ. Roma 1 "La Sapienza", Roma, Italy
Email address: `pierluigi.vellucci@sbai.uniroma1.it`

Contemporary Mathematics
Volume **706**, 2018
http://dx.doi.org/10.1090/conm/706/14214

On spectral sets of integers

Dorin Ervin Dutkay and Isabelle Kraus

ABSTRACT. Based on tiles and on the Coven-Meyerowitz property, we present some examples and some general constructions of spectral subsets of integers.

CONTENTS

1. Introduction

DEFINITION 1.1. Let G be a locally compact abelian group and let $\widehat{G}$ be its Pontryagin dual group. A finite subset A of G is called *spectral* (in G) if there exists a subset Λ of $\widehat{G}$ with $\#\Lambda = \#A$ such that

$$\frac{1}{\#A}\sum_{a\in A}\varphi(a)\overline{\varphi}'(a) = \delta_{\varphi\varphi'}, \quad (\varphi, \varphi' \in \Lambda) \tag{1.1}$$

In this case, Λ is called a *spectrum* for A (in the group $\widehat{G}$).

It is easy to check that the spectral property can be rephrased in the following ways.

2010 *Mathematics Subject Classification.* Primary 05B45, 65T50.
Key words and phrases. Spectral set, tiling, cyclotomic polynomial, Coven-Meyerowitz property.
Swedish Foundation for International Cooperation in Research and Higher Education (STINT) and Swedish Research Council (grant 2007-6338). This work was partially supported by a grant from the Simons Foundation (#228539 to Dorin Dutkay).

PROPOSITION 1.2. *Let A be a finite subset of a locally compact abelian group and Λ a finite subset of $\widehat{G}$ with $\#\Lambda = \#A$. The following statements are equivalent:*

(i) *Λ is a spectrum for A.*

(ii) *The matrix*

$$\frac{1}{\sqrt{\#A}}\left(\varphi(a)\right)_{a\in A,\varphi\in\Lambda} \tag{1.2}$$

is unitary.

(iii) *For every $a, a' \in A$,*

$$\frac{1}{\#\Lambda}\sum_{\varphi\in\Lambda}\varphi(a-a') = \delta_{aa'}. \tag{1.3}$$

We will be interested mainly in the spectral subsets of $\mathbb{Z}$ and the spectral subsets of $\mathbb{Z}_N$. Since the dual group of $\mathbb{Z}$ is the group $\mathbb{T} = \{z \in \mathbb{C} : |z| = 1\}$ which can be identified with $[0,1)$, a finite subset A of $\mathbb{Z}$ is spectral if and only if there exists a finite subset Λ in $[0,1)$ (or $\mathbb{R}$) such that $\#\Lambda = \#A$ and the matrix

$$\frac{1}{\sqrt{\#A}}\left(e^{2\pi i a\lambda}\right)_{a\in A,\lambda\in\Lambda}$$

is unitary.

Since the dual group of $\mathbb{Z}_N$ is $\mathbb{Z}_N$, a subset A of $\mathbb{Z}_N$ is spectral if and only if there exists a subset Λ of $\mathbb{Z}_N$ such that $\#A = \#\Lambda$ and the matrix

$$\frac{1}{\sqrt{\#A}}\left(e^{2\pi i a\lambda/N}\right)_{a\in A,\lambda\in\Lambda}$$

is unitary.

DEFINITION 1.3. For two subsets A and B of $\mathbb{Z}$, we write $A \oplus B$ to indicate that for each $c \in A + B$ there are unique numbers $a \in A$ and $b \in B$ such that $a + b = c$.

For a set A of non-negative integers we denote by $A(x)$ the *associated polynomial*

$$A(x) = \sum_{a\in A} x^a.$$

A subset A of $\mathbb{Z}$ is called a *tile* if there exists a subset C of $\mathbb{Z}$ such that $A \oplus C = \mathbb{Z}$.

In 1974 [**Fug74**], Fuglede proposed a conjecture that states that Lebesgue measurable spectral sets in $\mathbb{R}^n$ coincide with sets that tile $\mathbb{R}^n$. The conjecture was disproved by Tao [**Tao04**] in dimensions five and higher and later in dimensions three and higher [**FMM06, KM06b, KM06a, Mat05**]. All these counterexamples were based on some constructions in finite groups, so the Fuglede conjecture fails for groups of the form $\mathbb{Z}_{N_1} \times \mathbb{Z}_{N_2} \times \mathbb{Z}_{N_3}$. However, the conjecture is still open at this moment in dimensions one and two. It is known that the Fuglede conjecture in $\mathbb{R}$, under some additional hypotheses, can be reduced to the Fuglede conjecture for $\mathbb{Z}$, see [**DL14**].

CONJECTURE 1.4. [**Fuglede's conjecture for** $\mathbb{Z}$] *A finite subset of $\mathbb{Z}$ is spectral if and only if it is a tile.*

A basic result (see [**New77, CM99**]) shows that every tiling set C is *periodic*, i.e., there exists $N \in \mathbb{N}$ such that $C + N = C$. If B is any set consisting of one representative from C for each class modulo N, then $C = B \oplus N\mathbb{Z}$ and so

$A \oplus (B \oplus N\mathbb{Z}) = \mathbb{Z}$, and therefore $A \oplus B$ is a complete set of representatives modulo N.

PROPOSITION 1.5. [**CM99**] *Let N be a positive integer and A, B sets of non-negative integers. The following statements are equivalent:*

(i) $A \oplus (B \oplus N\mathbb{Z}) = \mathbb{Z}$.
(ii) *$A \oplus B$ is a complete set of representatives of $\mathbb{Z}_N$. In other words $A \oplus B = \mathbb{Z}_N$, where addition is understood modulo N.*
(iii) $A(x)B(x) \equiv 1 + x + \cdots + x^{N-1} \bmod (x^N - 1)$.
(iv) *$A(1)B(1) = N$ and for every factor $t > 1$ of N, the cyclotomic polynomial $\Phi_t(x)$ divides $A(x)$ or $B(x)$.*

Thus, tiles for $\mathbb{Z}$ coincide with tiles for the groups $\mathbb{Z}_N$.

In [**CM99**], Coven and Meyerowitz found a sufficient condition for a subset of $\mathbb{Z}$ to be a tile, formulated in terms of cyclotomic polynomials.

THEOREM 1.6. [**CM99**] *Let A be a finite set of non-negative integers with corresponding polynomial $A(x) = \sum_{a \in A} x^a$. Let S_A be the set of prime powers s such that the cyclotomic polynomial $\Phi_s(x)$ divides $A(x)$. Consider the following conditions on $A(x)$.*

(T1) $A(1) = \prod_{s \in S_A} \Phi_s(1)$.
(T2) *If $s_1, \ldots, s_m \in S_A$ are powers of distinct primes, then the cyclotomic polynomial $\Phi_{s_1 \ldots s_m}(x)$ divides $A(x)$.*

If $A(x)$ satisfies (T1) and (T2), then A tiles the integers with period $N := \operatorname{lcm}(S_A)$. The tiling set B can be obtained as follows: define $B(x) = \prod \Phi_s(x^{t(s)})$, where the product is taken over all prime power factors s of N which are not in S_A, and $t(s)$ is the largest factor of N relatively prime to s. Then $B(x)$ is the polynomial associated to a set of non-negative integers B.

DEFINITION 1.7. A finite set A of non-negative integers is said to satisfy the *Coven-Meyerowitz property* (or the CM-property), if it satisfies conditions (T1) and (T2) in Theorem 1.6. We call the tiling set B in Theorem 1.6, the *Coven-Meyerowitz (CM) tiling set* associated to A, and we denote it by $B = \mathrm{CM}(A)$.

The converse of the Coven-Meyerowitz theorem also seems to be true, but at the moment, it is just a conjecture. Coven and Meyerowitz showed that tiles satisfy the (T1) property.

THEOREM 1.8. [**CM99**] *Let A be a finite set of non-negative integers with corresponding polynomial $A(x) = \sum_{a \in A} x^a$ and let S_A be the set of prime powers s such that the cyclotomic polynomial $\Phi_s(x)$ divides $A(x)$. If A tiles the integers, then (T1) holds.*

Also, they proved that tiles with a cardinality that has only one or two prime factors satisfy the CM-property.

THEOREM 1.9. [**CM99**] *Let A be a finite set of non-negative integers with corresponding polynomial $A(x) = \sum_{a \in A} x^a$ such that $\#A$ has at most two prime factors and A tiles $\mathbb{Z}$. Then A satisfies (T2), and therefore it has the CM-property.*

REMARK 1.10. Note that the CM-tiling set B or $C_S = B \oplus \operatorname{lcm}(S_A)\mathbb{Z}$ does not depend on A, it depends only on $S = S_A$. Also, the proof shows that if A satisfies

(T1) and (T2), then it has a universal tiling of period $\operatorname{lcm}(S_A)$, which is a tiling set for all the sets A' with $S_{A'} = S_A$.

Note also that $B, C_S \subset p\mathbb{Z}$ for every prime factor $p \in S_A$ since p is a factor of N and every divisor $\Phi_s(x^{t(s)})$ of $B(x)$ is a polynomial in x^p. This is because either $s = p^{\alpha+1}$ with $\alpha \geq 1$, and then we use Proposition 5.4(iii), or s is a power of a prime different than p and then $t(s)$ is a multiple of p, being the largest factor of $\operatorname{lcm}(S_A)$ relatively prime to s.

Later, Łaba proved that sets with the CM-property are spectral.

THEOREM 1.11. [**Łab02**] *Let A be set of non-negative integers that satisfies the CM-property. Then A is a spectral set. A spectrum for A can be obtained as follows: consider the set Λ_A of all numbers of the form*

$$\sum_{s \in S_A} \frac{k_s}{s},$$

where $k_s \in \{0, 1, \dots, p-1\}$ if $s = p^\alpha \in S_A$, with p prime.

DEFINITION 1.12. With the notations as in Theorems 1.6 and 1.11, for $N := \operatorname{lcm}(S_A)$, we denote by $Ł_A := \Lambda_A$ and we call $Ł_A$ the *Łaba spectrum* of A.

Combining the results of Coven-Meyerowitz and Łaba, Dutkay and Haussermann showed that if a set has the CM-property, then the tiling sets and the spectra are in a nice complementary relation.

THEOREM 1.13. [**DH15**] *Let A be a finite set of non-negative integers with the CM-property. Let $N = \operatorname{lcm}(S_A)$, and let $B = \operatorname{CM}(A)$ be its CM-tiling set. Then B has the CM-property, and if $Ł_A$, $Ł_B$ are the corresponding Łaba spectra, then $(N \cdot Ł_A) \oplus (N \cdot Ł_B) = \mathbb{Z}_N$.*

Many examples of tiles are found in the literature. Many fewer spectral sets are known. In this paper, we gather some of the examples of tiling sets in the literature and show that they have the CM-property and explicitly describe the tiling sets and the spectra. In Section 2, we describe the tiles with cardinality of a prime power. In Section 3, we describe Szabó's examples and show that they have the CM-property and describe the tiling sets and spectra. In Section 4, we present some general constructions of spectral sets, tiling sets, and sets with the CM-property.

2. One prime power

THEOREM 2.1. *Let A be a set of non-negative integers with cardinality p^n, where p is prime and $n \in \mathbb{N}$. Then the set A tiles the integers if and only if there exist integers $1 \leq \alpha_1 < \dots < \alpha_n$ and for each $0 \leq k \leq n$, and each $1 \leq i_1, \dots, i_{k-1} \leq p-1$ there exists a complete set of representatives modulo p, $\{a_{i_1,\dots,i_{k-1},i_k} : 0 \leq i_k \leq p-1\}$, $a_{i_1,\dots,i_{k-1},0} = 0$, such that the set A is congruent modulo p^{α_n} to the set*

$$A' = \left\{ \sum_{k=1}^{n} p^{\alpha_k - 1} a_{i_1,\dots,i_k} : 0 \leq i_1, \dots, i_n \leq n-1 \right\}. \tag{2.1}$$

In this case $S_A = \{p^{\alpha_1}, \dots, p^{\alpha_n}\}$, *the CM-tiling set is*

(2.2)

$$B = \mathrm{CM}(A) = \left\{ \sum_{\substack{j=0,\dots,\alpha_n-1 \\ j\neq\alpha_1-1,\dots,\alpha_n-1}} b_j p^j : 0 \le b_j \le p-1, 0 \le j \le \alpha_n - 1, j \neq \alpha_1 - 1, \dots, \alpha_n - 1 \right\}.$$

The Łaba spectra of A *and* B *are, respectively*

$$Ł_A = \left\{ \sum_{i=1}^{n} \frac{k_i}{p^{\alpha_i}} : 0 \le k_i \le p-1, 1 \le i \le n \right\} \tag{2.3}$$

$$Ł_B = \left\{ \sum_{\substack{j=1,\dots,\alpha_n \\ j\neq\alpha_1,\dots,\alpha_n}} \frac{k_j}{p^j} : 0 \le k_j \le p-1, 1 \le j \le \alpha_n, j \neq \alpha_1, \dots, \alpha_n \right\} \tag{2.4}$$

REMARK 2.2. Let us explain a bit more the structure of the set A'. Think of the base p decomposition of a number. For the set A', we only use the digits corresponding to positions $\alpha_1 - 1, \alpha_2 - 1, \dots, \alpha_n - 1$. The rest of the digits are 0. In position $\alpha_1 - 1$ we use a complete set of representatives modulo p, $\{a_{i_1} : 0 \le i_1 \le p-1\}$ with $a_0 = 0$. Once the first digit a_{i_1} is chosen for the digit in position $\alpha_1 - 1$, we use another complete set of representatives modulo p, $\{a_{i_1,i_2} : 0 \le i_2 \le p-1\}$, with $a_{i_1,0} = 0$. Note that, this complete set of representatives is allowed to be different for different choices of i_1.

For $1 \le k \le n$, once the digits $a_{i_1}, a_{i_1,i_2}, \dots, a_{i_1,\dots,i_{k-1}}$ have been chosen for positions $\alpha_1 - 1, \alpha_2 - 1, \dots, \alpha_{k-1} - 1$ respectively, for the digit in position $\alpha_k - 1$ we pick a complete set of representatives, $\{a_{i_1,\dots,i_k} : 0 \le i_k \le p-1\}$, with $a_{i_1,\dots,i_{k-1},0} = 0$.

We will need some results from [**CM99**].

DEFINITION 2.3. Let S be a set of powers of at most two primes. Define $\mathcal{T}_S$ to be the collection of all subsets A of $\{0, 1, \dots, \mathrm{lcm}(S) - 1\}$ which tile the integers and satisfy $\min(A) = 0$ and $S_A = S$. Note that $\mathcal{T}_\emptyset = \{0\}$ because $\mathrm{lcm}(\emptyset) = 1$.

LEMMA 2.4. *Let* S *be a set of powers of at most two primes. A finite set* A' *with* $\min(A') = 0$ *and* $S_{A'} = S$ *tiles the integers if and only if* A' *is congruent modulo* $\mathrm{lcm}(S)$ *to a member of* $\mathcal{T}_S$.

PROOF. Let A be an element of $\mathcal{T}_S$ and $A' \equiv A (\mathrm{mod}\, \mathrm{lcm}(S))$. Let $N = \mathrm{lcm}(S)$. Since $A' \in \mathcal{T}_S$, there exists a set B such that $A' \oplus B = \mathbb{Z}_N$. Then, since $A \equiv A' (\mathrm{mod}\, N)$, it follows that $A \oplus B = \mathbb{Z}_N$.

Conversely, if A' tiles the integers and $S_A = S$, then by Lemma 2.5, $\#A'$ has at most two prime factors, so it has the CM-property, by Theorem 1.9. Therefore, it has a tiling set of period $\mathrm{lcm}(S)$, $B \oplus \mathrm{lcm}(S)\mathbb{Z}$, by Remark 1.10. Let A be the set obtained from A' by reducing modulo $\mathrm{lcm}(S)$. Then $\min(A) = 0$ and $A \subset \{0, 1, \dots, \mathrm{lcm}(S) - 1\}$. Also, A has the same tiling set $B \oplus \mathrm{lcm}(S)\mathbb{Z}$. Then, by Lemma 5.3, $S_{A'} = S_A$ as the complement of S_B in the set of all prime power factors of $\mathrm{lcm}(S)$. ☐

LEMMA 2.5. *Let A be a finite set of non-negative integers which is a tile. Then $\#A$ has at most two prime factors if and only if S_A consists of powers of at most two primes.*

PROOF. Since A is a tile, it satisfies the (T1) property, by Theorem 1.8. So, using Proposition 5.4(iv),

$$\#A = \prod_{s \in S_A} \Phi_s(1) = \prod_{p^\alpha \in S_A} p.$$

Thus $\#A$ has at most two prime factors if and only if S_A consists of powers of at most two primes. □

LEMMA 2.6. **[CM99]** *Suppose S contains powers of only one prime p. Let $\overline{S} = \{p^\alpha : p^{\alpha+1} \in S, \alpha \geq 1\}$.*

(i) *If $p \notin S$ then*

$$\mathcal{T}_S = \{p\overline{A} : \overline{A} \in \mathcal{T}_{\overline{S}}\}.$$

(ii) *If $p \in S$ then*

$$\mathcal{T}_S = \Big\{\cup_{i=0}^{p-1}(\{a_i\} \oplus p\overline{A}_i) : \overline{A}_i \in \mathcal{T}_{\overline{S}}, a_0 = 0, \{a_0, a_1, \dots, a_{p-1}\} \text{ a complete}$$
$$\text{set of representatives modulo } p \text{ and every } \{a_i\} \oplus p\overline{A}_i \subset \{0, 1, \dots, \operatorname{lcm}(S) - 1\}\Big\}.$$

PROPOSITION 2.7. **[CM99]** *Let p be a prime number. Then*

(i) *The only member of $\mathcal{T}_\emptyset$ is $\{0\}$.*
(ii) *For $\alpha \geq 0$, the only member of $\mathcal{T}_{\{p^{\alpha+1}\}}$ is $p^\alpha\{0, 1, \dots, p-1\}$.*

THEOREM 2.8. *Let p be a prime number. Let $S = \{p^{\alpha_1}, p^{\alpha_2}, \dots, p^{\alpha_n}\}$ with $1 \leq \alpha_1 < \alpha_2 < \cdots < \alpha_n\}$. The following statements are equivalent:*

(i) *$A \in \mathcal{T}_S$.*
(ii) *For $1 \leq k \leq n-1$, there exist numbers $a_{i_1,\dots,i_k}$ $i_1, \dots, i_k = 0, 1 \dots, p-1$ with the following properties*
   (a) *The set $\{a_{i_1} : 0 \leq i_1 \leq p-1\}$ is a complete set of representatives modulo p, $a_0 = 0$,*
   (b) *For each $2 \leq k \leq n-1$, and each $i_1, \dots, i_{k-1}$ in $\{0, 1 \dots, p-1\}$, the set $\{a_{i_1,\dots,i_{k-1},i_k} : 0 \leq i_k \leq p-1\}$ is a complete set of representatives modulo p, $a_{i_1,\dots,i_{k-1},0} = 0$,*
   (c) *For each $1 \leq k \leq n-1$*

$$a_{i_1,\dots,i_k} + p^{\alpha_{k+1}-\alpha_k} a_{i_1,\dots,i_{k+1}} + \cdots + p^{\alpha_{n-1}-\alpha_k} a_{i_1,\dots,i_{n-1}} \leq p^{\alpha_n - \alpha_k} - 1. \tag{2.5}$$

   (d)

$$A = \{p^{\alpha_1 - 1} a_{i_1} + p^{\alpha_2 - 1} a_{i_1,i_2} + \cdots + p^{\alpha_{n-1}-1} a_{i_1,\dots,i_{n-1}} + p^{\alpha_n - 1} j : 0 \leq i_1, \dots, i_n, j \leq p-1\}. \tag{2.6}$$

PROOF. We prove the equivalence of (i) and (ii) by induction on n. For $n = 1$, the result follows from Proposition 2.7. Assume now the statements are equivalent for n and take $S = \{p^{\alpha_1}, \dots, p^{\alpha_{n+1}}\}$. Using Lemma 2.6(i), we have that $A \in \mathcal{T}_S$ if and only if $A = p^{\alpha_1 - 1}\overline{A}$ with $\overline{A} \in \mathcal{T}_{\overline{S}_1}$ where $\overline{S} = \{p, p^{\alpha_2 - \alpha_1 + 1}, \dots, p^{\alpha_{n+1} - \alpha_1 + 1}\}$. Using Lemma 2.6(ii), we have that $\overline{A} \in \mathcal{T}_{\overline{S}}$ if and only if there exists a complete set of representatives modulo p, $\{a_{i_1} : 0 \leq i_1 \leq p-1\}$, $a_0 = 0$ and, for each

$0 \le i_1 \le p-1$, a set $\overline{A}_{i_1} \in \mathcal{T}_{\overline{S}'}$, where $\overline{S}' = \{p^{\alpha_2-\alpha_1}, \dots, p^{\alpha_{n+1}-\alpha_1}\}$ such that $a_{i_1} + p\overline{A}_{i_1} \subset \{0, \dots, \mathrm{lcm}(\overline{S}) - 1\} = \{0, \dots, p^{\alpha_{n+1}-\alpha_1+1} - 1\}$, and

$$\overline{A} = \cup_{i_1=0}^{p-1}(\{a_{i_1}\} + p\overline{A}_{i_1}).$$

Using the induction hypothesis for the set $\overline{S}'$, we get that, for each $0 \le i_1 \le p-1$ the set $\overline{A}_{i_1}$ must be of the form

$$\overline{A}_{i_1} = \{p^{\alpha_2-\alpha_1-1}a_{i_1,i_2} + p^{\alpha_3-\alpha_1-1}a_{i_1,i_2,i_3} + \dots + p^{\alpha_n-\alpha_1-1}a_{i_1,\dots,i_n} \\ +p^{\alpha_{n+1}-\alpha_1-1}j : 0 \le i_2, \dots, i_n, j \le p-1\},$$

where for each $0 \le i_2, \dots, i_{k-1} \le p-1$, the set $\{a_{i_1,\dots,i_k} : 0 \le i_k \le p-1\}$ is a complete set of representatives modulo p and $a_{i_1,\dots,i_{k-1},0} = 0$. Also

$$a_{i_1,i_2,\dots,i_k} + p^{(\alpha_{k+1}-\alpha_1)-(\alpha_k-\alpha_1)}a_{i_1,\dots,i_{k+1}} + \dots + p^{(\alpha_n-\alpha_1)-(\alpha_k-\alpha_1)} \\ \le p^{(\alpha_{n+1}-\alpha_1)-(\alpha_k-\alpha_1)} - 1$$

and this implies (c) for $k \ge 2$. We must also have

$$a_{i_1} + p^{\alpha_2-\alpha_1}a_{i_1,i_2} + \dots p^{\alpha_n-\alpha_1}a_{i_1,\dots,i_n} + p^{\alpha_{n+1}-\alpha_1}(p-1) \le p^{\alpha_{n+1}-\alpha_1+1},$$

and this implies (c) for $k = 1$.

Then

$$A = p^{\alpha_1-1} \cup_{i_1=0}^{p-1} (\{a_{i_1}\} + p\overline{A}_{i_1}),$$

and (d) follows.

□

PROOF OF THEOREM 2.1. Assume that A and A' have the given form. We show that $A' \oplus B = \mathbb{Z}_{p^{\alpha_n}}$. Note that $\#A' = p^n$ and $\#B = p^{\alpha_n-n}$ so $\#A' \cdot \#B = p^{\alpha_n}$. By Lemma 5.2, it is enough to show that $(A' - A') \cap (B - B) = \{0\}$ in $\mathbb{Z}_{p^{\alpha_n}}$. If we pick an element in the intersection, it can be written in both ways as

$$\sum_{k=1}^{n} p^{\alpha_k-1}(a_{i_1,\dots,i_k} - a_{i'_1,\dots,i'_k}) = \sum_{\substack{j=0,\dots,\alpha_n-1 \\ j \ne \alpha_1-1,\dots,\alpha_n-1}} p^j(b_j - b'_j).$$

Take the first index l such that $i_l \ne i'_l$. Then the left-hand side is divisible by p^{α_l-1} but not by p^{α_l}. Since p^{α_l-1} does not appear on the right hand side, it follows, by contradiction, that both sides are equal to 0.

For the converse, if A is a tile then the result follows from Theorem 2.8.

It remains to check that the CM-tiling set and the Łaba spectra are those given in (2.2),(2.3) and (2.4). By Lemma 5.3, we have that S_A and S_B are complementary, so

$$S_B = \{p^j : j \in \{1, \dots, \alpha_n\}, j \ne \alpha_1, \dots, \alpha_n\}.$$

Then the CM-tiling set is defined by the polynomial

$$\prod_{\substack{j=1\dots\alpha_n \\ j\ne\alpha_1,\dots,\alpha_n}} \Phi_j(x) = \prod_{\substack{j=1\dots\alpha_n \\ j\ne\alpha_1,\dots,\alpha_n}} (1 + x^{p^{j-1}} + x^{2p^{j-1}} + \dots + x^{(p-1)p^{j-1}}) \\ = \sum_{\substack{0\le b_j\le p-1 \\ 1\le j\le\alpha_n, j\ne\alpha_1,\dots,\alpha_n}} x^{\sum_{1\le j\le\alpha_n, j\ne\alpha_1,\dots,\alpha_n} b_j x^{p^{j-1}}}.$$

This implies (2.2). Since we have the form of S_B, (2.3) and (2.4) follow immediately.

□

REMARK 2.9. In [**New77**], Newman classifies the finite sets of integers which tile $\mathbb{Z}$ when the number of elements in the set is a prime power. The tiling condition is stated in Theorem 2.10. In Proposition 2.11, we determine the relation between the numbers e_{ij} in Newman's paper and the numbers α_i and the set S_A in Theorem 2.1.

THEOREM 2.10. [**New77**] *Let $a_1, a_2, \dots, a_k$ be distinct integers with $k = p^\alpha$, p a prime, α a positive integer. For each pair $a_i, a_j, i \neq j$, we denote by $p^{e_{ij}}$ the highest power of p which divides $a_i - a_j$. The set $\{a_1, a_2, \dots, a_k\}$ is a tile if and only if there are at most α distinct e_{ij}.*

PROPOSITION 2.11. *Let $A = \{a_1, \dots, a_k\}$ be a set of non-negative integers which tile $\mathbb{Z}$, with $\#A = p^n$, p a prime, n a positive integer. Then $S_A = \{p^{e_{ij}+1} : 1 \leq i, j \leq p^n\}$, where e_{ij} denotes the highest power of p which divides $a_i - a_j$, as in Theorem* 2.10.

PROOF. Let $S_A = \{p^{\alpha_1}, \dots, p^{\alpha_n}\}$. Since A is a tile, we have by Theorem 2.1 that A is congruent modulo p^{α_n} to the set

$$A' = \left\{ \sum_{k=1}^{n} p^{\alpha_k - 1} a_{i_1, \dots, i_k} : 0 \leq i_1, \dots, i_n \leq n - 1 \right\}.$$

Take $a, a' \in A$. Then

$$a = \sum_{k=1}^{n} p^{\alpha_k - 1} a_{i_1, \dots, i_k} + p^{\alpha_n} m,$$

$$a' = \sum_{k=1}^{n} p^{\alpha_k - 1} a_{i_{1'}, \dots, i_{k'}} + p^{\alpha_n} m'.$$

So

$$a - a' = \sum_{k=1}^{n} (a_{i_1, \dots, i_k} - a_{i_{1'}, \dots, i_{k'}}) p^{\alpha_k - 1} + p^{\alpha_n}(m - m').$$

If $i_1 = i_{1'}, \dots, i_n = i_{n'}$, then $a = a'$. So then $m = m'$. If there exists a k such that $i_k \neq i_{k'}$, then take the smallest such k. Let e be the largest power of p such that p^e divides $a - a'$. Then

$$a - a' = p^{\alpha_k - 1}(a_{i_1, \dots, i_{k-1}, i_k} - a_{i_{1'}, \dots, i_{k-1'}, i_{k'}}) + \sum_{j=k+1}^{n} (a_{i_1, \dots, i_j} - a_{i_{1'}, \dots, i_{j'}}) p^{\alpha_j - 1} + p^{\alpha_n}(m - m').$$

Since $\{a_{i_1, \dots, i_{k-1}, l} : 0 \leq l \leq p - 1\}$ is a complete set of representatives modulo p, this means that $a - a'$ is divisible by $p^{\alpha_k - 1}$, but not p^{α_k}. Therefore, $e = \alpha_k - 1$. Relabeling $a = a_i$ and $a' = a_j$, we get that $e_{ij} = \alpha_k - 1$. Doing this for all $a_i, a_j \in A$, we get that the set of all e_{ij} is $\{\alpha_1 - 1, \dots, \alpha_n - 1\}$ and the result follows. □

For the case when the cardinality of a tile has two prime factors, to describe the structure of the such tiles, one can use the following lemma from [**CM99**].

LEMMA 2.12. *Suppose S contains powers of both the primes p and q. Let*

$$\overline{S} = \{p^\alpha : p^{\alpha+1} \in S\} \cup \{q^\beta : q^\beta \in S\}, \quad \overline{S}' = \{p^\alpha : p^\alpha \in S\} \cup \{q^\beta : q^{\beta+1} \in S\}.$$

(i) *If $p \in S$, then*

$$\mathcal{T}_S = \Big\{ \cup_{i=0}^{p-1}(\{a_i\} \oplus p\overline{A}_i) : \overline{A}_i \in \mathcal{T}_{\overline{S}}, a_0 = 0, \{a_0, a_1, \dots, a_{p-1}\} \text{ a complete}$$

$$\text{set of representatives modulo } p \text{ and every } \{a_i\} \oplus p\overline{A}_i \subset \{0, 1, \dots, \mathrm{lcm}(S) - 1\} \Big\}.$$

(ii) *If $q \in S$, then*

$$\mathcal{T}_S = \Big\{ \cup_{i=0}^{q-1}(\{a_i\} \oplus q\overline{A}_i) : \overline{A}_i \in \mathcal{T}_{\overline{S}'}, a_0 = 0, \{a_0, a_1, \dots, a_{q-1}\} \text{ a complete}$$

$$\text{set of representatives modulo } q \text{ and every } \{a_i\} \oplus q\overline{A}_i \subset \{0, 1, \dots, \mathrm{lcm}(S) - 1\} \Big\}.$$

(iii) *If $p, q \notin S$, then $A \subset p\mathbb{Z}$ or $A \subset q\mathbb{Z}$ and*

$$\{A \in \mathcal{T}_S : A \subset p\mathbb{Z}\} = \{p\overline{A} : \overline{A} \in \mathcal{T}_{\overline{S}}\}, \quad \{A \in \mathcal{T}_S : A \subset q\mathbb{Z}\} = \{q\overline{A} : \overline{A} \in \mathcal{T}_{\overline{S}'}\}.$$

3. Szabó's examples

In [**Sza85**], Szabó constructed a class of examples to give a negative answer to two questions due to K. Corrádi and A.D. Sands respectively:

If G is a finite abelian group and $G = A_1 \oplus A_2$ is one of its normed factorizations, in the sense that both factors contain the zero, must one of the factors contain some proper subgroup of G?

If G is a finite abelian group and $G = A_1 \oplus A_2$ is a factorization, must one of the factors be contained in some proper subgroup of G?

We will present here Szabó's examples and prove that they have the CM-property and find their tiling sets and spectra. We begin with a general proposition which encapsulates the core ideas in Szabó's examples.

PROPOSITION 3.1. *Let G be a finite abelian group. Suppose we have a factorization $A \oplus B' = G$. Let $B_1, \dots, B_r$ be disjoint subsets of B' and $g_1, \dots, g_r$ in G with the property that*

$$A + B_i = A + B_i + g_i \text{ for all } i. \tag{3.1}$$

Define

$$B = B' \cup \left(\bigcup_{i=1}^{r} (B_i + g_i) \right) \setminus \left(\bigcup_{i=1}^{r} B_i \right). \tag{3.2}$$

Then $A \oplus B = G$.

PROOF. The sets $A + B_i$ are disjoint. Indeed, if not, then there are $a, a' \in A$, $b_i \in B_i$, $b_j \in B_j$, $i \neq j$ such that $a + b_i = a' + b_j$. Because of the unique factorization, $a = a'$ and $b_i = b_j$, which contradicts the fact that B_i and B_j are disjoint. We have the disjoint union

$$G = (A + (B' \setminus \cup_i B_i)) \cup \bigcup_i (A + B_i) = (A + (B' \setminus \cup_i B_i)) \cup \bigcup_i (A + B_i + g_i).$$

Take $b \neq b'$ in B and we want to prove that $A + b$ is disjoint from $A + b'$. If b or b' is in $B' \setminus \cup_i B_i$, this is clear from the hypothesis. If b and b' lie in different sets $B_i + g_i$ this is again clear since the union above is disjoint. If $b, b' \in B_i + g_i$, then $b = c + g_i$, $b' = c' + g_i$ with $c, c' \in B_i \subset B'$, and since $b \neq b'$, we have $c \neq c'$. Therefore $A + c$ is disjoint from $A + c'$ and the same is true for $A + b = A + c + g_i$ and $A + b' = A + c' + g_i$. Thus $A \oplus B = G$. □

Szabó constructed his examples in $\mathbb{Z}_{m_1} \times \cdots \times \mathbb{Z}_{m_r}$. We note here that working in the group $\mathbb{Z}_{m_1} \times \ldots \mathbb{Z}_{m_r}$ with $m_1, \ldots, m_r$ relatively prime is equivalent to working in $\mathbb{Z}_{m_1 \ldots m_r}$ because of the following isomorphism.

PROPOSITION 3.2. *Let $m_1, \ldots, m_r$ be relatively prime non-negative integers. Let $m = m_1 \ldots m_r$. The map $\Psi : \mathbb{Z}_{m_1} \times \cdots \times \mathbb{Z}_{m_r} \to \mathbb{Z}_m$,*

$$\Psi(k_1, \ldots, k_r) = \sum_{i=1}^{r} k_i \frac{m}{m_i} \tag{3.3}$$

is an isomorphism.

PROOF. It is clear that Ψ is a morphism. We check that it is injective. If $\Psi(k_1, \ldots, k_r) = 0$ then $\sum_{i=1}^r k_i \frac{m}{m_i} = 0$. Since m/m_i is divisible by m_j for all $i \neq j$, then $k_j \frac{m}{m_j} \equiv 0 (\operatorname{mod} m_j)$. Since $\frac{m}{m_j}$ is relatively prime to m_j, it is invertible in $\mathbb{Z}_{m_j}$ so $k_j \equiv 0 (\operatorname{mod} m_j)$. Thus $(k_1, \ldots, k_r) = (0, 0, \ldots, 0)$. Thus, Ψ is injective. Since the two sets have the same cardinality, Ψ is also surjective. $\square$

EXAMPLE 3.3. [**Sza85**] Let G be the direct product of the cyclic groups of orders $m_1, \ldots, m_r$ and generators $g_1, \ldots, g_r$ respectively. Assume $r \geq 3$. We can think of G as $\mathbb{Z}_{m_1} \times \cdots \times \mathbb{Z}_{m_r}$ and we can pick the generators $g_1 = (1, 0, \ldots, 0), \ldots, g_r = (0, \ldots, 0, 1)$. But we can also pick other generators, for example $g_1 = (q, 0 \ldots, 0)$ where q is some number which is relatively prime to m_1. Let π be a permutation of the set $\{1, \ldots, r\}$ that does not have cycles of length 1 or 2. (Szabó assumes π is cyclic, and since $r \geq 3$, our condition is satisfied, but we do not need π to be cyclic.)

If $g \in G \setminus \{0\}$ and m is a positive integer which is less than or equal to the order of g, we denote by $[g]_m = \{0, g, 2g, \ldots, (m-1)g\}$.

Assume now $m_i = u_i v_i$ where u_i, v_i are integers greater than one.

Obviously

$$G = \sum_{i=1}^{r} [g_i]_{m_i} = \sum_{i=1}^{r} ([g_i]_{u_i} + [u_i g_i]_{v_i}) = A \oplus B',$$

where

$$A = \sum_{i=1}^{r} [g_i]_{u_i} \text{ and } B' = \sum_{i=1}^{r} [u_i g_i]_{v_i}.$$

Then Szabó picks $B_i = [u_i g_i]_{v_i} + u_{\pi(i)} g_{\pi(i)}$ and $g_i = g_i$, as in Proposition 3.1. An easy check shows that the sets B_i are disjoint (here is where we need π to have only cycles of length ≥ 3) and $A + B_i + g_i = A + B_i$ (the main property used here is that $[g_i]_{u_i} + [u_i g_i]_{v_i} + g_i = [g_i]_{m_i} + g_i = [g_i]_{m_i} = [g_i]_{u_i} + [u_i g_i]_{v_i}$. Thus, the properties in Proposition 3.1 are satisfied and with

$$B = B' \cup \left(\bigcup_{i=1}^{r} ([u_i g_i]_{v_i} + u_{\pi(i)} g_{\pi(i)} + g_i) \right) \setminus \left(\bigcup_{i=1}^{r} ([u_i g_i]_{v_i} + u_{\pi(i)} g_{\pi(i)}) \right),$$

we have a new factorization $A \oplus B = G$.

Next, we construct spectra for the sets A, B' and B. For this, we regard G as we mentioned before: $G = \mathbb{Z}_{m_1} \times \cdots \times \mathbb{Z}_{m_r}$ and $g_1 = (1, 0, \ldots, 0), \ldots, g_r = (0, \ldots, 0, 1)$.

Then we have

$$A = \sum_{i=1}^{r} [g_i]_{u_i} = \{(k_1, \ldots, k_r) : 0 \leq k_i \leq u_i - 1 \text{ for all } i\},$$

$$B' = \sum_{i=1}^{r} [u_i g_i]_{v_i} = \{((u_1 k_1, \dots, u_r k_r) : 0 \le k_i \le v_i - 1\},$$

$$B = B' \cup \left(\bigcup_{i=1}^{r} ([u_i g_i]_{v_i} + u_{\pi(i)} g_{\pi(i)} + g_i) \right) \setminus \left(\bigcup_{i=1}^{r} ([u_i g_i]_{v_i} + u_{\pi(i)} g_{\pi(i)}) \right)$$
$$= \{((u_1 k_1, \dots, u_r k_r) : 0 \le k_i \le v_i - 1\} \cup$$
$$\bigcup_{i=1}^{r} \{(0, \dots, 0, k_i u_i + 1, 0, \dots, 0, u_{\pi(i)}, 0, \dots, 0) : 0 \le k_i \le v_i - 1 \text{ for all } i\}$$
$$\setminus \bigcup_{i=1}^{r} \{(0, \dots, 0, k_i u_i, 0, \dots, 0, u_{\pi(i)}, 0, \dots, 0) : 0 \le k_i \le v_i - 1 \text{ for all } i\},$$

where $k_i u_i + 1$ and $k_i u_i$ are on position i and $u_{\pi(i)}$ is on position $\pi(i)$.

PROPOSITION 3.4. *The set A has a spectrum (in G)*

$$\Lambda_A = \{(v_1 j_1, \dots, v_r j_r) : 0 \le j_i \le u_i - 1 \textit{ for all } i\}.$$

The sets B' and B have spectrum (in G)

$$\Lambda_B = \{(j_1, \dots, j_r) : 0 \le j_i \le v_i - 1 \textit{ for all } i\}.$$

Also $\Lambda_A \oplus \Lambda_B = G$.

PROOF. For the set A, we check that $\{0, v_i, 2v_i, \dots, (u_i - 1)v_i\}$ in $\mathbb{Z}_{m_i}$ has spectrum $\{0, 1, \dots, u_i - 1\}$. Indeed

$$\sum_{j=0}^{u_i - 1} e^{2\pi i \frac{(k-k')jv_i}{u_i v_i}} = u_i \delta_{kk'} \text{ for all } k, k' \in \{0, 1, \dots, u_i - 1\}.$$

Then, the result follows from Proposition 5.1. For B', a similar argument can be used.

For B we will use a lemma:

LEMMA 3.5. *Given $(k_1, \dots, k_r) \in \mathbb{Z}_{m_1} \times \dots \times \mathbb{Z}_{m_r}$, if one of the k_i is a non-zero multiple of u_i, then*

$$\sum_{l_1=0}^{v_1-1} \cdots \sum_{l_r=0}^{v_r-1} e^{2\pi i \frac{l_1 k_1}{m_1}} \dots e^{2\pi i \frac{l_r k_r}{m_r}} = 0.$$

PROOF. The sum splits into the product of the sums $\sum_{l_i=0}^{v_i-1} e^{2\pi i \frac{l_i k_i}{u_i v_i}}$, and if $k_i = k u_i$, $0 < k < v_i$ then we further get $\sum_{l_i=0}^{v_i-1} e^{2\pi i \frac{l_i k}{v_i}} = 0$. □

Now, take two distinct points $b = (b_1, \dots, b_r) \ne b' = (b'_1, \dots, b'_r)$ in B. We want to prove that

$$\sum_{(l_1, \dots, l_r) \in \Lambda_B} e^{2\pi i \frac{l_1 (b_1 - b'_1)}{m_1}} \dots e^{2\pi i \frac{l_r (b_r - b'_r)}{m_r}} = 0. \tag{3.4}$$

We denote

$$B_0 = \{(u_1 k_1, \dots, u_r k_r) : 0 \le k_i \le v_i - 1\} \setminus$$
$$\bigcup_{i=1}^{r} \{(0, \dots, 0, k_i u_i, 0, \dots, 0, u_{\pi(i)}, 0, \dots, 0) : 0 \le k_i \le v_i - 1 \text{ for all } i\},$$

$$B_i = \{(0,\dots,0,k_iu_i,0,\dots,0,u_{\pi(i)},0,\dots,0) : 0 \le k_i \le v_i - 1 \text{ for all } i\},$$
$$\tilde{B}_i = \{(0,\dots,0,k_iu_i+1,0,\dots,0,u_{\pi(i)},0,\dots,0) : 0 \le k_i \le v_i - 1 \text{ for all } i\}.$$

So $B = B_0 \cup \cup_{i=1}^r \tilde{B}_i$, disjoint union.

If $b, b' \in B_0$, then the result follows from the fact that Λ_B is a spectrum for B', so B' is a spectrum for Λ_B. If $b \in B_0$ and $b' \in \tilde{B}_i$, then b' is of the form $b' = (0,\dots,0,k_i'u_i+1,0,\dots,0,u_{\pi(i)},0,\dots,0)$ and b is of the form $b = (k_1u_1,\dots,k_ru_r)$.

If one of the $k_1,k_2,\dots,k_{i-1},k_{i+1},\dots,k_{\pi(i)-1},k_{\pi(i)+1},\dots,k_r$ is non-zero, then we apply Lemma 3.5 to $b-b'$ and obtain (3.4).

If all these are zero, then, if $k_{\pi(i)} \neq 1$, again we use Lemma 3.5 and obtain (3.4). If $k_{\pi(i)} = 1$, then that means that $b \in B_i$, a contradiction.

If $b' \in \tilde{B}_i$, and b is in a set $\tilde{B}_j$ with $j \neq i$, then b is of the form $b = (0,\dots,0,k_ju_j+1,0,\dots,0,u_{\pi(j)},0,\dots,0)$. If $\pi(i) = j$ then $\pi(j) \neq i$ (becasue π has no cycles of length 2), and $\pi(j) \neq \pi(i)$. Therefore, applying Lemma 3.5 to $b-b'$, using the $\pi(j)$ component, we get (3.4).

If $\pi(i) \neq j$, then since $\pi(i) \neq \pi(j)$, we can use the $\pi(i)$ component in Lemma 3.5 for $b-b'$ and again obtain (3.4).

Finally, if $b, b' \in \tilde{B}_i$, then $b-b' \in (B'-B') \setminus \{0\}$ and the result follows, since Λ_B is a spectrum for B'.

The factorization $\Lambda_A \oplus \Lambda_B = G$ is obvious. □

Note that the factorization $\Lambda_A \oplus \Lambda_B = G$ is completely analogous to the factorization $A \oplus B' = G$; the roles of u_i and v_i are interchanged. Therefore, we can perform the same type of operations on Λ_A as we did on B' and get new factorizations and spectra.

So, let us consider a permutation σ of $\{1,\dots,r\}$ which has only cycles of length ≥ 3. Define

$$\Lambda_A' = \Lambda_A \cup \left(\bigcup_{i=1}^r ([v_ig_i]_{u_i} + v_{\sigma(i)}g_{\sigma(i)}) + g_{\sigma(i)}\right) \setminus \left(\bigcup_{i=1}^r ([v_ig_i]_{u_i} + v_{\sigma(i)}g_{\sigma(i)})\right).$$

PROPOSITION 3.6. *Λ_A' is a spectrum of A and $\Lambda_A' \oplus \Lambda_B = G$.*

PROPOSITION 3.7. *Let $m = m_1 \dots m_r$. With the isomorphism Φ in Proposition 3.2, the image of the sets $A, B', B, \Lambda_A, \Lambda_A'$ and Λ_B in $\mathbb{Z}_m$ have the CM-property.*

PROOF. We have

$$\Psi(A) = \left\{\sum_{i=1}^r k_i \frac{m}{m_i} : 0 \le k_i \le u_i - 1\right\}.$$

We use the notation

$$Q_n(x) = 1 + x + \dots + x^{n-1}.$$

We have

$$A(x) = \prod_{i=1}^r Q_{u_i}\left(x^{\frac{m}{m_i}}\right).$$

Then, with Proposition 5.4(ii) and (vi) (since u_i and $\frac{m}{m_i}$ are mutually prime), we get

$$A(x) = \Psi(A)(x) = \prod_{i=1}^r \prod_{d|u_i, d>1} \Phi_{u_i}\left(x^{\frac{m}{m_i}}\right) = \prod_{i=1}^r \prod_{d|u_i,d>1} \prod_{t|\frac{m}{m_i}} \Phi_{dt}(x).$$

If p^α, $\alpha > 0$ is a prime power such that $\Phi_{p^\alpha}(x)$ divides $A(x)$, then there exist $i \in \{1, \ldots, r\}$, $d > 1$, $d|u_i$ and $t|\frac{m}{m_i}$ such that $p^\alpha = dt$. Since d and t are relatively prime and $d > 1$, we must have $p^\alpha = d$. Thus, S_A consists of the prime powers that divide one of the u_i.

Take $p_1^{\alpha_1}, \ldots, p_n^{\alpha_n}$ in S_A. Then $p_1^{\alpha_1}$ divides one of the u_i. By relabeling, let us assume $p_1^{\alpha_1}, \ldots, p_j^{\alpha_j}$ divide u_i and $p_{j+1}^{\alpha_{j+1}}, \ldots, p_n^{\alpha_n}$ do not, hence they divide some $u_{i_{j+1}}, \ldots, u_{i_n}$, respectively, different than u_i. Then $d = p_1^{\alpha_1} \ldots p_j^{\alpha_j} > 1$ and $d|u_i$. Also $t = p_{j+1}^{\alpha_{j+1}} \ldots p_n^{\alpha_n}$ divides $\frac{m}{m_i}$ therefore $\Phi_{dt}(x)$ divides $A(x)$ and so A has the CM-property.

For the sets B' and B, first note, from Lemma 5.3, that $S_B = S_{B'}$ consists of prime powers p^α, $\alpha > 0$ which divide m but do not divide any of the u_i.

Take $s_1, \ldots, s_n$ prime powers in S_B. We will show that, for $s = s_1 \ldots s_n$, $\Phi_s(x)$ does not divide $A(x)$, so, by Proposition 1.5, it has to divide $B(x)$ and $B'(x)$, which will mean that B and B' satisfy the CM-property.

Suppose not, so $\Phi_s(x)$ divides $A(x)$, then there exists $i \in \{1, \ldots, r\}$, $d > 1$, $d|u_i$ and $t|\frac{m}{m_i}$ such that $s = dt$. Each of the s_k divides exactly one of the m_j. By relabeling, suppose $s_1, \ldots, s_j$ are the prime powers that divide m_i and $s_{j+1}, \ldots, s_n$ do not. Then $d = s_1 \ldots s_j$. But, since $d|u_i$ it means that $s_1, \ldots, s_j$ divide u_i, a contradiction. Therefore, none of the s_k divide m_i so all of them divide $\frac{m}{m_i}$. But then, since $s = dt$, it follows that $d = 1$, again a contradiction. Thus, $\Phi_s(x)$ does not divide $A(x)$ so it has to divide $B(x)$ and $B'(x)$, which implies that B and B' have the CM-property.

To show that the Λ sets have the CM-property is completely analogous.

□

EXAMPLE 3.8. This example will show that it is possible for the cyclotomic polynomials to have multiplicity in a tile. They can appear with multiplicity in $A(x)$ and they can appear in both $A(x)$ and $B(x)$. However, from Lemma 5.3, we know that this is not possible for $\Phi_s(x)$ when s is a prime power.

Take $m_1 = 4$, $m_2 = 9$, $u_1 = v_1 = 2$, $u_2 = v_2 = 3$. Then

$$\Psi(A) = \{9a + 4b : a \in \{0,1\}, b \in \{0,1,2\}\},$$
$$\Psi(B') = \{18a + 12b : a \in \{0,1\}, b \in \{0,1,2\}\}.$$

As in the proof of Proposition 3.7, we have

$$A(x) = (1 + x^9)(1 + x^4 + x^8) = \Phi_2(x)\Phi_6(x)\Phi_{18}(x)\Phi_3(x)\Phi_6(x)\Phi_{12}(x).$$

Also, with Lemma 5.4,

$$B'(x) = (1 + x^{18})(1 + x^{12} + x^{24}) = Q_2(x^{18})Q_3(x^{12}) = Q_2((x^2)^9)Q_3((x^3)^4)$$
$$=\Phi_2(x^2)\Phi_6(x^2)\Phi_{18}(x^2)\Phi_3(x^3)\Phi_6(x^3)\Phi_{12}(x^3)=\Phi_4(x)\Phi_{12}(x)\Phi_{36}(x)\Phi_9(x)\Phi_{18}(x)\Phi_{36}(x).$$

4. Some general constructions

PROPOSITION 4.1. *Let A be a set of non-negative integers. If A is a complete set of representatives modulo N, then A has the CM-property.*

PROOF. We have $A \oplus \{0\} = \mathbb{Z}_N$. By Proposition 1.5, there is an integer polynomial such that

$$A(x) = 1 + x + \cdots + x^{N-1} + (x^N - 1)Q(x). \tag{4.1}$$

Since A is a tile, it satisfies (T1), by Theorem 1.8. So

$$N = \#A = \prod_{s \in S_A} \Phi_s(1).$$

Since $\Phi_s(1) = p$ if $s = p^\alpha$ (by Proposition 5.4), it follows that the prime numbers that appear in the prime powers in S_A divide N and also, all prime powers that divide N must appear in S_A, otherwise

$$N = \#A > \prod_{s \in S_A} \Phi_s(1).$$

Thus, S_A consists of all prime powers that divide N so (T2) is also satisfied, because if $s_1, \dots, s_n$ are powers of distinct primes in S_A then $s = s_1 \dots s_n$ divides N so $\Phi_s(x)$ divides $A(x)$, by (4.1). □

PROPOSITION 4.2. *Let A be a finite set of non-negative integers which has a spectrum Λ in $\frac{1}{N}\mathbb{Z}$. If r is relatively prime to N, then $r\Lambda$ is a spectrum for A.*

PROOF. Let $\frac{\lambda}{N} \neq \frac{\lambda'}{N}$ in Λ with $\lambda, \lambda' \in \mathbb{Z}$. Then $A(e^{2\pi i \frac{\lambda - \lambda'}{N}}) = 0$. Let $\frac{\lambda - \lambda'}{N} = \frac{k}{s}$ with $k, s \in \mathbb{Z}$, $(k, s) = 1$. Then $\Phi_s(z)$ divides $A(z)$ since it is the minimal polynomial for $e^{2\pi i \frac{k}{s}}$. Since r is relatively prime to N, it is also relatively prime to s (s divides N). Then $e^{2\pi i \frac{kr}{s}}$ is also a primitive root of order s so $\Phi_s(e^{2\pi i \frac{kr}{s}}) = 0$ so $A(e^{2\pi \frac{\lambda - \lambda'}{N} r}) = A(e^{2\pi i \frac{kr}{s}}) = 0$. This shows that $r\Lambda$ is a spectrum for A. □

For tiles, there is an analogous but more powerful result, due to Tijdeman [**Tij95**].

THEOREM 4.3. [**Tijdeman's theorem**] *If $A \oplus B = \mathbb{Z}_N$ and r is relatively prime to $\#A$, then $rA \oplus B = \mathbb{Z}_N$.*

QUESTION. Is Proposition 4.2 true if r is merely relatively prime to $\#A$? This would be a dual of Tijdeman's theorem.

THEOREM 4.4. *Let A be a tile in $\mathbb{Z}_N$, $A \oplus B = \mathbb{Z}_N$, let M be some postive integer and, for each $a \in A$, let A_a be a tile in $\mathbb{Z}_M$ with a common tiling set C, $A_a \oplus C = \mathbb{Z}_M$. Then the set $\tilde{A} = \cup_{a \in A}(\{a\} \oplus NA_a)$ is a tile in $\mathbb{Z}_{NM}$ with tiling set $\tilde{B} = B \oplus NC$. If, in addition, the sets A and A_a, $a \in A$ satisfy the CM-property, then the set $\tilde{A}$ satisfies the CM-property.*

PROOF. Take x in $\mathbb{Z}_{NM}$. It can be written uniquely as $x = k_1 + Nk_2$ with $k_1 \in \mathbb{Z}_N$ and $k_2 \in \mathbb{Z}_M$. Since $A \oplus B = \mathbb{Z}_N$, k_1 can be written uniquely as $k_1 = a + b$, with $a \in A$ and $b \in B$. Since $A_a \oplus C = \mathbb{Z}_M$, k_2 can be written uniquely as $k_2 = a' + c$ with $a' \in A_a$ and $c \in C$. Thus $x = (a + Na') + (b + Nc)$.

Assume now that the sets A and A_a, $a \in A$ satisfy the CM-condition. Since $\tilde{A}$ is a tile, by Theorem 1.8, $\tilde{A}$ satisfies the (T1) property. To check the (T2) property, first we have to compute the set $S_{\tilde{A}}$.

Note first that

$$\tilde{A}(x) = \sum_{a \in A} x^a A_a(x^N). \tag{4.2}$$

Since $A_a \oplus C = \mathbb{Z}_M$ for all a, it follows the the sets A_a have the same cardinality, and, from Lemma 5.3, that S_{A_a} is the complement of S_C in the set of all prime power factors of M. Therefore, the sets S_{A_a} are all equal. By Lemma 5.5, it follows that the sets S_{NA_a} are all equal, and the sets NA_a satisfy the CM-property.

We will prove that

$$S_{\tilde{A}} = S_A \cup S_{NA_a} \text{ disjoint union.} \tag{4.3}$$

If s is a prime power in S_A, then s divides N (by Lemma 5.3) and if $\omega = e^{2\pi i/s}$ then $\omega^N = 1$. Thus $(NA_a)(\omega) = A_a(\omega^N) = A_a(1) = \#A_a \neq 0$ so s is not in S_{NA_a}. So the sets S_A and S_{NA_a} are disjoint.

With (4.2), we have

$$\tilde{A}(\omega) = \sum_{a\in A} \omega^a A_a(1) = \#A_a \cdot A(\omega) = 0.$$

Thus $s \in S_{\tilde{A}}$.

If s is a prime power in S_{NA_a}, then $\Phi_s(x)$ divides $(NA_a)(x) = A_a(x^N)$ for all $a \in A$, and by (4.2), it follows that $\Phi_s(x)$ divides $\tilde{A}(x)$, so $s \in S_{\tilde{A}}$. This proves that $S_A \cup S_{NA_a} \subset S_{\tilde{A}}$.

Since the sets $\tilde{A}$, A and NA_a are tiles, they satisfy the (T1) property. Therefore, we have

$$\#\tilde{A} = \prod_{s\in S_{\tilde{A}}} \Phi_s(1) \geq \prod_{s\in S_A} \Phi_s(1) \prod_{s\in S_{NA_a}} \Phi_s(1) = \#A \cdot \#(NA_a) = \#\tilde{A}.$$

Thus, we have equality in the inequality, and with Proposition 5.4, since $\Phi_s(1) > 1$ for any prime power s, it follows that we cannot have more elements in $S_{\tilde{A}}$, so (4.3) is satisfied.

Now take $s_1, \dots, s_n$ powers of distinct primes in $S_{\tilde{A}}$. If all the s_i are in S_A, then since A satisfies the (T2) property, it follows that for $s = s_1 \dots s_n$, $\Phi_s(x)$ divides $A(x)$. Also, in this case, s divides N, so if $\omega = e^{2\pi i/s}$ then $\omega^N = 1$. Using (4.2), we obtain that $\tilde{A}(\omega) = 0$ so $\Phi_s(x)$ divide $\tilde{A}(x)$.

If all the s_i are in S_{NA_a}, then since NA_a satisfies the (T2) property, we get that $\Phi_s(x)$ divides $(NA_a)(x) = A_a(x^N)$, for all $a \in A$, and using (4.2), we obtain that $\Phi_s(x)$ divides $\tilde{A}(x)$.

Now assume $s_1, \dots, s_j$ are in S_A (and hence they divide N) and $s_{j+1}, \dots, s_n$ are in S_{NA_a}. Let $s' = s_{j+1} \dots s_n$. We can factor N as $N = N_1N_2N_3$ where N_1 contains all the prime factors that appear in $s_1, \dots, s_j$, N_2 contains all the prime factors of N that appear in s' and N_3 contains all the prime factors of N that do not appear in $s_1, \dots, s_n$, and N_1, N_2, N_3 are mutually prime. (The numbers $s_1 \dots s_j$ and s' are relatively prime, because the s_i are powers of *distinct* primes.)

Then $\frac{N}{s_1\dots s_n} = \frac{N_1N_2N_3}{s_1\dots s_j s'}$ can be reduced to $k_1 \frac{k_2}{s''} N_3$ where $k_1 = \frac{N_1}{s_1\dots s_j}$ is an integer which contains only prime factors that appear in $s_1, \dots, s_j$, $\frac{N_2}{s'} = \frac{k_2}{s''}$, with k_2, s'' relatively prime integers, and k_2 contains only prime factors that appear in s'.

Then we also have $\frac{N}{s'} = \frac{N_1N_2N_3}{s'} = N_1\frac{k_2}{s''}N_3$, with s' and $N_1k_2N_3$ relatively prime, so if $\omega' = e^{2\pi i/s}$ then $\omega'^N = e^{2\pi i \frac{N}{s'}}$ is a primitive root of unity of order s''. Since NA_a satisfies the (T2) property, we get that $\Phi_{s'}(x)$ divides $(NA_a)(x)$ so $A_a(\omega'^N) = 0$ which means that $\Phi_{s''}(x)$ divides $A_a(x)$. Then we also have $\frac{N}{s'} = \frac{N_1N_2N_3}{s'} = N_1\frac{k_2}{s''}N_3$, with s' and $N_1k_2N_3$ relatively prime, so if $\omega' = e^{2\pi i/s}$ then $\omega'^N = e^{2\pi i \frac{N}{s'}}$ is a primitive root of unity of order s''. Since NA_a satisfies the (T2) property, we get that $\Phi_{s'}(x)$ divides $(NA_a)(x)$ so $A_a(\omega'^N) = 0$ which means that $\Phi_{s''}(x)$ divides $A_a(x)$.

Since $k_1k_2N_3$ is also relatively prime with s'', it follows that, for $\omega = e^{2\pi i/s}$, $\omega^N = e^{2\pi i \frac{k_1k_2N_3}{s''}}$ is also a primitive root of unity of order s''. Therefore, $(NA_a)(\omega) = A_a(\omega^N) = 0$ which means that $\Phi_s(x)$ divides $(NA_a)(x) = A_a(x^N)$, for all $a \in A$. From (4.2), it follows that $\Phi_s(x)$ divides $\tilde{A}(x)$.

Thus, $\tilde{A}(x)$ satisfies the (T2) property.

□

REMARK 4.5. We see, from Lemma 2.12, that, in the case when A is a tile and $\#A$ has at most two prime factors p and q, then reducing modulo $\mathrm{lcm}(S_A)$, either A is contained in $p\mathbb{Z}$ or $q\mathbb{Z}$, or it is of the form given in Theorem 4.4. We show in the next example that this is not always the case.

EXAMPLE 4.6. Consider Szabó's example, with $m_1 = 2^2$, $m_2 = 3^2$, $m_3 = 5^2$, so $G = \mathbb{Z}_4 \times \mathbb{Z}_9 \times \mathbb{Z}_{25}$ which is isomorphic to $\mathbb{Z}_{900}$ by the isomorphism in Proposition 3.2. Let $u_1 = v_1 = 2$, $u_2 = v_2 = 3$, $u_3 = v_3 = 5$ and the permutation π of $\{1,2,3\}$, $\pi(1) = 2$, $\pi(2) = 3$, $\pi(3) = 1$. Then

$$A = \{0,1\} \times \{0,1,2\} \times \{0,1,2,3,4\} \subset \mathbb{Z}_4 \times \mathbb{Z}_9 \times \mathbb{Z}_{25},$$

$$B' = \{0,2\} \times \{0,3,6\} \times \{0,5,10,15,20\} \subset \mathbb{Z}_4 \times \mathbb{Z}_9 \times \mathbb{Z}_{25}.$$

To construct the set B we replace the subset $B_1 = \{0,2\} \times \{3\} \times \{0\}$ of B' with the set $\tilde{B}_1 = B_1 + g_1 = \{1,3\} \times \{3\} \times \{0\}$, the set $B_2 = \{0\} \times \{0,3,6\} \times \{5\}$ with the set $\tilde{B}_2 = B_2 + g_2 = \{0\} \times \{1,4,7\} \times \{5\}$, and the set $B_3 = \{2\} \times \{0\} \times \{0,5,10,15,20\}$ with the set $\tilde{B}_3 = B_3 + g_3 = \{2\} \times \{0\} \times \{1,6,11,16,21\}$. so

$$B = B' \setminus (B_1 \cup B_2 \cup B_3) \cup (\tilde{B}_1 \cup \tilde{B}_2 \cup \tilde{B}_3).$$

The isomorphism in Proposition 3.2 is given by

$$\mathbb{Z}_4 \times \mathbb{Z}_9 \times \mathbb{Z}_{25} \ni (a_1,a_2,a_3) \mapsto a_1 \cdot 3^2 \cdot 5^2 + a_2 \cdot 2^2 \cdot 5^2 + a_3 \cdot 2^2 \cdot 3^2 \in \mathbb{Z}_{900}.$$

We have that $\Psi(B)$, which has the CM-property and tiles $\mathbb{Z}_{900}$, according to Proposition 3.7, does not have the form in Theorem 4.4.

Note that if a set $\tilde{A}$ is of the form in Theorem 4.4, i.e.,

$$\tilde{A} = \cup_{a\in a}(\{a\} \oplus NA_a),$$

then, for $k \in \mathbb{Z}_N$, if $k \in A$, then $\#\{a \in \tilde{A} : a \equiv k (\mathrm{mod}\, N)\} = \#A_a = \#\tilde{A}/\#C$, a constant which does not depend on k, and which divides $\#\tilde{A}$, and it is 0 if $k \notin A$. We say that $\tilde{A}$ is *equidistributed* $\mathrm{mod}\, N$.

We will see that B is not equidistributed $\mathrm{mod}\, 2$. Indeed, let $B_0 := B' \setminus (B_1 \cup B_2 \cup B_3)$. Then,

$$\Psi(B_0) \equiv 0 (\mathrm{mod}\, 2), \Psi(\tilde{B}_1) \equiv 1 (\mathrm{mod}\, 2), \Psi(\tilde{B}_2) \equiv 0 (\mathrm{mod}\, 2), \Psi(\tilde{B}_3) \equiv 0 (\mathrm{mod}\, 2),$$

$$\Psi(B_0) \equiv 0 (\mathrm{mod}\, 3), \Psi(\tilde{B}_1) \equiv 0 (\mathrm{mod}\, 3), \Psi(\tilde{B}_2) \equiv 1 (\mathrm{mod}\, 3), \Psi(\tilde{B}_3) \equiv 0 (\mathrm{mod}\, 3),$$

$$\Psi(B_0) \equiv 0 (\mathrm{mod}\, 5), \Psi(\tilde{B}_1) \equiv 0 (\mathrm{mod}\, 5), \Psi(\tilde{B}_2) \equiv 0 (\mathrm{mod}\, 5), \Psi(\tilde{B}_3) \equiv 1 (\mathrm{mod}\, 5).$$

Thus, $\Psi(B_0) \equiv 0 (\mathrm{mod}\, 30)$, and $\Psi(B_0)$ has 20 elements, a number which does not divide $\#\Psi(B) = 30$. This means that $\Psi(B)$ cannot have the form in Theorem 4.4.

We have $\Psi(\tilde{B}_1) = \{75, 525\}$, $\Psi(\tilde{B}_2) = \{280, 580, 880\}$, $\Psi(\tilde{B}_3) = \{486, 666, 846, 126, 306\}$, $\Psi(B_0) = \{0, 30, 60, 120, 150, 210, 240, 330, 360, 390, 420, 510, 540, 570, 600, 660, 690, 720, 840, 870\}$. Thus

$$\Psi(B) = \{0, 30, 60, 75, 120, 126, 150, 210, 240, 280, 306, 330, 360, 390, 420, 486, 510, 525, 540, 570, 580, 600, 660, 666, 690, 720, 840, 846, 870, 880\}.$$

PROPOSITION 4.7. *Let A be a finite set of non-negative integers which has a spectrum $\Lambda_1 \subset \frac{1}{N}\mathbb{Z}$. Suppose $\{A_a : a \in A\}$ are finite sets of non-negative integers that have a common spectrum Λ_2. Then the set $\cup_{a\in A}(\{a\} \oplus NA_a)$ is spectral with spectrum $\Lambda_1 + \frac{1}{N}\Lambda_2$.*

PROOF. Let $(\lambda_1, \lambda_2) \neq (\lambda_1', \lambda_2')$ in $\Lambda_1 \times \Lambda_2$. Since $N(\lambda_1 - \lambda_1') \in \mathbb{Z}$, we have

$$\sum_{a\in A}\sum_{b\in A_a} e^{2\pi i(a+Nb)(\lambda_1-\lambda_1'+\frac{1}{N}(\lambda_2-\lambda_2'))} = \sum_{a\in A} e^{2\pi ia(\lambda_1-\lambda_1'+\frac{1}{N}(\lambda_2-\lambda_2'))}\sum_{b\in A_a} e^{2\pi ib(\lambda_2-\lambda_2')}.$$

Since Λ_2 is a spectrum for all sets A_a, we get that, if $\lambda_2 \neq \lambda_2'$, the last sum is 0, for all $a \in A$.

If $\lambda_2 = \lambda_2'$, then the above sum becomes

$$\sum_{a\in A} e^{2\pi ia(\lambda_1-\lambda_1')}\#A_a = \#\Lambda_2 \sum_{a\in A} e^{2\pi ia(\lambda_1-\lambda_1')}$$

and since Λ_1 is a spectrum for A, this sum is 0 if $\lambda_1 \neq \lambda_1'$. □

COROLLARY 4.8. *Let A be a spectral set in $\mathbb{Z}_N$ with spectrum Λ_1. Suppose $\{A_a : a \in A\}$ are subsets of $\mathbb{Z}_M$ that have a common spectrum Λ_2 in $\mathbb{Z}_M$. Then the set $\cup_{a\in A}(\{a\} \oplus NA_a)$ is spectral in $\mathbb{Z}_{NM}$ with spectrum $M\Lambda_1 + \Lambda_2$.*

Next, we present a result which generalizes and refines a result from Terence Tao's blog, due to I. Łaba.

THEOREM 4.9. *Suppose $A \oplus B = \mathbb{Z}_N$ and $\#A$ and $\#B$ are relatively prime. Then A consists of a single representative from each class modulo $N/\#B$, and B consists of a single representative from each class modulo $N/\#A$. In other words, A is a complete set of representatives modulo $N/\#B$. In particular, A and B have the CM-property and are spectral.*

PROOF. Let $r = \#B$. Since $\#A$ and r are relatively prime, by Theorem 4.3, $rA \oplus B = \mathbb{Z}_N$. So $B \oplus (rA \oplus N\mathbb{Z}) = \mathbb{Z}$. Then, by Lemma 5.6, if $B' = \{b \in B : b \equiv 0 (\bmod r)\}$, we have $B' \oplus (A \oplus N/r\mathbb{Z}) = \mathbb{Z}$ and $\#B' = \#B/r = 1$. Thus $A \oplus N/r\mathbb{Z} = \mathbb{Z}$, which means that A consists of a single representative from each class modulo N/r. By symmetry, the statement is also true for B. □

REMARK 4.10. If A is a complete set of representatives modulo k and N is a multiple of k, $N = kl$, then A always tiles $\mathbb{Z}_N$ with $B = \{ki : i = 0, \ldots, l-1\}$ which is a subgroup of $\mathbb{Z}_N$. Note however that the tile B does not have to be a subgroup of $\mathbb{Z}_N$, even when $\#A = k$ and $\#B = l$ are mutually prime. For example, $\{0, 5, 6, 11\} \oplus \{0, 2, 10\} = \mathbb{Z}_{12}$ and neither of the two sets is a subgroup of $\mathbb{Z}_{12}$.

5. Appendix

PROPOSITION 5.1. *If $A_1, \ldots, A_n$ are finite spectral sets in $G_1, \ldots, G_n$ respectively, with corresponding spectra $\Lambda_1, \ldots, \Lambda_n$, then $A_1 \times \cdots \times A_n$ is spectral in $G_1 \times \cdots \times G_n$ with spectrum $\Lambda_1 \times \cdots \times \Lambda_n$.*

PROOF. Let $(\varphi_1,\dots,\varphi_n),(\varphi_1',\dots,\varphi_n')\in\Lambda_1\times\dots\times\Lambda_n$. Then

$$\sum_{(a_1,\dots,a_n)\in A_1\times\dots\times A_n}\varphi_1(a_1)\dots\varphi_n(a_n)\overline{\varphi}_1'(a_1)\dots\overline{\varphi}_n'(a_n)$$

$$=\left(\sum_{a_1\in A_1}\varphi_1(a_1)\overline{\varphi}_1'(a_1)\right)\dots\left(\sum_{a_n\in A_n}\varphi_n(a_n)\overline{\varphi}_n'(a_n)\right)=\#A_1\delta_{\varphi_1\varphi_1'}\dots\#A_n\delta_{\varphi_n\varphi_n'}.$$

□

LEMMA 5.2. *Let A and B be subsets of $\mathbb{Z}_N$ such that $\#A\cdot\#B=N$. Then $A\oplus B=\mathbb{Z}_N$ if and only if $(A-A)\cap(B-B)=\{0\}$.*

PROOF. The direct implication is clear. For the converse, define the map $A\times B\to\mathbb{Z}_N$, $(a,b)\mapsto a+b$. The condition implies that the map is injective. Since the two sets have the same cardinality N, the map is also surjective, so $A\oplus B=\mathbb{Z}_N$. □

LEMMA 5.3. [**CM99**] *Let $A(x)$ and $B(x)$ be polynomials with coefficients 0 and 1, $n=A(1)B(1)$, and R the set of prime power factors of N. If $\Phi_t(x)$ divides $A(x)$ or $B(x)$ for every factor $t>1$ of N, then*

(i) *$A(1)=\prod_{s\in S_A}\Phi_s(1)$ and $B(1)=\prod_{s\in S_B}\Phi_s(1)$.*
(ii) *S_A and S_B are disjoint sets whose union is R.*

PROPOSITION 5.4. [**CM99**] *Let p be a prime.*

(i) *A polynomial $P(x)\in\mathbb{Z}[x]$ is divisible by $\Phi_s(x)$ if and only if $P(\omega)=0$ for a primitive s-th root of unity ω.*
(ii) *$1+x+\dots+x^{s-1}=\prod_{t>1,t|s}\Phi_t(x)$.*
(iii) *$\Phi_p(x)=1+x+\dots+x^{p-1}$ and $\Phi_{p^{\alpha+1}}(x)=\Phi_p(x^{p^\alpha})$.*
(iv)

$$\Phi_s(1)=\begin{cases}0 & \text{if } s=1\\ q & \text{if } s \text{ is a power of a prime } q\\ 1 & \text{otherwise.}\end{cases}$$

(v)

$$\Phi_s(x^p)=\begin{cases}\Phi_{ps}(x) & \text{if } p \text{ is a factor of } s\\ \Phi_s(x)\Phi_{ps}(x) & \text{if } p \text{ is not a factor } s.\end{cases}$$

(vi) *If s and t are relatively prime, then $\Phi_s(x^t)=\prod_{r|t}\Phi_{rs}(x)$.*

LEMMA 5.5. [**CM99**] *Let $k>1$ and let $A=k\overline{A}$ be a finite set of non-negative integers.*

(i) *A tiles the integers if and only if $\overline{A}$ tiles the integers.*
(ii) *If p is prime, then $S_{p\overline{A}}=\{p^{\alpha+1}:p^\alpha\in S_{\overline{A}}\}\cup\{q^\beta\in S_{\overline{A}}:q\text{ prime },q\neq p\}$.*
(iii) *$A(x)$ satisfies (T1) if and only if $\overline{A}(x)$ satisfies (T1).*
(iv) *$A(x)$ satisfies (T2) if and only if $\overline{A}(x)$ satisfies (T2).*

LEMMA 5.6. **[CM99]** *Suppose $A \oplus C = \mathbb{Z}$, where A is a finite set of non-negative integers, $k > 1$ and $C \subset k\mathbb{Z}$. For $i = 0, 1, \ldots, k-1$, let $A_i = \{a \in A : A \equiv i (\bmod k)\}$, $a_i = \min(A_i)$, and $\overline{A}_i = \{a - a_i : a \in A_i\}/k$. Then*

(i) $A(x) = x^{a_0}\overline{A}_0(x^k) + x^{a_1}\overline{A}_1(x^k) + \cdots + x^{a_{k-1}}\overline{A}_{k-1}(x^k)$.
(ii) *Every* $\overline{A}_i \oplus C/k = \mathbb{Z}$.
(iii) *The elements of A are equally distributed modulo k, i.e., every $\#\overline{A}_i = (\#A)/k$.*
(iv) $S_{\overline{A}_0} = S_{\overline{A}_1} = \cdots = S_{\overline{A}_{k-1}}$.
(v) *When k is prime $S_A = \{k\} \cup S_{k\overline{A}_0}$ and if every $\overline{A}_i(x)$ satisfies (T2), then $A(x)$ satisfies (T2).*

Acknowledgments

Most of the paper was written while Dorin Dutkay was visiting the Institute of Mathematics of the Romanian Academy, with the support of the Bitdefender Invited Professorship. We would like to thank Professors Lucian Beznea, Şerban Stratila and Dan Timotin for their kind hospitality.

References

[CM99] Ethan M. Coven and Aaron Meyerowitz, *Tiling the integers with translates of one finite set*, J. Algebra **212** (1999), no. 1, 161–174, DOI 10.1006/jabr.1998.7628. MR1670646

[DH15] Dorin Ervin Dutkay and John Haussermann, *Tiling properties of spectra of measures*, Math. Z. **280** (2015), no. 1-2, 525–549, DOI 10.1007/s00209-015-1435-6. MR3343918

[DL14] Dorin Ervin Dutkay and Chun-Kit Lai, *Some reductions of the spectral set conjecture to integers*, Math. Proc. Cambridge Philos. Soc. **156** (2014), no. 1, 123–135, DOI 10.1017/S0305004113000558. MR3144214

[FMM06] Bálint Farkas, Máté Matolcsi, and Péter Móra, *On Fuglede's conjecture and the existence of universal spectra*, J. Fourier Anal. Appl. **12** (2006), no. 5, 483–494, DOI 10.1007/s00041-005-5069-7. MR2267631

[Fug74] Bent Fuglede, *Commuting self-adjoint partial differential operators and a group theoretic problem*, J. Functional Analysis **16** (1974), 101–121. MR0470754

[KM06a] Mihail N. Kolountzakis and Máté Matolcsi, *Complex Hadamard matrices and the spectral set conjecture*, Collect. Math. **Vol. Extra** (2006), 281–291. MR2264214

[KM06b] Mihail N. Kolountzakis and Máté Matolcsi, *Tiles with no spectra*, Forum Math. **18** (2006), no. 3, 519–528, DOI 10.1515/FORUM.2006.026. MR2237932

[Łab02] I. Łaba, *The spectral set conjecture and multiplicative properties of roots of polynomials*, J. London Math. Soc. (2) **65** (2002), no. 3, 661–671, DOI 10.1112/S0024610702003149. MR1895739

[Mat05] Máté Matolcsi, *Fuglede's conjecture fails in dimension 4*, Proc. Amer. Math. Soc. **133** (2005), no. 10, 3021–3026, DOI 10.1090/S0002-9939-05-07874-3. MR2159781

[New77] Donald J. Newman, *Tesselation of integers*, J. Number Theory **9** (1977), no. 1, 107–111, DOI 10.1016/0022-314X(77)90054-3. MR0429720

[Sza85] Sándor Szabó, *A type of factorization of finite abelian groups*, Discrete Math. **54** (1985), no. 1, 121–124, DOI 10.1016/0012-365X(85)90071-8. MR787502

[Tao04] Terence Tao, *Fuglede's conjecture is false in 5 and higher dimensions*, Math. Res. Lett. **11** (2004), no. 2-3, 251–258, DOI 10.4310/MRL.2004.v11.n2.a8. MR2067470

[Tij95] R. Tijdeman, *Decomposition of the integers as a direct sum of two subsets*, Number theory (Paris, 1992), London Math. Soc. Lecture Note Ser., vol. 215, Cambridge Univ. Press, Cambridge, 1995, pp. 261–276, DOI 10.1017/CBO9780511661990.016. MR1345184

Department of Mathematics, University of Central Florida, 4000 Central Florida Blvd., P.O. Box 161364, Orlando, Florida 32816-1364
Email address: Dorin.Dutkay@ucf.edu

Department of Mathematics, University of Central Florida, 4393 Andromeda Loop N., Orlando, Florida 32816-1364
Email address: izzy.kraus@knights.ucf.edu

Contemporary Mathematics
Volume **706**, 2018
http://dx.doi.org/10.1090/conm/706/14208

Spectral fractal measures associated to IFS's consisting of three contraction mappings

Ian Long

ABSTRACT. In 1981, Hutchinson showed that for each iterated function system $\{\tau_1, \cdots, \tau_n\}$ on $[0, 1]$, there exists a unique probability measure, called a Hutchinson measure or fractal measure, whose support is the attractor of the of the iterated function system. A common question that has been asked about measures among this class is whether or not they are spectral. That is, for a given Hutchinson measure μ does there exists a set $\Lambda \subseteq \mathbb{R}$, called a spectrum for μ, such that the set of exponentials $\{e^{2\pi i \lambda x} : \lambda \in \Lambda\}$ is an orthonormal basis for $L^2(\mu)$. In the case where the measure is a Bernoulli convolution measure (that is, its associated iterated function system if of the form $\{\lambda(x-1), \lambda(x+1)\}$ with $0 < \lambda < 1$), it is known precisely whether or not the measure is spectral. For more general Hutchinson measures, less is known. We consider the question of whether or not a Hutchinson measure associated to an iterated function system $\{\frac{1}{R}(x+b_1), \frac{1}{R}(x+b_2), \frac{1}{R}(x+b_3)\}$ with $R, b_i \in \mathbb{Z}$ is spectral. We demonstrate that such a measure is spectral if and only if $3 \mid R$ and the elements of the set $B = \{b_1, b_2, b_3\}$ are "well-spaced."Furthermore, we find an explicit spectrum of each spectral measure.

1. Introduction: Hutchinson measures and determining when they are spectral

For over three decades, there has been a great deal of attention paid to a family of "fractal measures"first discovered by Hutchinson ([**Hut81**]). The definition of these measures depends on that of an iterated function system:

DEFINITION 1.1. Let (X, d) be a complete metric space. An **iterated function system** or **IFS** is a set of contraction mappings $\{\tau_1, \cdots, \tau_n\}$ on X.

Given an IFS $I = \{\tau_1, \cdots, \tau_n\}$ on a complete metric space it is well-known (see for example the previously given reference) that there is a unique compact subset $A \subseteq X$ which is invariant with respect to I. That is, there is such an A that satisfies

$$A = \bigcup_{i=1}^{n} \tau_i(A).$$

2010 *Mathematics Subject Classification.* Primary 43A99, 46C07.
Key words and phrases. Functional analysis, harmonic analysis, fractal measure.

This A is referred to as the **attractor** of the IFS.

To each IFS $I = \{\tau_1, \cdots, \tau_n\}$ on $\mathbb{R}$, Hutchinson showed that there exists a unique Borel regular probability measure μ which satisfies a self-similarity property:

$$\mu(\cdot) = \frac{1}{n}\sum_{j=1}^{n} \mu(\tau_j^{-1}(\cdot))$$

and has as its support the attractor of I. In virtually all cases, this attractor is a Cantor set. Thus, the vast majority of these **Hutchinson measures** are singular with respect to Lebesgue measure.

The study of these measures intensified with the detailed examination of the $\frac{1}{4}$-Cantor measure by Jorgensen and Pedersen in [**JP98**]. This measure is the Hutchinson measure associated to the IFS

$$\left\{x \mapsto \frac{1}{4}x, x \mapsto \frac{1}{4}x + \frac{1}{2}\right\}.$$

The support of this measure is a Cantor set formed by removing the second and fourth fourths at each iteration of its creation (as opposed to removing the second third at each iteration to create the classic Cantor set). Initial enthusiasm for $\mu_{\frac{1}{4}}$ was a result of the fact that Jorgensen and Pedersen showed that it was a spectral measure.

Definition 1.2. A measure μ on $X \subseteq \mathbb{R}$ is said to be **spectral** if there exists $\Lambda \subseteq \mathbb{R}$ such that

$$\{e_\lambda : X \to \mathbb{C}, e_\lambda(x) = e^{2\pi i \lambda x} | \ \lambda \in \Lambda\}$$

is an orthonormal basis for $L^2(\mu)$. The set Λ is referred to as a **spectrum** and $\{e_\lambda : \lambda \in \Lambda\}$ is referred to as a **Fourier basis** for μ.

Jorgensen and Pedersen were able to show that the $\frac{1}{4}$-Cantor measure has spectrum

$$\Lambda = \left\{\sum_{k=0}^{m'} 4^k \ell_k : \ell_k \in \{0,1\}, m' \in \mathbb{N}_0\right\}$$
$$= \{0, 1, 4, 5, 16, 17, 20, 21, \cdots\}.$$

They also demonstrated that there exist a wide variety of Hutchinson measures which are not spectral. For example, the $\frac{1}{3}$-Cantor measure $\mu_{\frac{1}{3}}$, associated with the IFS

$$\left\{x \mapsto \frac{1}{3}x, x \mapsto \frac{1}{3}x + \frac{2}{3}\right\},$$

is not spectral. In fact, they were able to show that a maximal orthogonal set of exponentials in $L^2(\mu_{\frac{1}{3}})$ contains at most two elements.

The method used by Jorgensen and Pedersen (and subsequently Laba and Wang in [**ŁW02**] and Dutkay, Haussermann, and Lai in [**DHL15**]) to determine that a given Hutchinson measure is spectral involves finding Hadamard triples:

DEFINITION 1.3. Let $R \in \mathbb{Z}$ with $|R| \geq 2$ and $B, L \subseteq \mathbb{R}$ with $0 < |B| = |L| = N < \infty$. Assume further that $0 \in L$ and $b \cdot \ell \in \mathbb{Z}$ for all $b \in B, \ell \in L$. Then we say that (R, B, L) is a **Hadamard triple** if the $N \times N$ matrix

$$H_{R,B,L} = \frac{1}{\sqrt{N}} \left[e^{\frac{2\pi i b \cdot \ell}{R}} \right]_{b \in B, \ell \in L}$$

is unitary.

We can relate the definition of a Hadamard triple to the definition of a Hutchinson measure in the following way: choose $R \in \mathbb{Z}$ with $|R| \geq 2$ and $B = \{b_1, \cdots, b_N\} \subseteq \mathbb{R}$. Since $|R| \geq 2$, we can define an IFS

$$I_{R,B} = \{\tau_i : \tau_i(x) = R^{-1}(x + b_i), 1 \leq i \leq N\}.$$

Therefore there is a unique Hutchinson measure $\mu_{R,B}$ on $\mathbb{R}$ satisfying the self-similarity property

$$\mu_{R,B}(\cdot) = \frac{1}{N} \sum_{i=1}^{N} \mu_{R,B}(R \cdot -b_i).$$

If R and B are chosen "correctly", then $\mu_{R,B}$ is spectral:

THEOREM 1.4. (**[ŁW02] Theorem 1.2)** *Let $R \in \mathbb{Z}$ with $|R| \geq 2$ and let $B \subseteq \mathbb{Z}$ be a finite set of integers. If there exists $L \subseteq \mathbb{Z}$ with $0 \in L$ such that (R, B, L) is a Hadamard triple, then $\mu_{R,B}$ is spectral.*

Defining Hutchinson measures by choosing an R and a B (as we shall do from now on), we are able to conveniently express the Fourier transform of $\mu_{R,B}$ as an absolutely convergent infinite product in terms of R and B by using the self-similarity condition of the Hutchinson measure:

$$\begin{aligned}
\widehat{\mu_{R,B}}(t) &= \left(\frac{1}{|B|} \sum_{b \in B} e^{\frac{2\pi i b t}{R}} \right) \widehat{\mu_{R,B}} \left(\frac{t}{R} \right) \\
&= \left(\frac{1}{|B|} \sum_{b \in B} e^{\frac{2\pi i b t}{R}} \right) \left(\frac{1}{|B|} \sum_{b \in B} e^{\frac{2\pi i b t}{R^2}} \right) \widehat{\mu_{R,B}} \left(\frac{t}{R^2} \right) \\
&= \prod_{d=1}^{\infty} \left(\frac{1}{|B|} \sum_{b \in B} e^{\frac{2\pi i b t}{R^d}} \right).
\end{aligned}$$

LEMMA 1.5. (**[JP98]**) *The infinite product expression for $\mu_{R,B}$ above is absolutely convergent for all $t \in \mathbb{R}$, thus is zero if and only if one of the terms in the product is zero.*

PROOF. It is known (for example, see **[Ahl78]** pg. 192) that the infinite product above will converge if and only if the series

$$\sum_{d=1}^{\infty} \left| \left(\frac{1}{|B|} \sum_{b \in B} e^{\frac{2\pi i b t}{R^d}} \right) - 1 \right|$$

converges. Noting that

$$0 \leq \sum_{d=1}^{\infty} \left| \left(\frac{1}{|B|} \sum_{b \in B} e^{\frac{2\pi i b t}{R^d}} \right) - 1 \right| \leq \sum_{b \in B} \frac{1}{|B|} \sum_{d=1}^{\infty} \left| -1 + e^{\frac{2\pi b t}{R^d}} \right|,$$

showing that this product converges will follow immediately if we can show that $\sum_{d=1}^{\infty}\left|-1+e^{\frac{2\pi bt}{R^d}}\right|$ converges for each $b \in B$. Since $|R| \geq 2$, this follows by the Ratio Test:

$$\begin{aligned}
\lim_{d\to\infty}\frac{\left|-1+e^{\frac{2\pi ibt}{R^{d+1}}}\right|}{\left|-1+e^{\frac{2\pi ibt}{R^{d}}}\right|} &= \left|\lim_{d\to\infty}\frac{e^{\frac{\pi ibt}{R^{d+1}}}}{e^{\frac{\pi ibt}{R^{d}}}}\right|\cdot\left|\lim_{d\to\infty}\frac{e^{\frac{\pi ibt}{R^{d+1}}}-e^{-\frac{\pi ibt}{R^{d+1}}}}{e^{\frac{\pi ibt}{R^{d}}}-e^{-\frac{\pi ibt}{R^{d}}}}\right| \\
&= \left|\lim_{d\to\infty}\frac{\sin\left(\frac{\pi bt}{R^{d+1}}\right)}{\sin\left(\frac{\pi bt}{R^{d}}\right)}\right| \\
&= \left|\lim_{d\to\infty}\frac{-\cos\left(\frac{\pi}{R^{d+1}}\right)\cdot\pi btR^{-d-1}\ln(R)}{-\cos\left(\frac{\pi}{R^{d}}\right)\cdot\pi btR^{-d}\ln(R)}\right| \\
&= \frac{1}{R}.
\end{aligned}$$

□

It is worth noting that higher-dimensional analogues to the Hutchinson measures defined here exist. In particular, if one chooses R to be an $N \times N$ integer matrix which is expansive (all of its eigenvalues have modulus greater than 1) and $B \subseteq \mathbb{R}^N$ with $|B| < \infty$, one can define a Hutchinson measure $\mu_{R,B}$ on a subset of $\mathbb{R}^N$ which satisfies an analogous self-similarity condition. Furthermore, if $L \subseteq \mathbb{R}^N$ exists such that $|L| = |B|$ and the matrix

$$H_{R,B,L} = \frac{1}{\sqrt{N}}\left[e^{2\pi i\langle R^{-1}b,\ell\rangle}\right]_{b\in B,\ell\in L}$$

is unitary, Dutkay and Lai proved a result analogous to Theorem 1.4 ([**DHL15**]). Recent efforts have been made to guarantee the spectrality or non-spectrality of higher-dimensional Hutchinson measures $\mu_{R,B}$, as well as to find explicit spectra of specific higher-dimensional Hutchinson measures ([**Li07**],[**DJ07**]). Additionally, there have been a number of publications in recent years dedicated to finding Fourier frames (as opposed to Fourier bases) for Hutchinson measures (see e.g. [**DHW14**],[**PW17**],[**DR16**]) and a number of publications dedicated to examining the set of bounded linear maps on the L^2-space of a Hutchinson measure ([**JKS12**], [**JKS14**], [**JKS14b**]). Finally, in the case of several Hutchinson measures, the question of when both a set $\Lambda \subseteq \mathbb{R}$ and one of its integer multiples are spectra for that measure has been considered ([**JKS11**], [**DH16**], [**DK16**], [**LX17**]). While these related areas of study will not be considered again in this publication, they are mentioned to demonstrate the numerous subjects related to Hutchinson measures which are studied.

If one knowns that a given Hutchinson measure is spectral, it is not an immediate consequence that one knows an explicit spectrum for that Hutchinson measure. However, if one has a Hutchinson measure $\mu_{R,B}$ and a set L such that (R, B, L) is Hadamard triple, one is immediately provided with a "likely candidate"for a spectrum of $\mu_{R,B}$:

DEFINITION 1.6. Choose $R \in \mathbb{Z}$ and L a finite subset of $\mathbb{R}$. We define the following sets:

$$X(L) = \left\{ \sum_{k=1}^{\infty} R^{-k}\ell_k : \ell_k \in L \right\}$$

$$\Gamma(L) = \left\{ \sum_{k=0}^{m'} R^{k}\ell_k : \ell_k \in L, m' \in \mathbb{N}_0 \right\}.$$

The aforementioned "likely candidate"is the set $\Gamma(L)$:

DEFINITION 1.7. ([**DJ12**]) Let (R, B, L) be a Hadamard triple with $R \in \mathbb{Z}$, $B, L \subseteq \mathbb{R}$, and $|B| = |L| = N$. Define $X(L)$ as above and define the IFS $\{\sigma_{\ell_1}, \cdots, \sigma_{\ell_N} : \ell_i \in L\}$ on $[0,1]$ such that $\sigma_{\ell_i}(x) = R^{-1}(x + \ell_i)$. Then an **$L$-cycle** C is a finite set $\{x_1, \cdots, x_n\} \in X(L)$ such that there exist $\ell_{x_1}, \cdots, \ell_{x_n} \in L$ (not necessarily distinct) satisfying $\sigma_{\ell_{x_i}}(x_i) = x_{i+1}$ if $1 \le i \le n-1$ and $\sigma_{\ell_{x_n}}(x_n) = x_1$. C is called **B-extreme** if $|\chi_B(x)| = 1$ for all $x \in C$, where

$$\chi_B(x) = \frac{1}{N}\sum_{b \in B} e^{2\pi i b x}.$$

THEOREM 1.8. ([**DJ12**] **Lemma 3.3**) *Let (R, B, L) be a Hadamard triple with $0 \in B \cap L$. Define $\Gamma(L)$ as above. Then $\Gamma(L)$ is a spectrum for $\mu_{R,B}$ if and only if the only B-extreme L-cycle is $\{0\}$.*

The following lemma is quite useful when attempting to find B-extreme L-cycles:

LEMMA 1.9. ([**DJ12**] **Lemma 3.4**) *If C is a B-extreme L-cycle (where (R, B, L) is a Hadamard triple) then for all $x \in C$ and all $b \in B$ it must be the case that $bx \in \mathbb{Z}$.*

EXAMPLE 1.10. Let $R = 4, B = \{0, 2\}$ and thus $\mu_{R,B}$ is the $\frac{1}{4}$-Cantor measure. Then if we choose $L = \{0, 1\}$ one can check that (R, B, L) is a Hadamard triple.

Our possible source of a B-extreme L-cycle is the set

$$\frac{1}{2}\mathbb{Z} \cap \left[\sum_{k=1}^{\infty} 4^{-k} \cdot 0, \sum_{k=1}^{\infty} 4^{-k} \cdot 1 \right] = \frac{1}{2}\mathbb{Z} \cap \left[0, \frac{1}{3}\right] = \{0\}.$$

It is immediate then that $\{0\}$ is the only B-extreme L-cycle and thus $\Gamma(L)$ is a spectrum for $\mu_{R,B}$.

It is not always the case that the likely candidate $\Gamma(L)$ is a spectrum for $\mu_{R,B}$, however, as is demonstrated in the following example:

EXAMPLE 1.11. Let $R = 9, B = \{-8, 0, 8\}$ and define $\mu_{R,B}$ to be the invariant measure associated to the IFS

$$I_{R,B} = \left\{ \tau_{-1}, \tau_0, \tau_1 : \mathbb{R} \to \mathbb{R}, \tau_k(x) = \frac{1}{9}x + \frac{8k}{9} \right\}.$$

Then if we choose $L = \{0, 3, 6\}$, one can check that (R, B, L) is a Hadamard triple. Define the IFS in terms of L as well

$$I_{R,L} = \left\{ \sigma_{\ell_0}, \sigma_{\ell_1}, \sigma_{\ell_2} : \mathbb{R} \to \mathbb{R}, \sigma_{\ell_k}(x) = \frac{1}{9}(x + 3k) \right\}.$$

Our possible source of a B-extreme L-cycle is the set

$$\frac{1}{8}\mathbb{Z} \cap \left[\sum_{k=1}^{\infty} 9^{-k} \cdot 0, \sum_{k=1}^{\infty} 9^{-k} \cdot 6\right] = \frac{1}{8}\mathbb{Z} \cap \left[0, \frac{3}{4}\right].$$

We have however that $\sigma_{\ell_2}\left(\frac{3}{4}\right) = \frac{3}{4} = \sum_{k=1}^{\infty} 9^{-k} \cdot 6 \in X(L)$. Thus, $\left\{\frac{3}{4}\right\}$ is a B-extreme L-cycle and $\Gamma(L)$ is not a spectrum for $\mu_{R,B}$. One can show (as we will in Section 4) that $\frac{1}{8}\Gamma(L)$ is a spectrum for $\mu_{R,B}$.

As this previous example illustrates, determining a spectrum of a Hutchinson measure can be a delicate process even when one is guaranteed of its existence by the existence of a corresponding Hadamard triple. As such, there are relatively few results which place conditions on R and B which guarantee that $\mu_{R,B}$ is spectral with a particular spectrum, even with the restrictions $R \in \mathbb{Z}$, $B \subseteq \mathbb{Z}$. The results which do exist apply predominantly to the case where $|B| = 2$. In particular, if it is assumed that $B = \{\pm 1\}$ (and in this case $\mu_{R,B}$ is called a Bernoulli convolution measure), it is known that $\mu_{R,B}$ is spectral if and only if R is an even integer. Furthermore, if $R = 2n$ specifically, $\mu_{R,B}$ has spectrum

$$\Gamma(L) = \left\{\sum_{j=0}^{m'} R^j \ell_j : \ell_j \in \left\{0, \frac{n}{2}\right\}, 0 \leq m' < \infty\right\}.$$

See [**Dai12**] for more information on these results.

This publication will be an attempt to add to these results by answering the following question: precisely which Hutchinson measures $\mu_{R,B}$ defined such that $R \in \mathbb{Z}, B \subseteq \mathbb{Z}$, and $|B| = 3$ are spectral? Furthermore, for those which are spectral, we will determine a canonical spectrum. We will refer to such measures as Hutchinson-3 measures for the sake of simplicity:

Definition 1.12. We say that a Hutchinson measure $\mu_{R,B}$ is a (one-dimensional) **Hutchinson-N measure** if $R \in \mathbb{Z}$ and $B \subseteq \mathbb{Z}$ with $|B| = N$.

By the end of the publication, we will have successfully determined a condition on R and B which tells us precisely which Hutchinson-3 measures are spectral. In particular, we will show first in Section 2 that it must be the case that $3 \mid R$ (Corollary 2.4). We will then show that the elements of B must be "almost" evenly-spaced. That is, if we choose $B = \{b_1, b_2, b_3\}$ with $b_1 < b_2 < b_3$, $b_3 - b_2 = b_2 - b_1$, and choose R divisible by 3, the measure $\mu_{R,B}$ will be spectral. Furthermore, we will show that if the elements of B satisfy a weaker condition called "well-spaced" (Definition 3.1), then the measure $\mu_{R,B}$ will be spectral. We show this in two steps: first we introduce and examine the stricter definition of "well-spaced about the origin" in Sections 3 and 4, and generalize it to define well-spacedness in Section 5. Furthermore, we will find explicit spectra of spectral Hutchinson-3 measures in Sections 3,4, and 5.

2. Spectral Hutchinson-3 measures: a necessary condition on R for the resulting measure to be spectral

The process of determining which Hutchinson-3 measures $\mu_{R,B}$ are spectral will require a number of steps. The first of these will be placing a condition on R: in the same way that a Bernoulli convolution measure $\mu_{R,B}$ requires R such that $2 \mid R$ in order to be spectral (as does a Hutchinson-2 measure more generally), it is a requirement that $3 \mid R$ for Hutchinson-3 measures.

LEMMA 2.1. *If $x_1, x_2 \in \mathbb{R}$ are chosen such that*

$$1 + e^{2\pi i x_1} + e^{2\pi i x_2} = 0,$$

then we must have that $x_1 = \frac{j}{3} + n_1$ and $x_2 = \frac{j'}{3} + n_2$ for some $n_1, n_2 \in \mathbb{Z}$ and one of j, j' being 1 and the other being 2.

PROOF. Under these assumptions, we have that $e^{2\pi i x_1} = -1 - e^{2\pi i x_2}$ so that

$$1 = |e^{2\pi i x_1}|^2 = |-1 - e^{2\pi i x_2}|^2 = 2 + 2\cos(2\pi x_2).$$

Thus $\cos(2\pi x_2) = -\frac{1}{2}$ and we must have that $x_2 = \frac{1}{3} + n$ or $\frac{2}{3} + n$ for some $n \in \mathbb{Z}$. Solving for x_1 using our first equation above, we find that $x_1 = \frac{2}{3} + m$ or $x_1 = \frac{1}{3} + m$ for some $m \in \mathbb{Z}$ (respectively). □

LEMMA 2.2. (**[DJ07] Theorem 3.1**) *Let $R \in \mathbb{Z}, B = \{b_1, b_2, b_3\} \subseteq \mathbb{Z}$ and define the function*

$$m_B(x) = \frac{1}{3} \sum_{b \in B} e^{2\pi i b x}$$

for $x \in \mathbb{R}$. Let Z be the set of zeroes of m_B in $[0, 1)$ and let

$$\mathcal{O}(Z) = \{R^n x \mod \mathbb{Z} | x \in Z, n \in \mathbb{N}\}.$$

Assume that the Euclidean distance from $\mathcal{O}(Z)$ to $\mathbb{Z}$ is $\delta > 0$. Then there exist at most $\frac{1}{\delta} + 1$ mutually orthogonal exponential functions in $L^2(\mu_{R,B})$.

THEOREM 2.3. *Let $R \in \mathbb{Z}$ and $B = \{0, b_1, b_2\} \subseteq \mathbb{Z}$ with $b_2 > b_1 > 0$ and assume that $\mu_{R,B}$ is spectral. Then it must be the case that $3 \mid R$. Furthermore, if $3 \nmid R$ and $b_2 - b_1 = M \in \mathbb{Z}$, then $L^2(\mu_{R,B})$ contains at most $3M + 1$ mutually orthogonal exponential functions. In particular, $\mu_{R,B}$ is not a spectral measure in this case.*

PROOF. By Lemma 2.1 above we have that $0 = m_B(x) = 3m_B(x) = 1 + e^{2\pi i b_1 x} + e^{2\pi i b_2 x}$ only if $b_1 x = \frac{1}{3} + n_1$ and $b_2 x = \frac{2}{3} + n_2$ or $b_1 x = \frac{2}{3} + n_1$ and $b_2 x = \frac{1}{3} + n_2$ for some $n_1, n_2 \in \mathbb{Z}$. In the former case we get that $(b_2 - b_1)x = \frac{1}{3} + (n_2 - n_1)$ so that $x = \frac{\frac{1}{3}+n}{b_2-b_1} = \frac{1+3n}{3(b_2-b_1)}$ for some $n \in \mathbb{Z}$ and in the latter case we similarly find that $x = \frac{-\frac{1}{3}+n}{b_2-b_1}$ so that $x = \frac{2+3n'}{3(b_2-b_1)}$ for some $n' \in \mathbb{Z}$. We cover both bases by assuming that $x = \frac{\ell+3n}{3(b_2-b_1)}$ for some $n \in \mathbb{Z}$ and $\ell \in \{1, 2\}$.

Regardless of our choice of ℓ, n, we have that $x \notin \mathbb{Z}$ since $3 \nmid \ell + 3n$. Furthermore, if we assume that $3 \nmid R$, then we have that $3(b_2 - b_1) \nmid R^k$ for any $k \in \mathbb{N}$. Thus, for all k we have that $\frac{R^k(\ell+3n)}{3(b_2-b_1)} = N + \frac{m}{3(b_2-b_1)}$ for some $N \in \mathbb{Z}$ and $0 < m < 3(b_2 - b_1)$. So for $x \in \mathcal{O}(Z)$ we have that the minimum possible Euclidean distance between the orbit of x under R and $\mathbb{Z}$ is $\frac{1}{3(b_2-b_1)}$ and therefore the minimum possible Euclidean distance from $\mathcal{O}(Z)$ to $\mathbb{Z}$ is $\frac{1}{3(b_2-b_1)}$. Applying Lemma 2.2, we see that $L^2(\mu_{R,B})$ contains at most $3(b_2 - b_1) + 1$ mutually orthogonal exponential functions as claimed. This proves the theorem. □

COROLLARY 2.4. *Let $R \in \mathbb{Z}$, $B = \{b_1, b_2, b_3\} \subseteq \mathbb{Z}$. If $3 \nmid R$, then $\mu_{R,B}$ is not spectral.*

PROOF. Assume without loss of generality that $b_1 < b_2 < b_3$ and let $B' = B - b_1$. Then $0 \in B' \subseteq \mathbb{Z}$ and by Theorem 2.3, $\mu_{R,B'}$ is not spectral. Furthermore

$$\widehat{\mu_{R,B}}(t) = \prod_{d=1}^{\infty} e^{\frac{2\pi i b_1 t}{R^d}} \left(\frac{1}{|B|} \sum_{b' \in B'} e^{\frac{2\pi i b' t}{R^d}} \right)$$

for all $t \in \mathbb{R}$. Thus $\widehat{\mu_{R,B}}(t) = 0$ if and only if $\widehat{\mu_{R,B'}}(t) = 0$ so $\mu_{R,B}$ is not spectral. □

3. Well-spacedness about the origin and the canonical spectrum

The fact that we know $3 \mid R$ is necessary for a Hutchinson-3 measure to be spectral is a significant step toward knowing exactly which Hutchinson-3 measures are spectral. Conditions on the set B must come next, and the following definition is the first step in the direction of determining said conditions:

DEFINITION 3.1. Let $R \in \mathbb{Z}$ satisfy $R = 3^n k$, where $n \geq 1$ and $k \in \mathbb{Z}$ with $3 \nmid k$. Let $B = \{-b+m, 0, b\}$ or $B = \{-b, 0, b-m\}$ where $b \in \mathbb{N}$ and $m \in [0, b) \cap \mathbb{N}_0$ and let n_1 be the largest nonnegative integer such that $3^{n_1} \mid b$. Then if additionally $3^{n_1+1} \mid m$, we say that (R, B) is **well-spaced about the origin**. If instead we have additionally that $3^{n_1+1} \nmid m$, we say that (R, B) is **poorly-spaced about the origin**.

This definition will be central to determining precisely which Hutchinson-3 measures are spectral. By the end of this section, we will have determined that well-spacedness about the origin of (R, B) is a sufficient condition for $\mu_{R,B}$ to be spectral in many cases. Furthermore, if $\mu_{R,B}$ is poorly-spaced about the origin, $L^2(\mu_{R,B})$ contains no two mutually orthogonal exponentials and therefore $\mu_{R,B}$ is not spectral.

THEOREM 3.2. *If (R, B) is well-spaced about the origin, there exists $L \subseteq \mathbb{R}$ such that (R, B, L) is a Hadamard triple. If the largest positive power of 3 which divides R is larger than the largest nonnegative power of 3 which divides b, then we may choose $L \subseteq \mathbb{Z}$.*

PROOF. Let n be the largest positive integer such that $3^n \mid R$, $k = \frac{R}{3^n}$, and let n_1 be the largest nonnegative integer such that $3^{n_1} \mid b$. By the assumption that $3 \nmid k$ we have that either $k \equiv 1 \pmod 3$ or $k \equiv 2 \pmod 3$. Thus, noting that a matrix that is equivalent via row and/or column interchanges to a matrix of the form

$$\frac{1}{\sqrt{3}} \begin{bmatrix} 1 & 1 & 1 \\ 1 & \omega & \omega^2 \\ 1 & \omega^2 & \omega \end{bmatrix},$$

(where ω is a primitive third root of unity) is a Hadamard matrix, it is a simple calculation to show that if $L = \{0, k3^{n-n_1-1}, 2k3^{n-n_1-1}\}$ and B is given as in the statement of the theorem, then $H_{R,B,L}$ is a equivalent via row and/or column interchanges matrix of this form as desired. Clearly $n > n_1$ would imply that $L \subseteq \mathbb{Z}$. □

THEOREM 3.3. *If (R, B) is well-spaced about the origin and the largest positive power of 3 which divides R is larger than the largest nonnegative power of 3 which divides b, then $\mu_{R,B}$ is spectral.*

PROOF. This follows immediately from Theorem 3.2 and Theorem 1.4. □

THEOREM 3.4. *If (R,B) is poorly-spaced about the origin, then there are no two mutually orthogonal exponentials in $L^2(\mu_{R,B})$. In particular, $\mu_{R,B}$ is not a spectral measure.*

PROOF. Let n_1 be the largest nonnegative integer such that $3^{n_1} \mid b$ so that $3^{n_1+1} \nmid m$ by assumption. We will show that there is no $t \in \mathbb{R}$ such that $\widehat{\mu_{R,B}}(t) = 0$. Indeed, we have using the self-similarity of $\mu_{R,B}$ that

$$\widehat{\mu_{R,B}}(t) = \frac{1}{3}\left(1 + e^{\frac{2\pi i(-b+m)t}{R}} + e^{\frac{2\pi ibt}{R}}\right)\widehat{\mu_{R,B}}\left(\frac{t}{R}\right)$$
$$= \prod_{d=1}^{\infty} \frac{1}{3}\left(1 + e^{\frac{2\pi i(-b+m)t}{R^d}} + e^{\frac{2\pi ibt}{R^d}}\right)$$

or

$$\widehat{\mu_{R,B}}(t) = \frac{1}{3}\left(1 + e^{\frac{-2\pi ibt}{R}} + e^{\frac{2\pi i(b-m)t}{R}}\right)\widehat{\mu_{R,B}}\left(\frac{t}{R}\right)$$
$$= \prod_{d=1}^{\infty} \frac{1}{3}\left(1 + e^{\frac{-2\pi ibt}{R^d}} + e^{\frac{2\pi i(b-m)t}{R^d}}\right).$$

Thus, we have by Lemma 2.1 that $\widehat{\mu_{R,B}}(t) = 0$ only if $\frac{(-b+m)t}{R^d} = \frac{1}{3} + m_1$ and $\frac{bt}{R^d} = \frac{2}{3} + m_2$ or $\frac{(-b+m)t}{R^d} = \frac{2}{3} + m_1$ and $\frac{bt}{R^d} = \frac{1}{3} + m_2$ for some $m_1, m_2 \in \mathbb{Z}$ and $d \geq 1$ in the first case. Similarly in the second case we have that $\widehat{\mu_{R,B}}(t) = 0$ only if $\frac{(b-m)t}{R^d} = \frac{1}{3} + m_1$ and $-\frac{bt}{R^d} = \frac{2}{3} + m_2$ or $\frac{(b-m)t}{R^d} = \frac{2}{3} + m_1$ and $-\frac{bt}{R^d} = \frac{1}{3} + m_2$ for some $m_1, m_2 \in \mathbb{Z}$ and $d \geq 1$. In either case, we may add to get that $\frac{mt}{R^d} \in \mathbb{Z}$. Thus, there is some $k_1 \in \mathbb{Z}$ such that $t = \frac{R^d k_1}{m}$.

But this implies that either $\frac{1}{3} + m_1 = \frac{bt}{R^d} = \frac{bk_1}{m}$ or $\frac{2}{3} + m_1 = \frac{bt}{R^d} = \frac{bk_1}{m}$. In either case, this is a contradiction as we assumed n_1 to be the largest nonnegative integer such that $3^{n_1} \mid b$ and also assumed that $3^{n_1+1} \nmid m$. Thus, $\frac{bk_1}{m}$ cannot be written as $\frac{1}{3} + m_1$ or $\frac{2}{3} + m_1$ for $m_1 \in \mathbb{Z}$ and $\widehat{\mu_{R,B}}$ is nonzero as claimed. □

COROLLARY 3.5. *If (R,B) is poorly-spaced about the origin and $b' \in \mathbb{Z}$, there are no two mutually orthogonal exponentials in $L^2(\mu_{R,B+b'})$. In particular, $\mu_{R,B+b'}$ is not a spectral measure.*

PROOF. Similarly to the previous theorem, in this case we find that

$$\widehat{\mu_{R,B}}(t) = \prod_{d=1}^{\infty} \frac{1}{3} e^{\frac{2\pi ib't}{R^d}}\left(1 + e^{\frac{2\pi i(-b+m)t}{R^d}} + e^{\frac{2\pi ibt}{R^d}}\right)$$

or

$$\widehat{\mu_{R,B}}(t) = \prod_{d=1}^{\infty} \frac{1}{3} e^{\frac{2\pi ib't}{R^d}}\left(1 + e^{\frac{2\pi i(b-m)t}{R^d}} + e^{\frac{-2\pi ibt}{R^d}}\right)$$

for some appropriately-chosen b, m. Thus $\widehat{\mu_{R,B+b'}}(t) = 0$ if and only if $\widehat{\mu_{R,B}}(t) = 0$ and this proof follows from the previous theorem. □

For a given pair (R,B) which is well-spaced about the origin, we have demonstrated that there is a canonical choice of a set L such that (R,B,L) is a Hadamard triple. This provides us both with a guarantee that $\mu_{R,B}$ is spectral when $L \subseteq \mathbb{Z}$ and gives us a canonical "guess"as to a possible spectrum of $\mu_{R,B}$. Namely, let R be written as the product of integers $R = 3^n k$ where $n \geq 1$ and $3 \nmid k$ and either choose $B = \{-b+m, 0, b\}$ or $B = \{-b, 0, b-m\}$ for $b \in \mathbb{N}, m \in \mathbb{N}_0 \cap [0, b)$ such that if n_1 is the largest nonnegative number satisfying $3^{n_1} \mid b$, then $3^{n_1+1} \mid m$. Then we may choose

$$L = \{0, k3^{n-n_1-1}, 2k3^{n-n_1-1}\}$$

to form a Hadamard triple, and as such we may use Theorem 1.8 to check whether or not

$$\Gamma(L) = \left\{ \sum_{j=0}^{m'} R^j \ell_j : \ell_j \in \{0, k3^{n-n_1-1}, 2k3^{n-n_1-1}\}, m' \in \mathbb{N}_0 \right\}$$

is a spectrum for $\mu_{R,B}$. As was demonstrated in Example 1.11, it is not immediately obvious whether or not this choice of $\Gamma(L)$ will be a spectrum for $\mu_{R,B}$. In fact, in order for $\Gamma(L)$ to be a spectrum for $\mu_{R,B}$, we must first choose R and B which do not satisfy the following condition:

DEFINITION 3.6. Let (R, B) be well-spaced about the origin with $B = \{-b + m, 0, b\}$ or $B = \{-b, 0, b-m\}$ and n_1 the largest nonnegative integer such that $3^{n_1} \mid b$. We say that (R, B) satisfies the **R-Divisibility Condition** if one of the following conditions are met:

(i) There exists $d \in \mathbb{N}$ such that $R^d - 1 \mid \frac{b}{3^{n_1}}$ and $R^d - 1 \mid \frac{m}{3^{n_1}}$.
(ii) There exists $d \in \mathbb{N}$ such that $R^d - 1 \mid \frac{2b}{3^{n_1}}$ and $R^d - 1 \mid \frac{2m}{3^{n_1}}$.

PROPOSITION 3.7. *If (R, B) is well-spaced about the origin and satisfies the* **R-Divisibility Condition**, *then*

$$\Gamma(L) = \left\{ \sum_{j=0}^{m'} R^j \ell_j : \ell_j \in \{0, k3^{n-n_1-1}, 2k3^{n-n_1-1}\}, m' \in \mathbb{N}_0 \right\}$$

is not a spectrum for $\mu_{R,B}$.

PROOF. We claim that if there exists $d \in \mathbb{N}$ satisfying (i) or (ii) above, then there exists a non-trivial B-extreme L-cycle. Toward showing this, we claim that either

$$\left\{ \frac{R^d}{R^d-1}\left(\frac{1}{3^{n_1+1}}\right), \frac{R^{d-1}}{R^d-1}\left(\frac{1}{3^{n_1+1}}\right), \dots, \frac{R}{R^d-1}\left(\frac{1}{3^{n_1+1}}\right) \right\} \subseteq X(L) \cap \frac{1}{b}\mathbb{Z} \cap \frac{1}{b-m}\mathbb{Z}$$

or

$$\left\{ \frac{R^d}{R^d-1}\left(\frac{2}{3^{n_1+1}}\right), \frac{R^{d-1}}{R^d-1}\left(\frac{2}{3^{n_1+1}}\right), \dots, \frac{R}{R^d-1}\left(\frac{2}{3^{n_1+1}}\right) \right\} \subseteq X(L) \cap \frac{1}{b}\mathbb{Z} \cap \frac{1}{b-m}\mathbb{Z}.$$

Indeed, we have that

$$\frac{R}{R^d-1} \cdot \left(\frac{1}{3^{n_1+1}}\right) = \sum_{j=1}^{\infty} R^{-dj} \cdot k3^{n-n_1-1} \in X(L)$$

by construction so that if $2 \leq d' \leq d$, we have that

$$\frac{R^{d'}}{R^d-1} \cdot \left(\frac{1}{3^{n_1+1}}\right) = \sum_{j=1}^{\infty} R^{-dj+d'-1} \cdot k3^{n-n_1-1} \in X(L)$$

as well. Similarly, we also have that $\frac{R^{d'}}{R^d-1}\cdot\left(\frac{2}{3^{n_1+1}}\right)\in X(L)$ if $1\le d'\le d$.

The assumption that (i) or (ii) holds implies that for all integers d' with $1\le d'\le d$, we have that either

$$\frac{R^{d'}}{R^d-1}\cdot\left(\frac{1}{3^{n_1+1}}\right)=\frac{\left(\frac{R^{d'}}{R^d-1}\cdot\frac{b}{3^{n_1+1}}\right)}{b}=\frac{\left(\frac{R^{d'}}{R^d-1}\cdot\frac{b-m}{3^{n_1+1}}\right)}{b-m}\in\frac{1}{b}\mathbb{Z}\cap\frac{1}{b-m}\mathbb{Z}$$

or

$$\frac{R^{d'}}{R^d-1}\cdot\left(\frac{2}{3^{n_1+1}}\right)=\frac{\left(\frac{R^{d'}}{R^d-1}\cdot\frac{2b}{3^{n_1+1}}\right)}{b}=\frac{\left(\frac{R^{d'}}{R^d-1}\cdot\frac{2b-2m}{3^{n_1+1}}\right)}{b-m}\in\frac{1}{b}\mathbb{Z}\cap\frac{1}{b-m}\mathbb{Z}.$$

Thus, defining the maps $\sigma_{\ell_1},\sigma_{\ell_2},\sigma_{\ell_3}$ on $\mathbb{R}$ by

$$\begin{aligned}\sigma_{\ell_1}(x)&=\frac{x}{R}\\ \sigma_{\ell_2}(x)&=\frac{1}{R}\left(x+k3^{n-n_1-1}\right)=\frac{x}{R}+\frac{1}{3^{n_1+1}}\\ \sigma_{\ell_3}(x)&=\frac{1}{R}\left(x+2k3^{n-n_1-1}\right)=\frac{x}{R}+\frac{2}{3^{n_1+1}},\end{aligned}$$

and noting that

$$\begin{aligned}\sigma_{\ell_2}\left(\frac{R}{R^d-1}\cdot\frac{1}{3^{n_1+1}}\right)&=\frac{R^d}{R^d-1}\cdot\frac{1}{3^{n_1+1}}\\ \sigma_{\ell_3}\left(\frac{R}{R^d-1}\cdot\frac{2}{3^{n_1+1}}\right)&=\frac{R^d}{R^d-1}\cdot\frac{2}{3^{n_1+1}},\end{aligned}$$

we have that

$$\left\{\frac{R^d}{R^d-1}\left(\frac{1}{3^{n_1+1}}\right),\frac{R^{d-1}}{R^d-1}\left(\frac{1}{3^{n_1-1}}\right),\dots,\frac{R}{R^d-1}\left(\frac{1}{3^{n_1+1}}\right)\right\}$$

or

$$\left\{\frac{R^d}{R^d-1}\left(\frac{2}{3^{n_1+1}}\right),\frac{R^{d-1}}{R^d-1}\left(\frac{2}{3^{n_1+1}}\right),\dots,\frac{R}{R^d-1}\left(\frac{2}{3^{n_1+1}}\right)\right\}$$

is a non-trivial B-extreme L-cycle as desired. □

There are several assumptions we can make which guarantee that $\Gamma(L)$ will be a spectrum for $\mu_{R,B}$ in the case that (R,B) is well-spaced about the origin, and they are outlined in the theorems and corollaries that conclude this section. Though some the conditions required for these theorems and corollaries could be more satisfying, the reader will take note that, beginning now, none of the conditions which follow require that we choose $L\subseteq\mathbb{Z}$.

THEOREM 3.8. *Let $R\in\mathbb{Z}$ be given by $R=3^nk$ for some $n\ge 1$ and $k\in\mathbb{Z}$ such that $3\nmid k$. Also assume $R\ne\pm 3$. Let $B=\{-b,0,b\}$ for some $b\in\mathbb{N}$ and let n_1 be the largest nonnegative integer such that $3^{n_1}\mid b$ and define $b'=\frac{b}{3^{n_1}}$. Furthermore, assume that (R,B) does not satisfy the R-divisibility condition. Then*

$$\Gamma(L)=\left\{\sum_{j=0}^{m'}R^j\ell_j:\ell_j\in\{0,k3^{n-n_1-1},2k3^{n-n_1-1}\},m'\in\mathbb{N}_0\right\}$$

is a spectrum for $\mu_{R,B}$ if either of the conditions below are satisfied.

(a) $\frac{12b'}{5} < R$.
(b) $k > 0$ *and there are no integers contained in any of the following intervals:*

$$I_{1,2} = \left(\frac{b'}{3R}, \frac{b'}{3R} + \frac{2b'}{3R(R-1)}\right]$$
$$I_{1,3} = \left(\frac{2b'}{3R}, \frac{2b'}{3R} + \frac{2b'}{3R(R-1)}\right]$$
$$I_{2,1} = \left(\frac{b'}{3}, \frac{b'}{3} + \frac{2b'}{3R(R-1)}\right]$$
$$I_{2,2} = \left(\frac{b'}{3R} + \frac{b'}{3}, \frac{b'}{3R} + \frac{b'}{3} + \frac{2b'}{3R(R-1)}\right]$$
$$I_{2,3} = \left(\frac{2b'}{3R} + \frac{b'}{3}, \frac{2b'}{3R} + \frac{b'}{3} + \frac{2b'}{3R(R-1)}\right]$$
$$I_{3,1} = \left(\frac{2b'}{3}, \frac{2b'}{3} + \frac{2b'}{3R(R-1)}\right]$$
$$I_{3,2} = \left(\frac{b'}{3R} + \frac{2b'}{3}, \frac{b'}{3R} + \frac{2b'}{3} + \frac{2b'}{3R(R-1)}\right]$$
$$I_{3,3} = \left(\frac{2b'}{3R} + \frac{2b'}{3}, \frac{2b'}{3R} + \frac{2b'}{3} + \frac{2b'}{3R(R-1)}\right]$$

or $k < 0$ *and there are no integers contained in each* $-I_{i,j}$ *above.*

PROOF. We assume $k > 0$ for simplicity, noting that the $k < 0$ case is nearly identical. Using Theorem 1.8, we will show that the only B-extreme L-cycle is the set $\{0\}$. By construction, we have that

$$X(L) \subseteq \left[0, 2k3^{n-n_1-1}\sum_{j=1}^{\infty} R^{-j}\right] = \left[0, \frac{2}{3^{n_1+1}}\left(\frac{R}{R-1}\right)\right].$$

Furthermore, by Lemma 1.9, we have that any $x \in X(L)$ which is an element of a B-extreme L-cycle must satisfy $x \in \frac{1}{b}\mathbb{Z}$. Thus, we must show that no B-extreme L-cycles besides $\{0\}$ can be formed from elements in the set $\frac{1}{b}\mathbb{Z} \cap \left[0, \frac{2}{3^{n_1+1}}\left(\frac{R}{R-1}\right)\right]$ to prove the given claim.

We define $\sigma_{\ell_1}, \sigma_{\ell_2}, \sigma_{\ell_3}$ on $\mathbb{R}$ by

$$\sigma_{\ell_1}(x) = \frac{x}{R}$$
$$\sigma_{\ell_2}(x) = \frac{1}{R}\left(x + k3^{n-n_1-1}\right) = \frac{x}{R} + \frac{1}{3^{n_1+1}}$$
$$\sigma_{\ell_3}(x) = \frac{1}{R}\left(x + 2k3^{n-n_1-1}\right) = \frac{x}{R} + \frac{2}{3^{n_1+1}}.$$

Also, we let $\frac{m}{b} \in \frac{1}{b}\mathbb{Z} \cap \left(0, \frac{2}{3^{n_1+1}}\left(\frac{R}{R-1}\right)\right]$ with $m \in \mathbb{Z}_{\geq 0}$ the first element of a B-extreme L-cycle. That is, assume there exist $\ell_{k_1}, \cdots, \ell_{k_d} \in L$ such that

$$\sigma_{\ell_{k_d}} \circ \sigma_{\ell_{k_{d-1}}} \circ \cdots \circ \sigma_{\ell_{k_1}}\left(\frac{m}{b}\right) = \frac{m}{b},$$
$$\sigma_{\ell_{k_{d'}}} \circ \cdots \circ \sigma_{\ell_{k_1}}\left(\frac{m}{b}\right) \in \frac{1}{b}\mathbb{Z} \cap \left(0, \frac{2}{3^{n_1+1}}\left(\frac{R}{R-1}\right)\right].$$

for all $1 \leq d' \leq d$. Note that we do not allow the choice $m = 0$ as $\sigma_{\ell_1}(0) = 0$ and we do not care about the extreme cycle $\{0\}$ and $\sigma_{\ell_2}(0), \sigma_{\ell_3}(0) \notin \frac{1}{b}\mathbb{Z}$ by construction. Also, the fact that we assume that (R, B) does not satisfy the R-divisibility condition implies that is cannot be the case that $k_d \neq 1$ while $k_{d'} = 1$ for all $d' < d$. Finally, it is worth noting that it cannot be the case that $\ell_{k_i} = \ell_1$ for all i, as then the finite sequence C would be decreasing and we could not possibly have that $\sigma_{\ell_{k_d}} \circ \sigma_{\ell_{k_{n-1}}} \circ \cdots \circ \sigma_{\ell_{k_1}}\left(\frac{m}{b}\right) = \frac{m}{b}$.

With these facts in mind, we note that it must be the case that there exists $i \in \{1, \cdots, d-1\}$ such that one of the following is true:

$$k_{i+1} = 2, k_i = 1$$
$$k_{i+1} = 3, k_i = 1$$
$$k_{i+1} = 1, k_i = 2$$
$$k_{i+1} = 1, k_i = 3,$$
$$k_{i+1} = 2, k_i = 2,$$
$$k_{i+1} = 2, k_i = 3,$$
$$k_{i+1} = 3, k_i = 2,$$
$$k_{i+1} = 3, k_i = 3.$$

Without loss of generality, then, we consider the image of $\frac{m}{b}$ under these possible compositions of two maps. If said image is not contained in $\frac{1}{b}\mathbb{Z} \cap \left[0, \frac{2}{3^{n_1+1}}\left(\frac{R}{R-1}\right)\right]$, we will have shown that no B-extreme L-cycles exist and therefore proven the theorem. To this end, we first claim that $\sigma_{\ell_1}\left(\frac{m}{b}\right), \sigma_{\ell_2}\left(\frac{m}{b}\right)$, or $\sigma_{\ell_3}\left(\frac{m}{b}\right) \in \frac{1}{b}\mathbb{Z}$ only if $k \mid m$ and $3^{n-1} \mid m$. Indeed, we have that

$$\sigma_{\ell_1}\left(\frac{m}{b}\right) = \frac{m}{Rb} = \frac{\frac{m}{R}}{b}$$
$$\sigma_{\ell_2}\left(\frac{m}{b}\right) = \frac{m}{Rb} + \frac{1}{3^{n_1+1}} = \frac{\frac{m}{R} + \frac{b}{3^{n_1+1}}}{b}$$
$$\sigma_{\ell_3}\left(\frac{m}{b}\right) = \frac{m}{Rb} + \frac{2}{3^{n_1+1}} = \frac{\frac{m}{R} + \frac{2b}{3^{n_1+1}}}{b}.$$

In the case of σ_{ℓ_1}, we must have that $R \mid m$ if $\frac{\frac{m}{R}}{b} \in \frac{1}{b}\mathbb{Z}$, and thus $3^{n-1}k \mid m$. And in the case of the other maps, since $\frac{b}{3^{n_1+1}} \in \mathbb{Z} + \left\{\frac{1}{3}\right\}$, we may assume that $3^{n-1} \mid m, k \mid m$, but $3^n \nmid m$. In either case, we may write $m = 3^{n-1}km'$ for some $m' \in \mathbb{Z}$. With this in mind, we have that

$$\sigma_{\ell_2} \circ \sigma_{\ell_1}\left(\frac{m}{b}\right) = \sigma_{\ell_2}\left(\frac{m'3^{n-1}k}{bR}\right) = \frac{\frac{m'}{3R} + \frac{b'}{3}}{b}$$

$$\sigma_{\ell_3} \circ \sigma_{\ell_1}\left(\frac{m}{b}\right) = \sigma_{\ell_3}\left(\frac{m'3^{n-1}k}{bR}\right) = \frac{\frac{m'}{3R} + \frac{2b'}{3}}{b}$$

$$\sigma_{\ell_1} \circ \sigma_{\ell_2}\left(\frac{m}{b}\right) = \sigma_{\ell_1}\left(\frac{m'3^{n-1}k}{bR} + \frac{1}{3^{n_1+1}}\right) = \frac{\frac{m'+b'}{3R}}{b}$$

$$\sigma_{\ell_1} \circ \sigma_{\ell_3}\left(\frac{m}{b}\right) = \sigma_{\ell_1}\left(\frac{m'3^{n-1}k}{bR} + \frac{2}{3^{n_1+1}}\right) = \frac{\frac{m'+2b'}{3R}}{b}$$

$$\sigma_{\ell_2} \circ \sigma_{\ell_2}\left(\frac{m}{b}\right) = \sigma_{\ell_2}\left(\frac{m'3^{n-1}k}{bR} + \frac{1}{3^{n_1+1}}\right) = \frac{\frac{m'+b'}{3R} + \frac{b'}{3}}{b}$$

$$\sigma_{\ell_2} \circ \sigma_{\ell_3}\left(\frac{m}{b}\right) = \sigma_{\ell_2}\left(\frac{m'3^{n-1}k}{bR} + \frac{1}{3^{n_1+1}}\right) = \frac{\frac{m'+2b'}{3R} + \frac{b'}{3}}{b}$$

$$\sigma_{\ell_3} \circ \sigma_{\ell_2}\left(\frac{m}{b}\right) = \sigma_{\ell_3}\left(\frac{m'3^{n-1}k}{bR} + \frac{1}{3^{n_1+1}}\right) = \frac{\frac{m'+b'}{3R} + \frac{2b'}{3}}{b}$$

$$\sigma_{\ell_3} \circ \sigma_{\ell_2}\left(\frac{m}{b}\right) = \sigma_{\ell_3}\left(\frac{m'3^{n-1}k}{bR} + \frac{1}{3^{n_1+1}}\right) = \frac{\frac{m'+2b'}{3R} + \frac{2b'}{3}}{b},$$

where $b' = \frac{b}{3^{n_1}}$. Thus, in order to ensure that any of the numbers above lies in $\frac{1}{b}\mathbb{Z}$, we must have (at the very least) that $R \mid m'$, $R \mid m' + b'$, or $R \mid m' + 2b'$, depending on the case above. However, our assumption that $\frac{m}{b} \in \left(0, \frac{2}{3^{n_1+1}}\left(\frac{R}{R-1}\right)\right]$, implies that

$$\frac{m'3^{n-1}k}{b} \leq \frac{2}{3^{n_1+1}}\left(\frac{R}{R-1}\right)$$

and equivalently that

$$m' \leq \frac{2b'}{3^n k - 1}.$$

Therefore, we have that

$$0 \leq m' \leq m' + b' \leq m' + 2b' \leq \frac{2b'}{3^n k - 1} + 2b' = \left(\frac{3^n k}{3^n k - 1}\right) \cdot 2b' \leq \frac{12}{5} \cdot b'$$

(this last inequality is due to the fact that one can easily show with calculus that $\frac{x}{x-1} < \frac{6}{5}$ for $x \geq 6$). Thus, if (a) holds, we do not have that $R \mid m'$, $R \mid m' + b'$, or $R \mid m' + 2b'$. This implies that there are no B-extreme L-cycles besides $\{0\}$ and the theorem is proven in this case.

In the case where (b) holds, we note that the interval $I_{i,j}$ is defined such that

$$I_{i,j} = b \cdot \sigma_{\ell_i} \circ \sigma_{\ell_j}\left(\left(0, \frac{2}{3^{n_1+1}} \cdot \frac{R}{R-1}\right]\right)$$

for each $i, j \in \{1, 2, 3\}$ not both 1. Thus, if no integers are contained in these intervals, there are no elements in $\frac{1}{b}\mathbb{Z}$ contained in the images of $\left(0, \frac{2}{3^{n_1+1}} \cdot \frac{R}{R-1}\right]$ under these combinations of maps. Coupled with the assumption that (R, B) does not satisfy the R-divisibility condition, this suffices to prove the theorem. $\square$

The conditions (a) and (b) above could certainly be more satisfactory, and the fact that they are not is a direct consequence of the fact that (R, B) is assumed well-spaced about the origin with $m = 0$. In the case where $m \neq 0$, the following results will show that there need not be as strong assumptions made to guarantee that the canonical choice $\Gamma(L)$ is a spectrum for $\mu_{R,B}$. This demonstrates an interesting juxtaposition: on the one hand, it is easier to guarantee that (R, B) is well-spaced about the origin when $m = 0$ (see Theorem 3.3). On the other hand, the spectrum of $\mu_{R,B}$ is more likely to be the canonical choice $\Gamma(L)$ when (R, B) is well-spaced about the origin with $m \neq 0$.

COROLLARY 3.9. *Let (R, B) we well-spaced about the origin, with $B = \{-b + m, 0, b\}$ or $B = \{-b, 0, b - m\}$ for some $b \in \mathbb{N}$ and $m \in \mathbb{N}_0 \cap [0, b)$. Let $B' = \{-b, 0, b\}$ and assume that (R, B') satisfies condition a) or b) given in the statement of Theorem 3.8 so that $\mu_{R,B'}$ has spectrum*

$$\Gamma(L) = \left\{ \sum_{j=0}^{m'} R^j \ell_j : \ell_j \in \{0, k3^{n-n_1-1}, 2k3^{n-n_1-1}\}, m' \in \mathbb{N}_0 \right\}.$$

Then we have that $\mu_{R,B}$ is spectral with spectrum $\Gamma(L)$ as well.

PROOF. By Theorem 3.8 we have that $\mu_{R,B'}$ is spectral with spectrum $\Gamma(L)$, which by Theorem 1.8 is true if and only if $\{0\}$ is the only B'-extreme L-cycle (note that both (R, B, L) and (R, B', L) are Hadamard triples by construction). By Lemma 1.9, the set of all B'-extreme L-cycles is contained in $\frac{1}{b}\mathbb{Z} \cap X(L)$, while the set of all B-extreme L cycles is contained in $\frac{1}{b-m}\mathbb{Z} \cap \frac{1}{b}\mathbb{Z} \cap X(L)$. Thus, $\{0\}$ is the only B-extreme L cycle and by Theorem 1.8 the corollary has been proved. □

THEOREM 3.10. *Let (R, B) be well-spaced about the origin and assume that (R, B) does not satisfy the R-divisibility condition. Specifically, let $B = \{-b + m, 0, b\}$ or $B = \{-b, 0, b - m\}$ with $b \in \mathbb{N}$ and $m \in \mathbb{N}_0 \cap [0, b)$ and let n_1 be the largest nonnegative integer satisfying $3^{n_1} \mid b$. Let $R = 3^n k$ for some $n \geq 1$, $k \in \mathbb{Z}$ with $3 \nmid k$. Then if $\gcd(b, b - m) \leq d$ and*

$$\left| \frac{2}{3^{n_1+1}} \left(\frac{R}{R-1} \right) \right| < \frac{1}{d},$$

then $\mu_{R,B}$ has spectrum

$$\Gamma(L) = \left\{ \sum_{j=0}^{m'} R^j \ell_j : \ell_j \in \{0, k3^{n-n_1-1}, 2k3^{n-n_1-1}\}, m' \in \mathbb{N}_0 \right\}.$$

PROOF. We assume $k > 0$, noting that the $k < 0$ case is very similar. We note that by Theorem 1.8, $\Gamma(L)$ is a spectrum for $\mu_{R,B}$ only if the set

$$\frac{1}{b-m}\mathbb{Z} \cap \frac{1}{b}\mathbb{Z} \cap \left(0, \frac{2}{3^{n_1+1}} \cdot \frac{R}{R-1}\right] = \frac{1}{\gcd(b, b-m)}\mathbb{Z} \cap \left(0, \frac{2}{3^{n_1+1}} \cdot \frac{R}{R-1}\right] \neq \emptyset.$$

However, if $\gcd(b, b - m) \leq d$, the smallest nonnegative number in $\frac{1}{\gcd(b,b-m)}\mathbb{Z}$ is at least $\frac{1}{d} > \frac{2}{3^{n_1+1}} \cdot \frac{R}{R-1}$. This proves the theorem. □

COROLLARY 3.11. *Let (R,B) be well-spaced about the origin and assume that (R,B) does not satisfy the R-divisibility condition. If $\gcd(b, b-m) = 1$, then $\mu_{R,B}$ has spectrum*

$$\Gamma(L) = \left\{ \sum_{j=0}^{m'} R^j \ell_j : \ell_j \in \{0, k3^{n-n_1-1}, 2k3^{n-n_1-1}\}, m' \in \mathbb{N}_0 \right\}.$$

PROOF. This corollary follows immediately from the statement of Theorem 3.10 with $d = 1$. □

EXAMPLE 3.12. Let $R = 9$ and $B = \{-b, 0, b\}$ for a given $b \in \mathbb{N}_0 \cap [0, 9)$. Then if n_1 is the largest nonnegative integer such that $3^{n_1} \mid b$, we have for each choice of b that if we choose $L = \{0, 3^{1-n_1}, 2 \cdot 3^{1-n_1}\} \subseteq \mathbb{Z}$, then (R, B, L) is a Hadamard triple so by Theorem 1.4 $\mu_{R,B}$ is spectral. In the case where $b = 1, 2, 3, 6$, we have by condition (a) from Corollary 4.3 that $\Gamma(L)$ is a spectrum for $\mu_{R,B}$. Condition (b) from Corollary 4.3 guarantees that $\Gamma(L)$ is spectrum for $\mu_{R,B}$ when $b = 5, 7$. But, when $b = 4, 8$ we have that (R, B) satisfies the R-divisibility condition. Thus, $\Gamma(L)$ is not a spectrum for $\mu_{R,B}$ in these cases. We will be able to find a spectrum for these measures in the next section.

4. Well-spacedness about the origin and an alternative spectrum

In the cases where (R, B) is well-spaced about the origin but our potential canonical spectrum $\Gamma(L)$ is not a spectrum for $\mu_{R,B}$, we are able to easily determine an alternative spectrum for this measure. By the end of this section, we will be able to say that any measure $\mu_{R,B}$ where (R, B) is well-spaced about the origin is spectral and determine an explicit spectrum for that measure. Most of the work required to show this is done by the following two lemmas.

LEMMA 4.1. **([JP98] Lemma 3.3)** *Let $R \in \mathbb{Z}$ and $B \subseteq \mathbb{Z}$ be a finite set. Let $\Lambda \subseteq \mathbb{R}$ be chosen such that $\{e_\lambda : \lambda \in \Lambda\}$ is orthonormal in $L^2(\mu_{R,B})$ and furthermore define*

$$Q_{R,B,\Lambda}(t) = \sum_{\gamma \in \Lambda} |\widehat{\mu_{R,B}}(t-\gamma)|^2, t \in \mathbb{R}.$$

Then $\{e_\lambda : \lambda \in \Lambda\}$ is an orthonormal basis for $L^2(\mu_{R,B})$ if and only if $Q_{R,B,\Lambda}(t) \equiv 1$.

LEMMA 4.2. *Let $R \in \mathbb{Z}$ and $B, L \subseteq \mathbb{Z}$. Then if $\alpha \neq 0$ and $\mu_{R,B}$ has spectrum $\Gamma(L)$, $\mu_{R,\alpha B}$ has spectrum $\alpha^{-1}\Gamma(L)$.*

PROOF. We note first that for a given $t \in \mathbb{R}$:

$$\begin{aligned}
\widehat{\mu_{R,\alpha B}}(\alpha^{-1}t) &= \left(\frac{1}{|B|} \sum_{\alpha b \in \alpha B} e^{\frac{2\pi i (\alpha b)(\alpha^{-1}t)}{R}} \right) \widehat{\mu_{R,\alpha B}}\left(\frac{\alpha^{-1}t}{R}\right) \\
&= \left(\frac{1}{|B|} \sum_{b \in B} e^{\frac{2\pi i b t}{R}} \right) \widehat{\mu_{R,\alpha B}}\left(\frac{\alpha^{-1}t}{R}\right) \\
&= \prod_{k=1}^{\infty} \left(\frac{1}{|B|} \sum_{b \in B} e^{\frac{2\pi i b t}{R^k}} \right) \\
&= \widehat{\mu_{R,B}}(t).
\end{aligned}$$

Thus the set $\{e_{\alpha^{-1}\lambda} : \alpha^{-1}\lambda \in \alpha^{-1}\lambda\}$ is orthogonal in $L^2(\mu_{R,\alpha B})$ and we have for a given $t \in \mathbb{R}$ that

$$Q_{R,\alpha B,\alpha^{-1}\Lambda}(t) = \sum_{\lambda\in\Lambda} |\widehat{\mu_{R,\alpha B}}(t-\alpha^{-1}\lambda)|^2 = \sum_{\lambda\in\Lambda} |\widehat{\mu_{R,B}}(\alpha t-\lambda)|^2 = Q_{R,B,\Lambda}(\alpha t) = 1.$$

By Lemma 4.1, we have proven the desired claim.

□

COROLLARY 4.3. *Let (R, B) be well-spaced about the origin with n the largest positive integer such that $3^n \mid R$, $k = \frac{R}{3^n}$, and let n_1 be the largest nonnegative integer such that $3^{n_1} \mid b$. Define $L = \{0, k3^{n-n_1-1}, 2k3^{n-n_1-1}\}$ so that*

$$\Gamma(L) = \left\{ \sum_{j=0}^{m'} R^j \ell_j : \ell_j \in \{0, k3^{n-n_1-1}, 2k3^{n-n_1-1}\}, m' \in \mathbb{N}_0 \right\}.$$

Then

$$\frac{3^{n_1}}{\gcd(b, b-m)}\Gamma(L)$$

is a spectrum for $\mu_{R,B}$.

PROOF. If $\alpha = \gcd(b, b-m)$, then the nonzero elements in $\alpha^{-1}B$ are relatively prime and therefore by Corollary 3.11 we have that $\mu_{R,\alpha^{-1}B}$ has

$$3^{n_1}\Gamma(L) = \left\{ \sum_{j=0}^{m'} R^j \ell_j : \ell_j \in \{0, k3^{n-1}, 2k3^{n-1}\}, m' \in \mathbb{N}_0 \right\}$$

as a spectrum. Thus, by Lemma 4.2 the result is proved. □

COROLLARY 4.4. *If (R, B) is well-spaced about the origin, then $\mu_{R,B}$ is spectral.*

EXAMPLE 4.5. It is entirely possible that a given Hutchinson measure has both of these possible spectra as spectra (and they are distinct spectra). An example of this phenomenon is given by $\mu_{R,B}$ where $R = 9, B = \{-2, 0, 2\}$. As we verified in Example 3.12, the set

$$\Gamma(L) = \left\{ \sum_{j=0}^{m'} 9^j \ell_j : \ell_j \in \{0, 3, 6\} \right\}$$

is a spectrum for $\mu_{R,B}$. However, by Corollary 4.3, we have that

$$\frac{1}{2}\Gamma(L) = \left\{ \sum_{j=0}^{m'} 9^j \ell_j : \ell_j \in \left\{0, \frac{3}{2}, 3\right\} \right\}$$

is a spectrum for $\mu_{R,B}$ as well.

5. Which Hutchinson-3 measures are spectral?

Finally, we are ready to generalize our previous results and state precisely which Hutchinson-3 measures are spectral. To do so, we will generalize the concept of well-spacedness about the origin.

DEFINITION 5.1. Let $R \in \mathbb{Z}$ satisfy $R = 3^n k$, where $n \geq 1$ and $k \in \mathbb{Z}$ with $3 \nmid k$. Let $B = \{b_1, b_2, b_3\} \subseteq \mathbb{Z}$ with $b_1 < b_2 < b_3$ and let n_1 be the largest nonnegative integer which divides $\max\{b_3 - b_2, b_2 - b_1\}$. Then if additionally

$$3^{n_1+1} \mid (\max\{b_3 - b_2, b_2 - b_1\} - \min\{b_3 - b_2, b_2 - b_1\}),$$

we say that (R, B) is **well-spaced**. If instead we have additionally that $3^{n_1+1} \nmid (\max\{b_3 - b_2, b_2 - b_1\} - \min\{b_3 - b_2, b_2 - b_1\})$, we say that (R, B) is **poorly-spaced**.

The following two facts are now worth noting:

(1) A pair (R, B) which is well-spaced about the origin is also well-spaced, and a pair which is poorly-spaced about the origin is also poorly-spaced.

(2) A Hutchinson-3 measure $\mu_{R,B}$ with $3 \mid R$ is either well-spaced or poorly-spaced.

With this in mind, the next two lemmas will help us prove our main result:

LEMMA 5.2. *Let $R \in \mathbb{Z}$ and $B \subseteq \mathbb{Z}$ be an arbitrary finite set and $b' \in \mathbb{Z}$ be arbitrary. Let $\mathcal{Z}(\widehat{\mu_{R,B}})$ be the set of all zeroes of $\widehat{\mu_{R,B}}$ and define $\mathcal{Z}(\widehat{\mu_{R,B+b'}})$ analogously. Then $\mathcal{Z}(\widehat{\mu_{R,B}}) = \mathcal{Z}(\widehat{\mu_{R,B+b'}})$.*

PROOF. The proof of this lemma follows from the infinite product representation of these Fourier transforms and Lemma 1.5. Specifically:

$$\widehat{\mu_{R,B}}(t) = \prod_{d=1}^{\infty} \left(\frac{1}{|B|} \sum_{b \in B} e^{\frac{2\pi i b t}{R^d}} \right)$$
$$\widehat{\mu_{R,B+b'}}(t) = \prod_{d=1}^{\infty} e^{\frac{2\pi i b' t}{R^d}} \left(\frac{1}{|B|} \sum_{b \in B} e^{\frac{2\pi i b t}{R^d}} \right)$$

Since either product is zero if and only if one of its terms is zero, the result follows. □

LEMMA 5.3. *Let $R \in \mathbb{Z}$, $B \subseteq \mathbb{Z}$ be a finite set, and choose $b' \in \mathbb{Z}$. Let $\Lambda \subseteq \mathbb{R}$ be chosen such that $\{e_\lambda : \lambda \in \Lambda\}$ is orthonormal in both $L^2(\mu_{R,B})$ and $L^2(\mu_{R,B+b'})$. Then $\{e_\lambda : \lambda \in \Lambda\}$ is an orthonormal basis in $L^2(\mu_{R,B})$ if and only if it is an orthonormal basis in $L^2(\mu_{R,B+b'})$.*

PROOF. By Lemma 4.1, it will suffice to show that $\left|\widehat{\mu_{R,B}}(t)\right| = \left|\widehat{\mu_{R,B+b'}}(t)\right|$ for an arbitrary $t \in \mathbb{R}$. To this end, we make use of the fact that $\widehat{\mu_{R,B}}, \widehat{\mu_{R,B+b'}}$ can be represented as an absolutely convergent infinite product by Lemma 1.5. Letting

$\chi_B(t) = \frac{1}{|B|}\sum_{b\in B} e^{2\pi i b t}$, we have that

$$
\begin{aligned}
\left|\widehat{\mu_{R,B+b'}}(t)\right| &= \left|\prod_{d=1}^{\infty} e^{\frac{2\pi i b' t}{R^d}} \chi_B\left(\frac{t}{R^d}\right)\right| \\
&= \left|\lim_{q\to\infty}\prod_{d=1}^{q} e^{\frac{2\pi i b' t}{R^d}} \chi_B\left(\frac{t}{R^d}\right)\right| \\
&= \lim_{q\to\infty}\prod_{d=1}^{q} \left|e^{\frac{2\pi i b' t}{R^d}} \chi_B\left(\frac{t}{R^d}\right)\right| \\
&= \lim_{q\to\infty}\prod_{d=1}^{q} \left|\chi_B\left(\frac{t}{R^d}\right)\right| \\
&= \lim_{q\to\infty}\left|\prod_{d=1}^{q} \chi_B\left(\frac{t}{R^d}\right)\right| \\
&= \left|\lim_{q\to\infty}\prod_{d=1}^{q} \chi_B\left(\frac{t}{R^d}\right)\right| \\
&= \left|\widehat{\mu_{R,B}}(t)\right|.
\end{aligned}
$$

□

We can now conclude this publication by proving our main theorem.

THEOREM 5.4. *Let $\mu_{R,B}$ be a Hutchinson-3 measure. Then it is spectral if and only if (R, B) is well-spaced. Furthermore, if $B = \{b_1, b_2, b_3\}$ with $b_1 < b_2 < b_3$ and (R, B) is well-spaced, then $\mu_{R,B}$ has the same spectra as $\mu_{R,B-b_2}$ (by construction $(R, B - b_2)$ is well-spaced about the origin).*

PROOF. Let (R, B) be well-spaced and let $B = \{b_1, b_2, b_3\}$ with $b_1 < b_2 < b_3$. Then by construction $B - b_2$ is well-spaced about the origin, and by the previous two lemmas we have that a set $\Lambda \subseteq \mathbb{R}$ is a spectrum for $\mu_{R,B}$ if and only if it is a spectrum for $\mu_{R,B-b_2}$. Thus, without loss of generality we may prove this corollary under the assumption that (R, B) is well-spaced about the origin rather than well-spaced. Also, the second statement in the theorem will follow immediately once we can prove the first statement of the theorem under this assumption.

With this in mind: if we assume that (R, B) is well-spaced about the origin, we have by Corollary 4.4 that $\mu_{R,B}$ is spectral. If (R, B) is not well-spaced about the origin but $b_2 = 0$, then either $3 \nmid R$ or (R, B) is poorly-spaced about the origin. Thus, by either Corollary 2.4 or Theorem 3.4, $\mu_{R,B}$ is not spectral and the theorem is proved. □

Acknowledgements

I would like to thank my advisor, Judy Packer, for proofreading these results and more importantly for the tremendous amount of guidance and support she has given me over the last several years. I would also like to thank the anonymous referee for time spent proofreading.

References

[Ahl78] L. V. Ahlfors, *Complex analysis*, 3rd ed., McGraw-Hill Book Co., New York, 1978. An introduction to the theory of analytic functions of one complex variable; International Series in Pure and Applied Mathematics. MR510197

[Dai12] X.-R. Dai, *When does a Bernoulli convolution admit a spectrum?*, Adv. Math. **231** (2012), no. 3-4, 1681–1693, DOI 10.1016/j.aim.2012.06.026. MR2964620

[DH16] D. E. Dutkay and J. Haussermann, *Number theory problems from the harmonic analysis of a fractal*, J. Number Theory **159** (2016), 7–26, DOI 10.1016/j.jnt.2015.07.009. MR3412709

[DHL15] D. Dutkay, J. Haussermann, and C.-K. Lai. *Hadamard triples generate self-affine spectral measures.* **arXiv:1506.01503** (2015)

[DHW14] D. E. Dutkay, D. Han, and E. Weber, *Continuous and discrete Fourier frames for fractal measures*, Trans. Amer. Math. Soc. **366** (2014), no. 3, 1213–1235, DOI 10.1090/S0002-9947-2013-05843-6. MR3145729

[DJ07] D. E. Dutkay and P. E. T. Jorgensen, *Analysis of orthogonality and of orbits in affine iterated function systems*, Math. Z. **256** (2007), no. 4, 801–823, DOI 10.1007/s00209-007-0104-9. MR2308892

[DJ12] D. E. Dutkay and P. E. T. Jorgensen, *Fourier duality for fractal measures with affine scales*, Math. Comp. **81** (2012), no. 280, 2253–2273, DOI 10.1090/S0025-5718-2012-02580-4. MR2945155

[DK16] D. Dutkay and I. Kraus. *Scaling of spectra of Cantor-type measures and some number theoretic considerations.* Advances in Mathematics **arXiv:1609.01928** (2016).

[DR16] D. E. Dutkay and R. Ranasinghe, *Weighted Fourier frames on fractal measures*, J. Math. Anal. Appl. **444** (2016), no. 2, 1603–1625, DOI 10.1016/j.jmaa.2016.07.042. MR3535778

[Hut81] J. E. Hutchinson, *Fractals and self-similarity*, Indiana Univ. Math. J. **30** (1981), no. 5, 713–747, DOI 10.1512/iumj.1981.30.30055. MR625600

[JKS11] P. E. T. Jorgensen, K. A. Kornelson, and K. L. Shuman, *Families of spectral sets for Bernoulli convolutions*, J. Fourier Anal. Appl. **17** (2011), no. 3, 431–456, DOI 10.1007/s00041-010-9158-x. MR2803943

[JKS12] P. E. T. Jorgensen, K. A. Kornelson, and K. L. Shuman, *An operator-fractal*, Numer. Funct. Anal. Optim. **33** (2012), no. 7-9, 1070–1094, DOI 10.1080/01630563.2012.682127. MR2966145

[JKS14a] P. E. T. Jorgensen, K. A. Kornelson, and K. L. Shuman, *Scalar spectral measures associated with an operator-fractal*, J. Math. Phys. **55** (2014), no. 2, 022103, 23, DOI 10.1063/1.4863897. MR3202868

[JKS14b] P. E. T. Jorgensen, K. A. Kornelson, and K. L. Shuman, *Scaling by 5 on a $\frac{1}{4}$-Cantor measure*, Rocky Mountain J. Math. **44** (2014), no. 6, 1881–1901, DOI 10.1216/RMJ-2014-44-6-1881. MR3310953

[JP98] P. E. T. Jorgensen and S. Pedersen, *Dense analytic subspaces in fractal L^2-spaces*, J. Anal. Math. **75** (1998), 185–228, DOI 10.1007/BF02788699. MR1655831

[Li07] J.-L. Li, *$\mu_{M,D}$-orthogonality and compatible pair*, J. Funct. Anal. **244** (2007), no. 2, 628–638, DOI 10.1016/j.jfa.2006.10.003. MR2297038

[ŁW02] I. Łaba and Y. Wang, *On spectral Cantor measures*, J. Funct. Anal. **193** (2002), no. 2, 409–420, DOI 10.1006/jfan.2001.3941. MR1929508

[LX17] J.-L. Li and D. Xing, *Multiple spectra of Bernoulli convolutions*, Proc. Edinb. Math. Soc. (2) **60** (2017), no. 1, 187–202, DOI 10.1017/S0013091515000565. MR3589848

[PW17] G. Picioroaga and E. S. Weber, *Fourier frames for the Cantor-4 set*, J. Fourier Anal. Appl. **23** (2017), no. 2, 324–343, DOI 10.1007/s00041-016-9471-0. MR3622655

Department of Mathematics, University of Colorado, Campus Box 395, Boulder, Colorado 80309

Current address: 3013 Acton Rd, Parkville, Maryland 21234

Email address: ian.long@colorado.edu

Contemporary Mathematics
Volume **706**, 2018
http://dx.doi.org/10.1090/conm/706/14211

A matrix characterization of boundary representations of positive matrices in the Hardy space

John E. Herr, Palle E. T. Jorgensen, and Eric S. Weber

ABSTRACT. Spectral measures give rise to a natural harmonic analysis on the unit disc via a boundary representation of a positive matrix arising from a spectrum of the measure. We consider in this paper the reverse: for a positive matrix in the Hardy space of the unit disc we consider which measures, if any, yield a boundary representation of the positive matrix. We introduce a potential characterization of those measures via a matrix identity and show that the characterization holds in several important special cases.

1. Introduction

1.1. The Szegő kernel. The classical Hardy space $H^2(\mathbb{D})$ consists of those holomorphic functions f defined on $\mathbb{D}$ satisfying

$$\|f\|_{H^2}^2 := \sup_{0<r<1} \int_0^1 |f(re^{2\pi ix})|^2\, dx < \infty. \tag{1.1}$$

It is well-known that an equivalent description of $H^2(\mathbb{D})$ is as the space of holomorphic functions on $\mathbb{D}$ with square-summable coefficients:

$$H^2(\mathbb{D}) = \left\{ \sum_{n=0}^{\infty} c_n z^n \;\middle|\; \sum_{n=0}^{\infty} |c_n|^2 < \infty \right\},$$

where the norm is then equivalently given by

$$\|f\|_{H^2}^2 = \sum_{n=0}^{\infty} |c_n|^2.$$

In addition, for each $f \in H^2(\mathbb{D})$, there exists a (unique) function $f^* \in L^2(\mathbb{T})$, which we shall call the Lebesgue boundary function of f, such that

$$\lim_{r\to 1^-} \int_0^1 |f(re^{2\pi ix}) - f^*(e^{2\pi ix})|^2\, dx = 0. \tag{1.2}$$

In fact, $\lim_{r\to 1^-} f(re^{2\pi ix}) = f^*(e^{2\pi ix})$ pointwise for almost every x. If $f(z) = \sum_{n=0}^{\infty} a_n z^n$ and $g(z) = \sum_{n=0}^{\infty} b_n z^n$ are two members of $H^2(\mathbb{D})$, the inner product

---

2010 *Mathematics Subject Classification.* Primary: 46E22, 30B30; Secondary 42B05, 28A25.

of f and g in $H^2(\mathbb{D})$ can be described in two ways:

$$\langle f, g\rangle_{H^2} = \sum_{n=0}^{\infty} a_n \overline{b_n} = \int_0^1 f^*(e^{2\pi ix})\overline{g^*(e^{2\pi ix})}\, dx.$$

Because the point-evaluation functionals on the Hardy space are bounded, the Hardy space is a reproducing kernel Hilbert space. Its kernel is the classical Szegő kernel $k(w, z) =: k_w$, defined by

$$k_w(z) := \frac{1}{1 - \overline{w}z}.$$

We then have

$$f(z) = \langle f, k_z\rangle_{H^2} = \int_0^1 f^*(e^{2\pi ix})\overline{k_z^*(e^{2\pi ix})}\, dx$$

for all $f \in H^2(\mathbb{D})$. In particular,

$$k(w, z) := \int_0^1 k_w^*(e^{2\pi ix})\overline{k_z^*(e^{2\pi ix})}\, dx. \tag{1.3}$$

Equation (1.3) shows that the Szegő kernel reproduces itself with respect to what is, by some definition, its boundary. The measure on the circle used to define k_z^* in (1.2) is Lebesgue measure, as is the measure in (1.3). The intent of this paper is to show that among the functions in the Hardy space, there are a host of other kernels that reproduce with respect to their boundaries. However, these boundary functions will not be taken with respect to Lebesgue measure, but with respect to some other measure on the circle $\mathbb{T}$, and the integration of these boundary functions will also be done with respect to this measure. Of interest are two main questions: which positive matrices does the Hardy space contain that reproduce themselves by boundary functions with respect to a given measure, and with respect to which measures will a positive matrix reproduce itself by boundary functions? In [**13**], we focused on the first question whereas in the present paper we focus on the second.

1.2. Boundary representations. The boundary behavior of functions in the Hardy space arises in a number of contexts, such as the spectral theory of the shift operator [**7**], de Branges-Rovnyak spaces [**8, 19**], and "pseudo-continuable" functions [**1, 18**]

The boundary behavior of kernels in reproducing kernel Hilbert spaces has been considered by others, see for example [**2–4**]. In those papers, much of the theory is developed under the a priori assumption that a reproducing kernel Hilbert space "sits isometrically" within an L^2 space, by which the authors mean that the elements of the Hilbert space possesses an L^2 boundary as we define below and the two norms coincide. However, the authors also begin with a measure on the boundary and define a reproducing kernel by integrating the Cauchy kernel against the measure [**3**, Lemma 6.4]. In the previous paper [**13**], we extend this result to construct many kernels which reproduce themselves with respect to a fixed singular measure using the Kaczmarz algorithm [**14, 16**]. Other boundary representations were introduced in [**11**] to understand the nature of spectral measures. In [**11**], and spectral measure gives rise to a positive matrix on $\mathbb{D}$ which reproduces itself on the boundary with respect to the spectral measure. As demonstrated in [**13**], however, the measure need not be spectral.

REMARK 1.1. In this paper, we will be dealing with measures μ on the unit circle. The unit circle $\mathbb{T} := \{z \in \mathbb{C} : |z| = 1\}$ and its topology shall be identified with $[0, 1)$ via the relation $\xi = e^{2\pi i x}$ for $\xi \in \mathbb{T}$ and $x \in [0, 1)$. We will regard the measures μ as being supported on $[0, 1)$. A function $f(\xi)$ defined on $\mathbb{T}$ (for example, a boundary function) may be regarded as being in $L^2(\mu)$ if $f(e^{2\pi i x}) \in L^2(\mu)$. For aesthetics, the inner product (norm) in $L^2(\mu)$ will be denoted $\langle \cdot, \cdot \rangle_\mu$ ($\|\cdot\|_\mu$) rather than $\langle \cdot, \cdot \rangle_{L^2(\mu)}$ ($\|\cdot\|_{L^2(\mu)}$). The subscript will be suppressed where context suffices. A measure μ will be called *singular* if it is a Borel measure that is singular with respect to Lebesgue measure.

DEFINITION 1.2. If μ is a finite Borel measure on $[0, 1)$ and $f(z)$ is an analytic function on $\mathbb{D}$, we say that $f^\star \in L^2(\mu)$ is an $L^2(\mu)$-boundary function of f if

$$\lim_{r \to 1^-} \left\| f^\star(x) - f(re^{2\pi i x}) \right\|_\mu = 0.$$

If a function possesses an $L^2(\mu)$-boundary function, then clearly that boundary function is unique. The $L^2(\mu)$-boundary function of a function $f : \mathbb{D} \to \mathbb{C}$ shall be denoted $f^\star_\mu$, but we omit the subscript when context precludes ambiguity.

DEFINITION 1.3. A positive matrix (in the sense of E. H. Moore) on a domain E is a function $K(z, w) : E \times E \to \mathbb{C}$ such that for all finite sequences $\zeta_1, \zeta_2, \ldots, \zeta_n \in E$, the matrix

$$(K(\zeta_j, \zeta_i))_{ij}$$

is positive semidefinite.

Our interest is in positive matrices on $E = \mathbb{D}$, and more specifically those residing in $H^2(\mathbb{D})$. Recall that the classical Hardy space is a reproducing kernel Hilbert space. We therefore desire to find subspaces of the Hardy space that not only are Hilbert spaces with respect to the $L^2(\mu)$-boundary norm, but are in fact reproducing kernel Hilbert spaces with respect to this norm. The classical Moore-Aronszajn Theorem connects positive matrices to reproducing kernel Hilbert spaces [**5**]:

THEOREM (Moore-Aronszajn). *To every positive matrix $K(w, z)$ on a domain E there corresponds one and only one class of functions on E with a uniquely determined quadratic form in it, forming a Hilbert space and admitting $K(w, z)$ as a reproducing kernel. This class of functions is generated by all functions of the form $\sum_{k=1}^n \xi_k K(w_k, z)$, with norm given by*

$$\left\| \sum_{k=1}^n \xi_k K(w_k, z) \right\|^2 = \sum_{i,j=1}^n K(z_j, z_i) \overline{\xi_i} \xi_j.$$

Conversely, every reproducing kernel of a Hilbert space of functions on a common domain is a positive matrix. Let us then define two sets of interest:

DEFINITION 1.4. Let μ be a Borel measure on $[0, 1)$. We define $\mathcal{K}(\mu)$ to be the set of positive matrices K on $\mathbb{D}$ such that for each fixed $z \in \mathbb{D}$, $K(w, \cdot)$ possesses an $L^2(\mu)$-boundary $K^\star(w, \cdot)$, and $K(w, z)$ reproduces itself with respect to integration of these $L^2(\mu)$-boundaries, i.e.

$$K(w, z) = \int_0^1 K^\star(w, x) \overline{K^\star(z, x)} \, d\mu(x) \tag{1.4}$$

for all $z, w \in \mathbb{D}$.

DEFINITION 1.5. Let K be a positive matrix on $\mathbb{D}$. We define $\mathcal{M}(K)$ to be the set of nonnegative Borel measures μ on $[0,1)$ such that for each fixed $z \in \mathbb{D}$, $K(w,\cdot)$ possesses an $L^2(\mu)$-boundary $K^\star(w,\cdot)$, and $K(w,z)$ reproduces itself with respect to integration of these $L^2(\mu)$-boundaries as in Equation (1.4).

DEFINITION 1.6. A sequence $\{x_n\}_{n=0}^\infty$ in a Hilbert space $\mathbb{H}$ is called a frame [**9**] if there exist positive constants A and B such that

$$A\|\phi\|^2 \leq \sum_{n=0}^{\infty} |\langle \phi, x_n\rangle|^2 \leq B\|\phi\|^2 \tag{1.5}$$

for all $\phi \in \mathbb{H}$. If $\{x_n\}_{n=0}^\infty$ satisfies (possibly only) the right-hand inequality in (1.5), it is called a Bessel sequence. If $A = B$, the frame is called tight, and if $A = B = 1$, it is called a Parseval frame.

The quaternary Cantor measure μ_4 is the restriction of the $\frac{1}{2}$-dimensional Hausdorff measure to the quaternary Cantor set. Likewise, the ternary Cantor measure μ_3 is the restriction of the $\frac{\ln(2)}{\ln(3)}$-dimensional Hausdorff measure to the ternary Cantor set. In [**15**], Jorgensen and Pedersen showed that the quaternary Cantor measure is spectral. That is, there exists a set $\Gamma \subset \mathbb{Z}$ such that the set of complex exponentials $\left\{e^{2\pi i n x}\right\}_{\lambda\in\Gamma}$ is an orthonormal basis of $L^2(\mu_4)$. From this, Dutkay and Jorgensen [**11**] constructed a positive matrix G_Γ inside H^2 that reproduces itself both in H^2 and with respect to $L^2(\mu_4)$-boundary integration. Thus $G_\Gamma \in \mathcal{K}(\mu_4)$.

In [**15**], it was also shown that μ_3 is not spectral. Thus, it is not possible to construct a positive matrix for μ_3 in the same way as for μ_4. However, it is sufficient for μ_3 to possess an exponential frame:

PROPOSITION. *If there exists a sequence $\{n_j\}_{j=0}^\infty$ of nonnegative integers such that $\{e^{2\pi i n_j x} : j \geq 0\} \subset L^2(\mu)$ is a frame, then $\mathcal{K}(\mu)$ is nonempty.*

We proved a generalization of this result in [**13**]. It is still unknown whether μ_3 possesses an exponential frame, which motivates our interest in understanding the boundary representations of positive matrices.

1.3. Kernels from a coefficient matrix. Let $C = (c_{mn})$ be a bi-infinite matrix, where $m, n \geq 0$. We consider the formal power series

$$K_C(w,z) = \sum_n \sum_m c_{mn}\overline{w}^m z^n. \tag{1.6}$$

We shall assume that $c_{nm} = \overline{c}_{mn}$; we shall make additional assumptions on C as needed. For example, if we assume that $\{c_{mn}\}$ is a bounded sequence, then the formal power series K_C converges absolutely on $\mathbb{D}\times\mathbb{D}$, and thus K_C is holomorphic on $\mathbb{D}$ in z and antiholomorphic on $\mathbb{D}$ in w. For the remainder of the paper, we shall assume the coefficient sequence is bounded.

Moreover, we wish K_C to be a positive matrix on $\mathbb{D}\times\mathbb{D}$, so we assume that C has this property. When the matrix C defines a bounded linear operator on $\ell^2(\mathbb{N}_0)$, then K_C is a positive matrix if and only if C is a positive operator. Indeed, for $z \in \mathbb{D}$, we denote by $\vec{z}$ the element of $\ell^2(\mathbb{N}_0)$ where $(\vec{z})_n = (z^n)_n$. Then, for $z_1, \dots, z_N \in \mathbb{D}$ and $\xi_1, \dots, \xi_N \in \mathbb{C}$,

$$\sum_{j=1}^N \sum_{k=1}^N \xi_j \overline{\xi}_k K_C(z_k, z_j) = \sum_{j=1}^N \sum_{k=1}^N \xi_j \overline{\xi}_k \langle C\vec{z}_j, \vec{z}_k\rangle_{\ell^2} = \left\langle C\sum_{j=1}^N \xi_j \vec{z}_j, \sum_{k=1}^N \xi_k \vec{z}_k \right\rangle_{\ell^2}$$

which is nonnegative if and only if C is a positive self-adjoint operator on $\ell^2(\mathbb{N}_0)$. Assuming that C is a bounded linear operator on $\ell^2(\mathbb{N}_0)$ has the additional virtue that for every $w \in \mathbb{D}$, $K_C(w, \cdot) \in H^2(\mathbb{D})$, since the coefficient sequence $(\overline{C\vec{w}})_m$ is square-summable. We have established the following:

LEMMA 1.7. *If $C = (c_{mn})_{mn}$ is a bounded, positive, self-adjoint operator on $\ell^2(\mathbb{N}_0)$, then the kernel K_C as given in Equation (1.6) is a positive matrix such that for each $w \in \mathbb{D}$, $K_C(w, \cdot) \in H^2(\mathbb{D})$. Moreover, for $w, z \in \mathbb{D}$,*

$$K_C(w, z) = \langle C\vec{z}, \vec{w} \rangle_{\ell^2}. \tag{1.7}$$

For a given C which defines a positive matrix as in Equation (1.6), we wish to determine which Borel measures on $\mathbb{T}$, if any, are in $\mathcal{M}(K_C)$. We shall approach the question via the following meta-theorem:

META-THEOREM. *A measure μ is in $\mathcal{M}(K_C)$ if and only if the matrix equation $C = CMC$ is satisfied, where the matrix $M = (\hat{\mu}(n - m))_{mn}$.*

We describe this as a meta-theorem for several reasons. First, even if C is a bounded operator on $\ell^2(\mathbb{N}_0)$, the expression CMC may not be well-defined. Indeed, *a priori* this product is only defined when C and M are bounded operators on $\ell^2(\mathbb{N}_0)$; we may have only one or neither of these matrices with that property. Second, the matrix equality does not *a priori* assure that the kernel functions $K_C(w, \cdot)$ have μ-boundaries. Our goal in the present paper is to establish the meta-theorem for two special cases: i) for diagonal matrices C, and ii) for C and μ for which M which are bounded operators on $\ell^2(\mathbb{N}_0)$. We have a description of which μ has the property that M is bounded [**6**, **10**], see also [**12**, **17**]:

LEMMA 1.8. *The matrix $M = (\hat{\mu}(n - m))_{mn}$ is a bounded operator on $\ell^2(\mathbb{N}_0)$ if and only if $\mu << \lambda$ and the Radon-Nikodym derivative $\dfrac{d\mu}{d\lambda} \in L^\infty(\lambda)$.*

2. Lebesgue measure: kernels in $H^2(\mathbb{D})$ with equal norms

We assume that the coefficient matrix C defines a bounded, positive, self-adjoint operator on $\ell^2(\mathbb{N}_0)$ and consider initially the special case of Lebesgue measure.

THEOREM 2.1. *Suppose $C = (c_{mn})$ defines a bounded, positive, self-adjoint operator on $\ell^2(\mathbb{N}_0)$. The following are equivalent:*

(1) *$\lambda \in \mathcal{M}(K_C)$;*
(2) *the coefficient matrix C is a projection;*
(3) *the norm induced by K_C is equal to the Hardy space norm in the following sense: for all $\xi_1, \ldots, \xi_N \in \mathbb{C}$ and $w_1, \ldots, w_N \in \mathbb{D}$,*

$$\left\| \sum_{j=1}^{N} \xi_j K_C(w_j, \cdot) \right\|_{K_C} = \left\| \sum_{j=1}^{N} \xi_j K_C(w_j, \cdot) \right\|_{H^2};$$

(4) *there exists a subspace M of the Hardy space such that the Parseval frame $g_n = P_M z^n$ is such that $c_{mn} = \langle g_n, g_m \rangle$;*
(5) *there exists a subspace M of the Hardy space such that the projection of the Szegö kernel onto M is K_C.*

REMARK 2.2. The equivalence of 1 and 2 would follow immediately from our meta-theorem. We will establish the meta-theorem for absolutely continuous measures with bounded Radon-Nikodym derivative in the next section–we present here a proof that uses only the equality of norms.

PROOF. ($1 \Leftrightarrow 3$) If $\lambda \in \mathcal{M}(K_C)$, then

$$\begin{aligned}
\left\| \sum_{j=1}^{N} \xi_j K_C(w_j, \cdot) \right\|_{H^2}^2 &= \int_0^1 \left(\sum_{j=1}^{N} \xi_j K_C^\star(w_j, \cdot) \right) \left(\overline{\sum_{k=1}^{N} \xi_k K_C^\star(w_k, \cdot)} \right) d\lambda \\
&= \sum_{j=1}^{N} \sum_{k=1}^{N} \xi_j \overline{\xi}_k \int_0^1 K_C^\star(w_j, \cdot) \overline{K_C^\star(w_k, \cdot)} d\lambda \\
&= \sum_{j=1}^{N} \sum_{k=1}^{N} \xi_j \overline{\xi}_k K_C(w_j, w_k) \\
&= \left\| \sum_{j=1}^{N} \xi_j K_C(w_j, \cdot) \right\|_{K_C}^2 .
\end{aligned}$$

Conversely, if the norms are equal, we have by the polarization identity

$$\begin{aligned}
K_C(w, z) &= \langle K_C(w, \cdot), K_C(z, \cdot) \rangle_{K_C} \\
&= \langle K_C(w, \cdot), K_C(z, \cdot) \rangle_{H^2} \\
&= \int_0^1 K_C^\star(w, \cdot) \overline{K_C^\star(z, \cdot)} d\lambda.
\end{aligned}$$

($2 \Leftrightarrow 3$) Consider the following calculations:

$$\begin{aligned}
\left\| \sum_{j=1}^{N} \xi_j K_C(w_j, \cdot) \right\|_{K_C}^2 &= \sum_{j=1}^{N} \sum_{k=1}^{N} \xi_j \overline{\xi}_k K_C(w_j, w_k) \\
&= \sum_{j=1}^{N} \sum_{k=1}^{N} \xi_j \overline{\xi}_k \langle C\vec{w}_k, \vec{w}_j \rangle_{\ell^2} \\
&= \left\langle C \left(\sum_{k=1}^{N} \overline{\xi}_k \vec{w}_k \right), \sum_{j=1}^{N} \overline{\xi}_j \vec{w}_j \right\rangle_{\ell^2} \qquad (2.1)
\end{aligned}$$

and

$$\begin{aligned}
\left\| \sum_{j=1}^{N} \xi_j K_C(w_j, \cdot) \right\|_{H^2}^2 &= \left\| \sum_{n=0}^{\infty} \left(\sum_{m=0}^{\infty} c_{mn} \sum_{j=1}^{N} \xi_j \overline{w}_j^m \right) z^n \right\|_{H^2}^2 \\
&= \left\| \sum_{n=0}^{\infty} \overline{\left(\sum_{m=0}^{\infty} c_{nm} \sum_{j=1}^{N} \overline{\xi}_j w_j^m \right)} z^n \right\|_{H^2}^2 \\
&= \left\langle C \left(\sum_{j=1}^{N} \overline{\xi}_j \vec{w}_j \right), C \left(\sum_{k=1}^{N} \overline{\xi}_k \vec{w}_k \right) \right\rangle_{\ell^2} . \qquad (2.2)
\end{aligned}$$

It follows that if C is projection, then the inner-products in Equations (2.1) and (2.2) are equal. Conversely, if the norms are equal, then by the polarization identity, we have that $C^2 = C$; since C is assumed self-adjoint, C is a projection.

($3 \Leftrightarrow 5$) If the norms are equal, then the RKHS generated by K_C is a closed subspace M of $H^2(\mathbb{D})$ (with equal norm), and hence the projection P_M of the Szegö kernel is the reproducing kernel for M, as is K_C. Conversely, if K_C is the projection of the Szegö kernel onto M, then the norms are equal.

($2 \Leftrightarrow 4$) If C is a projection, then we can define $\Phi : H^2(\mathbb{D}) \to H^2(\mathbb{D})$ by

$$\Phi f(z) = \sum_m (C\vec{f})_m z^m, \quad \text{where} \quad f(z) = \sum_n f_n z^n \quad \text{and} \quad \vec{f} = (f_n)_n.$$

It is readily verified that Φ is a projection on $H^2(\mathbb{D})$. We have

$$c_{mn} = \langle \Phi z^n, z^m \rangle = \langle \Phi z^n, \Phi z^m \rangle.$$

Conversely, if $g_n = P_M z^n$, then $\{g_n\}$ is a Parseval frame, thus its Grammian matrix is a projection. □

3. Diagonal coefficient matrices

We consider the special case of when C is a diagonal matrix. Let $\Gamma \subset \mathbb{N}_0$ and consider $K_\Gamma(w, z) = \sum_{\gamma \in \Gamma} (z\overline{w})^\gamma$. We will see that either a) there are many absolutely continuous measures in $\mathcal{M}(K_\Gamma)$, or b) only Lebesgue measure is in $\mathcal{M}(K_\Gamma)$. The determining factor of which possibility occurs is the difference set of Γ.

3.1. The kernels K_4 and K_3. Two specific kernels that fall into this category that we wish to understand are the kernels K_3 and K_4. Recall that a spectrum for μ_4 is

$$\Gamma_4 := \left\{ \sum_{j=0}^{N} l_j 4^j | l_j \in \{0, 1\} \right\} = \{0, 1, 4, 5, 16, 17, 20, 21, \dots\}.$$

Then

$$K_4(w, z) := \sum_{\gamma \in \Gamma_4} (\overline{w}z)^\gamma = \prod_{j=0}^{\infty} \left(1 + (\overline{w}z)^{4^j}\right).$$

An introduction to K_4 appears in [**11**], where it is shown that $\mu_4 \in \mathcal{M}(K_4)$. We show in Corollary 3.9 below that there are many (absolutely continuous) measures in $\mathcal{M}(K_4)$.

We also consider the kernel K_3, defined analogously to K_4:

$$K_3(w, z) := \prod_{j=0}^{\infty} \left(1 + (\overline{w}z)^{3^j}\right) = \sum_{n \in \Gamma_3} (\overline{w}z)^n,$$

where

$$\Gamma_3 = \left\{ \sum_{j=0}^{N} l_j 3^j | l_j \in \{0, 1\} \right\} = \{0, 1, 3, 4, 9, 10, 12, 13, \dots\}.$$

We shall show in Corollary 3.6 below that $\mathcal{M}(K_3)$ contains only Lebesgue measure.

Note that K_Γ corresponds to the diagonal coefficient matrix C with $c_{mm} = 1$ if and only if $m \in \Gamma$, and $c_{mm} = 0$ otherwise. Therefore C is a projection, and hence as a consequence of Theorem 2.1, we have:

COROLLARY 3.1. *For any* $\Gamma \subset \mathbb{N}_0$, $\lambda \in \mathcal{M}(K_\Gamma)$.

We shall also consider diagonal matrices C which are not projections–in fact we can consider diagonal matrices C which are not bounded operators on $\ell^2(\mathbb{N}_0)$. For example, the Bergmann kernel is given by

$$K_B(w,z) = \sum_{n=0}^{\infty}(n+1)(\overline{w}z)^n.$$

We shall show in Corollary 3.3 below that there are no representing measures for C which have distinct nonzero diagonal entries.

3.2. The meta-theorem for diagonal coefficient matrices. For two matrices $A = (a_{mn})$ and $B = (b_{mn})$, we say that AB is *defined in the matrix sense* if for every $m, n \in \mathbb{N}_0$, the sum $\sum_{k=0}^{\infty} = a_{mk}b_{kn}$ converges. Note that this holds if A, B are bounded operators on $\ell^2(\mathbb{N}_0)$. We say ABC is defined in the matrix sense if AB, BC, $(AB)C$, and $A(BC)$ are defined in the matrix sense and $(AB)C = A(BC)$.

THEOREM 3.2. *Suppose C is diagonal matrix such that $c_{nn} \geq 0$ and for every $0 < r < 1$, $\sum c_{nn}r^n < +\infty$. Let μ be a Borel probability measure on $[0,1)$ with $M = (\widehat{\mu}(n-m))_{mn}$. Then the following hold:*

(1) *$\sum_{n=0}^{\infty} c_{nn}\overline{w}^n e^{2\pi inx}$ converges in $L^2(\mu)$.*
(2) *CMC is defined in the matrix sense.*
(3) *$K_C(w,z)$ reproduces itself with respect to $L^2(\mu)$ boundaries if and only if the equation $C = CMC$ holds.*

PROOF. For the first part, we have for any $|w| < 1$,

$$\sum_{n=0}^{\infty}\left\|c_{nn}\overline{w}^n e^{2\pi inx}\right\|_{\mu} = \sum_{n=0}^{\infty}|c_{nn}\overline{w}^n| < \infty.$$

Thus, $\sum_{n=0}^{\infty} c_{nn}\overline{w}^n e^{2\pi inx}$ is absolutely summable in $L^2(\mu)$ and thus converges in $L^2(\mu)$.

For the second part, observe that

$$(CM)_{mn} = \sum_{k=0}^{\infty} c_{mk}\hat{\mu}(n-k) = c_{mm}\hat{\mu}(n-m)$$

$$(MC)_{mn} = \sum_{k=0}^{\infty}\hat{\mu}(k-m)c_{kn} = \hat{\mu}(n-m)c_{nn}.$$

Likewise,

$$(C(MC))_{mn} = \sum_{k=0}^{\infty} c_{mk}\hat{\mu}(n-k)c_{nn} = c_{mm}c_{nn}\hat{\mu}(n-m) = c_{mm}M_{mn}c_{nn}$$

$$((CM)C)_{mn} = \sum_{k=0}^{\infty} c_{mm}\hat{\mu}(k-m)c_{kn} = c_{mm}\hat{\mu}(n-m)c_{nn} = c_{mm}M_{mn}c_{nn}.$$

This shows that CM, MC, $C(MC)$, and $(CM)C$ are defined in the matrix sense, and that $C(MC) = (CM)C$.

Now, suppose $K_C(w,z)$ reproduces itself with respect to $L^2(\mu)$ boundary. The first part, together with Abel summability, shows that $K_C^{\star}(w,x) = \sum_{m=0}^{\infty} c_{mm}\overline{w}^n e^{2\pi imx}$.

By continuity of the inner product in $L^2(\mu)$, we have

$$\begin{aligned}\int_0^1 K_C^\star(w,x)\overline{K_C^\star(z,x)}\,d\mu(x) &= \sum_{n=0}^{\infty}\overline{c_{nn}}\left(\int_0^1 K_C^\star(w,x)e^{-2\pi inx}\,d\mu(x)\right)z^n\\ &= \sum_{n=0}^{\infty}\sum_{m=0}^{\infty} c_{nn}c_{mm}\left(\int_0^1 e^{2\pi imx}e^{-2\pi inx}\,d\mu(x)\right)\overline{w}^m z^n\\ &= \sum_{n=0}^{\infty}\sum_{m=0}^{\infty} c_{nn}c_{mm}\hat{\mu}(n-m)\overline{w}^m z^n\\ &= \sum_{n=0}^{\infty}\sum_{m=0}^{\infty} c_{mm}M_{mn}c_{nn}\overline{w}^m z^n\\ &= \sum_{n=0}^{\infty}\sum_{m=0}^{\infty}(CMC)_{mn}\overline{w}^m z^n\end{aligned}$$

Therefore, Equation (1.4) holds if and only if

$$\sum_{n=0}^{\infty}\sum_{m=0}^{\infty} c_{mn}\overline{w}^m z^n = \sum_{n=0}^{\infty}\sum_{m=0}^{\infty}(CMC)_{mn}\overline{w}^m z^n$$

holds, which by uniqueness of Taylor coefficients, holds if and only if $C = CMC$. □

COROLLARY 3.3. *Suppose C is a diagonal matrix which satisfies the hypotheses of Theorem 3.2 and which has two distinct nonzero diagonal entries. Then $\mathcal{M}(K_C) = \emptyset$.*

PROOF. Suppose $\mu \in \mathcal{M}(K_C)$. Then we must have $c_{mm} = (CMC)_{mm} = c_{mm}M_{mm}c_{mm} = \|\mu\|c_{mm}^2$ for all $m \in \mathbb{N}_0$. Thus, for any nonzero diagonal entry $c_{mm} = \|\mu\|$. □

DEFINITION 3.4. For a set $A \subset \mathbb{R}$, the difference set is

$$\mathcal{D}(A) = \{x - y \mid x, y \in A\}.$$

COROLLARY 3.5. *If $\Gamma \subset \mathbb{N}_0$ and μ is a probability measure, $\mu \in \mathcal{M}(K_\Gamma)$ if and only if*

$$\hat{\mu}(n) = 0 \tag{3.1}$$

for all $n \in \mathcal{D}(\Gamma) \setminus \{0\}$.

PROOF. We verify $C = CMC$ holds if and only if Equation (3.1) holds. We have $c_{mn} = 1$ if $m = n \in \Gamma$ and 0 otherwise. Thus by the calculation in Theorem 3.2,

$$(CMC)_{mn} = M_{mn}$$

if $m, n \in \Gamma$ and

$$(CMC)_{mn} = 0$$

otherwise. Thus, $C = CMC$ holds if and only if $M_{mn} = 0$ whenever $m, n \in \Gamma$ with $m \neq n$. The result now follows since

$$M_{mn} = \hat{\mu}(n - m).$$

□

3.3. The kernels K_3 and K_4, continued.

COROLLARY 3.6. *Suppose* $\Gamma \subset \mathbb{N}_0$ *is such that* $\mathcal{D}(\Gamma) = \mathbb{Z}$. *Then* $\mathcal{M}(K_\Gamma) = \{\lambda\}$. *In particular,* $\mathcal{M}(K_3) = \{\lambda\}$.

PROOF. By Corollary 3.1, we have $\lambda \in \mathcal{M}(K_\Gamma)$. Now, suppose $\mu \in \mathcal{M}(K_\Gamma)$. We must have that the matrix equation $C = CMC$ is satisfied. Thus, for $m \neq n \in \Gamma$, we have $0 = c_{mn} = c_{mm}M_{mn}c_{nn} = \hat{\mu}(n-m)$. Since the difference set of Γ is $\mathbb{Z}$, it follows that $\hat{\mu}(k) = 0$ for $k \in \mathbb{Z} \setminus \{0\}$, whence μ must be Lebesgue measure. The claim for K_3 is a consequence of Lemma 3.7. □

LEMMA 3.7. *The difference set* $\mathcal{D}(\Gamma_3) = \mathbb{Z}$.

PROOF. We prove that $\mathcal{D}(\Gamma_3)$ is invariant under the iterated functions $\varphi_0(x) = 3x$, $\varphi_1(x) = 3x+1$, and $\varphi_2(x) = 3x - 1$. Indeed, suppose that $n \in \mathcal{D}(\Gamma_3)$, then $n = \eta_1 - \eta_2$ for $\eta_k \in \Gamma_3$. Since Γ_3 is invariant under φ_0 and φ_1, we have

$$\begin{aligned}
\varphi_0(n) &= \varphi_0(\eta_1) - \varphi_0(\eta_2) \in \mathcal{D}(\Gamma_3) \\
\varphi_1(n) &= \varphi_1(\eta_1) - \varphi_0(\eta_2) \in \mathcal{D}(\Gamma_3) \\
\varphi_2(n) &= \varphi_0(\eta_1) - \varphi_1(\eta_2) \in \mathcal{D}(\Gamma_3).
\end{aligned}$$

Clearly $\{-1, 0, 1\} \subset \mathcal{D}(\Gamma_3)$, so since it is invariant under $\varphi_0, \varphi_1, \varphi_2$, our claim is established. □

REMARK 3.8. A consequence of Corollary 3.6 is that Γ_3 is *not* a spectrum of any measure. Likewise, any $\Gamma \subsetneq \mathbb{Z}$ whose difference set is $\mathbb{Z}$ is not a spectrum of any measure.

COROLLARY 3.9. *If* $\Gamma \subset \mathbb{N}_0$ *is such that* $\mathcal{D}(\Gamma) \neq \mathbb{Z}$, *then there exist absolutely continuous measures in* $\mathcal{M}(K_\Gamma)$. *In particular,* $\mathcal{M}(K_4)$ *contains many absolutely continuous measures.*

PROOF. We define μ by its Radon-Nikodym derivative: choose

$$\frac{d\mu}{d\lambda}(\theta) = 1 + \sum_{n \notin \mathcal{D}(\Gamma)} b_n \cos(2\pi n \theta)$$

subject to the constraint that $\sum_{n \notin \mathcal{D}(\Gamma)} |b_n| < 1$. It follows that μ is a probability measure such that

$$\widehat{\frac{d\mu}{d\lambda}}(n) = 0$$

for $n \in \mathcal{D}(\Gamma) \setminus \{0\}$, and so satisfies Corollary 3.5.

Now, we claim that $\mathcal{D}(\Gamma_4) \neq \mathbb{Z}$. Indeed, if we define

$$\Gamma_4' = \left\{ \sum_{j=0}^{N} l_j 4^j | l_j \in \{0, 2\} \right\},$$

then we claim that $\Gamma_4' \cap \mathcal{D}(\Gamma_4) = \{0\}$. To establish this, suppose we have

$$\sum_{j=0}^{N_1} l_j 4^j = \sum_{j=0}^{N_2} p_j 4^j - \sum_{j=0}^{N_3} q_j 4^j$$

with $l_j \in \{0,2\}$ and $p_j, q_j \in \{0,1\}$. We may assume $N_1 = N_2 = N_3$ by padding with 0's if necessary. Thus, we have

$$\sum_{j=0}^{N} p_j 4^j = \sum_{j=0}^{N} l_j 4^j + \sum_{j=0}^{N} q_j 4^j = \sum_{j=0}^{N} (l_j + q_j) 4^j,$$

where $l_j + q_j \in \{0,1,2,3\}$. Since the base 4 expansion is unique, we must have that $p_j = l_j + q_j$ for all j, which can only occur when $l_j = 0$ for all j. □

4. Absolutely continuous measures

We proceed now to prove the Meta-Theorem in the case that the Grammian matrix M is a bounded operator on $\ell^2(\mathbb{N}_0)$. As mentioned previously, this occurs when the measure μ is absolutely continuous with bounded Radon-Nikodym derivative.

THEOREM 4.1. *Suppose C is a bounded, positive, self-adjoint operator on $\ell^2(\mathbb{N})$, $\mu << \lambda$, and $\frac{d\mu}{d\lambda} \in L^\infty(\mathbb{T})$. Then $\mu \in \mathcal{M}(K_C)$ if and only if $C = CMC$, where M is the Grammian matrix of $\{e_n\}_{n=0}^\infty \subset L^2(\mu)$, i.e. $M_{mn} = \hat{\mu}(n-m)$.*

PROOF. Since $\frac{d\mu}{d\lambda} \in L^\infty(\mathbb{T})$, the sequence $\{e_n\}_{n=0}^\infty \subset L^2(\mu)$ is a Bessel sequence. Thus, M is a bounded operator on $\ell^2(\mathbb{N})$, and the matrix product CMC is defined. Moreover, for every $w \in \mathbb{D}$, we have that since

$$\left(C\vec{w}\right)_n \in \ell^2(\mathbb{N})$$

the series

$$\sum_{n=0}^\infty \left(\sum_{m=0}^\infty c_{mn} \overline{w}^m \right) e^{2\pi i n t}$$

converges in $L^2(\mu)$. Thus, we have that for every $w \in \mathbb{D}$, the $L^2(\mu)$ boundary is given by

$$K_C^\star(w,t) = \sum_{n=0}^\infty \left(\sum_{m=0}^\infty c_{mn} \overline{w}^m \right) e^{2\pi i n t}$$

by Abel summation. We calculate

$$\begin{aligned}
\int_0^1 K_C^\star(w,t)\overline{K_C^\star(z,t)}d\mu(t) &= \sum_{n=0}^\infty \left(\sum_{m=0}^\infty c_{mn} \overline{w}^m \right) \int_0^1 e^{2\pi i n t} \overline{K_C^\star(z,t)} d\mu(t) \\
&= \sum_{n=0}^\infty \left(\sum_{m=0}^\infty c_{mn} \overline{w}^m \right) \sum_{k=0}^\infty \left(\sum_{l=0}^\infty \overline{c}_{lk} z^l \right) \int_0^1 e^{2\pi i n t} e^{-2\pi i k t} d\mu(t) \\
&= \sum_{n=0}^\infty \left(\sum_{m=0}^\infty c_{mn} \overline{w}^m \right) \sum_{k=0}^\infty \left(\sum_{l=0}^\infty c_{kl} z^l \right) M_{nk}. \qquad (4.1)
\end{aligned}$$

We have by the Cauchy-Schwarz inequality:

$$\begin{aligned}\sum_l \sum_k |c_{kl} M_{nk} z^l| &\leq \sum_l |z^l| \sqrt{\sum_k |c_{kl}|^2} \sqrt{\sum_k |M_{nk}|^2} \\ &\leq \|C\| \|M\| \sum_l |z^l| \\ &< \infty.\end{aligned}$$

Therefore,

$$(4.2) \qquad (4.1) = \sum_{n=0}^{\infty} \left(\sum_{m=0}^{\infty} c_{mn} \overline{w}^m \right) \sum_{l=0}^{\infty} (MC)_{nl} \, z^l.$$

Again by the Cauchy-Schwarz inequality:

$$\begin{aligned}\sum_m \sum_n \sum_l |c_{mn} (MC)_{nl} \, z^l \overline{w}^n| &\leq \sum_m \sum_l |z^l \overline{w}^n| \sqrt{\sum_n |c_{mn}|^2} \sqrt{\sum_n |\,(MC)_{nl}|^2} \\ &\leq \|C\| \|MC\| \sum_m \sum_l |z^l \overline{w}^n| \\ &< \infty.\end{aligned}$$

Whence

$$(4.3) \qquad (4.2) = \sum_{l=0}^{\infty} \sum_{m=0}^{\infty} (CMC)_{ml} \overline{w}^m z^l.$$

Consequently, Equation (1.4) if and only if

$$(4.4) \qquad \sum_{n=0}^{\infty} \sum_{m=0}^{\infty} c_{mn} \overline{w}^m z^n = \sum_{n=0}^{\infty} \sum_{m=0}^{\infty} (CMC)_{mn} \overline{w}^m z^n.$$

Equation (4.4) holds if and only if $C = CMC$ by the uniqueness of Taylor series coefficients. □

5. Preservation of norms of subspaces of $L^2(\lambda)$

The proofs of Theorems 3.2 and 4.1 suggest that the property that K_C reproduces itself with respect to some μ on the boundary is related to the following question: given a closed subspace $V \subset L^2(\lambda)$, for which measures ν does the following norm preservation identity hold for all $f \in V$:

$$\int |f|^2 d\lambda = \int |f|^2 d\nu?$$

Of course, this is ill-defined, because for $f \in L^2(\lambda)$, the question of whether $f \in L^2(\nu)$ and subsequently norm equality may depend on the representative. However, this ambiguity can be made precise using the boundary behavior of kernels as in Theorems 5.3 and 5.6.

Definition 5.1. Suppose $V \subset L^2_+(\lambda)$ is a closed subspace and let $\widetilde{V} \subset H^2(\mathbb{D})$ be the space consisting of all functions whose $L^2(\lambda)$ boundaries are in V. We say the measure μ preserves the norm of V if for every $f \in V$, with $f(x) = \sum_{n=0}^{\infty} a_n e^{2\pi i n x}$, the corresponding function $F(z) = \sum_{n=0}^{\infty} a_n z^n$ has a $L^2(\mu)$ boundary $F^\star$ and

$$\|f\|_\lambda = \|F^\star\|_\mu.$$

LEMMA 5.2. *Suppose $\Gamma \subset \mathbb{N}_0$ and μ satisfies Equation 3.1, then for every f in the subspace generated by K_Γ, f possesses a $L^2(\mu)$-boundary, and the norm of the $L^2(\mu)$-boundary agrees with the $H^2(\mathbb{D})$ norm of f.*

PROOF. Consider $\Gamma_1 \subset \Gamma$ of finite cardinality, and $f(z) = \sum_{\gamma\in\Gamma_1} a_\gamma z^\gamma$. Let $f^\star(e^{2\pi i\theta}) = \sum_{\gamma\in\Gamma_1} a_\gamma e^{2\pi i\gamma\theta}$. We claim that $f^\star$ is the $L^2(\mu)$-boundary of f and that

$$\|f^\star\|_\mu = \|f\|_{H^2}.$$

Indeed, we have that the $L^2(\mu)$-boundary of the function z^γ is $e^{2\pi i\gamma\theta}$ by uniform convergence, and thus by linearity $f^\star$ is the $L^2(\mu)$-boundary of f. Moreover,

$$\begin{aligned}
\|f^\star\|_\mu^2 &= \int_0^1 \left(\sum_{\gamma\in\Gamma_1} a_\gamma e^{2\pi i\gamma\theta}\right)\left(\sum_{\gamma'\in\Gamma_1} \overline{a_{\gamma'}} e^{-2\pi i\gamma'\theta}\right) d\mu \\
&= \sum_{\gamma\in\Gamma_1}\sum_{\gamma'\in\Gamma_1} a_\gamma \overline{a_{\gamma'}} \int_0^1 e^{2\pi i(\gamma-\gamma')} d\mu \\
&= \sum_{\gamma\in\Gamma_1} |a_\gamma|^2 \\
&= \|f\|_{H^2}^2. \qquad (5.1)
\end{aligned}$$

Now, for $f(z) = \sum_{\gamma\in\Gamma} a_\gamma z^\gamma$, the series

$$f^\star(e^{2\pi i\theta}) = \sum_{\gamma\in\Gamma} a_\gamma e^{2\pi i\gamma\theta}$$

converges in $L^2(\mu)$, and $f^\star$ is the $L^2(\mu)$-boundary for f by Abel summability. The equality of norms follows from taking limits in Equation (5.1). $\square$

THEOREM 5.3. *Let $\Gamma \subset \mathbb{N}_0$ and let V be the closed span of $\{e^{2\pi i\gamma\theta}\}_{\gamma\in\Gamma}$ in $L^2(\lambda)$. The measure μ preserves the norm of V if and only if $C = CMC$, where C is the diagonal matrix $c_{nn} = 1$ if $n \in \Gamma$ and 0 otherwise, and $M = (\hat{\mu}(n-m))_{mn}$.*

PROOF. We have that $K_C (= K_\Gamma)$ is the reproducing kernel for the space $\widetilde{V}$, and thus, if μ preserves the norm of V, we have by the polarization identity that $\mu \in \mathcal{M}(K_C)$. By Theorem 3.2 we must have $C = CMC$.

Conversely, if $C = CMC$, then $\mu \in \mathcal{M}(K_C)$, and thus μ preserves the norms of finite linear combinations $\sum_{j=1}^N b_j K_C^\star(w_j,\cdot)$:

$$\begin{aligned}
\|\sum_{j=1}^N b_j K_C^\star(w_j,\cdot)\|_\mu^2 &= \sum_{j=1}^N\sum_{k=1}^N b_j\overline{b_k}\int_0^1 K_C^\star(w_j,\cdot)\overline{K_C^\star(w_k,\cdot)}d\mu \\
&= \sum_{j=1}^N\sum_{k=1}^N b_j\overline{b_k} K_C(w_j,w_k) \\
&= \|\sum_{j=1}^N b_j K_C(w_j,\cdot)\|_{H^2}^2 \\
&= \|\sum_{j=1}^N b_j K_C^\star(w_j,\cdot)\|_\lambda^2.
\end{aligned}$$

We see by the proof of Corollary 3.5 that if $C = CMC$, then μ satisfies Equation (3.1). By Lemma 5.2, every element of the space spanned by K_C possesses an $L^2(\mu)$ boundary, and by density, μ then preserves the norms of all elements of V. $\square$

LEMMA 5.4. *Suppose C is a projection on $\ell^2(\mathbb{N}_0)$ and $\mu << \lambda$ is such that $\frac{d\mu}{d\lambda} \in L^\infty(\lambda)$. Let $N = (\hat{\mu}(m-n))_{mn}$. If $C = CNC$, then for every sequence $(a_n)_n$ in the range of C*

$$\|\sum a_n e_n\|_\mu = \|\sum a_n e_n\|_\lambda.$$

Note that $N = M^T$ in our previous notation.

PROOF. Our hypotheses yield that the series $\sum_n a_n e^{2\pi i n\theta}$ converges in $L^2(\mu)$. We have

$$\begin{aligned}
\|\sum a_n e^{2\pi i n\theta}\|_\mu^2 &= \sum_{n,m} a_n \overline{a_m} \int_0^1 e^{2\pi i(n-m)\theta} d\mu(\theta) \\
&= \langle N(a_n)_n, (a_n)_n \rangle \\
&= \langle CNC(a_n)_n, (a_n)_n \rangle \\
&= \langle C(a_n)_n, (a_n)_n \rangle \\
&= \|\sum a_n e^{2\pi i n\theta}\|_\lambda^2.
\end{aligned}$$

$\square$

LEMMA 5.5. *Suppose V is a subspace of $L^2_+(\lambda)$. Let C be the projection on $\ell^2(\mathbb{N}_0)$ such that $f = \sum_n a_n e_n \in V$ if and only if $(a_n)_n$ is in the range of C. Then the reproducing kernel of $\widetilde{V}$ is K_{C^T}.*

PROOF. First note that for $w \in \mathbb{D}$,

$$K_{C^T}(w,z) = \sum_n \sum_m (C^T)_{mn} \overline{w}^m z^n = \sum_n \left(C\vec{\overline{w}}\right)_n z^n$$

is such that the coefficients are in the range of C. Thus, $K_C(w,\cdot) \in \widetilde{V}$. Now, for $f(z) = \sum_n a_n z^n \in \widetilde{V}$, we have that

$$\begin{aligned}
\langle f, K_{C^T} \rangle_{H^2} &= \langle (a_n)_n, C\vec{\overline{w}} \rangle_{\ell^2} \\
&= \langle (a_n)_n, \vec{\overline{w}} \rangle_{\ell^2} \\
&= \sum_n a_n w^n \\
&= f(w),
\end{aligned}$$

and thus K_{C^T} is the kernel as claimed. $\square$

THEOREM 5.6. *Suppose V is a subspace of $L^2_+(\lambda)$, and let $\mu << \lambda$ with $\frac{d\mu}{d\lambda} \in L^\infty(\lambda)$. Then μ preserves the norm of V if and only if $C^T = C^T M C^T$, where C is the projection on $\ell^2(\mathbb{N}_0)$ with the property that K_{C^T} is the reproducing kernel of $\widetilde{V} \subset H^2(\mathbb{D})$ and $M = (\hat{\mu}(n-m))_{mn}$.*

PROOF. For $w_1, w_2 \in \mathbb{D}$, $K_{C^T}(w_j, z) = \sum_n (C\vec{\overline{w}}_j)_n z^n$. By our assumptions, the $L^2(\mu)$ boundary of $K_{C^T}(w_j, \cdot)$ is $\sum_n (C\vec{\overline{w}}_j)_n e^{2\pi i n\theta}$.

Suppose that μ preserves the norm of V. We have by the Polarization Identity:

$$\begin{aligned}\int_0^1 K^\star_{C^T}(w_1,\theta)\overline{K^\star_{C^T}(w_2,\theta)}d\mu(\theta) &= \langle K^\star_{C^T}(w_1,\cdot), K^\star_{C^T}(w_2,\cdot)\rangle_\mu \\ &= \left\langle \sum_n (C\vec{\overline{w}}_1)_n e^{2\pi i n\theta}, \sum_n (C\vec{\overline{w}}_2)_n e^{2\pi i n\theta} \right\rangle_\lambda \\ &= \langle K_{C^T}(w_1,\cdot), K_C(w_2,\cdot)\rangle_{H^2} \\ &= K_{C^T}(w_1, w_2).\end{aligned}$$

Therefore, $\mu \in \mathcal{M}(K_{C^T})$, and hence by Theorem 4.1, $C^T = C^T M C^T$.

Conversely, if $C^T = C^T M C^T$, then $C = CNC$. Therefore, for every finite linear combination $\sum_{j=1}^N \xi_j K^\star_{C^T}(w_j, \cdot) \in V$, we have by Lemma 5.4 that

$$\|\sum_{j=1}^N \xi_j K^\star_{C^T}(w_j,\cdot)\|_\mu = \|\sum_{j=1}^N \xi_j K^\star_{C^T}(w_j,\cdot)\|_\lambda.$$

By density, μ preserves the norm of V. □

References

[1] A. B. Aleksandrov, *Inner functions and related spaces of pseudocontinuable functions* (Russian, with English summary), Zap. Nauchn. Sem. Leningrad. Otdel. Mat. Inst. Steklov. (LOMI) **170** (1989), no. Issled. Lineĭn. Oper. Teorii Funktsiĭ. 17, 7–33, 321, DOI 10.1007/BF01099304; English transl., J. Soviet Math. **63** (1989), no. 2, 115–129. MR1039571

[2] Daniel Alpay, Aad Dijksma, Heinz Langer, and Gerald Wanjala, *Basic boundary interpolation for generalized Schur functions and factorization of rational J-unitary matrix functions*, Interpolation, Schur functions and moment problems, Oper. Theory Adv. Appl., vol. 165, Birkhäuser, Basel, 2006, pp. 1–29, DOI 10.1007/3-7643-7547-7_1. MR2222518

[3] Daniel Alpay and Harry Dym, *Hilbert spaces of analytic functions, inverse scattering and operator models. I*, Integral Equations Operator Theory **7** (1984), no. 5, 589–641, DOI 10.1007/BF01195919. MR766625

[4] Daniel Alpay and Harry Dym, *Hilbert spaces of analytic functions, inverse scattering and operator models. II*, Integral Equations Operator Theory **8** (1985), no. 2, 145–180, DOI 10.1007/BF01202812. MR782616

[5] N. Aronszajn, *Theory of reproducing kernels*, Trans. Amer. Math. Soc. **68** (1950), 337–404, DOI 10.2307/1990404. MR0051437

[6] Peter G. Casazza, *The art of frame theory*, Taiwanese J. Math. **4** (2000), no. 2, 129–201, DOI 10.11650/twjm/1500407227. MR1757401

[7] Douglas N. Clark, *One dimensional perturbations of restricted shifts*, J. Analyse Math. **25** (1972), 169–191, DOI 10.1007/BF02790036. MR0301534

[8] Louis de Branges and James Rovnyak, *Square summable power series*, Holt, Rinehart and Winston, New York-Toronto, Ont.-London, 1966. MR0215065

[9] R. J. Duffin and A. C. Schaeffer, *A class of nonharmonic Fourier series*, Trans. Amer. Math. Soc. **72** (1952), 341–366, DOI 10.2307/1990760. MR0047179

[10] Dorin Ervin Dutkay, Deguang Han, Qiyu Sun, and Eric Weber, *On the Beurling dimension of exponential frames*, Adv. Math. **226** (2011), no. 1, 285–297, DOI 10.1016/j.aim.2010.06.017. MR2735759

[11] Dorin Ervin Dutkay and Palle E. T. Jorgensen, *Affine fractals as boundaries and their harmonic analysis*, Proc. Amer. Math. Soc. **139** (2011), no. 9, 3291–3305, DOI 10.1090/S0002-9939-2011-10752-4. MR2811284

[12] Dorin Ervin Dutkay and Chun-Kit Lai, *Uniformity of measures with Fourier frames*, Adv. Math. **252** (2014), 684–707, DOI 10.1016/j.aim.2013.11.012. MR3144246

[13] John E. Herr, Palle E.T. Jorgensen, and Eric S. Weber, *Positive matrices in the Hardy space with prescribed boundary representations via the Kaczmarz algorithm*, to appear in J. Anal. Math., 2017.

[14] John E. Herr and Eric S. Weber, *Fourier series for singular measures*, Axioms **6** (2017), no. 2:7, 13 ppg., http://dx.doi.org/10.3390/axioms6020007.

[15] Palle E. T. Jorgensen and Steen Pedersen, *Dense analytic subspaces in fractal L^2-spaces*, J. Anal. Math. **75** (1998), 185–228, DOI 10.1007/BF02788699. MR1655831

[16] Stanisław Kwapień and Jan Mycielski, *On the Kaczmarz algorithm of approximation in infinite-dimensional spaces*, Studia Math. **148** (2001), no. 1, 75–86, DOI 10.4064/sm148-1-7. MR1881441

[17] Chun Kit Lai, *Spectral Analysis on Fractal Measures and Tiles*, ProQuest LLC, Ann Arbor, MI, 2012. Thesis (Ph.D.)–The Chinese University of Hong Kong (Hong Kong). MR3130842

[18] A. G. Poltoratskiĭ, *Boundary behavior of pseudocontinuable functions* (Russian, with Russian summary), Algebra i Analiz **5** (1993), no. 2, 189–210; English transl., St. Petersburg Math. J. **5** (1993), no. 2, 389–406. MR1223178

[19] Donald Sarason, *Sub-Hardy Hilbert spaces in the unit disk*, University of Arkansas Lecture Notes in the Mathematical Sciences, vol. 10, John Wiley & Sons, Inc., New York, 1994. A Wiley-Interscience Publication. MR1289670

Department of Mathematics, Butler University, Indianapolis, Indiana 46208
Email address: jeherr@butler.edu

Department of Mathematics, University of Iowa, Iowa City, Iowa 52242
Email address: palle-jorgensen@uiowa.edu

Department of Mathematics, Iowa State University, 396 Carver Hall, Ames, Iowa 50011
Email address: esweber@iastate.edu

Contemporary Mathematics
Volume **706**, 2018
http://dx.doi.org/10.1090/conm/706/14209

Gibbs effects using Daubechies and Coiflet tight framelet systems

Mutaz Mohammad and En-Bing Lin

This paper is dedicated to our families.

ABSTRACT. In this article, we study the Gibbs phenomenon for compactly supported framelets, such as Daubechies and Coiflets framelets to illustrate the Gibbs effect. The tight framelets representation of a square integrable function is essentially a generalized wavelet representation. We show a numerical evidence that there is no Gibbs phenomenon when we exhibit the framelets expansion for a square integrable function by using 1st order Daubechies tight framelets for two generators. The investigation of Gibbs phenomenon in Daubechies tight framelets, however, shows that it exists for higher order. Also, we provide a numerical values of the overshoots and undershoots when we use Coiflets tight frame representation.

1. Introduction

Daubechies was first to construct compactly supported orthogonal wavelets with a preassigned degree of smoothness [**Da1**]. She intended to construct a wavelet with N vanishing moments and supported in $[0, 2N-1]$, where a wavelet function ψ is said to have N (≥ 2) vanishing moments if,

$$\int x^n \psi(x)\, dx = 0, \quad n = 0, 1, \cdots, N-1.$$

When $N = 1$, it is Haar wavelet and has only one vanishing moment. In fact, all systems built by using the unitary extension principle (UEP) of Ron and Shen [**RS**] have only one vanishing moment.

Daubechies looked for h_k $'s$ in a dilation equation,

$$\phi = \sqrt{2} \sum_{k\in\mathbb{Z}} h_k \phi(2x-k),$$

such that the orthonormal condition $\hat{\phi}(2p\pi) = \delta_{0,p}$, where $p = 0, 1, ..., N-1$, is satisfied, and h_k are as follows:

2010 *Mathematics Subject Classification.* Primary 42C15, 65T60; Secondary 65T40.

Key words and phrases. Gibbs phenomenon, unitary extension principle, tight wavelet frames, Daubechies wavelets, Coiflets.

The first author was supported by the PD Grant, Research Office, Zayed University.

$$h_0 = \frac{v(v-1)}{\sqrt{2}(v^2+1)}, h_1 = \frac{(v-1)}{\sqrt{2}(v^2+1)}, h_2 = \frac{(v+1)}{\sqrt{2}(v^2+1)}, h_3 = \frac{v(v+1)}{\sqrt{2}(v^2+1)}.$$

By choosing $v = \frac{\pm 1}{\sqrt{3}}$, a scaling function ϕ has support $[0,2]$.

Daubechies wavelet does not have a closed form, but instead, can be obtained recursively by

$$\begin{aligned}\phi_n(x) &= \sum_{k\in\mathbb{Z}} \sqrt{2} h_k \phi_{n-1}(2x-k), \\ \phi_0(x) &= \chi_{[0,1)}(x).\end{aligned}$$

It is known that the smoothness of the wavelets increases with N. For an application in numerical analysis, Coifman asked Daubechies to construct a family of wavelets ψ having N vanishing moments, minimum size support and

$$\int_{\mathbb{R}} \phi = 1,$$

in such a way we have a smooth orthogonal system [**Da1**]. Daubechies wavelet has vanishing moment for wavelet function ψ only. However, Daubechies designed, at that request, a wavelet (Coiflets) that has vanishing moment for both wavelet and scaling functions ψ and ϕ. The wavelet is near symmetric such that wavelet function ψ has $2N$ vanishing moments and the scaling function has $2N-1$ vanishing moments. The two functions have a support width of length $6N-1$. Also there is no closed form for Coiflets and it can be obtained recursively.

DEFINITION 1.1. A sequence $\{f_k\}_{k=1}^{\infty}$ of elements in $L^2(\mathbf{R})$ is a *frame* for $L^2(\mathbf{R})$ if there exist constants $A, B > 0$ such that

$$A\|f\|^2 \leq \sum_{k=1}^{\infty} |\langle f, f_k\rangle|^2 \leq B\|f\|^2, \forall f \in L^2(\mathbf{R}).$$

The numbers A, B are called frame bounds [**Ch**].

A frame is *tight* if the frame bounds A and B are equal, and it is *Parseval frame* if $A = B = 1$. All frames can be viewed as a generalization of an orthonormal basis. For any function $f \in L^2(\mathbf{R})$, the dyadic dilation operator D is defined by $Df(x) = \sqrt{2}f(2x)$, and the translation operator T is defined by $T_a f(x) = f(x-a)$ for $a \in \mathbf{R}$.

DEFINITION 1.2. A compactly supported function $\phi \in L^2(\mathbf{R})$ is said to be refinable if

$$\phi(x) = 2\sum_{k\in\mathbf{Z}} h_0[k]\phi(2x-k), \tag{1.1}$$

for some finite supported sequence $h_0[k] \in \ell_2(\mathbf{Z})$. The sequence h_0 is called the *low pass filter* of ϕ.

Let $\Psi = \{\psi_\ell\}_{\ell=1}^{r} \subset L^2(\mathbf{R})$ be of the form

$$\psi_\ell = 2\sum_{k\in\mathbf{Z}} h_\ell[k]\phi(2\cdot -k), \tag{1.2}$$

where $\{h_\ell[k],\ k \in \mathbf{Z}\}_{\ell=1}^{r}$ is a finitely supported sequence and is called a *high pass filter* of the system. Define the wavelet system

$$X(\Psi) = \{\psi_{\ell,j,k} : \quad 1 \le \ell \le r \quad ; j,k \in \mathbf{Z}\},$$

where $\psi_{\ell,j,k}(x){=}D^jT_k\psi_\ell$. The general setup is to construct a tight frame for $L^2(\mathbf{R})$ and of the form of $X(\Psi)$.

THEOREM 1.3 (Unitary Extension Principle [**RS**]). *Let $\phi \in L^2(\mathbf{R})$ be the compactly supported refinable function with its finitely supported low pass filter h_0. Let $\{h_\ell[k], k \in \mathbf{Z}\}_{\ell=1}^{r}$ be a set of finitely supported sequences, then the system*

$$X(\Psi) = \{\psi_{\ell,j,k} : 1 \le \ell \le r; j,k \in \mathbf{Z}\} \tag{1.3}$$

forms a tight frame for $L^2(\mathbf{R})$ provided the equalities

$$\sum_{\ell=0}^{r}\sum_{k\in\mathbf{Z}} \overline{h_\ell[k]}h_\ell[k-p] = \delta_{0,p} \tag{1.4}$$

and

$$\sum_{\ell=0}^{r}\sum_{k\in\mathbf{Z}} (-1)^{k-p}\,\overline{h_\ell[k]}h_\ell[k-p] = 0 \tag{1.5}$$

hold for all $p \in \mathbf{Z}$, where $\delta_{0,p}$ is the Kronecker delta.

When $X(\Psi)$ forms a (tight) frame of $L^2(\mathbf{R})$, each function ψ_ℓ is called (tight) framelet and the whole system $X(\Psi)$ is called (tight) framelet system. If Daubechies (Coiflets) scaling functions are used to generate framelets, we call $X(\Psi)$ Daubechies (Coiflets) tight framelets system. More details can be found in [**Be**, **Ch**, **Da1**, **Da2**, **DHRS**, **DS**] and the references therein.

2. Daubechies and Coiflets tight framelets for two generators

Assume that for a fixed N, the requirements of Theorem 1.3 hold for the Daubechies and Coiflet scaling functions db_{2N} and $coif6N$, respectively, where

$$\phi^N(x) = 2\sum_{k\in\mathbb{Z}} h_0[k]\phi^N(2x-k),\ {}_N\psi_\ell(x) \qquad = 2\sum_{k\in\mathbb{Z}} h_\ell[k]\phi^N(2x-k).$$

In this paper, we consider $r = 2$ in equations (1.4) and (1.5) to generate tight framelet system using Daubechies and Coiflet scaling functions. Note that, equations (1.4) and (1.5) provide $4N-1$ equations with $4N$ unknowns in finding high pass filters $h_\ell[k]$. We used the Levenberg-Marquardt algorithm (LMA) iterative technique [**Le**, **Lo**] to solve the nonlinear equations. The LMA significantly outperforms the Gauss-Newton algorithm and the method of gradient descent.

Before proceeding to present examples of framelets system, we will introduce a class of splines with very interesting features, the B-splines $N_m, m \in \mathbf{N}$.

DEFINITION 2.1 (**B-splines** [**Ch**, **Da2**]). The B-spline N_{m+1} is defined as follows by using the convolution

$$N_{m+1}(x) := (N_m * N_1)(x), x \in \mathbf{R},$$

where $N_1(x)$ is defined to be $\chi_{[0,1)}(x)$ (the characteristic function for the interval $[0,1)$).

The B-splines are well studied, and have many desirable properties [**Wa**]. In fact, $N_m(x)$ has the following dyadic dilation relation

$$N_m(x) = \sum_{k=0}^{m} 2^{-m+1}\binom{m}{k} N_m(2x-k). \tag{2.1}$$

LEMMA 2.2. *The low pass filter $h_0[k]$ of N_m is equal to $2^{-m}\binom{m}{k}$.*

PROOF. Obvious by looking to equation (2.1) and Definition 1.2 $\square$

Note that $db_2 = N_1(x)$. Therefore, the low pass filter of db_2 is equal to $[1/2, 1/2]$.

2.1. Examples. In this section we provide some examples of the construction of Daubechies and Coiflets tight framelets system. Note that, and as proposed before, the low and high pass filters $h_0[k]$ and $h_\ell[k]$ can be found by using the LMA and equations (1.4) and (1.5), where $k \in [0, 2N-1]$ in the case of db_{2N} and $k \in [0, 6N-1]$ for $coif6N$.

EXAMPLE 2.3. Let $h_0[k] = [1/2, 1/2]$, be the low pass filter of $db_2 = \chi_{[0,1)}(x)$. Define

$$\begin{aligned} h_1[k] &= [0.353373237958641, -0.353783252808043], \\ h_2[k] &= [-0.353733570306848, 0.353323497802083]. \end{aligned}$$

Then $h_0[k]$, $h_1[k]$, and $h_2[k]$ satisfy equations (1.4) and (1.5). Hence, the system $X(\Psi)$ defined in equation (1.3) is a tight framelet of $L^2(\mathbf{R})$. We illustrate db_2 and its framelet generators in Figure 1.

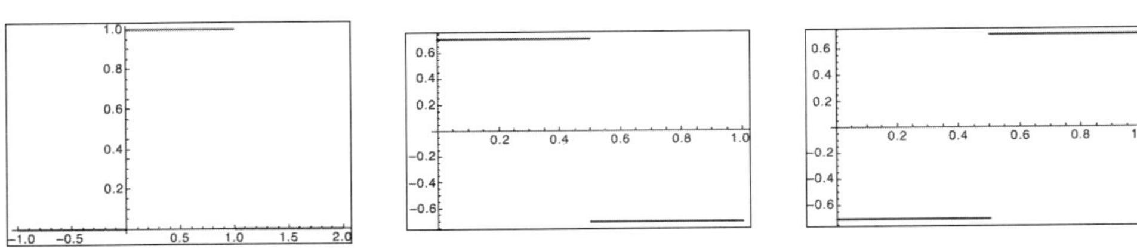

FIGURE 1. Daubechies scaling function of order 1 with its generators.

EXAMPLE 2.4. Let $h_0[k]=[\frac{1+\sqrt{3}}{8}, \frac{3+\sqrt{3}}{8}, \frac{3-\sqrt{3}}{8}, \frac{1-\sqrt{3}}{8}]$ be the low pass filter of db_4. Define

$$\begin{aligned} h_1[k] &= [-0.069459010353791, -0.120052106314821, 0.446985358193581, \\ &\quad -0.257977351573073], \\ h_2[k] &= [0.059574082608293, 0.103478971889253, -0.387406580578380, \\ &\quad 0.223772890970135]. \end{aligned}$$

Then $h_0[k]$, $h_1[k]$, and $h_2[k]$ satisfy equations (1.4) and (1.5). Hence, the system $X(\Psi)$ defined in equation (1.3) is a tight framelet of $L^2(\mathbf{R})$. We illustrate db_4 and its framelet generators ψ_1, ψ_2 in Figure 2.

We illustrate db_6 and db_8, and its Daubechies tight framelet generators, respectively, in Figures 3 and 4 respectively.

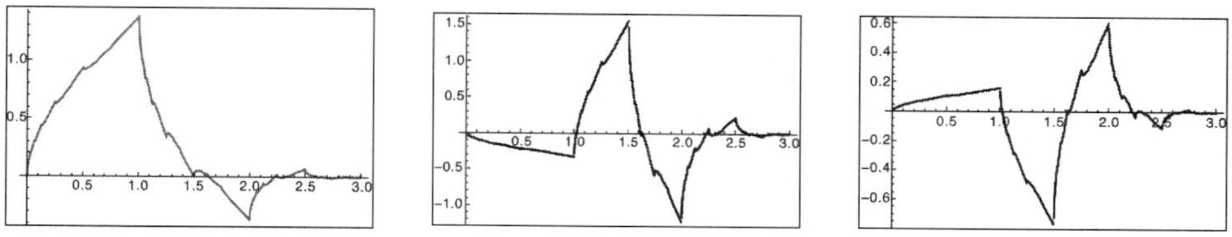

FIGURE 2. Daubechies scaling function of order 2 with its generators.

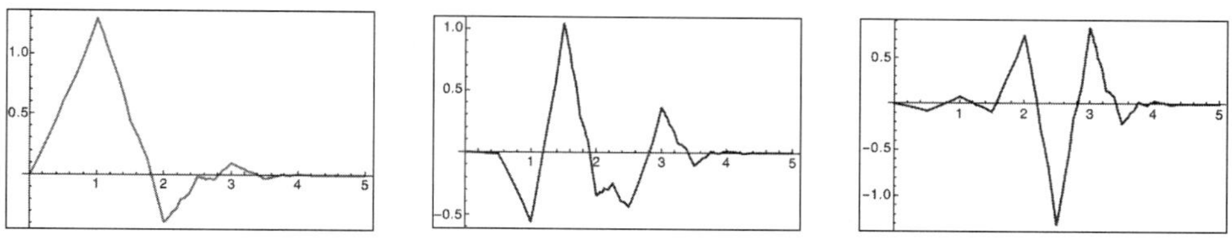

FIGURE 3. Daubechies scaling function of order 3 with its generators.

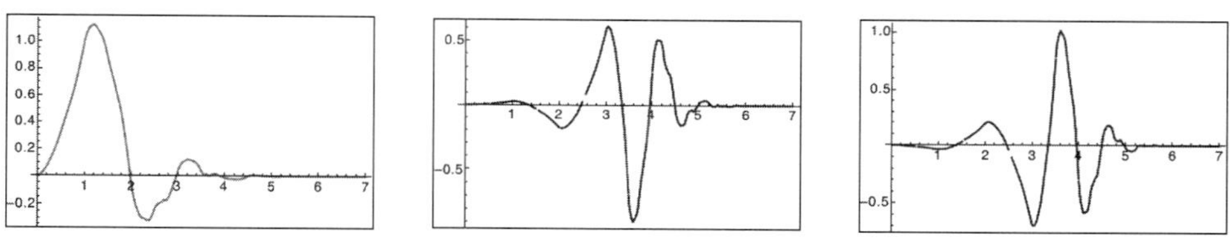

FIGURE 4. Daubechies scaling function of order 4 with its generators.

EXAMPLE 2.5. Using Coiflets scaling functions, let

$$\begin{aligned} h_0[k] &= [-0.051429728471, 0.238929728471, 0.602859456942, \\ &\quad 0.272140543058, -0.051429728471, -0.011070271529], \end{aligned}$$

be the low pass filter associated with $\phi^1 = coif6$. Then, by MATLAB, we obtain the set of generators as follows.

$$\begin{aligned} {}_1\psi_1(x) &= 2((-0.003121131787350)\phi^1[2x] + (0.014830692820540)\phi^1[2x-1] + \\ &\quad (0.080981568511578)\phi^1[2x-2] + (-0.179256226513660)\phi^1[2x-3] + \\ &\quad (0.070897582950706)\phi^1[2x-4] + (0.015348267448896)\phi^1[2x-5]), \\ {}_1\psi_2(x) &= 2((0.010622665113976)\phi^1[2x] + (-0.049247304653133)\phi^1[2x-1] + \\ &\quad (-0.259812371875876)\phi^1[2x-2] + (0.575592505759038)\phi^1[2x-3] + \\ &\quad (-0.228168721777402)\phi^1[2x-4] + (-0.049086165341934)\phi^1[2x-5]). \end{aligned}$$

Hence, the system $X(\Psi)$ defined in equation (1.3) forms a Coiflet tight framelet of $L^2(\mathbf{R})$. We illustrate $coif6$ and its Coiflets tight framelet generators in Figure 5.

EXAMPLE 2.6. For $N = 2$, we have $\phi^2 = coif12$. Then, the low pass filter is given by

$$\begin{aligned} h_0[k] &= [0.0116, -0.0293, -0.0476, 0.2730, 0.5747, 0.2949, -0.0541 \\ &\quad -0.0420, 0.0167, 0.0040, -0.0013, -5.0951e-004]. \end{aligned}$$

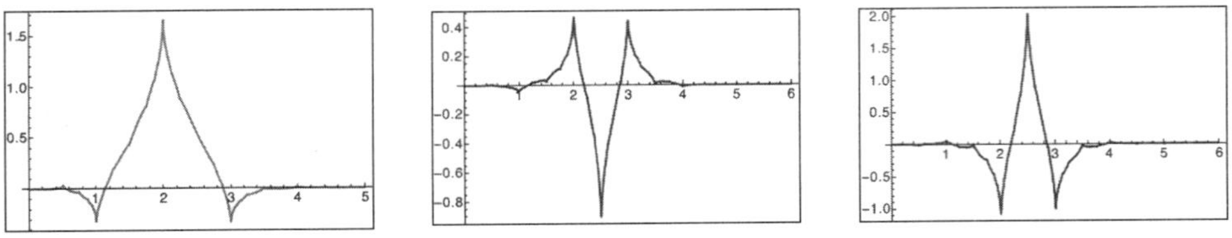

FIGURE 5. Coiflet scaling function of order 1 with its generators.

Again, by MATLAB, we obtain solutions for h_1 and h_2. Hence, we have

$$\begin{aligned}
{}_2\psi_1(x) \;=\; & 2((0.001172798035670)\phi^2[2x] + (-0.009000984913855)\phi^2[2x-1] + \\
& (-0.012596244814838)\phi^2[2x-2] + (0.009265925003131)\phi^2[2x-3] + \\
& (0.200090525277286)\phi^2[2x-4] + (-0.364663765944759)\phi^2[2x-5] + \\
& (0.179585296965332)\phi^2[2x-6] + (0.019874783895727)\phi^2[2x-7] + \\
& (-0.029019218150040)\phi^2[2x-8] + (0.020107729011366)\phi^2[2x-9] + \\
& (-0.015112081123584)\phi^2[2x-10] + (0.000083812662825)\phi^2[2x-11]).
\end{aligned}$$

$$\begin{aligned}
{}_2\psi_2(x) \;=\; & 2((-0.002149568366948)\phi^2[2x] + (0.011749982385303)\phi^2[2x-1] + \\
& (0.023344106542529)\phi^2[2x-2] + (-0.024024705737818)\phi^2[2x-3] + \\
& (-0.235941132147702)\phi^2[2x-4] + (0.439654397890551)\phi^2[2x-5] + \\
& (-0.182232874177506)\phi^2[2x-6] + (-0.081570734262437)\phi^2[2x-7] + \\
& (0.031082044158223)\phi^2[2x-8] + (0.035495389908025)\phi^2[2x-9] + \\
& (-0.014817198184965)\phi^2[2x-10] + (-0.000753121270391)\phi^2[2x-11]).
\end{aligned}$$

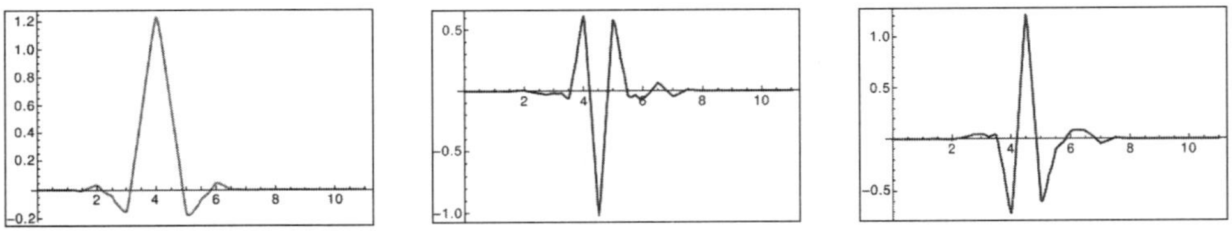

FIGURE 6. Coiflet scaling function of order 2 with its generators.

We illustrate $coif12$ and its Coiflets tight framelet generators ψ_1, ψ_2 in Figure 6.

3. The Gibbs phenomenon in Daubechies and Coiflets tight framelet expansions

It is known that the Gibbs phenomenon is not the failure of convergence at the discontinuity, nor the existence of overshoot and undershoot, but rather the failure of removing the overshoot by higher approximations. The overshoot at such a break is approximately 8.95% of the jump size. It appears a minimum, which undershoots the function by 4.84% of the total jump size there [**Je**, **Ri**].

As it is related to the behavior of truncated partial sums of a function with a jump discontinuity, the presence of this phenomenon is undesirable, and it is not so easily defeated. Therefore, examining series representations to avoid it or at least to reduce it is very important. This phenomenon is known as the Gibbs effect.

The first study of Gibbs phenomenon for framelet expansions by using B-spline functions has been examined in [**ML**] but it was somewhat predicted by Gribonval and Nielsen in [**GN**]. It has been stated there, the use of framelets generated by those scaling functions that have only one vanishing moment, may obtain less Gibbs effect than with wavelets.

In this paper, we see the absence of the phenomenon when we use the framelet series approximation obtained from Daubechies scaling function of order one but for two generators. This framelet system is obtained by the UEP.

To illustrate what is happening in the Gibbs effect, let us examine the partial sums of a framelet series. Assume that $X(\Psi)$ forms a tight frame for $L^2(\mathbf{R})$. For any function $f \in L^2(\mathbf{R})$, f can be expanded as follows,

$$f = \sum_{\ell=1}^{r} \sum_{j\in\mathbf{Z}} \sum_{k\in\mathbf{Z}} \langle f, \psi_{\ell,j,k}\rangle \psi_{\ell,j,k}. \tag{3.1}$$

We first provide a precise definition of the Gibbs phenomenon for framelet expansion of a given function $f : \mathbf{R} \longrightarrow \mathbf{R}$, such that $f \in L^2(\mathbf{R})$ which has a compact support and a jump discontinuity at the origin as follows.

Definition 3.1. Suppose a function f has a jump discontinuity at $x = 0$, i.e., limits $\lim_{x\to 0^+} f(x)$ and $\lim_{x\to 0^-} f(x)$ exist, and that $f(0^+) \neq f(0^-)$. Define $S_{n,m}f$ to be the truncated partial sum of equation (3.1) where j and k are up to n and m respectively. We say that the framelet expansion of f exhibits the Gibbs phenomenon at the right hand side of $x = 0$ if there is a sequence $x_s > 0$ converging to 0, and

$$\lim_{n,m\to\infty} S_{n,m}f(x_s) \begin{cases} > f(0^+), & if \quad f(0^+) > f(0^-) \\ < f(0^+), & if \quad f(0^+) < f(0^-) \end{cases}.$$

Similarly, we can define the Gibbs phenomenon on the left-hand side of $x = 0$.

To study the Gibbs effect, some researchers consider the function D [**Je**], where

$$D(x) = \begin{cases} 1, & x \geq 0 \\ 0, & x < 0 \end{cases}.$$

Whereas, some other researchers consider the function f as follows [**Ke, Ra**],

$$f(x) = \begin{cases} 1 - x, & 0 < x \leq 1 \\ -1 - x, & -1 \leq x < 0 \\ 0, & \text{else} \end{cases}.$$

Note that D and f are very useful for expressing all sorts of functions with (jump) discontinuities. As an example of:

- a removable discontinuity, we have

$$|ax + b| = (ax + b) D(ax + b) + (-ax - b) D(-ax - b);$$

- and, as a non-removable discontinuity, define

$$g(x) = \begin{cases} x^2, & x \geq 4 \\ -x, & x < 4 \end{cases}$$

Then, $g(x) = -x + (x^2 + x)D(x - 4)$.

In our study, we will use f. Note that if

$$J(x) = \left\{ \begin{array}{cc} a+1-x, & a < x \leq a+1 \\ a-1-x, & a-1 \leq x < a \\ 0, & \text{else} \end{array} \right\}$$

then, J has a jump discontinuity at the point a and $J(x) = f(x-a)$. More precisely, we have the following general statement.

PROPOSITION 3.2. *Any function with finitely many jump discontinuities can be written in terms of J plus a continuous function at the origin.*

PROOF. Let g be a function with a jump discontinuity, say at $x = a$, of magnitude M. We could put several of these together for g but we would likely only be looking at one such function at a time. Suppose that J and g in the same direction of the given jump (i.e., if $g(a^+) > g(a^-)$, then $J(a^+) > J(a^-)$, and similarly for $g(a^+) < g(a^-)$) or multiply J by (- or +)M to create the needed jump in the same direction. Define

$$F(x) = (+ \text{ or } -)MJ(x) + d,$$

so that d is a constant that makes the jump endpoints of F and g matched at $x = a$. Our continuous function in the neighborhood of the point a will then be $g(x) - F(x)$. □

To examine Gibbs effects, the approximation formula can be written using scaling functions, here we use Daubechies and Coiflet functions. That is, the Daubechies approximation formula (the partial sum of order N) is given by

$$S_n^{r,2N} f(\cdot) = \int_{\mathbb{R}} f(y)\, K_n^{r,2N}(\cdot, y)\, dy, \tag{3.2}$$

where

$$K_n^{r,2N}(x,y) = 4\sum_{\ell=1}^{r}\sum_{j=-n}^{n}\sum_{k\in\mathbf{Z}}\sum_{m,s\in\mathbf{Z}} 2^j h_\ell[m] h_\ell[s] \phi^N(2^{j+1}x - 2k - s)\phi^N(2^{j+1}y - 2k - m)$$

is defined to be the reproducing Daubechies tight framelet kernel constructed from the UEP. Note that, in the case of Coiflet scaling functions, we just replace $2N$ by $6N$.

The investigation of Gibbs effects using Daubechies tight framelet system shows that it exists for higher order (i.e., $N \geq 2$). Whereas the case $N = 1$ showed that equation (3.2) do not exhibit Gibbs effect. We believe that the reasoning for the fact only 1st-order Daubechies tight framelets beat the Gibbs phenomenon is that the quasi-affine system generated by db_2 is clearly not in the Schwartz space, the space of smooth functions whose derivatives (including the function itself) decay at infinity faster than any power [**Ra**]. We provide a numerical evidence and some graphs to showcase the absence of the effects in Table 1 and Figure 7.

Tables 2, 3, and 4 present an approximation of the maximum undershoots and overshoots of the truncated function $S_n^{2,2N}$ near the jump $x = 0$ for $N = 2, 3,$ and 4, respectively.

Figures 8, 9, and 10 illustrate the phenomenon using Daubechies tight framelet of order 2, 3, and 4, respectively.

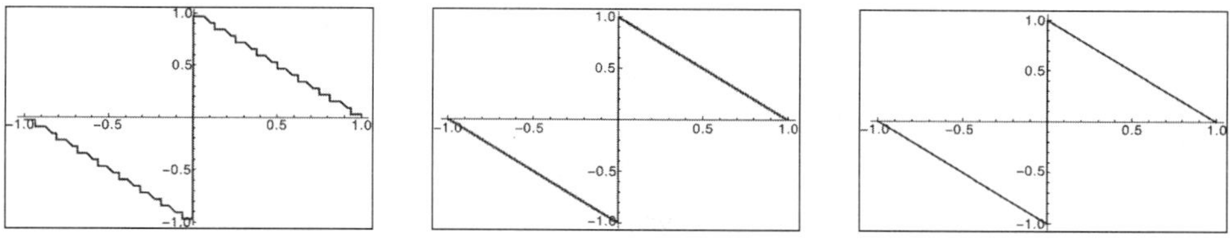

FIGURE 7. Illustration for the absence of Gibbs effects using $S_n^{2,2} f(x)$, $n = 3, 5$ and 10.

TABLE 1. Approximate maximum overshoot and undershoot near $x = 0$ using $S_n^{2,2} f(x)$.

Level	*Maximum*	*Minimum*
$n = 3$	0.968731	-0.968731
$n = 5$	0.992189	-0.992189
$n = 10$	0.999759	-0.999759

TABLE 2. Approximate maximum overshoot and undershoot near $x = 0$ using $S_n^{2,4} f(x)$.

Level	*Maximum*	*Minimum*
$n = 3$	1.58563	-0.95046
$n = 4$	1.60152	-1.00166
$n = 5$	1.60968	-1.02474

TABLE 3. Approximate maximum overshoot and undershoot near $x = 0$ using $S_n^{2,6} f(x)$.

Level	*Maximum*	*Minimum*
$n = 3$	1.18157	-1.19841
$n = 4$	1.21593	-1.22986
$n = 5$	1.23249	-1.24476

As we did for the truncated Daubechies tight framelet, we define the truncated Coiflets tight framelet expansion as

$$S_n^{r,6N} f = \int_{\mathbb{R}} f(y) K_n^{r,6N}(\cdot, y)\, dy$$

where $K_n^{r,6N}(x, y)$ is the reproducing Coiflet tight framelet kernel constructed from the UEP.

TABLE 4. Approximate maximum overshoot and undershoot near $x = 0$ using $S_n^{2,8} f(x)$.

Level	*Maximum*	*Minimum*
$n = 2$	0.886196	-1.24064
$n = 3$	1.01008	-1.28193
$n = 4$	1.06619	-1.30510

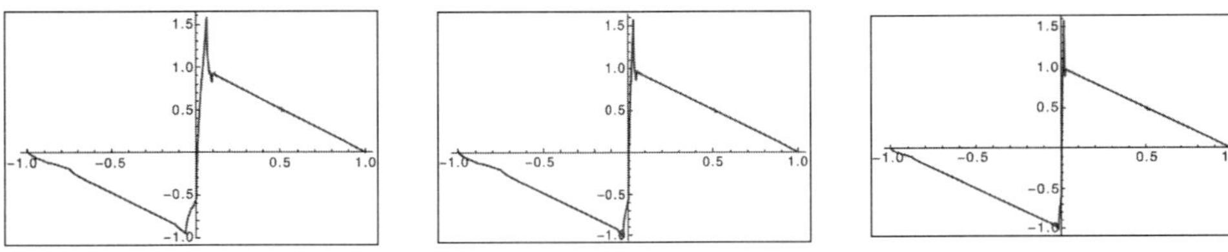

FIGURE 8. The overshoot and undershoot for $S_n^{2,4} f(x)$, $n = 3, 4$ and 5.

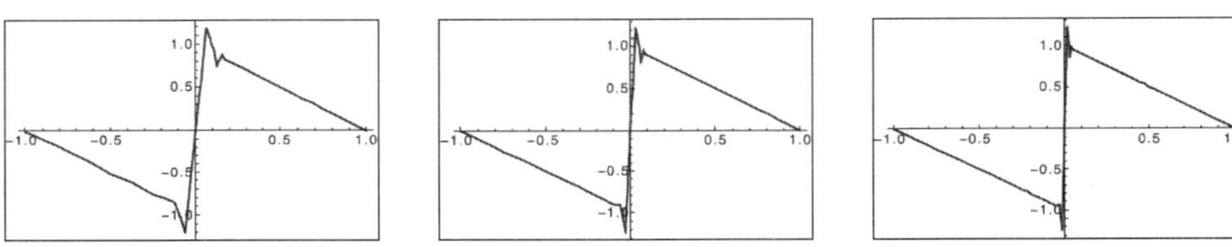

FIGURE 9. The overshoot and undershoot for $S_n^{2,6} f(x)$, $n = 3, 4$ and 5.

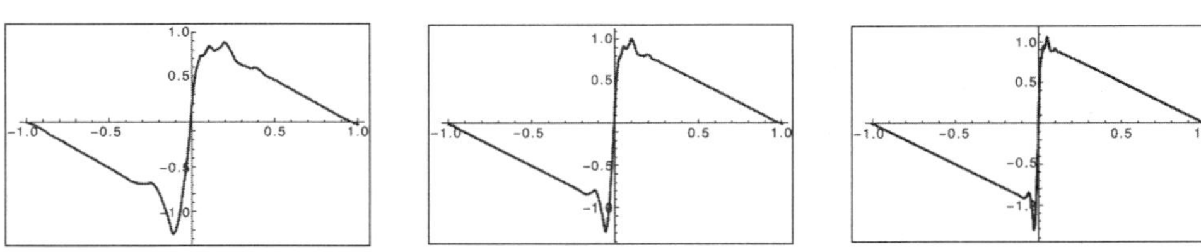

FIGURE 10. The overshoot and undershoot for $S_n^{2,8} f(x)$, $n = 2, 3$ and 4.

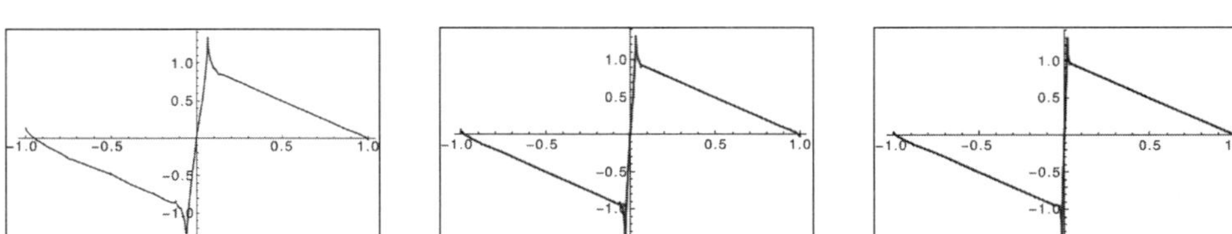

FIGURE 11. The overshoot and undershoot for the $S_n^{2,6} f(x)$

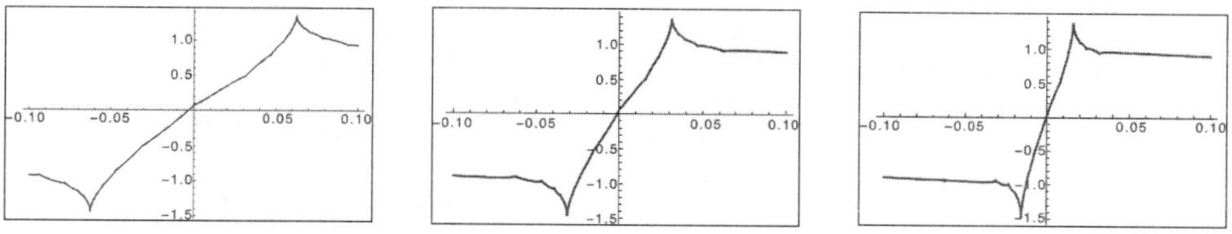

FIGURE 12. A closer view to the overshoot and undershoot of $S_n^{2,6}f\,(x)$

TABLE 5. Approximate maximum overshoot and undershoot near $x = 0$ using the truncated Coiflets sum $S_n^{2,6}f\,(x)$.

Level	*Maximum*	*Minimum*
$n = 3$	1.35422	-1.45369
$n = 4$	1.38898	-1.48304
$n = 5$	1.40574	-1.49756

Table 5 presents the values of the overshoot and undershoot for the jump $x = 0$. Figures 11 and 12 illustrate the Gibbs effect near 0, using $S_n^{2,6}f\,(x)$.

4. Conclusion

In this article, we showed numerically that Gibbs phenomenon didn't occur when the 1st Daubechies tight framelet system was used to expand a standard function having a jump discontinuity at the point $x = 0$. Also, for higher order cases we provided a numerical experiment of the existence of Gibbs effect using Coiflets and Daubechies tight framelet systems. This may provide some insights to future practical applications.

Acknowledgment

. The authors would like to thank Professor Susan Kelly from the University of Wisconsin-La Crosse for her helpful comments. We are very grateful to the anonymous referee's valuable comments and suggestions.

References

[Be] J. J. Benedetto and S. Li, *The theory of multiresolution analysis frames and applications to filter banks*, Appl. Comput. Harmon. Anal. **5** (1998), no. 4, 389–427, DOI 10.1006/acha.1997.0237. MR1646534

[Ch] O. Christensen, *An introduction to frames and Riesz bases*, Applied and Numerical Harmonic Analysis, Birkhäuser Boston, Inc., Boston, MA, 2003. MR1946982

[Da1] I. Daubechies, *Orthonormal bases of compactly supported wavelets*, Comm. Pure Appl. Math. **41** (1988), no. 7, 909–996, DOI 10.1002/cpa.3160410705. MR951745

[Da2] I. Daubechies, *Ten lectures on wavelets*, CBMS-NSF Regional Conference Series in Applied Mathematics, vol. 61, Society for Industrial and Applied Mathematics (SIAM), Philadelphia, PA, 1992. MR1162107

[DHRS] I. Daubechies, B. Han, A. Ron, and Z. Shen, *Framelets: MRA-based constructions of wavelet frames*, Appl. Comput. Harmon. Anal. **14** (2003), no. 1, 1–46, DOI 10.1016/S1063-5203(02)00511-0. MR1971300

[DS] R. J. Duffin and A. C. Schaeffer, *A class of nonharmonic Fourier series*, Trans. Amer. Math. Soc. **72** (1952), 341–366, DOI 10.2307/1990760. MR0047179

[GN] R. Gribonval and M. Nielsen, *On approximation with spline generated framelets*, Constr. Approx. **20** (2004), no. 2, 207–232, DOI 10.1007/s00365-003-0532-7. MR2036641

[Je] A. J. Jerri, *The Gibbs phenomenon in Fourier analysis, splines and wavelet approximations*, Mathematics and its Applications, vol. 446, Kluwer Academic Publishers, Dordrecht, 1998. MR1650415

[Ke] S. E. Kelly, *Gibbs phenomenon for wavelets*, Appl. Comput. Harmon. Anal. **3** (1996), no. 1, 72–81, DOI 10.1006/acha.1996.0006. MR1374397

[Le] K. Levenberg, *A method for the solution of certain non-linear problems in least squares*, Quart. Appl. Math. **2** (1944), 164–168, DOI 10.1090/qam/10666. MR0010666

[Lo] M.I.A. Lourakis. *A brief description of the Levenberg-Marquardt algorithm implemented by levmar,* Technical Report, Institute of Computer Science, Foundation for Research and Technology - Hellas, 2005.

[ML] M. Mohammad and E.-B. Lin, *Gibbs phenomenon in tight framelet expansions*, Commun. Nonlinear Sci. Numer. Simul. **55** (2018), 84–92, DOI 10.1016/j.cnsns.2017.06.029. MR3693373

[Ri] F. B. Richards, *A Gibbs phenomenon for spline functions*, J. Approx. Theory **66** (1991), no. 3, 334–351, DOI 10.1016/0021-9045(91)90034-8. MR1122292

[Ra] K. Raeen, *A Study of the Gibbs Phenomenon in Fourier Series and Wavelets*, M.S. Thesis, Mathematics Department, The University of New Mexico. (2008).

[RS] A. Ron and Z. Shen, *Affine systems in $L_2(\mathbf{R}^d)$: the analysis of the analysis operator*, J. Funct. Anal. **148** (1997), no. 2, 408–447, DOI 10.1006/jfan.1996.3079. MR1469348

[Wa] D. F. Walnut, *An introduction to wavelet analysis*, Applied and Numerical Harmonic Analysis, Birkhäuser Boston, Inc., Boston, MA, 2002. MR1854350

Department of Mathematics, Zayed University, P.O. Box 144534, Abu Dhabi, United Arab Emirates

Email address: Mutaz.Mohammad@zu.ac.ae

Department of Mathematics, Central Michigan University, Mt. Pleasant, Michigan 48859

Email address: lin1e@cmich.edu

Contemporary Mathematics
Volume **706**, 2018
http://dx.doi.org/10.1090/conm/706/14206

Conditions on shape preserving of stationary polynomial reproducing subdivision schemes

Yeon Hyang Kim

ABSTRACT. A stationary subdivision scheme reproducing polynomials is called a quasi-interpolatory scheme. In this paper, we study local conditions on the initial control points which guarantee preservation of monotonicity and convexity for a class of quasi-interpolatory subdivision schemes.

1. Introduction

A subdivision scheme is a set of rules by which a series of refinements are obtained from the initial control points. Subdivision schemes are a powerful tool for generating smooth curves and surfaces. These methods are simple, have fast algorithms and provide efficient mathematical representations. Thus subdivision schemes are widely used in modeling, image processing, computer-aided geometric design, and in constructing wavelet bases, [**3**, **5**, **9**]. They continue to gain increased interest because of their relation to multiresolution analysis and wavelets. As a general principle, these methods of construction are required to be local so that changes in a control point affect only a limited local area.

Given a set of initial control points $f^0 = \{ f_n^0 : n \in \mathbb{Z} \}$, let

$$f^k = \{ f_n^k : n \in \mathbb{Z} \}$$

be the set of control points at the kth level of refinement:

$$f^k = S^k f^0, \quad k \geq 0,$$

where $S : \ell(\mathbb{Z}) \to \ell(\mathbb{Z})$ is the subdivision operator. We say that the scheme S converges if for any initial data f^0, there exists $f \in C^m$, for some $m \geq 0$, such that

$$\lim_{k\to\infty} \sup_{j\in\mathbb{Z}} |f(2^{-k}j) - f_j^k| = 0,$$

and $f \neq 0$ for some initial data f^0. In this case we denote $S^\infty f^0 = f$, and we say that the scheme is C^m. If the refinement rule is the same at all levels and positions of iteration, the scheme is called a stationary subdivision scheme. A stationary subdivision scheme reproducing polynomials is called a quasi-interpolatory scheme.

2010 *Mathematics Subject Classification.* Primary 39B12.

Key words and phrases. Subdivision scheme, quasi-interpolant, polynomial reproduction, monotonicity preservation, convexity preservation.

This work was supported by the Central Michigan University FRCE Research Type A Grant #C48143.

A quasi-interpolatory scheme is an ordinary subdivision scheme whose mask has certain linear-phase moments, [**7**, **8**]. Much of the theory of subdivision schemes consists in deriving the convergence conditions of a given scheme. Choi et al. introduced binary five-point quasi-interpolatory subdivision schemes, which reproduce polynomials up to a certain degree. These schemes extend and unify not only the well-known Deslauriers-Dubuc interpolatory schemes but also the quadratic and cubic B-spline schemes, [**4**].

Besides convergence and smoothness, shape preserving is also an important topic in curves design, [**1**, **6**, **10**, **11**]. In many applications, one often encounters the problems of monotonicity preservation and convexity preservation. These problems concern fundamental features considered in many applications such as in generating a limit curve without introducing oscillations in a monotonic or convex portion of the initial data. Generally, a four-point interpolating subdivision scheme cannot preserve convexity. Dyn et al. [**6**] discussed conditions in which convexity is preserved, Cai [**1**] gave a convex preserving algorithm for a four-point binary scheme and published convexity preserving conditions for four-point ternary interpolatory subdivision schemes,[**2**]. In [**4**], the authors introduced the following binary five-point quasi-interpolatory subdivision scheme:

$$\begin{cases} f_{2i}^{k+1} = -\omega f_{i-2}^k + 4\omega f_{i-1}^k + (1-6\omega) f_i^k + 4\omega f_{i+1}^k - \omega f_{i+2}^k \\ f_{2i+1}^{k+1} = -\frac{1}{16} f_{i-1}^k + \frac{9}{16} f_i^k + \frac{9}{16} f_{i+1}^k - \frac{1}{16} f_{i+2}^k \end{cases} \tag{1}$$

The limit of this subdivision scheme is C^2 for the parameter $\omega \in (0.020262, 0.044039)$, [**4**]. In this paper we study local conditions imposed on the initial control points for this scheme in order to preserve monotonicity and convexity.

2. Conditions for shape preservaton

Throughout this paper, we let $\omega_1 := 0.020262$, $\omega_2 := 0.044039$. For the monotonicity preservation, we denote the first-order difference by

$$D_i^k := f_{i+1}^k - f_i^k.$$

Then for each $k \geq 0$, $i \in Z$ we have

$$D_{2i}^{k+1} = -\omega D_{i-2}^k + \left(\frac{1}{16} + 3\omega\right) D_{i-1}^k + \left(\frac{1}{2} - 3\omega\right) D_i^k + \left(-\frac{1}{16} + \omega\right) D_{i+1}^k,$$

$$D_{2i+1}^{k+1} = \left(-\frac{1}{16} + \omega\right) D_{i-1}^k + \left(\frac{1}{2} - 3\omega\right) D_i^k + \left(-\frac{1}{16} + 3\omega\right) D_{i+1}^k - \omega D_{i+2}^k.$$

The main result for monotonicity preservation is as follows:

THEOREM 2.1. *Assume that the initial control points are all strictly monotone, i.e., $D_i^0 > 0$. Denote*

$$q_i^k = \frac{D_{i+1}^k}{D_i^k}, \quad Q^k = \max_i \left\{ q_i^k, \frac{1}{q_i^k} \right\}, \quad k \geq 0,\ i \in \mathbb{Z},$$

and let $\omega \in (\omega_1, \omega_2)$. If $\frac{1}{\mu} \leq Q^0 \leq \mu$, where $\mu = \min\left\{3 + \frac{1}{16w}, 3 + 2\sqrt{2}\right\}$, then the binary five-point quasi-interpolatory subdivision scheme (1) preserves monotonicity.

PROOF. We use mathematical induction to prove $D_i^k > 0$ and $\frac{1}{\mu} \leq Q^k \leq \mu$. Suppose that $D_i^k > 0$ and $\frac{1}{\mu} \leq Q^k \leq \mu$.

$$\begin{aligned}
&\frac{D_{2i}^{k+1}}{D_{i-1}^k} \\
= \;& -\omega \frac{1}{q_{i-2}^k} + \left(\frac{1}{16} + 3\omega\right) + \left(\frac{1}{2} - 3\omega\right) q_{i-1}^k + \left(-\frac{1}{16} + \omega\right) q_i^k q_{i-1}^k \\
\geq \;& -\omega\mu + \left(\frac{1}{16} + 3\omega\right) + \left[\left(\frac{1}{2} - 3\omega\right) + \left(-\frac{1}{16} + \omega\right) q_i^k\right] q_{i-1}^k \\
\geq \;& -\omega\mu + \left(\frac{1}{16} + 3\omega\right) + \left[\left(\frac{1}{2} - 3\omega\right) + \left(-\frac{1}{16} + \omega\right) \mu\right] q_{i-1}^k \\
> \;& 0,
\end{aligned}$$

since $\mu \leq 3 + \frac{1}{16w}$ and $\left(\frac{1}{2} - 3\omega\right) + \left(-\frac{1}{16} + \omega\right)\mu > 0$. This implies that $D_{2i}^{k+1} > 0$. Similarly, we have

$$\begin{aligned}
&\frac{D_{2i+1}^{k+1}}{D_{i+1}^k} \\
= \;& \left(-\frac{1}{16} + \omega\right) \frac{1}{q_i^k q_{i-1}^k} + \left(\frac{1}{2} - 3\omega\right) \frac{1}{q_i^k} + \left(\frac{1}{16} + 3\omega\right) - \omega q_{i+1}^k > 0.
\end{aligned}$$

Now we prove $q_{2i}^{k+1} \leq \mu$.

$$\begin{aligned}
&\frac{D_{2i}^{k+1}}{D_{i-1}^k} \left(q_{2i}^{k+1} - \mu\right) \\
= \;& \omega\mu \frac{1}{q_{i-2}^k} + \left(-\frac{1}{16} + \omega\right) - \mu\left(\frac{1}{16} + 3\omega\right) + \left(\frac{1}{2} - 3\omega\right)(1 - \mu)\, q_{i-1}^k \\
& + \left[\left(\frac{1}{16} + 3\omega\right) - \mu\left(-\frac{1}{16} + \omega\right)\right] q_i^k q_{i-1}^k - \omega q_{i+1}^k q_i^k q_{i-1}^k \\
\leq \;& \omega\mu^2 + \left(-\frac{1}{16} + \omega\right) - \mu\left(\frac{1}{16} + 3\omega\right) + \left(\frac{1}{2} - 3\omega\right)(1 - \mu)\, q_{i-1}^k \\
& + \left[\left(\frac{1}{16} + 3\omega\right) - \mu\left(-\frac{1}{16} + \omega\right) - \omega q_{i+1}^k\right] q_i^k q_{i-1}^k =: I_1.
\end{aligned}$$

Since $q_{i+1}^k \leq \mu \leq 3 + \frac{1}{16w}$ and $\omega \in (\omega_1, \omega_2)$,

$$\left(\frac{1}{16} + 3\omega\right) - \omega q_{i+1}^k - \mu\left(-\frac{1}{16} + \omega\right) > 0,$$

which implies that

$$\begin{aligned}
I_1 \leq \;& \omega\mu^2 + \left(-\frac{1}{16} + \omega\right) - \mu\left(\frac{1}{16} + 3\omega\right) + \left(\frac{1}{2} - 3\omega\right)(1 - \mu)\, q_{i-1}^k \\
& + \left[\left(\frac{1}{16} + 3\omega\right) - \mu\left(-\frac{1}{16} + \omega\right) - \omega\frac{1}{\mu}\right] \mu q_{i-1}^k \\
= \;& \omega\mu^2 - \left(\frac{1}{16} + 3\omega\right)\mu + \left(-\frac{1}{16} + \omega\right) \\
& + \left[\left(\frac{1}{16} - \omega\right)\mu^2 + \left(-\frac{7}{16} + 6\omega\right)\mu + \frac{1}{2} - 4\omega\right] q_{i-1}^k =: I_2.
\end{aligned}$$

Since $\mu \le 3 + \frac{1}{16w}$ and $\omega \in (\omega_1, \omega_2)$,

$$\omega\mu^2 - \left(\frac{1}{16} + 3\omega\right)\mu + \left(-\frac{1}{16} + \omega\right) < 0.$$

Thus if $\left(\frac{1}{16} - \omega\right)\mu^2 + \left(-\frac{7}{16} + 6\omega\right)\mu + \frac{1}{2} - 4\omega \le 0$, then $q_{2i}^{k+1} \le \mu$. Let $\left(\frac{1}{16} - \omega\right)\mu^2 + \left(-\frac{7}{16} + 6\omega\right)\mu + \frac{1}{2} - 4\omega > 0$, then

$$\begin{aligned} I_2 &\le \omega\mu^2 - \left(\frac{1}{16} + 3\omega\right)\mu + \left(-\frac{1}{16} + \omega\right) \\ &\quad + \left[\left(\frac{1}{16} - \omega\right)\mu^2 + \left(-\frac{7}{16} + 6\omega\right)\mu + \frac{1}{2} - 4\omega\right]\mu \\ &= \left(\frac{1}{16} - \omega\right)(\mu - 1)(\mu^2 - 6\mu + 1) \le 0 \end{aligned}$$

since $1 \le \mu \le 3 + 2\sqrt{2}$, which implies that $q_{2i}^{k+1} \le \mu$. We next prove $q_{2i+1}^{k+1} \le \mu$.

$$\begin{aligned} &\frac{D_{2i+1}^{k+1}}{D_i^k}\left(q_{2i+1}^{k+1} - \mu\right) \\ &= \left[-\omega - \mu\left(-\frac{1}{16} + \omega\right)\right]\frac{1}{q_{i-1}^k} + \frac{1}{16} + 3\omega - \mu\left(\frac{1}{2} - 3\omega\right) \\ &\quad + \left[\frac{1}{2} - 3\omega - \mu\left(\frac{1}{16} + 3\omega\right) + \left(-\frac{1}{16} + \omega + \mu\omega\right)q_{i+1}^k\right]q_i^k =: I_3. \end{aligned}$$

We note that $-\omega - \mu\left(-\frac{1}{16} + \omega\right) \ge 0$ if $\omega \in (\omega_1, 1/32]$ and $-\frac{1}{16} + \omega + \mu\omega \ge 0$ if $\omega \in [1/32, \omega_2)$. If $-\frac{1}{16} + \omega + \mu\omega \ge 0$, then

$$\begin{aligned} &\frac{1}{2} - 3\omega - \mu\left(\frac{1}{16} + 3\omega\right) + \left(-\frac{1}{16} + \omega + \mu\omega\right)q_{i+1}^k \\ &\le \frac{1}{2} - 3\omega - \mu\left(\frac{1}{16} + 3\omega\right) + \left(-\frac{1}{16} + \omega + \mu\omega\right)\mu \end{aligned}$$

which is negative for $\omega \in (\omega_1, \omega_2)$. Thus if $-\omega - \mu\left(-\frac{1}{16} + \omega\right) \ge 0$, then

$$\begin{aligned} I_3 &\le \left[-\omega - \mu\left(-\frac{1}{16} + \omega\right)\right]\mu + \frac{1}{16} + 3\omega - \mu\left(\frac{1}{2} - 3\omega\right) \\ &\quad + \left[\frac{1}{2} - 3\omega - \mu\left(\frac{1}{16} + 3\omega\right) + \left(-\frac{1}{16} + \omega + \mu\omega\right)\mu\right]\mu \\ &= (\mu + 1)(\mu - 1)\left(\mu - 3 - \frac{1}{16\omega}\right) \le 0 \end{aligned}$$

since $1 \le \mu \le 3 + \frac{1}{16\omega}$.
If $-\omega - \mu\left(-\frac{1}{16} + \omega\right) < 0$, equivalently $\mu \le \frac{16\omega}{1-16\omega}$, then

$$\begin{aligned} I_3 &\le \left[-\omega - \mu\left(-\frac{1}{16} + \omega\right)\right]\frac{1}{\mu} + \frac{1}{16} + 3\omega - \mu\left(\frac{1}{2} - 3\omega\right) \\ &\quad + \left[\frac{1}{2} - 3\omega - \mu\left(\frac{1}{16} + 3\omega\right) + \left(-\frac{1}{16} + \omega + \mu\omega\right)\mu\right]\mu \\ &= \frac{1}{\mu}(\mu + 1)(\mu - 1)\left(\mu^2 - \left(2 + \frac{1}{8\omega}\right)\mu + \frac{1}{128\omega^2} + 1\right) \le 0 \end{aligned}$$

since $1 \le \mu \le \frac{16\omega}{1-16\omega}$. Finally, let $-\frac{1}{16}+\omega+\mu\omega<0$, then

$$\begin{aligned}&\frac{1}{2}-3\omega-\mu\left(\frac{1}{16}+3\omega\right)+\left(-\frac{1}{16}+\omega+\mu\omega\right)q_{i+1}^k\\ &\le \frac{1}{2}-3\omega-\mu\left(\frac{1}{16}+3\omega\right)+\left(-\frac{1}{16}+\omega+\mu\omega\right)\frac{1}{\mu}\end{aligned}$$

which is positive for $\omega \in (\omega_1, 1/32]$. Thus

$$\begin{aligned}I_3 &\le \left[-\omega-\mu\left(-\frac{1}{16}+\omega\right)\right]\mu+\frac{1}{16}+3\omega-\mu\left(\frac{1}{2}-3\omega\right)\\ &\quad+\left[\frac{1}{2}-3\omega-\mu\left(\frac{1}{16}+3\omega\right)+\left(-\frac{1}{16}+\omega+\mu\omega\right)\frac{1}{\mu}\right]\mu\\ &= -4\omega(\mu+1)(\mu-1)\le 0\end{aligned}$$

since $\mu \ge 1$. A consequence of all this is $q_{2i+1}^{k+1}\le\mu$ if $1\le\mu\le\min\left\{3+\frac{1}{16w}, 3+2\sqrt{2}\right\}$. A similar calculation shows that

$$q_{2i}^{k+1} \ge \frac{1}{\mu} \text{ and } q_{2i+1}^{k+1} \ge \frac{1}{\mu}.$$

This completes the proof. □

For convexity preservation, we denote the second-order difference by

$$d_i^k := f_{i+1}^k - 2f_i^k + f_{i+1}^k.$$

The main result for convexity preservation is as follows:

THEOREM 2.2. *Assume that the initial control points are all strictly convex, i.e.,* $d_i^0>0$. *Denote*

$$r_i^k = \frac{d_{i+1}^k}{d_i^k}, \quad R^k = \max_i\left\{r_i^k, \frac{1}{r_i^k}\right\}, \quad k\ge 0,\ i\in\mathbb{Z},$$

and let $\omega\in(\omega_1,\omega_2)$. *If* $\frac{1}{\mu}\le R^0\le\mu$, *where*

$$\mu = \begin{cases}\frac{1}{1-32\omega}, & \omega_1<\omega<\frac{-1+\sqrt{17}}{128}\\ \frac{3-48\omega+\sqrt{4352\omega^2-352\omega+9}}{32\omega}, & \frac{-1+\sqrt{17}}{128}\le\omega<\omega_2\end{cases},$$

then the binary five-point quasi-interpolatory subdivision scheme (1) *preserves convexity.*

PROOF. For simplicity, we denote

$$B(\omega) := \begin{cases}\frac{1}{1-32\omega}, & \omega_1<\omega<\frac{-1+\sqrt{17}}{128}\\ \frac{3-48\omega+\sqrt{4352\omega^2-352\omega+9}}{32\omega}, & \frac{-1+\sqrt{17}}{128}\le\omega<\omega_2\end{cases}.$$

We use mathematical induction to prove $d_i^k>0$ and $\frac{1}{\mu}\le R^k\le\mu$. Suppose that $d_i^k>0$ and $\frac{1}{\mu}\le R^k\le\mu$ then

$$d_{2i}^{k+1} = \left(-\frac{1}{16}+2\omega\right)d_{i+1}^k+\left(\frac{3}{8}-4\omega\right)d_i^k+\left(-\frac{1}{16}+2\omega\right)d_{i-1}^k.$$

For $\frac{1}{32} \le \omega < \omega_2$, since $-\frac{1}{16} + 2\omega \ge 0$ and $\frac{3}{8} - 4\omega > 0$, we have that $d_{2i}^{k+1} > 0$. Let $\omega \in (\omega_1, \frac{1}{32})$, then since $\frac{1}{\mu} \le R^k \le \mu$,

$$d_{2i}^{k+1} \ge d_i^k \left[\left(-\frac{1}{16} + 2\omega \right) 2\mu + \left(\frac{3}{8} - 4\omega \right) \right],$$

which is positive if $\mu < B(\omega)$. Also

$$\begin{aligned} d_{2i+1}^{k+1} &= d_i^k \left[-\omega r_{i+1}^k r_i^k + \left(\frac{1}{8} + \omega \right) r_i^k + \frac{1}{8} + \omega - \omega \frac{1}{r_{i-1}^k} \right] \\ &\ge d_i^k \left[r_i^k \left(-\omega\mu + \frac{1}{8} + \omega \right) + \frac{1}{8} + \omega - \omega\mu \right] > 0 \end{aligned}$$

since $\frac{1}{8} + \omega - \omega\mu > 0$ if $\mu < B(\omega)$.

Now we prove $r_{2i}^{k+1} \le \mu$. We have

$$\begin{aligned} &\frac{d_{2i}^{k+1}}{d_i^k} \left(r_{2i}^{k+1} - \mu \right) \\ = \ & r_i^k \left[-\omega r_{i+1}^k + \frac{1}{8} + \omega - \mu \left(-\frac{1}{16} + 2\omega \right) \right] \\ & + \frac{1}{8} + \omega - \mu \left(\frac{3}{8} - 4\omega \right) + \left[-\omega - \mu \left(-\frac{1}{16} + 2\omega \right) \right] \frac{1}{r_{i-1}^k} =: I_4. \end{aligned}$$

Since $-\omega \frac{1}{\mu} + \frac{1}{8} + \omega - \mu \left(-\frac{1}{16} + 2\omega \right) > 0$ because of our assumption on μ, we have

$$\begin{aligned} I_4 \le \ & \mu \left[-\omega \frac{1}{\mu} + \frac{1}{8} + \omega - \mu \left(-\frac{1}{16} + 2\omega \right) \right] \\ & + \frac{1}{8} + \omega - \mu \left(\frac{3}{8} - 4\omega \right) + \left[-\omega - \mu \left(-\frac{1}{16} + 2\omega \right) \right] \frac{1}{r_{i-1}^k}. \end{aligned}$$

When $-\omega - \mu \left(-\frac{1}{16} + 2\omega \right) \le 0$,

$$\begin{aligned} I_4 &\le \mu \left[-\omega \frac{1}{\mu} + \frac{1}{8} + \omega - \mu \left(-\frac{1}{16} + 2\omega \right) \right] \\ & \quad + \frac{1}{8} + \omega - \mu \left(\frac{3}{8} - 4\omega \right) + \left[-\omega - \mu \left(-\frac{1}{16} + 2\omega \right) \right] \frac{1}{\mu} \\ &= \frac{\mu - 1}{16\mu} \left[(-32\omega + 1)\mu^2 + (48\omega - 3)\mu + 16\omega \right], \end{aligned}$$

which is less than or equal to zero because of our assumption on μ. Let $-\omega - \mu \left(-\frac{1}{16} + 2\omega \right) > 0$, then $\omega_1 < \omega < \frac{-3+\sqrt{33}}{96}$ and

$$\begin{aligned} I_4 &\le \mu \left[-\omega \frac{1}{\mu} + \frac{1}{8} + \omega - \mu \left(-\frac{1}{16} + 2\omega \right) \right] \\ & \quad + \frac{1}{8} + \omega - \mu \left(\frac{3}{8} - 4\omega \right) + \left[-\omega - \mu \left(-\frac{1}{16} + 2\omega \right) \right] \mu \\ &= \frac{\mu - 1}{8} \left(-32\omega\mu + \mu - 1 \right), \end{aligned}$$

which is less than or equal to zero because of our assumption on μ for $\omega \in (\omega_1, \frac{-3+\sqrt{33}}{96})$. Thus in either case, $r_{2i}^{k+1} \le \mu$.

Next we show that $r_{2i+1}^{k+1} \le \mu$. We have

$$\begin{aligned} & \frac{d_{2i+1}^{k+1}}{d_i^k}\left(r_{2i+1}^{k+1}-\mu\right) \\ = \;& r_i^k\left[\left(-\frac{1}{16}+2\omega+\mu\omega\right)r_{i+1}^k+\frac{3}{8}-4\omega-\mu\left(\frac{1}{8}+\omega\right)\right] \\ & -\frac{1}{16}+2\omega-\mu\left(\frac{1}{8}+\omega\right)+\mu\omega\frac{1}{r_i^k} \\ \le \;& r_i^k\left[\left(-\frac{1}{16}+2\omega+\mu\omega\right)r_{i+1}^k+\frac{3}{8}-4\omega-\mu\left(\frac{1}{8}+\omega\right)\right] \\ & -\frac{1}{16}+2\omega-\mu\left(\frac{1}{8}+\omega\right)+\mu^2\omega =: I_5. \end{aligned}$$

There are four cases to consider, depend on the positivities and negativities of

$$-\frac{1}{16}+2\omega+\mu\omega \text{ and } \left(-\frac{1}{16}+2\omega+\mu\omega\right)r_{i+1}^k+\frac{3}{8}-4\omega-\mu\left(\frac{1}{8}+\omega\right).$$

Since $\mu \ge 1$, we note that $-\frac{1}{16}+2\omega+\mu\omega \ge 0$ if $\omega > \frac{1}{48}$.

Case 1. $-\frac{1}{16}+2\omega+\mu\omega \ge 0$ and $\left(-\frac{1}{16}+2\omega+\mu\omega\right)\mu+\frac{3}{8}-4\omega-\mu\left(\frac{1}{8}+\omega\right) \ge 0$. Then

$$\begin{aligned} I_5 \;\le\;& \left(-\frac{1}{16}+2\omega+\mu\omega\right)\mu^2+\left[\frac{3}{8}-4\omega-\mu\left(\frac{1}{8}+\omega\right)\right]\mu \\ & -\frac{1}{16}+2\omega-\mu\left(\frac{1}{8}+\omega\right)+\mu^2\omega \\ =\;& \frac{1}{16}(\mu-1)\left[16\omega\mu^2+(48\omega-3)\mu-32\omega+1\right] \le 0, \end{aligned}$$

if $1 \le \mu \le \frac{3-48\omega+\sqrt{4352\omega^2-352\omega+9}}{32\omega}$.

Case 2. $-\frac{1}{16}+2\omega+\mu\omega \ge 0$ and $\left(-\frac{1}{16}+2\omega+\mu\omega\right)\mu+\frac{3}{8}-4\omega-\mu\left(\frac{1}{8}+\omega\right) < 0$. Note that $\omega \ge \frac{1}{48}$ implies that $\left(-\frac{1}{16}+2\omega+\mu\omega\right)\mu+\frac{3}{8}-4\omega-\mu\left(\frac{1}{8}+\omega\right) \ge 0$. Thus we only consider for $\omega_1 < \omega < \frac{1}{48}$.

$$\begin{aligned} I_5 \;\le\;& \left(-\frac{1}{16}+2\omega+\mu\omega\right)+\left[\frac{3}{8}-4\omega-\mu\left(\frac{1}{8}+\omega\right)\right]\frac{1}{\mu} \\ & -\frac{1}{16}+2\omega-\mu\left(\frac{1}{8}+\omega\right)+\mu^2\omega \\ =\;& \frac{1}{8\mu}(\mu-1)\left[8\omega\mu^2+(8\omega-1)\mu+32\omega-3\right] \le 0, \end{aligned}$$

since $8\omega\mu^2+(8\omega-1)\mu+32\omega-3 \le 0$ when $\omega_1 < \omega < \frac{1}{48}$.

Case 3. $-\frac{1}{16}+2\omega+\mu\omega < 0$ and $\left(-\frac{1}{16}+2\omega+\mu\omega\right)\mu+\frac{3}{8}-4\omega-\mu\left(\frac{1}{8}+\omega\right) \ge 0$. Then

$$\begin{aligned} I_5 \;\le\;& \left(-\frac{1}{16}+2\omega+\mu\omega\right)+\left[\frac{3}{8}-4\omega-\mu\left(\frac{1}{8}+\omega\right)\right]\mu \\ & -\frac{1}{16}+2\omega-\mu\left(\frac{1}{8}+\omega\right)+\mu^2\omega \\ =\;& \frac{1}{8}(\mu-1)(-\mu-32\omega+1) \le 0, \end{aligned}$$

since $\mu \geq 1$.

Case 4. $-\frac{1}{16}+2\omega+\mu\omega<0$ and $\left(-\frac{1}{16}+2\omega+\mu\omega\right)\mu+\frac{3}{8}-4\omega-\mu\left(\frac{1}{8}+\omega\right)<0$. Since $-\frac{1}{16}+2\omega+\mu\omega<0$, we only consider for $\omega_1<\omega<\frac{1}{48}$.

$$\begin{aligned} I_5 &\leq \left(-\frac{1}{16}+2\omega+\mu\omega\right)\frac{1}{\mu^2}+\left[\frac{3}{8}-4\omega-\mu\left(\frac{1}{8}+\omega\right)\right]\frac{1}{\mu} \\ &\quad -\frac{1}{16}+2\omega-\mu\left(\frac{1}{8}+\omega\right)+\mu^2\omega \\ &= \frac{1}{16\mu^2}(\mu-1)\left[16\omega\mu^3-2\mu^2+(16\omega-5)\mu-32\omega+1\right]=:I_6. \end{aligned}$$

Let $f(\mu)=16\omega\mu^3-2\mu^2+(16\omega-5)\mu-32\omega+1$. Then f has its maxima at $\frac{1-\sqrt{-192\omega^2+60\omega+1}}{24\omega}$ and minima at $\frac{1+\sqrt{-192\omega^2+60\omega+1}}{24\omega}$. However, for $\omega_1<\omega<\frac{1}{48}$, $\frac{1-\sqrt{-192\omega^2+60\omega+1}}{24\omega}<0$ and $\frac{1+\sqrt{-192\omega^2+60\omega+1}}{24\omega}>5$. This together with $f(1)=-6$ implies that $I_6\leq 0$ in the interval $[1,5]$. A consequence of all this is $r_{2i+1}^{k+1}\leq\mu$ if $1\leq\mu\leq B(\omega)$. A similar calculation shows that

$$r_{2i}^{k+1}\geq\frac{1}{\mu},\text{ and } r_{2i+1}^{k+1}\geq\frac{1}{\mu}.$$

This completes the proof. □

Acknowledgments

The author thanks to Dr. B. Jeong for many helpful discussions during his visit to Central Michigan University. The author also thanks the referee for valuable comments.

References

[1] Z. Cai, *Four-point scheme and convexity-preserving algorithm*, Chin. J. CAD & CG, **6(1)** (1994), 33–36.

[2] Z. Cai, *Convexity preservation of the interpolating four-point C^2 ternary stationary subdivision scheme*, Comput. Aided Geom. Design **26** (2009), no. 5, 560–565, DOI 10.1016/j.cagd.2009.02.004. MR2526011

[3] A. S. Cavaretta, W. Dahmen, and C. A. Micchelli, *Stationary subdivision*, Mem. Amer. Math. Soc. **93** (1991), no. 453, vi+186, DOI 10.1090/memo/0453. MR1079033

[4] S. W. Choi, B.-G. Lee, Y. J. Lee, and J. Yoon, *Stationary subdivision schemes reproducing polynomials*, Comput. Aided Geom. Design **23** (2006), no. 4, 351–360, DOI 10.1016/j.cagd.2006.01.003. MR2219936

[5] N. Dyn, J. A. Gregory, and D. Levin, *Analysis of uniform binary subdivision schemes for curve design*, Constr. Approx. **7** (1991), no. 2, 127–147, DOI 10.1007/BF01888150. MR1101059

[6] N. Dyn, F. Kuijt, D. Levin, and R. van Damme, *Convexity preservation of the four-point interpolatory subdivision scheme*, Comput. Aided Geom. Design **16** (1999), no. 8, 789–792, DOI 10.1016/S0167-8396(99)00019-9. MR1706763

[7] B. Han, *Vector cascade algorithms and refinable function vectors in Sobolev spaces*, J. Approx. Theory **124** (2003), no. 1, 44–88, DOI 10.1016/S0021-9045(03)00120-5. MR2010780

[8] B. Han, *Properties of discrete framelet transforms*, Math. Model. Nat. Phenom. **8** (2013), no. 1, 18–47, DOI 10.1051/mmnp/20138102. MR3022977

[9] M. F. Hassan, I. P. Ivrissimitzis, N. A. Dodgson, and M. A. Sabin, *An interpolating 4-point C^2 ternary stationary subdivision scheme*, Comput. Aided Geom. Design **19** (2002), no. 1, 1–18, DOI 10.1016/S0167-8396(01)00084-X. MR1879678

[10] J.-L. Merrien and P. Sablonnière, *Monotone and convex C^1 Hermite interpolants generated by an adaptive subdivision scheme* (English, with English and French summaries), C. R. Acad. Sci. Paris Sér. I Math. **333** (2001), no. 5, 493–497, DOI 10.1016/S0764-4442(01)02072-9. MR1859243

[11] A. Le Méhauté and F. I. Utreras, *Convexity-preserving interpolatory subdivision*, Comput. Aided Geom. Design **11** (1994), no. 1, 17–37, DOI 10.1016/0167-8396(94)90023-X. MR1257702

Department of Mathematics, Central Michigan University, Mt. Pleasant, Michigan 48859

Email address: kim4y@cmich.edu

Contemporary Mathematics
Volume **706**, 2018
http://dx.doi.org/10.1090/conm/706/14219

W-Markov measures, transfer operators, wavelets and multiresolutions

Daniel Alpay, Palle Jorgensen, and Izchak Lewkowicz

ABSTRACT. In a general setting we solve the following inverse problem: Given a positive operator R, acting on measurable functions on a fixed measure space $(X, \mathcal{B}_X)$, we construct an associated Markov chain. Specifically, starting with a choice of R (the transfer operator), and a probability measure μ_0 on $(X, \mathcal{B}_X)$, we then build an associated Markov chain $T_0, T_1, T_2, \ldots$, with these random variables (r.v) realized in a suitable probability space $(\Omega, \mathcal{F}, \mathbb{P})$, and each r.v. taking values in X, and with T_0 having the probability μ_0 as law. We further show how spectral data for R, e.g., the presence of R-harmonic functions, propagate to the Markov chain. Conversely, in a general setting, we show that every Markov chain is determined by its transfer operator. In a range of examples we put this correspondence into practical terms: (i) iterated function systems (IFS), (ii) wavelet multiresolution constructions, and (iii) IFSs with random "control." Our setting for IFSs is general as well: a fixed measure space $(X, \mathcal{B}_X)$ and a system of mappings τ_i, each acting in $(X, \mathcal{B}_X)$, and each assigned a probability, say p_i which may or may not be a function of x. For standard IFSs, the p_i's are constant, but for wavelet constructions, we have functions $p_i(x)$ reflecting the multi-band filters which make up the wavelet algorithm at hand. The sets $\tau_i(X)$ partition X, but they may have overlap, or not. For IFSs with random control, we show how the setting of transfer operators translates into explicit Markov moves: Starting with a point $x \in X$, the Markov move to the next point is in two steps, combined yielding the move from $T_0 = x$ to $T_1 = y$, and more generally from T_n to T_{n+1}. The initial point x will first move to one of the sets $\tau_i(X)$ with probability p_i, and once there, it will "choose" a definite position y (within $\tau_i(X)$), now governed by a fixed law (a given probability distribution). For Markov chains, the law is the same in each move from T_n to T_{n+1}.

CONTENTS

2010 *Mathematics Subject Classification.* 37C30, 46L55, 47B65, 60J05, 60J10, 65T60.
Key words and phrases. Transfer operator, Markov chains, solenoid, wavelet multiresolution.
D. Alpay thanks the Foster G. and Mary McGaw Professorship in Mathematical Sciences, which supported his research.

1. Introduction

The purpose of our paper is to explore in two directions the interconnection between positive operators R defined in certain function spaces, on the one hand, and associated discrete time-random processes on the other. The direction back from R to the discrete time-random process, we refer to as "the inverse problem." It includes the construction of the process itself. By contrast, the direct problem starts with a given discrete-time random process, and then computes the associated transfer operator, or sequence of transfer operators, and then finally uses the latter in order to determine properties of the given random process under consideration.

Our second purpose is a list of applications of our results in the general setting, the applications ranging from homogeneous Markov chains with white noise-input, dynamics of endomorphisms, including logistics maps, encoding mappings, invariant measures, wavelets in a general setting of multi-resolutions and associated transfer operators, also called Ruelle operators. In the case of a single positive operators R, we obtain, via a solution to the inverse problem, an associated generalized Markov processes, but its detailed properties will depend on a prescribed weight function W, hence the term "W-Markov processes." In the case of a prescribed sequence of positive operators, we still obtain associated discrete time-random processes, now with each operator R_n accounting for the transfer of information from time n to time $n+1$. But these processes will not be Markov. Hence the Markov property is equivalent to $R_n = R$ for all n.

Returning to the case of our study of dynamics of endomorphisms, say σ in X, if the transfer operator R is σ-homogeneous, we show that the associated Markov processes will be of a special kind: when realized in the natural probability space of an associated solenoid $\mathrm{Sol}_\sigma(X)$ (see Definition 3.7 for the latter), we arrive at multi-scale resolutions in $\mathbf{L}_2(\mathrm{Sol}_\sigma(X), \mathcal{F}, \mathbb{P})$ (see Definition 3.16), with the scale of resolutions in question defined from the given endomorphism σ. In the case when σ is the scale endomorphism of a wavelet construction, we show that the wavelet multi-scale resolution will agree with that of the associated solenoid analysis. The latter framework is much more general, and covers a variety of multiresolution models.

TABLE 1. Increasing level of generality (each with its transfer operator and multiresolution; see Tables 2 and 4)

Case	$\mathbf{L}_2(\mathbb{R}, dx)$ $\longrightarrow$	$\mathbf{L}_2(\mathrm{Sol}_\sigma(X), \mathbb{P})$ $\longrightarrow$	$\mathbf{L}_2(\Omega, \mathbb{P})$

Before turning to the third theme in our paper, a few words on terminology: by a measure space $(X, \mathcal{B}_X)$ we mean a set X and a sigma-algebra $\mathcal{B}_X$ of subsets, each specified at the outset, usually with some additional technical restrictions. By a probability space, we mean a triple $(\Omega, \mathcal{F}, \mathbb{P})$, sample space Ω, sigma-algebra of events $\mathcal{F}$, and probability measure $\mathbb{P}$. We shall consider systems of random variables with values in measure spaces $(X, \mathcal{B}_X)$; different random variables may take values in different measure spaces. Our first order of business is to show that for any pair of random variables, say A and B, each taking values in a measure space, there is an associated transfer operator R, depending only on A and B, which "transfers"" information from one to the other. If A and B are independent, the associated operator R will be of rank-one, while if the sigma algebra generated by A is contained in that of B, then R will be the inclusion operator of the L_2-spaces of the respective distributions, the distribution of A and that of B.

One source of motivation for our present work is a number of recent papers dealing with generalized wavelet multiresolutions, see e.g., [**7, 34, 40, 41, 48, 55, 63**], and harmonic analysis on groupoids. While these themes may seem disparate, they are connected via a set of questions in operator algebra theory; see e.g., [**28, 45, 46**]. The positive operators considered here are in a general measure theoretic setting, but we stress that there is also a rich theory of positive integral operators is the metric space setting, often called Mercer operators, and important in the approach of Smale and collaborators to learning theory, see e.g., [**22, 61, 68**]. However for our present use, the setting of the Mercer operators is too restrictive.

While various aspects of our settings may have appeared in special cases in anyone or the other of existing treatments of Markov chains, the level of generality, the questions addressed, and the specific and detailed interconnections, some surprising, revealed below, we believe have not. Relevant references include [**20, 36, 38**] and the papers cited therein.

Aside from the Introduction, the paper is divided into three sections. Since our approach to the applications involves some issues of a general nature, we found it best to begin with general theory, Section 2, covering a number of new results, all based on several intriguing operator theoretic features of general systems of random variables, and their associated transfer operators. This is developed first, and its relevance to discrete-time random processes is then covered in the remaining of Section 2. From there, we then turn to Markov chains, developed in this rather general and operator theoretic framework, and with an emphasis on transfer operator related issues. It is our hope that this will be of interest to readers both in operator theory, and in random dynamical systems and their harmonic analysis. We have thus postponed the applications to the last section. This is dictated in part by our focus on those Markov chains and associated dynamical systems which are induced by endomorphisms in measure spaces. In Section 3 we show that this setting can be realized in probability spaces over solenoids. Each endomorphism induces a solenoid, and a Markov chain of a special kind. The usefulness of this point of view is then documented with a host of applications and detailed examples which we have included in several subsections in Section 4. We believe that our results in both the general theory and in our applications sections are of independent interest.

2. General theory

In this section, we consider the following general setting of random variables systems (r.v.s) on a prescribed probability space $(\Omega, \mathcal{F}, \mathbb{P})$, each r.v. taking values in a measure space $(X, \mathcal{B}_X)$; different random variables may take values in different measure spaces. Our aim is to make precise transfer between the different r.v.s making up the system. For this purpose we concentrate on the case of a pair of r.v.s, say A and B. There is then an associated transfer operator $R = R_{A,B}$, depending only on A and B, which "transfers"" information from one to the other.

The transfer operator makes precise the "intertwining" of the two random variables. Indeed, if A and B are in fact given to be independent, then the associated operator R will be of rank-one, or zero in the case of zero means. On the other hand, if the sigma algebra generated by A is contained in that of B, then R will be the inclusion operator of the L_2-spaces of the respective distributions, i.e., the distribution of A and that of B. We further show, in the general setting, that the product of the respective conditional expectations (the one for A and the one for B) are linked, via a factorization formula, by the transfer operator $R_{A,B}$. See Table 3 below.

While Section 2 is somewhat long and technical, it serves two important purposes: one, it offers lemmas to be used in the proofs of our main theorems later. The second purpose is to develop the tools we need in several inductive limit constructions to be used in our analysis of inverse problems, the inductive limits here concern the step of realizing infinite-dimensional discrete time-random processes as inductive limits of finite systems. For the finite systems themselves we develop here (the first five lemmas in Section 2) a new kernel analysis which will then be used later when we build the infinite dimensional probability models needed in the main theorems. As mentioned, a key tool is the notion of a transfer operator for a pair (or a finite number of) random variables. We shall include an analysis of the special case when one of the two r.v.s takes values in a discrete measure space. There are two reasons for this, one the interest in Markov chains with discrete state space, and the other is the study of such random variables as stopping time (see Definition 2.26).

Our approach to the analysis of finite systems of r.v.s is operator theoretic, relying on systems of isometries, co-isometries and projections, the latter in the form of conditional expectations. Of independent interest is our Corollary 2.36 which offers a representation of some operator relations known as the Cuntz-Krieger relations in operator algebra theory. Lemmas 2.2, 2.4, 2.9, 2.17, and 2.38 prepare the ground for what is to follow. Main results in the section includes Theorems 2.7, 2.19, 2.29, 2.30, and 2.39, as well as their corollaries and applications. Theorem 2.19 offers a model for the analysis of Markov processes in the general setting of our paper, Theorem 2.29 is a result which supplies a model for Markov chains driven by white noise. In this case we also compute an explicit invariant measure. This in turn is applied (Theorem 2.39) to a new random process realized naturally in a probability space over the Schur functions from complex analysis. Background references on calculus of random variables include [**26, 38, 47, 53, 60, 62**]; on classes of positive operators (Ruelle operators) [**14, 29, 43, 44**]; and on algebras of operators in Hilbert space [**8, 23, 24, 45, 50, 54, 55, 58, 64**].

2.1. Pairs of random variables and transfer operators. Let $(X, \mathcal{B}_X)$ be a measurable space. In this section, we define a transfer operator associated with two X-valued random variables, say A and B, defined on some probability space $(\Omega, \mathcal{F}, \mathbb{P})$. The distribution probability of A (also called "law") is defined by

$$\mu_A(L) = \mathbb{P}(A^{-1}(L)), \quad L \in \mathcal{B}_X,$$

and so, with $\mathcal{M}(X,\mathcal{B}_X)$ denoting the space of real-valued measurable functions defined on X,

$$\int_\Omega f(A(\omega))d\mathbb{P}(\omega) = \int_X f(x)d\mu_A(x), \quad \forall f \in \mathcal{M}(X,\mathcal{B}_X),$$

(and similarly for B).

DEFINITION 2.1. We denote by $\mathcal{F}_A$ the sub sigma-algebra of $\mathcal{F}$ defined by

$$\mathcal{F}_A = \left\{A^{-1}(L)\,;\, L \in \mathcal{B}\right\}. \tag{2.1}$$

By definition of μ_A, and with $\mathcal{F}_A$ introduced in Definition 2.1, the map

$$V_A f = f \circ A \tag{2.2}$$

is an isometry from $\mathbf{L}_2(X,\mathcal{B}_X,\mu_A)$ onto $\mathbf{L}_2(\Omega,\mathcal{F}_A,\mathbb{P})$. For the adjoint operator V_A^* we have the following covariance (in a sense analogue to the one in mathematical physics and representation theory).

In the following lemma and in the sequel, by $\mathbb{E}(\cdot \,|\, \mathcal{F}_A)$ we mean the orthogonal projection of $\mathbf{L}_2(\Omega,\mathcal{F},\mathbb{P})$ onto $\mathbf{L}_2(\Omega,\mathcal{F}_A,\mathbb{P})$. It can also be defined as

$$\mathbb{E}\left(\psi \,|\, \mathcal{F}_A\right) = \frac{d(\psi d\mathbb{P})}{d\mu_A} \tag{2.3}$$

in terms of Radon-Nikodym derivatives; see [**59**]. Furthermore, in (2.4), the symbol $\mathbb{E}_{A=x}(\psi \,|\, \mathcal{F}_A)$ stands for the unique function $g \in \mathcal{M}(X,\mathcal{B}_X)$ such that $E\left(\psi \,|\, \mathcal{F}_A\right) = g \circ A$ (existence and uniqueness are justified in the proof of the lemma).

LEMMA 2.2. *It holds that*

$$(V_A^*\psi)(x) = \mathbb{E}_{A=x}(\psi \,|\, \mathcal{F}_A), \quad \psi \in \mathbf{L}_2(\Omega,\mathcal{F},\mathbb{P}). \tag{2.4}$$

PROOF. We take $\psi \in \mathbf{L}_2(\Omega,\mathcal{F},\mathbb{P})$ and $f \in \mathbf{L}_2(X,\mu_A)$. We have

$$\begin{aligned}\langle V_A^*\psi, f\rangle_{\mu_A} &= \langle \psi, V_A f\rangle_{\mathbb{P}} \\ &= \langle \psi, f\circ A\rangle_{\mathbb{P}} \\ &= \int_\Omega \psi(\omega) f(A(\omega))d\mathbb{P}(\omega) \\ &= \int_\Omega f(A(\omega))\mathbb{E}\left(\psi \,|\, \mathcal{F}_A\right) d\mathbb{P}(\omega).\end{aligned}$$

But $\mathcal{F}_A$ is generated by the functions of the form

$$\chi_{A^{-1}(\Delta)} = \chi_\Delta \circ A, \quad \Delta \in \mathcal{B}_X,$$

and so there is a uniquely determined function $g \in \mathcal{M}(X,\mathcal{B}_X)$ such that $E\left(\psi \,|\, \mathcal{F}_A\right) = g \circ A$. (Uniqueness of g follows from the fact that $V_A : \mathbf{L}_2(X,\mathcal{B}_X,\mu_A) \longrightarrow \mathbf{L}_2(\Omega,\mathcal{F}_A,\mathbb{P})$ is an isometry). Hence

$$\langle V_A^*\psi, f\rangle_{\mu_A} = \int_\Omega f(x)g(x)d\mu_A(x),$$

and hence the formula,

$$(V_A^*\psi)(x) = g(x) = \mathbb{E}_{A=x}\left(\psi \,|\, \mathcal{F}_A\right).$$

$\square$

COROLLARY 2.3. *The measure* $(\psi d\mathbb{P})\circ A^{-1}$ *is absolutely continuous with respect to* μ_A*, and for* $\psi \in \mathbf{L}_2(\Omega, \mathcal{F}, \mathbb{P})$ *we have*

$$(\psi d\mathbb{P}) \circ A^{-1} = g d\mu_A, \tag{2.5}$$

and

$$V_A^* \psi = \frac{(\psi d\mathbb{P}) \circ A^{-1}}{d\mu_A}. \tag{2.6}$$

PROOF. From the previous proof we have on the one hand

$$\begin{aligned} \langle V_A^*\psi, f\rangle_{\mu_A} &= \int_\Omega f(A(\omega))\,(\psi(\omega)d\mathbb{P}(\omega)) \\ &= \int_X f(x)\left((\psi d\mathbb{P})\circ A^{-1}\right)(x) \end{aligned}$$

and on the other hand,

$$\begin{aligned} \langle V_A^*\psi, f\rangle_{\mu_A} &= \int_\Omega f(A(\omega))\,(\psi(\omega)d\mathbb{P}(\omega)) \\ &= \int_\Omega f(x)g(x)d\mu_A(x) \\ &= \int_\Omega f(A(\omega))g(A(\omega))d\mathbb{P}(\omega), \end{aligned}$$

and the claim follows by comparing these two computations. $\square$

With the above random variables A, B, we associate the positive operator $R_{A,B}$, which we call the transfer operator from A to B, defined by

$$R_{A,B} = V_A^* V_B, \tag{2.7}$$

see the figure below:

$$\begin{array}{ccc} \mathbf{L}_2(\mu_B) & \xrightarrow{\quad R_{A,B} \quad} & \mathbf{L}_2(\mu_A) \\ {\scriptstyle V_B}\searrow & & \nearrow{\scriptstyle V_A^*} \\ & \mathbf{L}_2(\Omega, \mathbb{P}) & \end{array}.$$

Note that both V_A^* and $R_{A,B}$ are positive operators in the following sense:

$$\psi \geq 0 \implies V_A^*\psi \geq 0$$

and

$$f \geq 0 \implies V_A^* V_B f \geq 0.$$

The following result shows that $R_{A,B}$ is a conditional expectation.

LEMMA 2.4. *We have:*

$$\begin{aligned} &(2.8)\ \mathbb{E}\left(f\circ A \,\middle|\, \mathcal{F}_B\right) &= (R_{A,B}^* f)\circ B = (R_{B,A}f)\circ B, \quad & f \in \mathbf{L}_2(X, \mathcal{B}_X, \mu_A) \\ &(2.9)\ \mathbb{E}\left(g\circ B \,\middle|\, \mathcal{F}_A\right) &= (R_{A,B}g)\circ A = (R_{B,A}^* g)\circ A, \quad & g \in \mathbf{L}_2(X, \mathcal{B}_X, \mu_B). \end{aligned}$$

PROOF. We prove (2.9). The proof of (2.8) is similar and follows from $(V_A^*V_B)^* = V_B^*V_A$. Let $f_1 \in \mathbf{L}_2(X, \mathcal{B}_X, \mu_B)$, and $f_2 \in \mathbf{L}_2(X, \mathcal{B}_X, \mu_A)$. On the one hand, we have

$$\langle V_A^*V_Bf_1, f_2\rangle_{\mu_A} = \int_\Omega ((R_{A,B}f_1)\circ A)(\omega)(f_2\circ A)(\omega)d\mathbb{P}(\omega).$$

On the other hand,

$$\begin{aligned}\langle V_A^*V_Bf_1, f_2\rangle_{\mu_A} &= \langle V_Bf_1, V_Af_2\rangle_{\mathbb{P}}\\ &= \int_\Omega (f_1\circ B)(\omega)(f_2\circ A)(\omega)d\mathbb{P}(\omega)\\ &= \int_\Omega \left(\mathbb{E}\left(f_1\circ B\right)\middle|\,\mathcal{F}_A)\right)(\omega)(f_2\circ A)(\omega)d\mathbb{P}(\omega)\end{aligned}$$

by definition of the conditional expectation, and the result follows. □

COROLLARY 2.5. *Let A, B and C be three random variables with transfer functions*

$$R_{A,B} : \mathbf{L}_2(X, \mathcal{B}_X, \mu_B) \longrightarrow \mathbf{L}_2(X, \mathcal{B}_X, \mu_A)$$

and

$$R_{B,C} : \mathbf{L}_2(X, \mathcal{B}_X, \mu_C) \longrightarrow \mathbf{L}_2(X, \mathcal{B}_X, \mu_B).$$

Then the following chain rule holds for all $f \in \mathbf{L}_2(X, \mathcal{B}_X, \mu_C)$ and $x \in X$:

$$(R_{A,B}R_{B,C}f)(x) = \mathbb{E}_{A=x}\left((R_{B,C}(f))\circ B\,\middle|\,\mathcal{F}_A\right). \tag{2.10}$$

PROOF. We have

$$\begin{aligned}(R_{A,B}R_{B,C}f)(x) &= (V_A^*V_BV_B^*V_Cf)(x)\\ &= (V_A^*\mathbb{E}_BV_Cf)(x)\\ &= \left(V_A^*\mathbb{E}_B\left(f\circ C\,\middle|\,\mathcal{F}_B\right)\right)(x)\\ &= (V_A^*\left((R_{B,C}(f))\circ B\right))(x),\end{aligned}$$

and the result follows from Lemma 2.2. □

In the following lemma, X is assumed locally compact, and $C_c(X)$ denotes the space of continuous functions on X with compact support.

LEMMA 2.6. *Assume that X is a locally compact topological space, and that $\mathcal{B}$ is the associated Borel sigma-algebra. Assume moreover that $R_{A,B}$ sends $C_c(X)$ into $C(X)$. Then it holds that*

$$\mathbb{E}\left(f\circ B\,\middle|\,A=x\right) = (R_{A,B}(f))(x). \tag{2.11}$$

PROOF. We denote by $\mathcal{F}_{A,x}$ the sigma-algebra generated by the set $\{A=x\}$. We have

$$\mathbb{E}(f\circ B\,|\,\mathcal{F}_{A,x}) = \mathbb{E}\left(\mathbb{E}\left(f\circ B\,\middle|\,\mathcal{F}_A\right)\,\middle|\,\mathcal{F}_{A,x}\right).$$

Using the previous lemma, we can then write

$$\int_{A=x}(R_{A,B}f\circ A)(\omega)d\mathbb{P}(\omega) = \int_{A=x}\mathbb{E}(f\circ B\,|\,\mathcal{F}_{A,x})d\mathbb{P}(\omega), \quad \text{and}$$
$$\int_{A\neq x}(R_{A,B}f\circ A)(\omega)d\mathbb{P}(\omega) = \int_{A\neq x}\mathbb{E}(f\circ B\,|\,\mathcal{F}_{A,x})d\mathbb{P}(\omega),$$

from which we get (2.11). □

THEOREM 2.7. *Let the following be as above: The probability space* $(\Omega, \mathcal{F}, \mathbb{P})$, *the random variables* A *and* B, *and the respective measures* μ_A *and* μ_B). *Let also* $R_{A,B}$ *be the corresponding transfer operator. Then the following are equivalent:*
(i) $\mu_B << \mu_A$, *with* $W = \frac{d\mu_B}{d\mu_A}$

(ii) It holds that

$$\int_X R_{A,B}(f)(x)d\mu_A(x) = \int_X f(x)W(x)d\mu_A(x), \tag{2.12}$$

that is, $\frac{d\mu_A R_{A,B}}{d\mu_A} = W$.

PROOF. Let $f_1 \in \mathbf{L}_2(X, \mathcal{B}_X, \mu_A)$ and $f_2 \in \mathbf{L}_2(X, \mathcal{B}_X, \mu_B)$. From the proof of Lemma 2.6 we have:

$$\int_X f_1(x)\left(R_{A,B}f_2\right)(x)d\mu_A(x) = \int_\Omega (f_1 \circ A)(\omega)(f_2 \circ B)(\omega)d\mathbb{P}(\omega)$$

Setting $f_1(x) \equiv 1$, we obtain

$$\int_X \left(R_{A,B}f_2\right)(x)d\mu_A(x) = \int_X f_2(x)d\mu_B(x),$$

so that $d(\mu_A R_{A,B}) = d\mu_B$. By definition of W, we obtain (2.12). The converse is clear. □

One can associate with the transfer operator $R_{A,B}$ two extreme cases: On the one end, if $\operatorname{rank} R_{A,B} = 1$, this corresponds to having A and B independent, see Proposition 2.8. No information is passed from A to B. On the other end, if $R_{A,B} = I$, it corresponds to the sub-Markovian case.

PROPOSITION 2.8. *The random variables* A *and* B *are independent if and only if the transfer operator* $R_{A,B}$ *has rank* 1.

PROOF. Indeed, assume first A and B independent, and let $f \in \mathbf{L}_2(X, \mathcal{B}_X, \mu_A)$ and $g \in \mathbf{L}_2(X, \mathcal{B}_X, \mu_B)$. We have:

$$\begin{aligned}
\langle f, R_{A,B}g\rangle_{\mu_A} &= \langle V_A f, V_B g\rangle_{\mathbb{P}} \\
&= \int_\Omega ((f \circ A)(\omega))((g \circ B)(\omega))d\mathbb{P}(\omega) \\
&= \left(\int_\Omega (f \circ A)(\omega)\right)\left(\int_\Omega (g \circ B)(\omega)d\mathbb{P}(\omega)\right) \\
&= \left(\int_X f(x)d\mu_A(x)\right)\left(\int_X g(x)d\mu_B(x)\right) \\
&= \langle f, 1\rangle_{\mu_A}\langle 1, g\rangle_{\mu_B},
\end{aligned}$$

the product of the means of the respective random variables $f(A)$ and $g(B)$, and hence $R_{A,B}$ has rank one. In Dirac's notation, (ket-bra) we can write

$$R_{A,B} = |1>_{\mu_A}<1|_{\mu_B}.$$

□

Given two projections P_1 and P_2 on a Hilbert space, we recall (see [**8**, p. 376] that the sequence $(P_2P_1)^m$ converges strongly to the projection on the intersection of the corresponding spaces. Applied to $P_1 = \mathbb{E}_B$ and $P_2 = \mathbb{E}_A$ we obtain that

$\lim_{m\to\infty}(\mathbb{E}_A\mathbb{E}_B)^m$ is the projection onto $\mathbb{E}_A(\mathbf{L}_2(\Omega,\mathcal{F},\mathbb{P}))\cap\mathbb{E}_B(\mathbf{L}_2(\Omega,\mathcal{F},\mathbb{P}))$, that is the orthogonal projection onto $\mathbf{L}_2(\Omega,\mathcal{F}_A\cap\mathcal{F}_B,\mathbb{P})$.

Here we have a more precise formula:

LEMMA 2.9. *With A, B and $\mathbb{E}_A, \mathbb{E}_B$ as above, let P denote the orthogonal projection onto the eigenspace corresponding to the eigenvalue1of* $\lim_{m\to\infty}(R_{A,B}R_{A,B}^*)^m$. *Then,*

$$\lim_{m\to\infty}(\mathbb{E}_A\mathbb{E}_B)^m\,\psi=\mathbb{E}\left(\psi\,|\,\mathcal{F}_A\cap\mathcal{F}_B\right)=V_APR_{A,B}V_B^*\psi,\quad\forall\psi\in\mathbf{L}_2(\Omega,\mathcal{F},\mathbb{P}).\tag{2.13}$$

PROOF. The proof follows from the formula

$$(\mathbb{E}_A\mathbb{E}_B)^{m+1}=V_A(R_{A,B}R_{A,B}^*)^mR_{A,B}V_B^*,\quad m=0,1,\ldots\tag{2.14}$$

which is true for $m=0$ and proved by induction as follows:

$$\begin{aligned}(\mathbb{E}_A\mathbb{E}_B)^{m+1}&=(\mathbb{E}_A\mathbb{E}_B)^m\,V_AR_{A,B}V_B^*\\&=\overbrace{\left(V_A(R_{A,B}R_{A,B}^*)^{m-1}R_{A,B}V_B^*\right)}^{\text{induction at rank } m}\overbrace{\left(V_AR_{A,B}V_B^*\right)}^{\mathbb{E}_A\mathbb{E}_B}\\&=V_A(R_{A,B}R_{A,B}^*)^{m-1}R_{A,B}\underbrace{V_B^*V_A}_{R_{A,B}^*}R_{A,B}V_B^*\\&=V_A(R_{A,B}R_{A,B}^*)^{m-1}R_{A,B}R_{A,B}^*R_{A,B}V_B^*\\&=V_A(R_{A,B}R_{A,B}^*)^mR_{A,B}V_B^*.\end{aligned}$$

To conclude we remark that $\lim_{m\to\infty}(\mathbb{E}_A\mathbb{E}_B)^m$, being a projection, has spectrum consisting of the eigenvalues 0 and 1. Indeed, let $S=R_{A,B}R_{A,B}^*$. By the assumptions, the projection-valued spectral resolution $E^{(S)}$ of the self-adjoint operator S satisfies

$$S=\int_0^1 tE^{(S)}(dt),$$

and so $\lim_{m\to\infty}S^m=E^{(S)}(\{1\})$, where $E^{(S)}(\{1\})$ (denoted by P in (2.13)) is the spectral projection onto

$$\{f\in\mathbf{L}_2(X,\mathcal{B}_X,\mu_A)\,:\,Sf=f\}.$$

As a result we get

$$\mathbb{E}\left(\cdot\,\middle|\,\mathcal{F}_A\cap\mathcal{F}_B\right)=V_AE^{(S)}(\{1\})R_{A,B}V_B^*.$$

□

For a related result, see [**64**].

As a corollary we have (where here and in the sequel we denote by $\mathbb{E}_A$ the conditional expectation onto $\mathcal{F}_A$):

COROLLARY 2.10. *In the notation of the previous proposition and of its proof, let $S=R_{A,B}R_{A,B}^*$, and let $f\in\mathbf{L}_2(X,\mathcal{B}_X,\mu_A)$ and $\psi=V_Af$. The following are equivalent:*

(1) $Sf = f$, *i.e.*, $E^{(S)}(\{1\})f = f$.

(2) ψ *satisfies* $\mathbb{E}_A\mathbb{E}_B\psi = \psi$.

(3) ψ *satisfies* $\mathbb{E}_B\mathbb{E}_A\psi = \psi$.

(4) $\mathbb{E}\left(\psi \,|\, \mathcal{F}_A \cap \mathcal{F}_B\right) = \psi$.

PROOF. If T is a contraction from a Hilbert space $\mathcal{H}$ into itself and $T\psi = \psi$ for some $\psi \in \mathcal{H}$, then we also have $T^*\psi = \psi$. Indeed, using $T\psi = \psi$ we obtain

$$\|\psi - T^*\psi\|^2 = \|T^*\psi\|^2 - \|\psi\|^2,$$

which is negative since T^* is also a contraction. Hence $\|\psi - T^*\psi\| = 0$ and $T^*\psi = \psi$. The proof of the corollary follows then by applying the above fact to $T = \mathbb{E}_A\mathbb{E}_B$. □

COROLLARY 2.11. *In the notation of the previous proposition, the following are equivalent for pairs of random variables A and B:*

(1) $\mathcal{F}_A \subset \mathcal{F}_B$, *(that is containment of the sigma-algebras of subsets of Ω)*

(2) $\mathbb{E}_A(\mathbf{L}_2(\Omega, \mathbb{P})) \subset \mathbb{E}_B(\mathbf{L}_2(\Omega, \mathbb{P}))$.

(3) $\mathbb{E}_A\mathbb{E}_B = \mathbb{E}_A$, *or equivalently* $\mathbb{E}_A \leq \mathbb{E}_B$, *where $\leq$ denotes the standard ordering of projections.*

(4) $\mathbb{E}_B\mathbb{E}_A = \mathbb{E}_A$, *equivalently* $\mathbb{E}_A \leq \mathbb{E}_B$.

(5) $R_{A,B}V_B^* = V_A^*$

(6) $V_BR_{B,A} = V_A$.

PROOF. This is essentially from the above, but see also the arguments outlined in Table 3 below. □

2.2. A formula for the conditional expectation. We are in the setting of Section 2.1. Let A be a X-valued random variable. For $f \in \mathcal{M}(X, \mathcal{B}_X)$ we denote by $M_{f\circ A}$ the operator of multiplication by $f \circ A$, from $\mathbf{L}_2(\Omega, \mathcal{F}, \mathbb{P})$ into itself. The space of all these operators when f runs through $\mathbf{L}^\infty(X, \mathcal{B}_X)$ is a commutative von Neumann algebra, denoted $\mathcal{M}_A$. By Stone's theorem (see [**58**]), there exists a $\mathcal{M}_A$-valued measure $\mathscr{E}_A$ on $(X, \mathcal{B}_X)$ such that

$$M_{f\circ A} = \int_X f(x)\mathscr{E}_A(dx). \tag{2.15}$$

For every $L \in \mathcal{B}_X$, the operator $\mathscr{E}(L) \in \mathcal{M}_A$, and so is of the form $f \circ A$ for some $f \in \mathbf{L}^\infty(X, \mathcal{B}_X)$, namely $f = \chi_L$. From the equality

$$\mathscr{E}_A(L) = \chi_{\{A\in L\}} = \chi_{A^{-1}(L)}, \quad L \in \mathcal{B}_X, \tag{2.16}$$

we shall use the notation (after identifying the function and the corresponding multiplier)

$$\mathscr{E}_A(dx) = M_{\chi_{\{A\in dx\}}} = \chi_{\{A\in dx\}} \tag{2.17}$$

and rewrite (2.15) as

$$f \circ A = \int_X f(x)\chi_{\{A\in dx\}}, \quad \text{or} \quad (f \circ A)(\omega) = \int_X f(x)\chi_{\{A(\omega)\in dx\}}. \tag{2.18}$$

REMARK 2.12. While $\chi_{\{A\in dx\}}$ is a heuristic notation, we stress that it is made precise via the spectral theorem in the form (2.15), and also by the conclusion of the following proposition.

PROPOSITION 2.13. *Let* $f \in \mathbf{L}_2(X, \mathcal{B}_X), \mu_A)$. *Then,*

$$\int_X f(x)^2 d\mu_A(x) = \int_\Omega \left(\int_X f(x)\chi_{\{A(\omega)\in dx\}} \right)^2 d\mathbb{P}(\omega). \tag{2.19}$$

PROOF. Consider finite partitions $\pi = \{L_i,\, i = 1, \ldots, m\}$ of X into sets of $\mathcal{B}$ such that $L_i \cap L_j = 0$ for $i \neq j$, and for every i chose $x_i \in L_i$. Let $|\pi| = \max_{i=1,\ldots,m} |\mu_A(L_i)|$. We obtain a filter of $\mathcal{B}$-partitions along which limits are taken. By definition of the integral with respect with a measure we have:

$$\int_X f(x)\chi_{\{A(\omega)\in dx\}} = \lim_{|\pi|\to 0} \sum_{i=1}^m f(x_i)\chi_{\{A(\omega)\in L_i\}}.$$

But

$$\mathbb{E}\left(\sum_{i=1}^m f(x_i)\chi_{\{A(\omega)\in L_i\}} \right)^2 = \sum_{i=1}^m f(x_i)^2 \mu_A(L_i)$$
$$\to \int_X f(x)^2 d\mu_A(x), \quad \text{as } |\pi| \to 0,$$

and the result follows. □

We now consider the case of a discrete random variable. We shall assume that $A : \Omega \longrightarrow \mathbb{N}_0$. So, $X = \mathbb{N}_0$, and the space $\mathbf{L}_2(X, \mathcal{B}_X, \mu_A)$ is the Hilbert space of $\ell_2(\mu_A)$ real-valued sequences $(\xi_n)_{n\in\mathbb{N}_0}$ such that

$$\sum_{n=0}^\infty \xi_n^2\, \mathbb{P}(\{A = n\}) < \infty. \tag{2.20}$$

We have

(2.21)

$$(M_{f\circ A}\psi)(\omega) = \sum_{n=0}^\infty \xi_n \chi_{\{A=n\}}(\omega)\psi(\omega),\ \forall \psi \in \mathbf{L}_2(\Omega, \mathcal{F}, \mathbb{P}),\ \forall f = (\xi_n)_{n\in\mathbb{N}_0} \in \ell_2(\mu_A),$$

and

$$\mathscr{E}_A(\{n\}) = M_{\chi_{\{A=n\}}}, \quad n = 0, 1, \ldots. \tag{2.22}$$

THEOREM 2.14. *The following formulas hold:*

(2.23)

$$\mathbb{E}(\psi \mid \mathcal{F}_A)(\omega) = \begin{cases} \int_X (V_A^*\psi)(x)\chi_{\{A\in dx\}}(\omega) & (\textit{continuous case}) \\ \sum_{k=0}^\infty \frac{1}{\mathbb{P}(\{A=k\})} \left(\int_{\{A=k\}} \psi(\omega) d\mathbb{P}(\omega) \right) \chi_{\{A=k\}}(\omega) & (\textit{discrete case}). \end{cases}$$

The proof of these formulas follow from Lemma 2.2 and (2.18).

2.3. Markov processes. We follow the notation of the previous section, but our starting point is now a sequence of X-valued random variables $T_0, T_1, \ldots$ defined on the given probability space $(\Omega, \mathcal{F}, \mathbb{P})$.

AXIOMS 2.15. *(i) Let $\mathcal{G}_n \subset \mathcal{F}$ be the smallest sigma-algebra for which the variables $T_0, \ldots, T_n$ are measurable. We assume that $V_{T_n}^* V_{T_{n+1}}$ does not depend on n, and that*

$$\mathbb{E}\left(f \circ T_{n+1} \middle| \mathcal{G}_n\right) = \mathbb{E}\left(f \circ T_{n+1} \middle| \mathcal{F}_n\right) = R(f) \circ T_n, \quad n = 0, 1, \ldots \tag{2.24}$$

(ii) The measures μ_0 and μ_1 are equivalent.

We refer to (2.24) as the Markov property in the present setting.

REMARK 2.16. If in the expression $R_{A,B} f = V_A^* V_B f$, to $A = T_n$ and $B = T_{n+1}$ and if moreovoer R_{T_{n+1}, T_n} is independent of n we get

$$\mathbb{E}\left(f \circ T_{n+1} \,\middle|\, \mathcal{F}_n\right) = (R(f)) \circ T_n$$

as a special case of (2.8). Iterating we get

$$\mathbb{E}\left(f \circ T_{n+k} \,\middle|\, \mathcal{F}_n\right) = \left(R^k(f)\right) \circ T_n.$$

LEMMA 2.17. *Condition (ii) from Axioms 2.15 holds if and only if $\mu_0\left(\{x : W(x) = 0\}\right) = 0$.*

PROOF. Since $\mu_1 << \mu_0$ we can write

$$\mu_1(\Delta) = \int_\Delta W(x) d\mu_0(x), \quad \forall \Delta \in \mathcal{B}_X, \tag{2.25}$$

where $W = \frac{d\mu_1}{d\mu_0}$. Let $\Delta_0 = \{x \in X \,;\, W(x) = 0\}$. By (2.25) we have $\mu_1(\Delta_0) = 0$. Assume that $\mu_0(\Delta_0) > 0$. Then $\mu_0 << \mu_1$ will not hold.

Conversely, if $\mu_0(\Delta_0) = 0$, then W^{-1} is well defined μ_0 a.e., and $\mu_0 << \mu_1$ with $\frac{d\mu_0}{d\mu_1} = \frac{1}{W}$. □

DEFINITION 2.18. Assume the previous axioms in force, and set $W = \frac{d\mu_1}{d\mu_0}$. The sequence $T_0, T_1 \ldots$ is called a W-Markov process.

Given $\prod_0^\infty X$, we denote by π_n the n-th coordinate function:

$$\pi_n(x_0, x_1, \ldots) = x_n.$$

In the present work, this product is always endowed with the cylinder sigma-algebra $\mathcal{C}$.

THEOREM 2.19. *Let $(\Omega, \mathcal{F}, \mathbb{P}, (T_n)_{n \in \mathbb{N}_0})$ satisfy axioms (i) and (ii) above. Then there is a probability measure $\mathbb{P}^\times$ on the Cartesian product $\prod_0^\infty X$, and an isomorphism $\widehat{T}$ between $(\Omega, \mathcal{F}, \mathbb{P}, (T_n)_{n \in \mathbb{N}_0})$ and $(\prod_{n=0}^\infty X, \mathcal{C}, \mathbb{P}^\times, (\pi_n)_{n \in \mathbb{N}_0})$, meaning that*

$$\pi_n \circ \widehat{T} = T_n, \quad n = 0, 1, \ldots \tag{2.26}$$

PROOF. We define

$$\widehat{T}(\omega) = (T_0(\omega), T_1(\omega), \ldots) \tag{2.27}$$

and

$$\mathbb{P}^\times(\Delta) = \mathbb{P}\left(\widehat{T}^{-1}(\Delta)\right), \quad \forall \Delta \in \mathcal{C}, \tag{2.28}$$

in other words, $\mathbb{P}^\times$ is the distribution of $\widehat{T}$. □

2.4. Discrete case. We first compute the transfer operator (see (2.7)) for a pair of random variables A and B on $(\Omega, \mathcal{F}, \mathbb{P})$ when A is discrete. See Section 3.3 for the notation. We denote by δ_n the Dirac function on $\mathbb{N}_0$, that is

$$\delta_n(m) = \begin{cases} 1, & \text{if} \quad m = n, \\ 0, & \text{if} \quad m \neq n. \end{cases}$$

PROPOSITION 2.20. *Let $(\Omega, \mathcal{F}, \mathbb{P})$ and A, B as above. Then the transfer operator $R_{B,A} : \ell_2(\mu_A) \longrightarrow \mathbf{L}_2(X, \mathcal{B}_X, \mu_B)$ is given by:*

$$(R_{B,A}(\delta_n))(x) = \mathbb{E}_{B=x}\left(\{A = n\} \mid \mathcal{F}_B\right). \tag{2.29}$$

PROOF. The result is immediate from Lemmas 2.2 and 2.4. We get for $x \in X$ and $n \in \mathbb{N}_0$

$$\begin{aligned}((V_B^* V_A)\delta_n)(x) &= \left(V_B^*\left(\chi_{\{A=n\}}\right)\right)(x) \\ &= \mathbb{E}_{B=x}\left(\{A = n\} \mid \mathcal{F}_B\right)\end{aligned}$$

where we have identified the indicator function $\chi_{\{A=n\}}$ on Ω with the subset

$$\{A = n\} = \{\omega \in \Omega;\, A(\omega) = n\}.$$

□

Note that, since $R_{B,A}^* = R_{A,B}$, by Lemma 2.4 we get for $f \in \mathbf{L}_2(X, \mathcal{B}_X, \mu_B)$

$$(R_{A,B}(f))(n) = \int_X f(x)\mathbb{E}_{B=x}\left(\{A = n\} \mid \mathcal{F}_B\right) d\mu_B(x). \tag{2.30}$$

PROPOSITION 2.21. *On $\mathbb{N}_0 \times \mathbb{N}_0$ we have the following positive definite kernel:*
(2.31)

$$k_B(n, m) = \langle R_{A,B}\delta_n, R_{A,B}\delta_m\rangle_{\mathbf{L}_2(X,\mathcal{B}_X,\mu_B)} = \int_\Omega \chi_{\{A=n\}}(\omega)\mathbb{E}_B\left(\chi_{\{A=m\}}\right)(\omega)d\mathbb{P}(\omega).$$

PROOF. This follows from (2.30) and from the formula $R_{B,A}^* R_{B,A} = V_A^* \mathbb{E}_B V_A$ □

The last result concerns the case where both A and B are discrete.

PROPOSITION 2.22. *Let $(\Omega, \mathcal{F}, \mathbb{P})$ and A, B as above, and assume that both A and B have discrete laws, with μ_A and μ_B both supported on the same countable discrete set, say S. Denote by $\delta_j^{(B)}$ the Dirac function viewed as a vector in $\ell_2(\mu_B)$. Then*

$$\left(R_{A,B}\left(\delta_j^{(B)}\right)\right)(i) = \mathbb{P}\left(B = j \mid A = i\right). \tag{2.32}$$

PROOF. This is immediate from Lemma 2.6. □

2.5. Martingales.

DEFINITION 2.23. Let $T_0, T_1, \ldots$ be a Markov chain, with each T_n taking values in the space $(X, \mathcal{B}_X)$. Let $M_0, M_1, \ldots$ be another X-valued random process. We say that $(M_n)_{n\in\mathbb{N}_0}$ is a martingale with respect to $(T_n)_{n\in\mathbb{N}_0}$ if the condition

$$\mathbb{E}\left(M_{n+k} \mid \bigvee_{j=0}^{n} \mathcal{F}_{T_j}\right) = M_n, \quad \forall n, k \in \mathbb{N}_0 \tag{2.33}$$

holds.

TABLE 2. Markov chains

Case	General state space $\mathcal{M}(X, \mathcal{B}_X)$	Discrete state space
Transition	$\mathbb{E}\left(f \circ T_{n+1} \,\middle|\, \bigvee_{j=0}^{n} \mathcal{F}_j\right)$ $= \mathbb{E}\left(f \circ T_{n+1} \,\middle|\, \mathcal{F}_n\right)$	$\mathbb{E}\left(T_{n+1} = j \,\middle|\, T_0 = i_0, \ldots T_n = i_n\right)$ $= \mathbb{E}\left(T_{n+1} = j \,\middle|\, T_n = i_n\right)$
Transfer operator	$(Rf)(x) = \mathbb{E}\left(f \circ T_{n+1} \,\middle|\, T_n = x\right)$ $\forall f \in \mathcal{M}(X, \mathcal{B}_X)$	$p_{i,j} = \mathbb{P}\left(T_{n+1} = j \,\middle|\, T_n = i\right)$
Harmonic functions	$(Rh)(x) = h(x)$	$\sum_j p_{i,j} h_j = h_i$

PROPOSITION 2.24. *Let $(X, \mathcal{B}_X)$ be a measure space, and let $T_0, T_1, \ldots$ be a X-valued Markov chain, with transition operator R acting on $\mathcal{M}(X, \mathcal{B}_X)$, and let h positive on X and such that $Rh = h$. Then, the process $M_n = h \circ T_n$, $n = 0, 1, \ldots$ is a martingale (see Definition 2.23 for the latter) with respect to $(T_n)_{n \in \mathbb{N}_0}$.*

PROOF. Since $(T_n)_{n \in \mathbb{N}_0}$ is a Markov chain we have for every $f \in \mathcal{M}(X, \mathcal{B}_X)$

$$\mathbb{E}\left(f \circ T_{n+k} \,\middle|\, \bigvee_{j=0}^{n} \mathcal{F}_j\right) = \mathbb{E}\left(f \circ T_{n+k} \,\middle|\, \mathcal{F}_n\right) = \left(R^k(f)\right) \circ T_n, \tag{2.34}$$

(by Lemma 2.8 with $B = T_{n+k}$ and T_n instead of A; see also (3.40) for the solenoid).

Setting $f = h$ in (2.34) we get

$$\begin{aligned}\mathbb{E}\left(M_{n+k} \,\middle|\, \bigvee_{j=0}^{n} \mathcal{F}_j\right) &= (R^k(h)) \circ T_n \\ &= h \circ T_n \\ &= M_n,\end{aligned}$$

which is the desired conclusion. □

REMARK 2.25. The same argument will hold for a positive function f such that $Rf = \lambda f$ for some $\lambda \neq 0$. Then, $M_n^{(\lambda)} = \lambda^{-n} f \circ T_n$ is a martingale with respect to $(T_n)_{n \in \mathbb{N}_0}$.

DEFINITION 2.26. *(stopping time)* A stopping time K for a random process $(T_n)_{n \in \mathbb{N}_0}$ on the probability space $(\Omega, \mathcal{F}, \mathbb{P})$ is a random variable $K : \Omega \longrightarrow \mathbb{N}_0$ such that

$$K^{-1}(\{n\}) \in \bigvee_{j=0}^{n} \mathcal{F}_{T_j}, \quad \forall n \in \mathbb{N}_0. \tag{2.35}$$

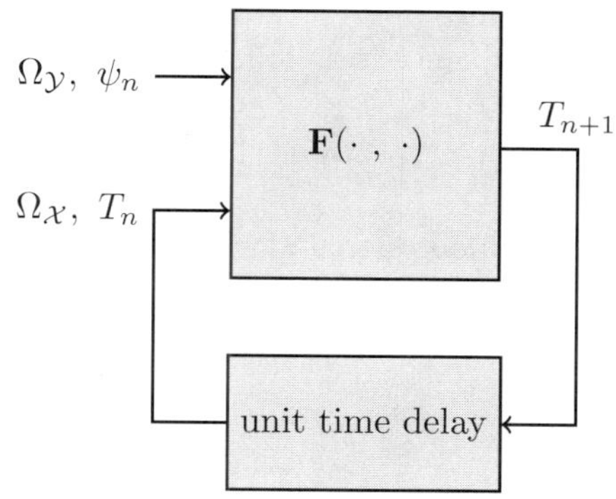

FIGURE 1. Feedback loop: illustration of the class of i.i.d. feedback processes from Definition 2.27.

Applying our previous analysis to the pair $A = (T_0, \ldots, T_n)$ (with associated space X^{n+1}) and $B = K$, we have the following stopping time formula:

$$(2.36)\qquad \left(R_{\{T_0,\ldots,T_n\},K} f\right)(m) = \mathbb{E}_{\{K=m\}}\left(f(T_0, \ldots, T_n) \,\middle|\, \mathcal{F}_K\right), \\ n, m \in \mathbb{N}_0, \quad \forall f \in \mathbf{L}_2\left(\left(\prod_{n=0}^{n} X\right), \mu_n\right),$$

where μ_n denotes the joint distribution of $\{T_0, \ldots, T_n\}$.

DEFINITION 2.27. Let $(X, \mathcal{B}_X)$ and $(Y, \mathcal{B}_Y)$ be two measure spaces, and let $F : X \times Y \longrightarrow X$ be a measurable function, where $X \times Y$ as a measure space is given the product sigma-algebra. Let $(T_n)_{n\in\mathbb{N}_0}$ be a Markov chain with values in X, and let $(\psi_n)_{n\in\mathbb{N}_0}$ be a system of independent identically distributed (i.i.d.) random variables with values in Y. If

$$(2.37)\qquad T_{n+1} = F(T_n, \psi_n), \quad n \in \mathbb{N}_0,$$

then one says that $(T_n)_{n\in\mathbb{N}_0}$ is a homogeneous Markov chain (HMC). In details, the requirement is that

$$T_{n+1}(\omega) = F(T_n(\omega), \psi_n(\omega)), \quad \omega \in \Omega,$$

where Ω refers to the sample space in the probability space $(\Omega, \mathcal{F}, \mathbb{P})$ which realizes the two processes; see also Theorems 2.29 and 2.30 below.

Recursion (2.37) is a feedback loop in the language of (non-linear) system theory. See Figure 1 below, and see e.g. [**35**] for information on feedback. We plan to explore these connections in a future publication.

There are many applications of these Markov processes, including to control, see [**36**], to feedback, see [**38**], and to Monte Carlo simulation, see e.g., [**20**] and [**62**, §5.5].

2.6. Homogeneous Markov chains (HMC). As above $(X, \mathcal{B}_X)$ is a set with a fixed sigma-algebra $\mathcal{B}_X$. We consider another measure-space $(Y, \mathcal{D})$. Let $\psi_0, \psi_1, \ldots$ be a sequence of independent identically distributed (i.i.d.) Y-valued random variables defined on $(\Omega, \mathcal{F}, \mathbb{P})$, with probability distribution ν. Such a sequence is called a white noise or a driving sequence; see Remark 2.31. One way to construct such a sequence is as follows. One takes

$$\Omega_Y = \prod_{n=0}^{\infty} Y \ (= Y^{\mathbb{N}_0})$$

endowed with the cylinder sigma-algebra $\mathcal{C}$ (see for instance [**47**] for the latter), and the infinite product measure $\nu_\infty = \nu \times \nu \cdots$, and set

$$\psi_n(y_0, y_1, \ldots) = y_n$$

Thus

$$\nu(D) = \nu_\infty(\psi_n^{-1}(D)), \quad D \in \mathcal{D}, \text{ and } n = 0, 1, \ldots \tag{2.38}$$

In particular,

$$\int_\Omega F(\cdot, \psi_0(\omega))d\nu_\infty(\omega) = \int_Y F(\cdot, y)d\nu(y) \tag{2.39}$$

We now consider a measurable map F from $X \times Y$ into X, where Y is another measure-space. We define

$$(R_F f)(x) = \int_Y f(F(x, y))d\nu(y), \qquad f \in \mathcal{M}(X, \mathcal{B}). \tag{2.40}$$

Let $\Omega_Y = \prod_{n=0}^{\infty} Y$, and for $\omega = (y_0, y_1, \ldots) \in \Omega_Y$, we define (with $F_y = F(\cdot, y)$)

$$\begin{aligned} \omega|n &= (y_0, \ldots, y_n) \\ F_{\omega|_n} &= F_{y_n} F_{y_{n-1}} \cdots F_{y_1} F_{y_0}. \end{aligned} \tag{2.41}$$

We assume that

$$\cap_{n=1}^{\infty} F_{\omega|_n}(X) = \{x_\omega\} \tag{2.42}$$

is a singleton. We then set

$$V(\omega) = x_\omega \quad \text{(see (2.42))}. \tag{2.43}$$

LEMMA 2.28. *Let $F : X \times Y \longrightarrow X$. The corresponding transfer operator in (2.40) is of the form*

$$(R_F f)(x) = \int_X f(t)\mu(dt, |x) \tag{2.44}$$

where $\mu(\cdot|x) \overset{\text{def.}}{=} (d\nu) \circ F_x^{-1}$, and where $F_x(\cdot) = F(x, \cdot) : Y \longrightarrow X$.

PROOF. We have

$$\begin{aligned} \int_Y f(F(x, y))d\nu(y) &= \int_Y (f \circ F_x)(y)d\nu(y) \\ &= \int_X f(t) \left(d\nu \circ F_x^{-1}\right)(t). \end{aligned}$$

Hence $\mu(\cdot|x) = d\nu \circ F_x^{-1}$ as claimed. □

THEOREM 2.29. *Let ν be a probability measure on $(Y, \mathcal{D})$, and let $\times_{n=0}^{\infty}\nu$ be the corresponding infinite product measure on Ω. Assume that (2.42) is in force, and let V be defined by (2.43). The formula*

$$\mu(B) = (\times_{n=0}^{\infty}\nu)(V^{-1}(B)), \quad B \in \mathcal{B}_X, \tag{2.45}$$

then defines a measure on $(X, \mathcal{B})$ *which satisfies*

$$\mu R_F = \mu, \tag{2.46}$$

that is

$$\iint_{X\times Y} f(F(x,y))d\nu(y)d\mu(x) = \int_X f(x)d\mu(x), \quad \forall f \in \mathcal{M}(X, \mathcal{B}_X) \tag{2.47}$$

holds.

PROOF. We define on Ω

$$\ell(y)(y_0, y_1, \ldots) = (y, y_0, y_1, \ldots), \quad \text{with } y \in Y \text{ and } \omega = (y_0, y_1, \ldots) \in \Omega.$$

Then it is clear from (2.42) and (2.43) that

$$F_y V = V\ell(y), \tag{2.48}$$

since

$$F_y F_{\omega|n} = F_{\ell(\omega)|n+1}.$$

Note that (2.48) means that the following commutative diagram is in force:

$$\begin{array}{ccccc} & \Omega_Y & \xrightarrow{\quad V \quad} & X & \\ \ell(y) & \downarrow & & \downarrow & F_y \\ & \Omega_Y & \xrightarrow{\quad V \quad} & X & \end{array}$$

We now prove (2.47). Let $f \in \mathcal{M}(X, \mathcal{B})$. Then R_F in (2.40) can be rewritten as

$$R_F f = \int_Y (f \circ F_y) d\nu(y).$$

With $\mathbb{Q} = \times_{n=0}^{\infty}\nu$ and $\mu = \mathbb{Q}\circ V^{-1}$ we get

$$\begin{aligned}
\int_X (R_F f)(x)d\mu(x) &= \int_X (R_F f)(x)(d\mathbb{Q}\circ V^{-1})(x)\\
&= \int_{\Omega_Y} ((R_F f)\circ V)(\omega)d\mathbb{Q}(\omega)\\
&= \int_{\Omega_Y}\int_Y ((f\circ F_y)\circ V)(\omega)d\mathbb{Q}(\omega)d\nu(y)\\
&\text{(and using (2.48))}\\
&= \int_{\Omega_Y}\int_Y (f\circ V\circ \ell(y))d\mathbb{Q}(\omega)d\nu(y)\\
&= \int_{\Omega_Y}\int_Y (f\circ V)(\omega)d\mathbb{Q}\circ \ell(y)^{-1}d\nu(y)\\
&\text{(and since } \mathbb{Q} \text{ is an infinite product measure)}\\
&= \int_{\Omega_Y}\int_Y (f\circ V)(\omega)d\mathbb{Q}(\omega)d\nu(y)\\
&= \int_X f(x)\int_Y (d\mathbb{Q}\circ V^{-1})(x)d\nu(y)\\
&= \int_X f(x)(d\mathbb{Q}\circ V^{-1})(x)\\
&\text{(since } d\nu \text{ is a probability measure)}\\
&= \int_X f(x)d\mu(x)\quad (\text{with } \mu = \mathbb{Q}\circ V^{-1})
\end{aligned}$$

□

THEOREM 2.30. *Let Y and F be as above. Let λ be a probability measure on $(X,\mathcal{B})$, and let $\psi_0,\psi_1,\dots$ be a sequence of i.i.d. Y-valued random variables with probability distribution ν. Then there exists a probability measure $\mathbb{P}$ on $(\Omega_Y,\mathcal{C})$ and a sequence of X-valued random variables $T_0,T_1,\dots$ on Ω_Y such that:*

(1) *λ is the distribution of T_0, that is*

$$\int_\Omega F(T_0(\omega),\cdot)d\mathbb{P}(\omega) = \int_X F(x,\cdot)d\lambda(x). \tag{2.49}$$

(2) *We have*

$$T_{n+1} = F(T_n,\psi_n),\quad n=0,1,\dots \tag{2.50}$$

(3) *It holds that*

$$\mathbb{E}\left(f\circ T_{n+1}\,\middle|\,\mathcal{F}_n\right) = \mathbb{E}\left(f\circ T_{n+1}\middle|\mathcal{G}_n\right) = (R_F(f))\circ T_n \tag{2.51}$$

where $\mathcal{F}_n = T_n^{-1}(\mathcal{B})$, where $\mathcal{G}_n$ is the smallest sigma-algebra for which the variables $T_0,\dots,T_n$ are measurable, and where R_F is given by (2.40).

(4) *We have*

(2.52)

$$\int_{\Omega_Y}(f_0\circ T_0)(f_1\circ T_1)\cdots(f_n\circ T_n)d\mathbb{P} = \int_X f_0(x)R_F(f_1R_F(f_2\cdots R_F(f_nh)\cdots))d\lambda(x)$$

with $f_0,f_1,\dots,f_n\in\mathcal{M}(X,\mathcal{B})$.

REMARK 2.31. Equation (2.50) is called an homogeneous Markov chain driven by white noise. The sequence $\psi_0, \psi_1, \ldots$ is called the driving sequence. See [**20**, p. 56]. For general background on time-homogeneous state equation and homogeneous Markov chains, [**36, 38**]. See also Theorem 3.9 below.

REMARK 2.32. When R is not normalized,and assuming the existence of a fixed point h for R, one defines

$$R'(f) = \frac{R(fh)}{h}.$$

Then, $R'1 = 1$. The above construction applied to the pair $(R', hd\lambda)$ will lead to the same probability measure $\mathbb{P}$. This is because

$$\int_X f_0(x)R(f_1R(f_2\cdots R(f_nh)\cdots))d\lambda(x)$$
$$= \int_X f_0(x)R'(f_1R'(f_2\cdots R'(f_n)\cdots))h(x)d\lambda(x).$$

PROOF OF THEOREM 2.30. The proof is divided into three steps, which we outline.

STEP 1: *Let* $\omega = (y_0, y_1, \ldots) \in \Omega_Y$ *and define* $\psi_n(\omega) = y_n$, *and* $T_0, T_1, \ldots$ *via*

$$T_n(\omega) = F(\cdots(F(F(F(T_0(\omega), y_0), y_1), \cdots), y_{n-1}), \ldots). \tag{2.53}$$

STEP 2: *Formula (2.52) defines a unique probability measure* $\mathbb{P}$ *on* Ω_Y *endowed with its cylinder sigma-algebra.*

The existence of $\mathbb{P}$ is an application of Kolmogorov's consistency principle; see for instance [**53**].

STEP 3: *The above probability, and the random functions* $T_0, T_1, \ldots$ *have the desired properties.*

See also Lemma 2.2 and 2.4.

□

COROLLARY 2.33. *The probability distribution of* T_n *is* $\mu_n = \lambda R^n$, $n = 1, 2, \ldots$

PROOF. This follows from Theorem 2.30. See also Theorem 2.19. □

COROLLARY 2.34.

$$\mu_n(B) = (\lambda \times \underbrace{\nu \times \cdots \times \nu}_{n\ times})(F_n^{-1}(B)), \quad B \in \mathcal{B}_X. \tag{2.54}$$

where

$$\begin{aligned} F_n(x, y_1, y_2, \ldots, y_n) &= (F_{y_n} \cdots F_{y_1})(x) \\ &= F(\cdots F(F(F(x, y_1), y_2) \cdots, y_{n-1}), y_n). \end{aligned} \tag{2.55}$$

We set

$$\pi_1(x, y) = x.$$

COROLLARY 2.35. *Let* $B \in \mathcal{B}$ *and* $x \in X$. *Then*

$$\mathbb{E}\left(T_{n+1} \in B \,\middle|\, T_n = x\right) = \nu(\pi_1^{-1}(x) \cap F^{-1}(B)) \tag{2.56}$$

PROOF. The result follows from Lemmas 2.2 and 2.4. □

Background references on multiresolutions include [**10, 12, 19, 25, 25**].

2.7. Multiresolutions and Cuntz-Krieger relations. As a corollary of the previous analysis we now consider the case where possibly more than two random variables are given.

COROLLARY 2.36. *Given N random variables $A_1, \ldots, A_N$ with values in X, ($N = \infty$ is allowed), the following hold:*

$$V_{A_u}^* V_{A_v} = R_{u,v} \quad \textit{(definition of the transfer operator from } A_u \textit{ to } A_v\textit{)} \tag{2.57}$$

$$V_{A_u}^* V_{A_u} = I_{\mathbf{L}_2(\mu_u)}, \quad u = 1, 2, \ldots, N \tag{2.58}$$

$$V_{A_u} V_{A_u}^* = \mathbb{E}\left(\cdot \,|\, \mathcal{F}_{A_u}\right), \quad u = 1, 2, \ldots, N \tag{2.59}$$

$$\sum_{u=1}^{N} V_{A_u} V_{A_u}^* = \mathbb{E}\left(\cdot \,|\, \cup_{u=1}^{N} \mathcal{F}_u\right), \tag{2.60}$$

$$\sum_{u=1}^{N} V_{A_u} V_{A_u}^* = \mathbb{E}\left(\cdot \,|\, \mathcal{F}\right) = I \quad \textit{if} \quad \mathcal{F} = \cup_{n=1}^{N} \mathcal{F}_{A_n}. \tag{2.61}$$

If $N = \infty$, the latter sums (2.60)-(2.61) *converge in the strong operator topology.*

PROOF. This follows from Corollary 2.5 and Table 3. □

REMARK 2.37. Relations (2.57)-(2.61) can be seen as a generalization of the Cuntz-Krieger relations (see [**23, 24**] for the latter), and they lead to a multiresolution decomposition of the probability space $\mathbf{L}_2(\Omega, \mathcal{F}, \mathbb{P})$; see Section 4.3. A practical interpretation of formulae (2.60) and (2.61), is the assertion that certain random variables may be reconstructed by samples. In this case, the sampling is performed with the use of random variables as specified in the premise in Corollary 2.36. For a practical use of related sampling formulas in learning theory, see e.g., [**68**].

2.8. The Schur algorithm and homogeneous Markov chains (HMC). The Schur algorithm provides an application of the above analysis. We first recall the following (see [**1, 13, 21, 31, 37, 67**]). Let s be a function analytic and strictly contractive in the open unit disk $\mathbb{D}$ (we will call such functions Schur functions, and denote their set by $\mathcal{S}$). Then, the functions $s_1, \ldots$ defined recursively by $s_0(z) = s(z)$ and

$$s_{n+1}(z) = \frac{s_n(z) - s_n(0)}{z(1 - \overline{s_n(0)} s_n(z))}, \quad n = 0, 1, \ldots \tag{2.62}$$

belong to $\mathcal{S}$ as long as $|s_n(0)| < 1$. The recursion stops at rank n if $|s_n(0)| = 1$. As already proved by Schur, this will happen if and only if s is a finite Blaschke product. The numbers $\rho_n = s_n(0)$, $n = 0, 1, \ldots$ are called the Schur parameters of s, and determine uniquely the function s in terms of a partial fraction exansion

$$s(z) = \rho_0 + \cfrac{z(1 - |\rho_0|^2)}{\overline{\rho_0} z - \cfrac{1}{\rho_1 + \cfrac{z(1 - |\rho_1|^2)}{\overline{\rho_1} - \cdots}}} \tag{2.63}$$

See [**70**, p. 285]. When s is a finite Blaschke product, the sequence is finite, and its last element is of modulus 1.

Set $X = \mathcal{S} \setminus \{\text{unitary constants and finite Blaschke products}\}$ and $Y = \mathbb{D}$. We define

$$F(s,\rho)(z) = \frac{s(z)-\rho}{z(1-s(z)\overline{\rho})} \tag{2.64}$$

which maps $X \times Y$ into X. We will also use the notations $F_\rho(s)$ and $(F_\rho(s))(z)$. We set $\Omega = \prod_{n=0}^{\infty} \mathbb{D}$ and,

$$\omega = (\rho_0, \rho_1, \ldots) \quad \text{and} \quad \omega|_n = (\rho_0, \ldots, \rho_n).$$

Furthermore, we define (see (2.41))

$$(F_{\omega|_n})(s) = \left(F_{\rho_n} F_{\rho_{n-1}} \cdots F_{\rho_1} F_{\rho_0}\right)(s), \quad \text{see (2.63)}.$$

We denote by V the map

$$V(\omega) = s_\omega \tag{2.65}$$

where $s_\omega \in \mathcal{S}$ is uniquely defined element from ω via (2.63).

LEMMA 2.38. *For every $\omega \in \Omega$ we have*

$$\cap_{n=0}^{\infty} F_{\omega|_n}(\mathcal{S}) = \{s_\omega\} \tag{2.66}$$

where $s_w = V(\omega)$, see (2.42).

PROOF. This follows from the fact that a given Schur function is uniquely determined by the sequence of Schur coefficients when the latter is infinite. See [**67**]. □

As a consequence of Theorem 2.29 we have the following result. In the proof the transfer operator now takes the form as in (2.40), that is

$$((R_F f)(s))(z) = \int_{\mathbb{D}} f((F(s,\rho))(z)) d\nu(\rho). \tag{2.67}$$

THEOREM 2.39. *Let ν be a probability measure on $\mathbb{D}$ endowed with its Borel sigma-algebra, and let $\mathbb{Q} = \mathbb{Q}_\nu = \nu \times \nu \times \cdots$ be the corresponding infinite product measure on $\Omega = \prod_{n=0}^{\infty} \mathbb{D}$ endowed with the cylinder sigma-algebra. Then $\mu = \mathbb{Q}_\nu \circ V^{-1}$, where V is defined by* (2.65)*, is a positive measure on $\mathcal{S}$ (or, more precisely, on the set $\mathcal{S}$ from which the unitary constants and finite Blaschke products have been removed) satisfying*

$$\mu R_F = \mu. \tag{2.68}$$

2.9. A summary of formulas. We now summarize some of the formulas obtained in this section, pertaining to two given X-valued random variables A and B

The proofs of these various formulas are given in the section. See in particular Lemmas 2.2, 2.4 and Corollary 2.5.

REMARK 2.40. In the case when B is discrete, say $B : \Omega \longrightarrow \mathbb{N}_0$, the formula for $R_{A,B}$ simplifies as follows:

$$\left(R_{A,B} f\right)(x) = \sum_{n=0}^{\infty} \frac{f(n)}{\mathbb{P}(\{B=n\})} \mathbb{E}_{\{A=x\}}\left(\chi_{\{B=n\}} \,\middle|\, \mathcal{F}_A\right), \quad \forall x \in X$$

TABLE 3. Summary

Hilbert spaces	$\mathbf{L}_2(\mu_A) \longrightarrow \mathbf{L}_2(\Omega, \mathbb{P})$	$\mathbf{L}_2(\Omega, \mathbb{P}) \longrightarrow \mathbf{L}_2(\mu_A)$
Operators	$V_A f = f \circ A$	$(V_A^* \psi)(x) = \mathbb{E}_{A=x}(\psi \mid \mathcal{F}_A)$
Hilbert spaces	$\mathbf{L}_2(\Omega, \mathbb{P}) \longrightarrow \mathbf{L}_2(\Omega, \mathbb{P})$	$\mathbf{L}_2(\mu_B) \longrightarrow \mathbf{L}_2(\mu_A)$
Operators	$V_A V_B^* \psi = \mathbb{E}(\psi \mid \mathcal{F}_B \cap \{A = B\})$ (need $V_B^* \psi \in \mathbf{L}_2(\mu_A) \cap \mathbf{L}_2(\mu_B)$)	$(V_A^* V_B f)(x) = \mathbb{E}_{A=x}(f \circ B \mid \mathcal{F}_A)$ the tranfer operator $R_{A,B}$
Special case $A = B$	$V_A V_A^* = \mathbb{E}(\cdot \mid \mathcal{F}_A)$	$V_A^* V_A = I_{\mathbf{L}_2(\mu_A)}$
Product of the conditional expectations $\mathbb{E}_{\mathcal{F}_A}$ and $\mathbb{E}_{\mathcal{F}_B}$	$\mathbb{E}_{\mathcal{F}_A} \mathbb{E}_{\mathcal{F}_B} = V_A R_{A,B} V_B^*$	$\mathbf{L}_2(\Omega, \mathbb{P}) \longrightarrow \mathbf{L}_2(\mu_B) \longrightarrow$ $\longrightarrow \mathbf{L}_2(\mu_A) \longrightarrow \mathbf{L}_2(\Omega, \mathbb{P})$

for functions $f : \mathbb{N}_0 \to \mathbb{R}$ such that

$$\sum_{n=0}^{\infty} |f(n)|^2 \mathbb{P}(\{B = n\}) < \infty, \quad i.e., f \in \ell_2(\mu_B).$$

3. Solenoid probability spaces

Why the solenoids? A number of reasons. Given an endomorphism σ in a measure space, the associated solenoid Sol_σ is then a useful tool for the study of scales of multiresolutions (see Definitions 3.7 and 3.16). The latter includes those resolutions arising naturally from discrete wavelet algorithms, as well as from the study of non-reversible dynamics in ergodic theory in and physics. In fact it is not so much Sol_σ itself that is central in this program, but rather probability spaces $(\mathrm{Sol}_\sigma, \mathcal{F}, \mathbb{P})$ where the solenoid is the sample space. It is the pair $(\mathcal{F}, \mathbb{P})$ which carries the information about the relevant scales of multiresolutions for the problem at hand, and the nature and the details of $(\mathcal{F}, \mathbb{P})$ change from one algorithm to the next; much like traditional wavelet analysis depend on scaling functions, father function, mother functions etc in $\mathbf{L}_2(\mathbb{R}^d)$. But the latter is too restrictive a framework; see e.g. Section 4 and [**9, 19, 25**]. See also Tables 1 and 4.

By "discrete wavelet algorithms" we mean recursive algorithms with selfsimilarity given by a scaling matrix. In one dimension, this may be just the N-adic scaling, but in general we allow for "discrete time" to be modelled by higher rank lattices, by more general discrete abelian groups, or even by infinite discrete sets with some given structure. For a given time-series, even in this general form, we may always introduce an associated generating function. This will be a function in "dual frequency variables" in one or more complex variables, and called the frequency response function (see e.g., [**19**]). In many classical wavelet settings the given discrete wavelet algorithms may be realized in $\mathbf{L}_2(\mathbb{R}^d)$ for some d, but such a realization places very strong restrictions and limitations on the given multi-band filters making up the discrete wavelet algorithm at hand. We show that with the Hilbert space $\mathbf{L}_2(\mathrm{Sol}_\sigma, \mathcal{F}, \mathbb{P})$, we can get around this difficulty, and still retain the useful features of multi-scale resolutions and selfsimilarity which makes the wavelet realizations so useful.

Motivated by multiresolutions in statistical computations, in many applications, and in particular in generalized wavelet algorithms, we study here a setting of dynamics of endomorphisms of measure spaces, denoting a given endomorphism by σ, say acting in X (see [**14, 16, 42–44**]). If the associated transfer operator R is further given to be σ-homogeneous (see Definition 3.1 below), we show that the associated R-Markov processes will be of a special kind: when realized in the natural probability space of an associated solenoid Sol_σ computed from σ, we then arrive at natural multi-scale resolutions inside the Hilbert space $\mathbf{L}_2(\mathrm{Sol}_\sigma, \mathcal{F}, \mathbb{P})$, with the scale of resolutions in question defined from the given endomorphism σ. In the case when σ is the scale endomorphism of a wavelet construction, we show that the multi-scale resolution at hand will agree with that of the associated solenoid analysis. And when a wavelet is realizable in Euclidean space, for example on the real line $\mathbb{R}$, then we show that then $\mathbb{R}$ is naturally embedded as a "curve" in the solenoid. Moreover, we identify the analogous multivariable setting with endomorphism and solenoid. Background references on analysis on solenoids and related multiresolutions include [**12, 19, 29, 40, 44, 49**].

In our discussion of solenoids and multiresolutions, we have here restricted the discussion to the commutative case, as our motivation is from stochastic processes. But in the recent literature, there is also an exciting, and somewhat parallel non-commutative theory of solenoids and their multiresolutions. It too is motivated (at least in part) by developments in the analysis of wavelet-multiresolutions, and the corresponding scaling operators. However, the relevant questions in the non-commutative theory are quite different from those addressed here. The relevant questions are simply different in the non-commutative theory. The differences between the two in fact reflect the dichotomy for two different notions of probability theory, the difference between (classical) commutative, versus non-commutative probability theory. Among the recent papers on the non-commutative theory, we mention [**10–12, 49, 50**], and the literature cited there.

3.1. Definitions. Consider a locally compact Hausdorff space X, with associated Borel sigma-algebra $\mathcal{B}$, let σ be a measurable endomorphism of X, which is onto. We denote by $\mathcal{M}(X, \mathcal{B})$ the space of all measurable functions from X into $\mathbb{R}$.

DEFINITION 3.1. A map R from $\mathcal{M}(X, \mathcal{B})$ into itself is called a σ-transfer operator (or a Ruelle operator) if

$$Rf \geq 0, \quad \forall f \in \mathcal{M}(X, \mathcal{B}) \text{ satisfying } f(x) \geq 0,\ \forall x \in X, \tag{3.1}$$

and if the pull-out property

$$R\left((f \circ \sigma)g\right) = fR(g), \quad \forall f, g \in \mathcal{M}(X, \mathcal{B}) \tag{3.2}$$

holds.

As a first example we have:

LEMMA 3.2. *Let $(X, \mathcal{B})$ be a measure-space and let A and B be two X-valued random variables on the probability space $(\Omega, \mathcal{F}, \mathbb{P})$, with transfer operator $R_{A,B}$ given by* (2.7). *Let σ be an endomorphism of X which is onto and such that*

$$\sigma \circ B = A. \tag{3.3}$$

Then $R_{A,B}$ satisfies the pull-out property (3.2). *Moreover*

$$R_{A,B}^* f = f \circ \sigma. \tag{3.4}$$

PROOF. Indeed

$$\begin{aligned} (R_{A,B}((f \circ \sigma)g)) \circ A &= \mathbb{E}\left[((f \circ \sigma)g) \circ B \,\middle|\, \mathcal{F}_A\right] \\ &= \mathbb{E}\left[(f \circ A)(g \circ B) \,\middle|\, \mathcal{F}_A\right] \quad (\text{since } \sigma \circ B = A) \\ &= (f \circ A)\mathbb{E}\left[g \circ B \,\middle|\, \mathcal{F}_A\right] \\ &= (f R_{A,B}(g)) \circ A. \end{aligned}$$

We now prove (3.4). We have

$$\begin{aligned} \langle g, f \circ \sigma \rangle_{\mu_B} &= \int_X g(x)((f \circ \sigma)(x)) d\mu_B(x) \\ &= \int_X R(g(f \circ \sigma))(x) d\mu_A(x) \\ &= \int_X f(x)(R(g)(x)) d\mu_A(x) \\ &= \langle R(g), f \rangle_{\mu_A}. \end{aligned}$$

□

We note that $\mathcal{F}_A \subset \mathcal{F}_B$ when (3.3) is in force. We now present an example of pairs of random variables for which neither $\mathcal{F}_A \subset \mathcal{F}_B$ nor $\mathcal{F}_B \subset \mathcal{F}_A$ hold. In particular they cannot be connected by an endomorphism of X.

EXAMPLE 3.3. Consider the space $\{-1, 1\}$ with probability distribution

$$p(\{1\}) = p(\{-1\}) = \frac{1}{2}.$$

We take $\Omega = \prod_{n=1}^{\infty} \{-1, +1\}$, and $\mathbb{P}$ the corresponding infinite product measure on the cylinder sigma-algebra. Let $a \in (0, 1)$ and define

$$E_a(\omega) = \sum_{k=1}^{\infty} \omega_k a^k, \tag{3.5}$$

where $\omega = (\omega_1, \omega_2, \ldots) \in \Omega$ and thus $\omega_k \in \{-1, 1\}$. The random variable E_a takes values in $\mathbb{R}$, and its distribution, defined by

$$\alpha_a(x_1, x_2) = \mathbb{P}(\omega \in \Omega,\ x_1 < E_a(\omega) < x_2)$$

has Fourier transform

$$\widehat{\alpha_a}(t) = \prod_{k=1}^{\infty} \cos(a^k t). \tag{3.6}$$

It is known that (see [**27, 29, 69**]:

(1) When $a < \frac{1}{2}$ the distributions α_a are singular, and mutually singular.
(2) When $a = \frac{1}{2}$ we obtain the Lebesgue measure.

(3) When $a \in \left(\frac{1}{2}, 1\right)$ the corresponding α_a are absolutely continuous with respect to Lebesgue measure, for almost all values of a. This is called the Erdös conjecture (see [**32, 33**]), and was proved in [**69**]. The only known value of $a > \frac{1}{2}$ for which α_a is known not to be absolutely continuous with respect to Lebesgue measure is the reciprocal of the golden ratio $a = \frac{\sqrt{5}-1}{2}$. See also [**26**, p. 48] for further references and information.

Taking a_1 and a_2 such that the corresponding distributions are mutually singular leads to random variables E_{a_1} and E_{a_2} which cannot be related by an endomorphism of X.

DEFINITION 3.4. The solenoid $\mathrm{Sol}_\sigma(X)$ associated with σ is the subset of sequences $(x_k)_{k\in\mathbb{N}_0}$ in $X^{\mathbb{N}_0}$ such that

$$\sigma(x_{k+1}) = x_k, \quad k = 0, 1, \ldots \tag{3.7}$$

REMARK 3.5. We think of a "point" in $\mathrm{Sol}_\sigma(X)$ as a path-governed by σ, and hence $\mathrm{Sol}_\sigma(X)$ as a path-space.

We set

$$\pi_k(x_0, x_1, \ldots) = x_k, \quad k \in \mathbb{N}_0 \text{ and } (x_k)_{k\in\mathbb{N}_0} \in X^{\mathbb{N}_0}, \tag{3.8}$$

and (3.7) can be rewritten as

$$\sigma \circ \pi_{n+1} = \pi_n, \quad n = 1, 2, \ldots \tag{3.9}$$

The endomorphism σ is (in general) neither one-to-one nor onto. But:

PROPOSITION 3.6. *The induced map $\widehat{\sigma}$ defined by*

$$\widehat{\sigma}(x_0, x_1, \ldots) = (\sigma(x_0), x_0, x_1 \ldots) \tag{3.10}$$

is one-to-one from $\mathrm{Sol}_\sigma(X)$ onto itself, with inverse

$$\widehat{\sigma}^{-1}(x_0, x_1, \ldots) = (x_1, x_2, \ldots) \tag{3.11}$$

PROOF. One-to-oneness is clear. Let τ denote the map in (3.11). Then,

$$\widehat{\sigma} \circ \tau(x_0, x_1, \ldots) = \widehat{\sigma}(x_1, \ldots) = (\sigma(x_1), x_1, \ldots) = (x_0, x_1, \ldots)$$

since $\sigma(x_1) = x_0$ and

$$\tau \circ \widehat{\sigma}(x_0, x_1, \ldots) = \tau(\sigma(x_0), x_0, x_1, \ldots) = (x_0, x_1, \ldots).$$

□

We note that

$$\pi_0 \circ \widehat{\sigma} = \sigma \circ \pi_0 \quad \text{and} \quad \pi_{n+1} \circ \widehat{\sigma} = \pi_n, \quad n = 0, 1, \ldots \tag{3.12}$$

Recall that the notation $\mathcal{F}_A$ was introduced in (2.1). We set

$$\mathcal{F}_{\pi_n} = \mathcal{F}_n,$$

DEFINITION 3.7. $\mathcal{F}_n = \pi_n^{-1}(\mathcal{B})$ is the sigma-algebra generated by the random variables $f \circ \pi_n$, where f runs through the measurable functions on $(X, \mathcal{B})$.

As an immediate consequence of (3.12) we have:

LEMMA 3.8. *In the notation of Definition 3.7, we have*

$$\mathcal{F}_0 \subset \mathcal{F}_1 \subset \cdots \subset \mathcal{F}_n \subset \mathcal{F}_{n+1} \subset \cdots$$

and $\cup_{n=0}^{\infty} \mathcal{F}_n = \mathcal{F}$.

PROOF. Let $f \in \mathcal{M}(X, \mathcal{B}_X)$ and $n \in \mathbb{N}$. We have

$$f \circ \pi_n = (f \circ \sigma) \circ \pi_{n+1}$$

and so $\mathcal{F}_n \subset \mathcal{F}_{n+1}$. Since the sigma-algebra $\mathcal{F}$ on $\mathrm{Sol}_\sigma(X)$ is the cylinder sigma-algebra obtained from $\prod_{n=0}^\infty X$ we have that $\mathcal{F} = \vee_{n=0}^\infty \mathcal{F}_n$, where $\vee$ denotes the lattice operation on sigma-algebras. □

THEOREM 3.9. *Let $(X, \mathcal{B}, \sigma, R, h, \lambda)$ be as above, and assume $Rh = h$. Then there exists a unique probability measure $\mathbb{P}$ defined on the cylinder sigma-algebra on the associated solenoid $\mathrm{Sol}_\sigma(X)$ such that*

$$\begin{aligned}(3.13)\quad \int_{\mathrm{Sol}_\sigma(X)} & (f_0\pi_0)(\omega)(f_1\pi_1)(\omega)\cdots(f_n\pi_n)(\omega)d\mathbb{P}(\omega)\\ &= \int_X (f_0(x)R(f_1R(f_2\cdots R(f_nh))))(x)d\lambda(x)\end{aligned}$$

for all $n \in \mathbb{N}_0$ and $f_0, \ldots, f_n \in \mathcal{M}(X, \mathcal{B})$. In the normalized case, we use $h \equiv 1$

PROOF. We first remark that, in view of (3.9), $\mathbb{P}$ (if it exists) is uniquely determined by

$$(3.14)\qquad \int_{\mathrm{Sol}_\sigma} (f \circ \pi_0)(\omega)d\mathbb{P}(\omega) = \int_X f(x)d\lambda(x),$$

and

$$(3.15)\qquad \int_{\mathrm{Sol}_\sigma(X)} (f \circ \pi_n)(\omega)d\mathbb{P}(\omega) = \int_X R^n(fh)(x)d\lambda(x).$$

For every $n \in \mathbb{N}_0$, there exists a measure $\mathbb{P}_n$ on $\mathcal{F}_n$ such that

$$(3.16)\qquad \int_{\mathrm{Sol}_\sigma(X)} (f \circ \pi_n)(\omega)d\mathbb{P}_n(\omega) = \int_X R^n(fh)(x)d\lambda(x).$$

Setting, in (3.15), $n+1$ instead of n, and taking into account (3.9), we have

$$\begin{aligned}\int_{\mathrm{Sol}_\sigma(X)} (f \circ \pi_n)(\omega)d\mathbb{P}_{n+1}(\omega) &= \int_{\mathrm{Sol}_\sigma(X)} (f \circ \pi_{n+1})(\omega)d\mathbb{P}_{n+1}(\omega)\\ &= \int_X R^{n+1}(f \circ \sigma)(x)d\lambda(x)\\ &= \int_X R^n\left(R(f \circ \sigma)\right)(x)d\lambda(x)\\ &= \int_X \left(R^n f\right)(x)d\lambda(x)\\ &\text{(by the pull-out property (3.2) since } R \text{ is normalized)}\\ &= \int_{\mathrm{Sol}_\sigma(X)} (f \circ \pi_n)(\omega)d\mathbb{P}_n(\omega).\end{aligned}$$

By Kolmogorov's extension theorem (see e.g. [**60**]), the family $(\mathbb{P}_n)_{n\in\mathbb{N}}$ extends to a probability measure $\mathbb{P}$ on the cylinder sigma-algebra. □

We define the measure (λR) by

$$(3.17)\qquad \int_X f(x)d(\lambda R)(x) = \int_X R(f)(x)d\lambda(x).$$

DEFINITION 3.10. We say that σ is ergodic if

$$\cap_{n=1}^{\infty}\sigma^{-n}\left(\mathcal{B}_X\right)=\{\emptyset, X\},$$

modulo sets of λ-measure zero.

The examples of endomorphisms σ which we consider here are ergodic.

THEOREM 3.11. *Let W be a positive measurable function on $(X,\mathcal{B})$. The following are equivalent:*
(1) $\lambda R << \lambda$, *and*

$$\frac{d\lambda R}{d\lambda}=W. \tag{3.18}$$

(2) $\mathbb{P}\circ\widehat{\sigma} << \mathbb{P}$ *and*

$$\frac{d\mathbb{P}\circ\widehat{\sigma}}{d\mathbb{P}}=W\circ\pi_0 \tag{3.19}$$

PROOF. Assume that (1) is in force. To prove (2) we will show that (3.19) holds, or equivalently, that

$$\int_{\mathrm{Sol}_\sigma(X)}\psi\mathbb{P}=\int_{\mathrm{Sol}_\sigma(X)}(\psi\circ\widehat{\sigma})\left(W\circ\pi_0\right)d\mathbb{P} \tag{3.20}$$

for all measurable functions ψ on $\mathrm{Sol}_\sigma(X)$. It suffices to take ψ of the form $\psi=f\circ\pi_n$ for $n=0,1,\ldots$. We first consider the case $n=0$. Let $\omega=(x_0,x_1,\ldots)$. Then $\widehat{\sigma}(\omega)=(\sigma(x_0),x_0,x_1,\ldots)$, and so with $\psi=f\circ\pi_0$, the right-hand side of (3.20) is equal to

$$\begin{aligned}
\int_{\mathrm{Sol}_\sigma(X)}(f\circ\pi_0\circ\widehat{\sigma})(\omega)(W\circ\pi_0)(\omega)d\mathbb{P}(\omega) &= \int_{\mathrm{Sol}_\sigma(X)}(f\circ\sigma)(x_0)(W\circ\pi_0)(\omega)d\mathbb{P}(\omega)\\
&= \int_{\mathrm{Sol}_\sigma(X)}(f\circ\sigma\circ\pi_0)(\omega)(W\circ\pi_0)(\omega)d\mathbb{P}(\omega)\\
&= \int_{\mathrm{Sol}_\sigma(X)}\left(((f\circ\sigma)W)\circ\pi_0\right)(\omega)d\mathbb{P}(\omega)\\
&= \int_X(f\circ\sigma)(x)W(x)d\lambda(x)\\
&= \int_X\left(R(f\circ\sigma)\right)(x)d\lambda(x)\quad \text{(using (3.17)}\\
&\qquad\qquad \text{and (3.18))}\\
&= \int_X f(x)d\lambda(x)\quad \text{since } R \text{ is normalized:}\\
&\qquad\qquad R1=1 \text{ and using the}\\
&\qquad\qquad \text{pullout property (3.2)}\\
&= \int_{\mathrm{Sol}_\sigma(X)}(f\circ\pi_0)(\omega)d\mathbb{P}(\omega),
\end{aligned}$$

which is the left-hand side of (3.20).

We now consider the case $n > 0$. The right-hand side of (3.20) is equal to:

$$\begin{aligned}\int_{\mathrm{Sol}_\sigma(X)} (f\circ\pi_n\circ\widehat{\sigma})(\omega)(W\circ\pi_0)(\omega)d\mathbb{P}(\omega) &= \int_{\mathrm{Sol}_\sigma(X)} f(x_{n-1})(W\circ\pi_0)(\omega)d\mathbb{P}(\omega)\\ &= \int_{\mathrm{Sol}_\sigma(X)} (f\circ\pi_{n-1})(\omega)(W\circ\pi_0)(\omega)d\mathbb{P}(\omega)\\ &= \int_X \left(R^{n-1}(f)\right)(x)W(x)d\lambda(x)\\ &\quad\text{(where we have used (3.15))}\\ &= \int_X \left(R^{n}(f)\right)(x)d\lambda(x)\\ &\quad\text{(by definition of } W)\\ &= \int_{\mathrm{Sol}_\sigma(X)} (f\circ\pi_n)(\omega)d\mathbb{P}(\omega)\\ &\quad\text{(by (3.15))}\\ &= \int_{\mathrm{Sol}_\sigma(X)} \psi(\omega)d\mathbb{P}(\omega),\end{aligned}$$

by definition of $\psi = f\circ\pi_n$.

Conversely, we now assume that $\mathbb{P}\circ\widehat{\sigma} << \mathbb{P}$, with Radon-Nikodym derivative given by (3.19). Let $\psi = f\circ\pi_1$. We have:

$$\int_{\mathrm{Sol}_\sigma(X)} (\psi\circ\widehat{\sigma})(\omega)(W\circ\pi_0)(\omega)d\mathbb{P}(\omega) = \int_{\mathrm{Sol}_\sigma(X)} \psi(\omega)d\mathbb{P}(\omega). \tag{3.21}$$

Since $\pi\circ\widehat{\sigma} = \pi_0$, this latter equality is equivalent to:

$$\int_{\mathrm{Sol}_\sigma(X)} \left((fW)\circ\pi_0\right)(\omega)d\mathbb{P}(\omega) = \int_{\mathrm{Sol}_\sigma(X)} (f\circ\pi_1)(\omega)d\mathbb{P}(\omega),$$

that is

$$\int_X (fW)(x)d\lambda(x) = \int_X (R(f))(x)d\lambda(x) \tag{3.22}$$

where we used (3.15). But (3.22) means that $\lambda R << \lambda$ with Radon-Nikodym derivative equal to W. □

REMARK 3.12. It follows from (3.15) that the probability distribution μ_n of the random variable π_n is equal to

$$d\mu_n(x) = W(x)((W\circ\sigma)(x))\cdots((W\circ\sigma^{n-1})(x))d\lambda(x),$$

and that

$$\frac{d\mu_{n+1}}{d\mu_n} = (W\circ\sigma^n)(x). \tag{3.23}$$

We will assume that there the Radon-Nikodym derivative $W = \frac{d(\lambda R)}{d\lambda}$ exists, that is:

$$\int_X R(f)(x)d\lambda(x) = \int_X f(x)W(x)d\lambda(x), \quad \forall f\in\mathcal{M}(X,\mathcal{B}), \tag{3.24}$$

and that, furthermore, the Ruelle operator R in equations (4.13) and (3.1) is of Perron-Frobenius type in the sense that there exists $h \geq 0$, $h \in \mathcal{M}(X, \mathcal{B})$ such that

$$Rh = h, \tag{3.25}$$

and normalized to

$$\int_X h(x) d\lambda(x) = 1. \tag{3.26}$$

When R is not normalized one can replace R with the operator R' defined by

$$R'f = \frac{R(fh)}{h}.$$

It satisfies $R'1 = 1$. See Remark 2.32.

DEFINITION 3.13. We will call $(X, \mathcal{B}, \sigma, R, h, \lambda)$ a generator for a path space when $\lambda R << \lambda$ and when R is normalized.

As a consequence of (3.9) we have (recall that σ is not one-to-one in general):

PROPOSITION 3.14. *The distributions μ_k and μ_{k+1} are related by*

$$\mu_{k+1} \circ \sigma^{-1} = \mu_k, \tag{3.27}$$

meaning that

$$\mu_{k+1}\left(\sigma^{-1}(B)\right) = \mu_k(B), \quad \forall B \in \mathcal{B}. \tag{3.28}$$

PROOF. By definition of μ_k we have:

$$\int_{\mathrm{Sol}_\sigma(X)} (f \circ \pi_k)(\omega) d\mathbb{P}(\omega) = \int_X f(x) d\mu_k(x).$$

Hence,

$$\begin{aligned}\int_X (f \circ \sigma)(x) d\mu_{k+1}(x) &= \int_{\mathrm{Sol}_\sigma(X)} ((f \circ \sigma) \circ \pi_{k+1})(\omega) d\mathbb{P}(\omega) \\ &= \int_{\mathrm{Sol}_\sigma(X)} (f \circ \pi_k)(\omega) d\mathbb{P}(\omega) \quad \text{(using (3.9))} \\ &= \int_X f(x) d\mu_k(x).\end{aligned}$$

It suffices to take $f(x) = \chi_B(x)$ to obtain (3.28). □

REMARK 3.15. (3.28) is independent of the given probability measure on the cylinder sigma-algebra.

3.2. The multiresolution associated with a solenoid. We begin with a table relative to the wavelet realization by unitary operators; the third column, related to the classical $\mathbf{L}_2(\mathbb{R}, dx)$ wavelets is elaborated upon in Section 4.3.

DEFINITION 3.16. Let $\mathcal{H}$ be a Hilbert space, let $U : \mathcal{H} \longrightarrow \mathcal{H}$ be a unitary operator and let $(\mathcal{H}_n)_{n \in \mathbb{Z}}$ be an indexed family of closed subspaces such that:
(i) $\mathcal{H}_{n+1} \subset \mathcal{H}_n$, $n \in \mathbb{Z}$,
(ii) $U^{-k}\mathcal{H}_0 = \mathcal{H}_k$, $k \in \mathbb{Z}$,
(iii) $\bigwedge_{k \in \mathbb{Z}} \mathcal{H}_k$ is at most one dimensional,
and
(iv) $\bigvee_{k \in \mathbb{Z}} \mathcal{H}_k = \mathcal{H}$.
Here $\bigwedge$ and $\bigvee$ refer to the lattice operations applied to closed subspaces in $\mathcal{H}$.

When $(\mathcal{H}, U, (\mathcal{H}_n)_{n\in\mathbb{Z}})$ satisfy (i)-(iv), then we say that it is a multiresolution (or multi-scale resolution), and that U is the associated scaling operator.

REMARK 3.17. Let U be a unitary operator which is part of a multiresolution, then it can be shown that the spectrum of U must be as follows: Except for the point $\lambda = 1$ occurring with at most multiplicity one, the spectrum of U must be absolutely continuous with uniform multiplicity infinity. This is an application of ideas of Wold, Lax-Phillips, and Stone-von Neumann; see [**52, 54**]. See also Remark 3.22 below.

We shall outline below a number of examples of multiresolutions, in wavelet theory and in dynamics more generally. This will make use of the theory we already developed in Section 2 above.

TABLE 4. Wavelets realization by unitary operators

The case	$\mathbf{L}_2(\mathbb{R}, dx)$	Fourier transform of $\mathbf{L}_2(\mathbb{R}, dx)$	General (solenoid) $\mathbf{L}_2(\mathrm{Sol}_\sigma, \mathcal{C}, \mathbb{P})$
The unitary operator	$(Ug)(x) = \frac{1}{\sqrt{N}} g(x/N)$	$(U\gamma)(t) = \sqrt{N}\gamma(Nt)$	$(U\psi)(\omega)$ $= (\psi \circ \widehat{\sigma})(m_0 \circ \pi_0)$
(1) Map onto the zero resolution subspace.	$K\xi = \sum_{k\in\mathbb{Z}} \xi_k \varphi(x-k)$ belongs to $\mathbf{L}_2(\mathbb{R}, dx)$ for $(\xi_n) \in \ell_2(\mathbb{Z})$.	$(Kf)(t) = f(t)\widehat{\varphi_0}(t)$ belongs to $\mathbf{L}_2(\mathbb{R}, dx)$ for $f \in \mathbf{L}_2(\mathbb{T})$.	$V_{\pi_0} f = f \circ \pi_0$ belongs to $\mathbf{L}_2(\mathrm{Sol}_\sigma, \mathbb{P})$ for $f \in \mathbf{L}_2(X, \lambda)$.
(2) Average operator.	$(S\xi)_j = \sum_{k\in\mathbb{Z}} \xi_k a_{Nk-j}$	$(Sf)(t) = m_0(t) f(Nt)$	$Sf = m_0 \cdot f \circ \sigma$
Level zero resolution Invariant subspace for S (using (1) above)	$\xi \in \ell_2(\mathbb{Z}) \simeq \mathcal{H}_0$	$f \in \mathbf{L}_2(\mathbb{T}) \simeq \mathcal{H}_0$	$f \in \mathbf{L}_2(X, \lambda) \simeq \mathcal{H}_0^{\mathrm{Sol}_\sigma}$

REMARK 3.18. In the above table, in the second column, the map m_0 is a continuous function $m_0(z)$ on the unit circle, and the coefficients a_n are the Fourier coefficients of $|m_0(e^{it})|^2$. In the third column, we are in the special case where $W = |m_0|^2$, then W satisfies the conditions of Theorem 3.11.

Recall from Lemma 2.17 that if μ_0 and μ_1 are equivalent then, the set where W vanishes has measure zero.

PROPOSITION 3.19. *Assume that μ_0 and μ_1 are equivalent. Then, the map*

$$U\psi = \sqrt{W \circ \pi_0}\,(\psi \circ \widehat{\sigma}) \tag{3.29}$$

is unitary from $\mathbf{L}_2(\mathrm{Sol}_\sigma(X), \mathbb{P})$ onto itself, and its inverse is given by

$$U^{-1}\psi = \frac{1}{\sqrt{W \circ \pi_1}} \psi \circ \widehat{\sigma}^{-1}, \tag{3.30}$$

where $\widehat{\sigma}^{-1}$ is given by (3.11).

PROOF. The first claim is a consequence of (3.21) with $|\psi^2|$ instead of ψ. We now check that $UU^{-1}\psi = \psi$. We have

$$\begin{aligned} UU^{-1}\psi &= \sqrt{W \circ \pi_0}\,(U^{-1}\psi) \circ \widehat{\sigma}^{-1} \\ &= \sqrt{W \circ \pi_0}\frac{1}{\sqrt{W \circ \pi_1 \circ \widehat{\sigma}}}\psi \circ \widehat{\sigma} \circ \widehat{\sigma}^{-1} \\ &= \psi \end{aligned}$$

since $\pi_1 \circ \widehat{\sigma} = \pi_0$. The proof that $U^{-1}U\psi = \psi$ is similar, and omitted. □

DEFINITION 3.20. Let

$$\mathcal{H}_0 = \{f \circ \pi_0 \mid f \in \mathbf{L}_2(X, hd\lambda)\}\,, \tag{3.31}$$

the resolution subspace. The family $\mathcal{H}_n = U^{-n}\mathcal{H}_0$, $n \in \mathbb{Z}$, is called the multiresolution associated with the solenoid, and will be denoted by MR_σ.

Note that

$$\int_{\mathbf{L}_2(\mathbb{P})} |f \circ \pi_0|^2 d\mathbb{P} = \int_X |f(x)|^2 h(x) d\lambda(x). \tag{3.32}$$

PROPOSITION 3.21. *Let $f \in L^\infty(X)$. The multiplication map*

$$M_{f\circ\pi_0} \tag{3.33}$$

sends $\mathbf{L}_2(\mathrm{Sol}_\sigma, \mathbb{P})$ into itself, and $\mathcal{H}_n$ into itself for all n. We have

$$\begin{array}{ccc} \mathbf{L}_2(\mathrm{Sol}_\sigma, \mathbb{P}) & \xrightarrow{\quad U \quad} & \mathbf{L}_2(\mathrm{Sol}_\sigma, \mathbb{P}) \\ M_{f\circ\pi_0}\Big\downarrow & & \Big\downarrow M_{f\circ\sigma\circ\pi_0} \\ \mathbf{L}_2(\mathrm{Sol}_\sigma, \mathbb{P}) & \xrightarrow{\quad U \quad} & \mathbf{L}_2(\mathrm{Sol}_\sigma, \mathbb{P}) \end{array}.$$

and the following covariance relation holds (see also Remarks 3.22 and 3.23)

$$UM_{f\circ\pi_0}U^{-1} = M_{f\circ\sigma\circ\pi_0}. \tag{3.34}$$

The map $f \mapsto M_f$ defines a representation of $L^\infty(X)$ by bounded operators on $\mathbf{L}_2(\mathrm{Sol}_\sigma, \mathbb{P})$.

PROOF. For $\psi \in \mathcal{M}(\Omega)$ we have

$$\begin{aligned} UM_{f\circ\pi_0}\psi &= \sqrt{W \circ \pi_0}(f \circ \pi_0 \circ \widehat{\sigma})(\psi \circ \widehat{\sigma}) \\ &= \sqrt{W \circ \pi_0}(f \circ \sigma \circ \pi_0)(\psi \circ \widehat{\sigma}) \text{ (where we use } \pi_0 \circ \widehat{\sigma} = \sigma \circ \pi_0\text{; see (3.12))} \\ &= M_{f\circ\sigma\circ\pi_0}U\psi. \end{aligned}$$

□

REMARK 3.22. Note that (3.34) is an instance of a covariance relation: It states that the representation M is unitarily equivalent to the representation obtained from it by substitution with the endomorphism σ. As a result, the projection valued measure determining M will satisfy the analogous covariance. This is outlined in (3.35) below. For the convenience of the reader, let us give the following analogy: Consider the two canonical variables P and Q in the canonical commutation relation from quantum mechanics in the Weyl exponentiated form. If E_Q denotes the projection valued spectral measure of Q, then the unitary one-parameter group $U(t)$, generated by P, satisfies a covariance in the form

$$U(t)E_Q(B)U(-t) = E_Q(B+t),$$

all for all Borel sets B, and all $t \in \mathbb{R}$. Here we use the word "covariance" in the same general context, but now for endomorphisms, also now instead for a single unitary operator. Many covariance relations have solutions that are unique up to unitary equivalence, for example the canonical $P - Q$ relation does; this is a form of the Stone-von Neumann uniqueness theorem. See [**51, 52, 54**].

REMARK 3.23. The commutative von Neumann algebra $\mathcal{M}_{\pi_0}$ of the multiplication operators $M_{f\circ\pi_0}$ with $f \in \mathbf{L}^\infty(X, \mathcal{B}_X)$ has the spectral representation (see Section 2.2 and equation (2.18))

$$M_{f\circ\pi_0} = \int_X f(x)\chi_{\{\pi_0\in dx\}}$$

where $\chi_{\{\pi_0\in dx\}}$ (also denoted by $\mathscr{E}_{\pi_0}(dx)$ is the projection-valued measure given by (2.16) and arising from the Stone theorem applied to $\mathcal{M}_{\pi_0}$; see [**58**].

Define

$$\mathscr{E}_{\pi_0}^{(\sigma)}(\omega) = M_{\chi_{\{\pi_0\in\sigma^{-1}(L)\}}}$$

As in (3.34) we arrive at the following selfsimilarity property for $\mathscr{E}_{\pi_0}$ with U given by (3.29):

$$U\mathscr{E}_{\pi_0}(L)U^{-1} = \mathscr{E}_{\pi_0}\left(\sigma^{-1}(L)\right), \quad \forall L \in \mathcal{B}_X, \tag{3.35}$$

which we also rewrite as $U\mathscr{E}_{\pi_0}U^{-1} = \mathscr{E}_{\pi_0}^{(\sigma)}$.

We now give another interpretation of the resolution subspace $\mathcal{H}_n$. For $\mathcal{F}_n$, see Definition 3.7.

PROPOSITION 3.24. *We have:*

$$\mathcal{H}_n = \mathbf{L}_2(\mathrm{Sol}_\sigma, \mathcal{F}_n, \mathbb{P}) = \mathbb{E}_{\mathcal{F}_n}(\mathbf{L}_2(\mathrm{Sol}_\sigma, \mathcal{F}, \mathbb{P})). \tag{3.36}$$

PROOF. The proof follows from Corollary 2.11. □

Equation (3.37) below is a generalization of the classical notion of martingale.

PROPOSITION 3.25. *Assume R normalized,* i.e. $R1 = 1$. *Then*

$$\mathbb{E}\left(f\circ\pi_{n+1}\middle|\mathcal{F}_n\right) = R(f)\circ\pi_n, \quad n = 0, \ldots \tag{3.37}$$

PROOF. This follows from Lemma 2.4. □

DEFINITION 3.26. The sequence $(T_n)_{n\in\mathbb{N}_0}$ of random variables from the probability space $(\Omega, \mathcal{F}, \mathbb{P})$ into the measurable space $(X, \mathcal{B})$ is called a Markov chain if

$$\mathbb{E}(T_{n+1}\in B\big|T_0, \ldots, T_n) = \mathbb{E}(T_{n+1}\in B\big|T_n) = (R(\chi_B))\circ T_n \tag{3.38}$$

PROPOSITION 3.27. *We have*

$$\mathbb{E}\left(f\circ\pi_{n+1}\middle|\mathcal{F}_0, \ldots, \mathcal{F}_n\right) = \mathbb{E}\left(f\circ\pi_{n+1}\middle|\mathcal{F}_n\right), \quad n = 0, \ldots \tag{3.39}$$

PROOF. The result follows from Theorem 2.30. □

THEOREM 3.28. *Assume that R is normalized, and let $(\pi_n)_{n\in\mathbb{N}_0}$ be the stochastic process on $\mathrm{Sol}_\sigma(X)$ defined by the coordinates. Then:*

$$\mathbb{E}\left(f\circ\pi_{n+k}\middle|\mathcal{F}_n\right) = R^k(f)\circ\pi_n, \quad n=0,\ldots, \quad k=1,2,\ldots \tag{3.40}$$

PROOF. The proof uses the chain rule for conditional expectation and induction. It is enough to consider the case $k=2$. We then have:

$$\begin{aligned}\mathbb{E}\left(f\circ\pi_{n+2}\middle|\mathcal{F}_n\right) &= \mathbb{E}\left(\mathbb{E}\left(f\circ\pi_{n+2}\middle|\mathcal{F}_{n+1}\right)\middle|\mathcal{F}_n\right)\\ &= \mathbb{E}\left(R(f)\circ\pi_{n+1}\middle|\mathcal{F}_n\right)\\ &= R^2(f)\circ\pi_n,\end{aligned}$$

where we used twice (3.37). □

3.3. Conditional expectations associated with a solenoid.

PROPOSITION 3.29. *The map $Vf = f\circ\pi_0$ is an isometry from $\mathbf{L}_2(X, hd\lambda)$ into $\mathbf{L}_2(\mathrm{Sol}_\sigma, \mathbb{P})$.*

PROOF. This is a corollary of Lemma 2.2. □

DEFINITION 3.30. The projection $\mathbb{E}_0 = VV^*$ in $\mathbf{L}_2(\mathrm{Sol}_\sigma, \mathbb{P})$ is called the conditional expectation onto $\mathcal{F}_0$ of the multiresolution $(\mathrm{Sol}_\sigma, \mathbb{P})$.

3.4. A general setting and an inverse problem. We now present a general setting, which includes the preceding analysis. We start from a probability space $(\Omega, \mathcal{F}, \mathbb{P})$, and a measurable space $(X, \mathcal{B})$. We assume given a sequence of random variables $(T_n)_{n\in\mathbb{N}_0}$ from Ω to X, and an endomorphism σ from X into X. We assume that

$$\sigma\circ T_{n+1} = T_n, \quad n=0,1,\ldots, \tag{3.41}$$

or, equivalently,

$$\forall\omega\in\Omega,\ T_{n+1}(\omega)\in\sigma^{-1}(T_n(\omega)). \tag{3.42}$$

The map

$$\widetilde{T}(\omega) = (T_0(\omega), T_1(\omega), \ldots) \tag{3.43}$$

is measurable from Ω into $\mathrm{Sol}_\sigma(X)$. It induces a probability measure $\mathbb{P}^*$ on the cylinder sigma-algebra of $X^{\mathbb{N}}$ via the formula

$$\mathbb{P}^*(A) = \mathbb{P}(\widetilde{T}^{-1}(A)). \tag{3.44}$$

The sequence $(T_n)_{n\in\mathbb{N}_0}$ generates a family of sigma-algebras, namely

$$\mathcal{F}_n = \left\{T_n^{-1}(A)\,;\, A\in\mathcal{B}\right\}. \tag{3.45}$$

In view of (3.4) we have $\mathcal{F}_n\subset\mathcal{F}_{n+1}$.

We now recall a technical lemma, to be used in the proof of Theorem 3.32

LEMMA 3.31. *Let $(\Omega, \mathcal{F})$ and $(X, \mathcal{B})$ be two measurable spaces and let T be a map from Ω into X. Let*

$$\mathcal{F}_T = \left\{T^{-1}(B)\,;\, B\in\mathcal{B})\right\}.$$

Then a real valued function ψ defined on Ω is $\mathcal{F}_T$-measurable if and only if it can be written in the form

$$\psi = f\circ T$$

for a uniquely defined $\mathcal{B}$-measurable function f.

THEOREM 3.32. *There exists a positive operator defined on the space of measurable functions from X to $\mathbb{R}$ such that*

$$\mathbb{E}\left(f \circ T_{n+1} \middle| \mathcal{F}_n\right) = R(f) \circ T_n. \tag{3.46}$$

PROOF. The existence of R follows from Lemma 3.31, and the positivity of R follows from the fact that a conditional expectation is an orthogonal projection. □

COROLLARY 3.33.

$$R(1) = 1, \tag{3.47}$$

$$R((f \circ \sigma)g) = fR(g). \tag{3.48}$$

PROOF. The first equation follows from setting $n = 0$ and $f \equiv 1$ in (3.46). The second equation is proved as follows. We have

$$\begin{aligned}
\mathbb{E}\left(((f \circ \sigma)g) \circ T_1 \,\middle|\, \mathcal{F}_0\right) &= \mathbb{E}\left((f \circ \sigma \circ T_1)(g \circ T_1) \,\middle|\, \mathcal{F}_0\right) \\
&= \mathbb{E}\left((f \circ T_0)(g \circ T_1) \,\middle|\, \mathcal{F}_0\right) \\
&= (f \circ T_0)\mathbb{E}\left((g \circ T_1) \,\middle|\, \mathcal{F}_0\right) \\
&= (f \circ T_0)(R(g) \circ T_0) \\
&= (fR(g)) \circ T_0.
\end{aligned}$$

□

4. Examples and applications: Transfer operators and Markov moves

While in the abstract, as we showed, Markov chains are derived from positive operators R, acting on functions on a fixed measure space $(X, \mathcal{B}_X)$. Starting with a choice of R (the transfer operator), we then build a Markov chain $T_0, T_1, T_2, \ldots$, with these random variables (r.v) realized in a suitable probability space $(\Omega, \mathcal{F}, \mathbb{P})$, and each r.v. taking values in X, measurable of course with respect to the respective sigma algebras, $\mathcal{F}$ on Ω, and $\mathcal{B}_X$ on X. Conversely, every Markov chain is determined by its transfer operator.

The purpose of the examples below is to put this correspondence into more practical terms. The range of the examples we give will cover (i) iterated function systems (IFS), (ii) wavelet multiresolution constructions, and (iii) IFSs with random "control."

An IFS on a fixed measure space $(X, \mathcal{B}_X)$ is a system of mappings τ_i, each acting in $(X, \mathcal{B}_X)$, and each assigned a probability, say p_i which may or may not be a function of x. For standard IFSs it is not, but for wavelet constructions it is. In the latter, the functions $p_i(x)$ reflect the multi-band filters making up the wavelet algorithm. Moreover, the sets $\tau_i(X)$ partition X, but they may have overlap, or not. The Markov chains for the non-overlapping IFSs are simpler.

Returning to the general case, we now briefly sketch the idea behind Markov moves in IFSs with random control in a bit more detail. The examples below will supply hands-on cases, serving to illustrate the general idea.

The Markov move: Starting with a point x in X, the Markov move to the next point is in two steps, as follows, the combined two steps describing the move from T_0 to T_1, and more generally from T_n to T_{n+1}. The initial point x will first move to one of the sets $\tau_i(X)$ with probability p_i, and once there, it will "choose" a definite position (within $\tau_i(X)$), and this second move will be prescribed by a fixed law (a given probability distribution); for example, the law could be the uniform distribution, or something different. However, for Markov chains, the law is the same in the move from T_n to T_{n+1}, for all n.

4.1. First examples.

EXAMPLE 4.1. In the first example, $X = [0,1]$ and $\sigma(x) = 4x(1-x)$, called the logistic map. von Neumann and Ulam proved that an invariant measure is

$$d\mu(x) = \frac{dx}{\pi\sqrt{x(1-x)}},$$

the Beta $B(\frac{1}{2},\frac{1}{2})$ distribution, i.e. $\mu \circ \sigma^{-1} = \mu$. See [**38**, pp. 87-91]. The corresponding transfer operator is

$$(Rf)(x) = \frac{1}{2}\left(f\left(\frac{1+\sqrt{1-x}}{2}\right) + f\left(\frac{1-\sqrt{1-x}}{2}\right)\right) \tag{4.1}$$

We note that

$$\mu R \neq \mu.$$

We now turn to an example of a transfer operator $R_F : X \times Y \longrightarrow X$

$$(R_F f)(x) = \int_Y f(F(x,y))d\nu(y)$$

in which

$$\mu R_F = \mu \tag{4.2}$$

for the $B(\frac{1}{2},\frac{1}{2})$ law μ. As a consequence of (4.2), we have that the corresponding probability measure $\mathbb{P}$ in $(\prod_{n=0}^{\infty} X, \mathcal{F}, \mathbb{P})$ will be shift-invariant.

EXAMPLE 4.2. We take $X = (0,1)$. The endomorphism σ will depend on a parameter $u \in (0,1)$, and is defined as follows. Set

$$\tau_0^{(u)}(x) = ux, \tag{4.3}$$

$$\tau_1^{(u)}(x) = u + (1-u)x. \tag{4.4}$$

Then,

$$\sigma^{(u)}(x) = \begin{cases} \frac{x}{u}, & 0 < x \leq u, \\ -\frac{u}{1-u} + \frac{x}{1-u}, & u < x < 1. \end{cases} \tag{4.5}$$

Then,

$$\sigma^{(u)} \circ \tau_i^{(u)}(x) = x, \quad \text{for } i = 1,2, \quad \text{and} \quad x \in (0,1).$$

Then,

$$R^{(u)}f(x) = \frac{1}{2}\left(f(\tau_0^{(u)})(x) + f(\tau_1^{(u)})(x)\right). \tag{4.6}$$

Let λ be the Lebesgue measure on $[0,1]$. Then

$$d(\lambda R^{(u)}) = W^{(u)} d\lambda, \quad u \in (0,1) \tag{4.7}$$

with

$$W^{(u)}(x) = \begin{cases} \frac{1}{2u}, & 0 \leq x < u, \\ \frac{1}{2(1-u)}, & u \leq x < 1. \end{cases}$$

Note that $W^{(u)}(x) \equiv 1$ if and only if $u = \frac{1}{2}$. For every $u \in (0,1)$ we have a quasi-invariant measure $\mathbb{P}^{(u)}$ such that

$$\frac{\mathbb{P}^{(u)} \circ \widehat{\sigma^{(u)}}}{d\mathbb{P}^{(u)}} = W^{(u)} \circ \pi_0.$$

Let $Y = \{0,1\} \times (0,1)$ and $d\nu = p_1 \times p_2$ be the product measure with $p_1(0) = p_1(1) = \frac{1}{2}$ and p_2 the uniform probability distribution on $(0,1)$. Let furthermore

$$F(x,(i,u)) = \begin{cases} ux, & \text{if} \quad i = 0, \\ (1-u)x + u, & \text{if} \quad i = 1, \end{cases}$$

and

$$\begin{aligned} (R_F f)(x) &= \int_Y f(F(x,(i,u)))d\nu(i,u) \\ &= \frac{1}{2}\int_0^1 (f(ux) + f((1-u)x+u))\,du \\ &= \frac{1}{2}\left(\frac{1}{x}\int_0^x f(t)dt + \frac{1}{1-x}\int_x^1 f(t)dt\right). \end{aligned} \tag{4.8}$$

Now we show that the transfer operator which we just introduced has an invariant measure with absolutely continuous density.

PROPOSITION 4.3. *Let R_F denote the transfer operator defined in* (4.8), *and set*

$$d\mu(x) = \frac{dx}{\pi\sqrt{x(1-x)}}, \qquad x \in (0,1). \tag{4.9}$$

We then have

$$\mu R_F = \mu,$$

that is,

$$\int_0^1 (R_F f)(x)d\mu(x) = \int_0^1 f(x)dx, \quad \forall f \in \mathcal{M}((0,1),\mathcal{B}). \tag{4.10}$$

PROOF. For $d\mu(x) = G(x)dx$ to satisfy (4.10) we must have

$$G(y) = \frac{1}{2}\left(\int_y^1 \frac{G(x)}{x}dx + \int_0^y \frac{G(x)}{1-x}dx\right).$$

Hence

$$\frac{G'(y)}{G(y)} = \frac{1}{2}\left(-\frac{1}{y} + \frac{1}{1-y}\right),$$

and hence the result. □

DEFINITION 4.4. We define the backward shift s on sequences of $\prod_{n=0}^{\infty} X$ by

$$s(x_0, x_1, x_2, \ldots) = (x_1, x_2, \ldots). \tag{4.11}$$

PROPOSITION 4.5. *In the setting of Theorem 2.29, let μ be an invariant measure for the transfer operator, and let π_0 be endowed with μ as probability law. Then the corresponding probability measure is shift-invariant:*

$$\mathbb{P} \circ s^{-1} = \mathbb{P}.$$

PROOF. $\mathbb{P}$ is built from the Kolmogorov construction by

$$\int_{\mathrm{Sol}_\sigma(X)} (f_0\pi_0)(\omega)(f_1\pi_1)(\omega)\cdots(f_n\pi_n)(\omega)d\mathbb{P}(\omega) = \int_X (f_0(x)R(f_1R(f_2\cdots R(f_nh))))(x)d\mu(x). \tag{4.12}$$

□

4.2. Cases where σ is not onto. When the endomorphism σ is not onto, the solenoid satisfies

$$\mathrm{Sol}_\sigma(X) \subset \prod_{n=1}^{\infty} X_\infty^{(\sigma)}, \quad \text{where} \quad X_\infty^{(\sigma)} \overset{\text{def.}}{=} \cap_{n=1}^{\infty}\sigma^n(X),$$

and the latter can be a very small set, as we now illustrate.

EXAMPLE 4.6. Take $X = [0,1]$ and $\sigma(x) = 2x(1-x)$. Then

$$\sigma(X) = [0, \frac{1}{2}] \quad \text{and} \quad X_\infty^{(\sigma)} = \left\{0, \frac{1}{2}\right\}.$$

The solenoid consists of the two points

$$(0,0,\ldots) \quad \text{and} \quad (\frac{1}{2}, \frac{1}{2}, \ldots).$$

EXAMPLE 4.7. This example is from complex dynamics. We take $X = \mathbb{C}$ and for a pre-assigned $c \in \mathbb{C}$,

$$\sigma_c(z) = z^2 + c.$$

Then $X_\infty^{(\sigma)}$ is the Julia set, see [**56**].

4.3. Solenoids associated with the unit circle. In the period since the mid 1990ties, the term "wavelet" has come to have a broader meaning: From referring to systems of bases in $\mathbf{L}_2(\mathbb{R})$ with dyadic scale symmetry, "wavelet" now typically refers to finite systems of functions on a suitable measure space that can be used in order to construct either an orthonormal basis, or frame basis by means of operators connected to algebraic and geometric information involving a notion of "scaling function." The latter often in the form of a probability measure on a solenoid-measure space. In the case of fractals, there are natural choices of finite systems of functions yielding very well-behaved orthonormal bases, and thus giving direct information about the topological structure of the particular fractal involved. Our framework below makes use of solenoids (from endomorphisms) in order to offer an even more inclusive framework for wavelet bases and multiresolutions. Background references for the present section include [**12, 15, 19, 29, 30, 40, 44, 45, 49**].

4.3.1. *Definition.* Starting from a continuous function $m_0(z)$ on the unit circle $\mathbb{T}$ and $N \in \{2,3,\ldots\}$ one can construct (at least) two representations of the algebra of operators generated by two operators T, U such that U is unitary and $UTU^{-1} = T^N$ (such an algebra is an algebra generated by a group of the kind studied in [**15**] by Baumslag and Solitar). To be more precise let $R = R_{m_0}$ denote the corresponding Ruelle operator:

$$(Rf)(z) = \frac{1}{N}\sum_{w^N=z} |m_0(w)|^2 f(w). \tag{4.13}$$

When $R1 = 1$ the infinite product $\prod_{u=1}^{\infty} \frac{m_0(e^{\frac{2\pi it}{N^u}})}{\sqrt{N}}$ belongs to $\mathbf{L}_2(\mathbb{R})$, and is the Fourier transform of the scaling function φ_0. The space

$$\mathcal{H}_0 = \{\widehat{\varphi_0}(t)f(t)\,;\, f \text{ function on } \mathbb{R}/\mathbb{Z} \text{ measurable and } f(t) = f(t+1)\} \tag{4.14}$$

is the 0-resolution subspace of the multiresolution

$$\mathcal{H}_k = \left\{2^{-k/2}\widehat{\varphi_0}(t/2^k)f(t)\,;\, f \text{ measurable function on } \mathbb{R}/\mathbb{Z} \text{ i.e.} f(t) = f(t+1)\right\}, k \in \mathbb{Z}. \tag{4.15}$$

One defines a representation ρ of $\mathbf{L}_\infty(\mathbb{T})$ into $\mathbf{B}(\mathbf{L}_2(\mathbb{R}))$ as follows: If $f \in \mathbf{L}_\infty(\mathbb{T})$ with associated Fourier series $f(e^{it}) = \sum_{n\in\mathbb{Z}} \widehat{f}(n)e^{int}$, one sets

$$\rho(f)(g) = \sum_{n\in\mathbb{Z}} \widehat{f}(n)g(x-n), \quad g \in \mathbf{L}_2(\mathbb{R}). \tag{4.16}$$

In this paper we remove the $\mathbf{L}_2(\mathbb{R}, dx)$ requirement (which we assumed in [**2**, **3**, **5**, **6**]) from the wavelet setting. Now wavelet multiresolutions may be viewed as a special case of a probability space multiresolution. In the latter, the resolution subspaces will be specified by a system of conditional expectations. In the classical wavelet application, the solenoid becomes the real line, realized as a dense curve in $\mathrm{Sol}_\sigma(X)$, and the solenoid measure P becomes Lebesgue measure.

We now consider the special case where X is equal to the unit circle $\mathbb{R}/\mathbb{Z} = \mathbb{T}$ and $\sigma(z) = z^N$. When using the notation $z = e^{2\pi it}$ we have $\sigma(t) = Nt \pmod 1$.

The solenoid G_N is a compact group, included in $\prod_{k=0}^{\infty}\mathbb{T}$, and consists of the sequences $z = (z_0, z_1, z_2, \ldots) \in \prod_{k=0}^{\infty}\mathbb{T}$ such that

$$z_{k+1}^N = z_k, \quad k = 1, 2, \ldots$$

See [**44**]. We define

$$\sigma(z_0, z_1, z_2, \ldots) = (z_0^N, z_1^N, z_2^N, \ldots) = (z_0^N, z_0, z_1, \ldots) \tag{4.17}$$

and

$$\tau(z_0, z_1, z_2, \ldots) = (z_1, z_2, \ldots) \tag{4.18}$$

We have

$$\sigma \circ \tau = \tau \circ \sigma = I.$$

It is the dual of the discrete group $\mathbb{Z}[1/N]$, with characters $\chi\left(\frac{\ell}{N^k}\right)$ given by

$$\langle \chi\left(\frac{\ell}{N^k}\right), z\rangle = z_k^\ell, \quad k, \ell = 0, 1, \ldots \tag{4.19}$$

See [**16**, **19**]. Note that (4.19) is well defined since

$$\langle \chi\left(\frac{N\ell}{N^{k+1}}\right), z\rangle = z_{k+1}^{N\ell} = (z_{k+1}^N)^\ell = z_k^\ell = \langle \chi\left(\frac{\ell}{N^k}\right), z\rangle.$$

4.3.2. *Ruelle operators and wavelets.* We use the term Ruelle operator consistent with [**4, 14, 19, 65**] to indicate a transfer operator which governs branching in a number of different context. Every filter in the family we have can be realized as a wavelet filter on the solenoid. Fix a low-pass filter m_0 with the usual properties, and assume that the function identically equal to 1 (denoted in this paper by $\mathbf{1}$) is an eigenvalue of R (defined by (4.13)) with eigenvalue 1, that is,

$$\frac{1}{N}\sum_{\substack{w\in\mathbb{T}\\ w^N=z}} |m_0(w)|^2 \equiv 1.$$

Then, one can construct φ_0 and use the space $\mathbf{L}_2(\mathbb{R}, dx)$. We study the representations of the algebra generated by (U, T) such that

$$UTU^{-1} = T^N \tag{4.20}$$

We take

$$U\left(\sum \xi_n\varphi_0(\cdot - n)\right) = m_0(z)f(z^N) \quad \text{with} \quad f(z) = \sum \xi_n z^n.$$

Thus we have a slanted Toeplitz matrix

$$(S\xi)_n = \sum_{j\in\mathbb{Z}} a_{n-jN}\xi_j.$$

The following result reflects the scaling law for the father function φ_0 of the wavelet under consideration,

$$\varphi_0(x) = \sqrt{N}\sum_{k\in\mathbb{Z}} a_k\varphi_0(Nx - k), \quad x \in \mathbb{R}, \tag{4.21}$$

where

$$m_0(x) = \sum_{k\in\mathbb{Z}} a_k e^{2\pi i k x}. \tag{4.22}$$

LEMMA 4.8. *For the operators K and S (see* (1) *and* (2) *in Table* 4 *above) we define*

$$K : \ell_2(\mathbb{Z}) \longrightarrow \mathcal{H}_0 \ (\textit{the zero resolution subspace in } \mathbf{L}_2(\mathbb{R}))$$

by

$$(K\xi)(x) = \sum_{n\in\mathbb{Z}} \xi_n\varphi_0(x - n).$$

Then

$$KS = UK \tag{4.23}$$

holds, that is the following diagram is commutative:

$$\begin{array}{ccc} \mathbf{L}_2(\mathbb{R}) & \xrightarrow{\quad U \quad} & \mathbf{L}_2(\mathbb{R}) \\ K\uparrow & & \uparrow K \\ \ell_2(\mathbb{Z}) & \xrightarrow{\quad S \quad} & \ell_2(\mathbb{Z}), \end{array}$$

where

$$(U\gamma)(x) = \frac{1}{\sqrt{N}}\gamma(x/N), \quad \gamma \in \mathbf{L}_2(\mathbb{R}, dx).$$

PROOF. We have for $\xi \in \ell_2(\mathbb{Z})$:

$$(KS\xi)(x) = \sum_{n\in\mathbb{Z}}\sum_{j\in\mathbb{Z}} a_{n-jN}\xi_j\varphi_0(x-n)$$

and

$$\begin{aligned}(UK\xi)(x) &= \sum_{j\in\mathbb{Z}} \xi_j \frac{1}{\sqrt{N}}\varphi_0\left(\frac{x-jN}{N}\right)\\ &= \sum_{j\in\mathbb{Z}}\sum_{k\in\mathbb{Z}} \xi_j a_k \varphi_0(x-jN-k)\\ &\text{and, with the change of variable } n = jN+k,\\ &= \sum_{n\in\mathbb{Z}}\sum_{j\in\mathbb{Z}} \xi_j a_{n-jN}\varphi_0(x-n),\end{aligned}$$

and the result follows. □

More generally for many choices of filters (see (4.22)) there are no $\mathbf{L}_2(\mathbb{R}, dx)$-solution to (4.21), and then one leaves the setting of $\mathbf{L}_2(\mathbb{R})$. We still get counterparts of (3.34) and (3.48) using the solenoid.

PROPOSITION 4.9. *The operator R in (4.13) is bounded from $\mathbf{L}_2(\mathbb{T}, d\lambda)$ into itself, and its adjoint is given by the formula*

$$(R^*f)(z) = |m_0(z)|^2 f(z^N). \tag{4.24}$$

We now discuss the multiresolution associated with m_0 and its relationships with the multiresolution MR_σ. We first note that the space $\mathcal{H}_0$ defined by (4.14) is equal to the closed linear span of the functions $x \mapsto \widehat{\varphi_0(x+k)}$, when k runs through $\mathbb{Z}$. In general the family of functions $x \mapsto \varphi_0(x+k)$ $(k \in \mathbb{Z})$ is not orthogonal in $\mathbf{L}_2(\mathbb{R}, dx)$.

PROPOSITION 4.10. *Let $W = |m_0|^2$, let*

$$h_{\varphi_0}(t) = \sum_{n\in\mathbb{Z}} |\widehat{\varphi_0}(t+n)|^2, \tag{4.25}$$

and let

$$(Rf)(t) = \frac{1}{N}\left(\sum_{k=0}^{N-1}(Wf)\left(\frac{t+k}{N}\right)\right). \tag{4.26}$$

Then

$$Rh_{\varphi_0} = h_{\varphi_0}. \tag{4.27}$$

PROOF. We have

$$\begin{aligned}
(Rh_{\varphi_0})(t) &= \frac{1}{N}\sum_{k\in\mathbb{Z}_N} W\left(\frac{t+k}{N}\right)\sum_{n\in\mathbb{Z}}|\widehat{\varphi_0}\left(\frac{t+k+nN}{N}\right)|^2 \\
&= \sum_{k\in\mathbb{Z}_N}\sum_{n\in\mathbb{Z}}|\widehat{\varphi_0}\left(t+\overbrace{k+nN}^{m}\right)|^2 \\
&= \sum_{m\in\mathbb{Z}}|\widehat{\varphi_0}(t+m)|^2 \\
&= h_{\varphi_0}(t),
\end{aligned}$$

where we wrote $\mathbb{Z}_N$ for the cyclic group $\mathbb{Z}/N\mathbb{Z}$, and we used the Euclidean algorithm on $\mathbb{Z}$, mod N, in the last step ($m = k+nN$). The first step used the scaling identity for φ_0 and $W = |m_0|^2$. □

As an application of Proposition 4.10 we get the following results for wavelets on solenoids.

COROLLARY 4.11. *Let $W = |m_0|^2$ and h_{φ_0} be as in Proposition 4.10; then $M_n \overset{\text{def.}}{=} h_{\varphi_0}(\pi_n)$ is a $(\pi_n)_{n\in\mathbb{N}_0}$-martingale, where $(\pi_n)_{n\in\mathbb{N}_0}$ denotes the $\mathrm{Sol}_N(\mathbb{T})$-Markov chain.*

COROLLARY 4.12. *Consider the wavelet filter m_0 with scaling function $\varphi_0 \in \mathbf{L}_2(\mathbb{R})$. Let h_{φ_0} be the corresponding harmonic function: $R_{m_0}h_{\varphi_0} = h_{\varphi_0}$, see (4.27). Then the level-0 isometry*

$$V_0 : f \in \mathbf{L}_2(\mathbb{T}, h_{\varphi_0}(t)dt) \;\mapsto\; (f(t)\widehat{\varphi_0}(t)) \in \mathbf{L}_2(\mathbb{R})$$

has the following explicit adjoint V_0^* computed on $\mathbf{L}_2(\mathbb{R})$:

$$(V_0^*\gamma)(t) = \frac{1}{h_{\varphi_0}(t)}\sum_{n\in\mathbb{Z}}\gamma(t+n)\overline{\widehat{\varphi_0}(t+n)}, \quad t\in\mathbb{R}, \quad \gamma\in\mathbf{L}_2(\mathbb{R},dx).$$

REMARK 4.13. For functions k defined on $[0,1]$ (or, equivalently, on $\mathbb{R}/\mathbb{Z}$) we introduce the Fourier coefficients

$$\widehat{k}(n) = \int_0^1 e^{-2\pi int}k(t)dt, \quad n\in\mathbb{Z}.$$

With $k = h_{\varphi_0}$ from (4.25) we can then compute the inner products $\int_{\mathbb{R}}\varphi_0(x+n)\overline{\varphi_0(x)}dx$ for $n\in\mathbb{Z}$. See the following proposition.

PROPOSITION 4.14. *Let h_{φ_0} be the harmonic function associated to a scaling function $\varphi_0 \in \mathbf{L}_2(\mathbb{R},dx)$. Then the following hold:*
(i)

$$\int_{\mathbb{R}}\varphi_0(x+n)\overline{\varphi_0}(x)dx = \widehat{h_{\varphi_0}}(n), \quad n\in\mathbb{Z}.$$

(ii) The generating function

$$\zeta\in\mathbb{C}\mapsto G_{\varphi_0}(\zeta) = \sum_{n\in\mathbb{Z}}\zeta^n\left(\int_{\mathbb{R}}\varphi_0(x+n)\overline{\varphi_0}(x)dx\right)$$

is an analytic extension of h_{φ_0} to an open neighborhood of $\mathbb{T}$.
(iii) The scaling function φ_0 is compactly supported on $\mathbb{R}$ if and only if G_{φ_0} is a polynomial.

PROOF. We need only to prove (i). The other two claims follow then easily. Using Parseval's equality in $\mathbf{L}_2(\mathbb{R})$ we have

$$\begin{aligned}\int_{\mathbb{R}} \varphi_0(x+n)\overline{\varphi_0}(x)dx &= \int_{\mathbb{R}} e^{-2\pi int}|\widehat{\varphi_0}(t)|^2dt\\ &= \sum_{m\in\mathbb{Z}}\int_0^1 e^{-2\pi int}|\widehat{\varphi_0}(t+m)|^2dt\end{aligned}$$

and using the dominated convergence theorem,

$$\begin{aligned}&= \int_0^1 e^{-2\pi int}\left(\sum_{m\in\mathbb{Z}}|\widehat{\varphi_0}(t+m)|^2\right)dt\\ &= \int_0^1 e^{-2\pi int}h_{\varphi_0}(t)dt\\ &= \widehat{h_{\varphi_0}}(n), \quad \forall n\in\mathbb{Z}.\end{aligned}$$

□

COROLLARY 4.15. *Orthogonality of the family* $\{\varphi_0(\cdot+n)\}_{n\in\mathbb{Z}}$ *in* $\mathbf{L}_2(\mathbb{R},dx)$ *is equivalent to the condition* $h_{\varphi_0}\equiv 1$.

In the next example we show that the Fejér kernels arise as h_{φ_0} for a family of scaling functions $\varphi_0\in\mathbf{L}_2(\mathbb{R},dx)$. We first recall that the Dirichlet kernel and Fejér kernels are defined respectively by

$$D_k(\zeta) = \sum_{j=-k}^{k}\zeta^j$$

and

$$F_k(\zeta) = \frac{\sum_{u=0}^k D_u(\zeta)}{k+1}.$$

EXAMPLE 4.16. We take $\varphi_0(x) = \frac{1}{\sqrt{2m+1}}\chi_{[0,2m+1]}(x)$, where $m\in\mathbb{N}$ is fixed. Then

$$\int_{\mathbb{R}}\varphi_0(x)\varphi_0(x-n)dx = \begin{cases} 0, & \text{if } |n|\geq 2m+1,\\ \frac{2m+1-n}{2m+1} & \text{for } n\in\{0,\ldots,2m\}.\end{cases}$$

Thus

$$(2m+1)h_{\varphi_0}(\zeta)=\zeta^{-2m}+2\zeta^{1-2m}+\cdots+(2m)\zeta^{-1}+(2m+1)+(2m)\zeta+\cdots+2\zeta^{2m-1}+\zeta^{2m},$$

which is the Fejér kernel F_{2m}.

4.3.3. *Realization using the solenoid.* We set $e_n(z)=z^n$, $n\in\mathbb{Z}$.

THEOREM 4.17. *Let* $z\in\mathbb{T}$. *The function*

$$L\left(\frac{n}{N^k}\right) = \left(R^k\left(e_nh\right)\right)(z) \tag{4.28}$$

is positive definite on $\mathbb{Z}[1/N]$, *and there exists a positive finite measure* $d\mu_z$ *on* $\mathrm{Sol}_N(\mathbb{T})$ *such that*

$$L\left(\frac{n}{N^k}\right) = \int_{\mathrm{Sol}_N(\mathbb{T})}\chi\left(\frac{n}{N^k}\right)(x)dP_z(x) \tag{4.29}$$

Proof: We first check that L is well defined. We have

$$\begin{aligned} L\left(\frac{Nn}{N^{k+1}}\right) &= \left(R^{k+1}\left(e_{Nn}h\right)\right)(z) \\ &= \left(R^{k}\left(Re_{nN}h\right)\right)(z) \\ &= R^{k}\left(R((e_n\circ\sigma)\cdot h)\right)(z) \\ &= \left(R^{k}e_nRh\right)(z) \\ &= \left(R^{k}e_nh\right)(z) \\ &= L\left(\frac{n}{N^k}\right) \end{aligned}$$

since R satisfies the pull-out property with $\sigma(z)=z^N$.

We now prove that L is positive definite on $\mathbb{Z}[1/N]$. Let $M\in\mathbb{N}$, $c_1,\ldots,c_M\in\mathbb{C}$ and $\frac{n_1}{N^{k_1}},\ldots,\frac{n_M}{N^{k_M}}\in\mathbb{Z}[1/N]$. In view of the first part of the proof, we assume all the denominators equal, say to k. We have

$$\begin{aligned} \sum_{u,v=1}^{M}\overline{c_u}c_vL\left(\frac{n_u}{N^k}-\frac{n_v}{N^k}\right) &= \sum_{u,v=1}^{M}\overline{c_u}c_vR^k\left((e_{n_u-n_v})h\right)(z) \\ &= \sum_{u,v=1}^{M}\overline{c_u}c_vR^k\left((e_{n_u-n_v})h\right)(z) \\ &= \sum_{u,v=1}^{M}\overline{c_u}c_vR^k\left((e_{n_u}\overline{e_{n_v}})h\right)(z) \\ &= \left(R\left(|g|^2h\right)\right)(z)\geq 0, \end{aligned}$$

with $g=\sum_{u=1}^{M}\overline{c_u}e_u$.

The second claim comes from Bochner's theorem for compact groups.

□

Let

$$\pi_0(z)=z_0,$$

and

$$U(\psi)=m_0(\pi_0(z))(\psi\circ\sigma_N)(z).$$

PROPOSITION 4.18. *U is unitary and its adjoint is given by the formula*

$$U^*\psi=\frac{1}{m\circ\pi_1}\psi\circ\sigma_N^{-1}. \tag{4.30}$$

Proof: The results follow from the previous considerations; see also [**44**].

□

4.3.4. *Multiresolutions.* We set

$$\mathcal{L}_k=\text{closed linear span}\left\{\chi\left(\frac{n}{N^k}\right),\ n\in\mathbb{Z}\right\},\quad k\in\mathbb{Z}.$$

4.3.5. *Embedding the real line into the solenoid.* We define

$$\gamma_N(t) = \left(e^{2\pi i[t]}, e^{2\pi i[t/N]}, e^{2\pi i[t/N^2]}, \ldots\right) \in \prod_{n=0}^{\infty} (\mathbb{R}/\mathbb{Z}), \tag{4.31}$$

where $[x]$ denotes the value of $x \in \mathbb{Z}$ modulo 1.

LEMMA 4.19. *The map γ_N is one-to-one from $\mathbb{R}$ into $\mathrm{Sol}_N(\mathbb{T})$, meaning that*

$$\gamma_N(t) = (1,1,1,\ldots) \iff t = 0.$$

Proof: See [**44**].

□

4.3.6. *Probability.* Let $h \geq 0$ be such that $Rh = h$, and assume that $\int_{\mathbb{T}} h(\lambda)d\lambda = 1$.

PROPOSITION 4.20. *The distribution of the random variable*

$$\pi_k(\mathbf{z}) = z_k$$

is $|m^{(k)}(z)|^2 h(z)d\lambda$, where

$$m^{(k)}(z) = m_0(z)m_0(z^N)\cdots m_0(z^{N^{k-1}}). \tag{4.32}$$

Proof: We want to show that for every bounded measurable function f on $\mathbb{T}$ we have

$$\int_{\mathbb{T}} f(e^{it})d\lambda(e^{it}) = \int_{\mathrm{Sol}_N(\mathbb{T})} f(\pi_k(\mathbf{z}))dP(\mathbf{z}).$$

We prove this equality for $f(z) = z^n$ (which we denoted by $e_n(z)$) with $n \in \mathbb{Z}$, that is

$$\int_{\mathbb{T}} e^{int}d\lambda(e^{it}) = \int_{\mathrm{Sol}_N(\mathbb{T})} e_n(\pi(\mathbf{z}))dP(\mathbf{z}). \tag{4.33}$$

But recall that

$$\chi\left(\frac{n}{N^k}\right)(\mathbf{z}) = z_k^n = (\pi_k(\mathbf{z}))^n = e_n(\pi_k(\mathbf{z})).$$

Hence

$$\begin{aligned}
\int_{\mathrm{Sol}_N(\mathbb{T})} \chi\left(\frac{n}{N^k}\right)(\mathbf{z})dP(\mathbf{z}) &= \int_{\mathbb{T}} R^k(e_n h)(e^{it})d\lambda(e^{it}) \\
&= \int_{\mathbb{T}} (R^{*k}(1))(e^{it})e_n(e^{it})h(e^{it})d\lambda(e^{it}) \\
&= \int_{\mathbb{T}} |m^{(k)}(e^{it})|^2 h(e^{it})d\lambda(e^{it}),
\end{aligned}$$

where we have used (4.24) to compute $R^{*k}1$. □

REMARK 4.21. We note that

$$\begin{aligned}
\int_{\mathbb{T}} |m^{(k)}(e^{it})|^2 h(e^{it})d\lambda(e^{it}) &= \int_{\mathbb{T}} (R^{*k}(1))(e^{it})h(e^{it})d\lambda(e^{it}) \\
&= \int_{\mathbb{T}} (R^k(h))(e^{it})d\lambda(e^{it}) \\
&= \int_{\mathbb{T}} h(e^{it})d\lambda(e^{it}) = 1,
\end{aligned}$$

as it should be.

4.3.7. *The martingale property.* As in the previous section h denotes a positive function such that $Rh = h$ and $\int_{\mathbb{T}} h(\lambda)d\lambda = 1$. Let $z, w \in \mathbb{T}$ be such that $w^N = z$, and set

$$P(z \mapsto w) = \frac{1}{N}|m(w)|^2 h(w).$$

The Markov property now reads

$$\sum_{\substack{w\in\mathbb{T}\\ w^N=z}} P(z \mapsto w) = 1. \tag{4.34}$$

The martingale property is now

$$R(\xi_{n+1}h) = \xi_n h,$$

and the following formulas hold:

$$\frac{R(fh)}{h} \leq 1 \quad \text{(conditional expectation)}$$

$$\frac{R(\xi(z^n))h}{h} = \xi$$

$$\sum_{w^N=z} \xi(w)h(w) = \xi(z)\frac{Rh}{h}$$

$$M_{z_0z_1}M_{z_1z_2}\cdots = R^k,$$

with

$$M_{z_0z_1}M_{z_1z_2} = \frac{1}{N}\sum_{z_1^n=z_0} |m(z_1)|^2 \sum_{z_2^N=z_1} |m(z_1)|^2.$$

4.4. Fractal examples. We here consider X to be the set of numbers of the form

$$x = \sum_{n=1}^{\infty} \frac{b_n}{3^n}, \quad \text{where} \quad b_n \in \{0, 2\}$$

and $\sigma(x) = 3x \pmod 1$. In symbolic form we have

$$(b_1, b_2, b_3, \ldots) \xrightarrow{\ \sigma\ } (b_2, b_3, b_4, \ldots)$$

More generally, let $d \in \mathbb{N}$ and let $A \in \mathbb{Z}^{d\times d}$ with all eigenvalues of modulus strictly bigger than 1, and let $m < |\det A|$. Fix d residue classes $b_1, \ldots, b_d$ in $\mathbb{Z}^d/A\mathbb{Z}^d$. We set $Y = \{b_1, \ldots, b_n\}$. We consider the set X of vectors in $\mathbb{R}^d$ of the form

$$x = \sum_{n=1}^{\infty} A^{-n}c_n$$

where to make connections with homogeneous Markov chains (see (2.50) for the latter) we define

$$F_{b_j}(x) = A^{-1}(x + b_j), \quad j = 1, \ldots, d \tag{4.35}$$

and $\sigma(x) = Ax$ modulo $\mathbb{Z}^d$.

Since Y is a finite set, a probability measure on Y is given by a finite number of positive numbers $p_1, \ldots, p_d$ adding up to 1, and the transfer operator is now given by

$$Rf = \sum_{j=1}^{d} p_j f \circ F_{b_j}. \tag{4.36}$$

Consider the set $\operatorname{Prob}(X)$ of probabilities on X, and let $\nu, \mu \in \operatorname{Prob}(X)$. Consider the Hausdorff distance between ν and μ:

$$d_H(\nu, \mu) = \sup\left\{\int_X f(x)(d\nu(x) - d\mu(x))\right\}$$

where the supremum is on the set of all Lipschitz functions:

$$|f(x) - f(y)| \leq \|x - y\|, \quad \text{(with } \|x-y\| \text{ being the usual distance in } \mathbb{R}^d\text{)}.$$

Define a measure νR on X via

$$\int_X f(d(\nu R)) = \int_X (Rf) d\nu.$$

A theorem of Hutchinson (see [**19, 39**]) states that the map $\nu \mapsto \nu R$ is then strictly contractive. There exists $\alpha \in (0,1)$ such that

$$d_H(R\nu, R\mu) \leq \alpha d_H(\nu, \mu).$$

Existence and uniqueness of a solution to the equation $\nu R = \nu$ follows from Banach fixed point theorem.

4.5. The Gauss operator. The present example is related to number theory and has links with information theory; see [**17, 18, 57, 66**]. We take $X = (0,1)$ and $d\lambda(x) = dx$, and

$$\sigma(x) =< \frac{1}{x} >, \tag{4.37}$$

where $< \cdot >$ denotes the "fractional part", defined as follows: If $x \in (\frac{1}{k+1}, \frac{1}{k})$ then $\sigma(x) = \frac{1}{x} - k$. We also define $\tau_k(x) = \frac{1}{x+k}$ with $k = 1, 2, \ldots$. Note that

$$\sigma \circ \tau_k(x) = x, \quad k = 1, 2, \ldots$$

The solenoid (see Definition 3.4) associated with X is described as follows:

$$\operatorname{Sol}_\sigma(0,1) = \left\{(x_0, x_1, \ldots) \in \prod_{n=0}^{\infty}(0,1) \text{ such that } x_k =< \frac{1}{x_{k+1}} >\right\}. \tag{4.38}$$

We thus obtain the continued fraction associated with x_0.

$$\begin{aligned} x_0 &= k_1 + \frac{1}{\sigma(x_1)} \\ &= k_1 + \cfrac{1}{k_2 + \cfrac{1}{\sigma(x_2)}} \\ &= k_1 + \cfrac{1}{k_2 + \cfrac{1}{k_3 + \cfrac{1}{\ddots}}} \\ &\vdots \end{aligned}$$

Now the transfer operator is given by:

$$(Rf)(x) = \sum_{n=1}^{\infty} \frac{1}{(n+x)^2} f\left(\frac{1}{n+x}\right). \tag{4.39}$$

PROPOSITION 4.22. *Let* $h(x) = \frac{1}{\ln 2} \frac{1}{1+x}$. *Then* $\int_0^1 h(x)dx = 1$ *and* $\lambda R = \lambda$.

PROOF. We have

$$\begin{aligned} \lambda R = \lambda &\iff \int_0^1 \left(\sum_{n=1}^{\infty} \frac{1}{(n+x)^2} f\left(\frac{1}{n+x}\right)\right) dx = \int_0^1 f(x)dx \\ &\iff \sum_{n=1}^{\infty} \int_0^1 \frac{1}{(n+x)^2} f\left(\frac{1}{n+x}\right) dx = \int_0^1 f(x). \end{aligned}$$

Note that the change of variable $y = \frac{1}{n+x}$ leads to

$$\int_0^1 \frac{1}{(n+x)^2} f\left(\frac{1}{n+x}\right) dx = \int_{\frac{1}{n+1}}^{\frac{1}{n}} f(y)dy,$$

and hence

$$\sum_{n=1}^{\infty} \int_0^1 \frac{1}{(n+x)^2} f\left(\frac{1}{n+x}\right) dx = \sum_{n=1}^{\infty} \int_{\frac{1}{n+1}}^{\frac{1}{n}} f(x)dx = \int_0^1 f(x)dx,$$

and the result follows. □

Acknowledgments

P. Jorgensen wishes to thank the department of mathematics at Ben-Gurion University for hospitality during a 4 weeks research visit in the Spring of 2016, allowing for collaboration, and completion of this research. The authors are very grateful for the constructive suggestions made by the referee.

References

[1] D. Alpay, *Algorithme de Schur, espaces à noyau reproduisant et théorie des systèmes* (French, with English and French summaries), Panoramas et Synthèses [Panoramas and Syntheses], vol. 6, Société Mathématique de France, Paris, 1998. MR1638044

[2] D. Alpay, P. Jorgensen, and I. Lewkowicz, *Extending wavelet filters: infinite dimensions, the nonrational case, and indefinite inner product spaces*, Excursions in harmonic analysis. Volume 2, Appl. Numer. Harmon. Anal., Birkhäuser/Springer, New York, 2013, pp. 69–111, DOI 10.1007/978-0-8176-8379-5_5. MR3050315

[3] D. Alpay, P. Jorgensen, and I. Lewkowicz, *Parametrizations of all wavelet filters: input-output and state-space*, Sampl. Theory Signal Image Process. **12** (2013), no. 2-3, 159–188. MR3285409

[4] D. Alpay, P. Jorgensen, and I. Lewkowicz, *Realizations of infinite products, Ruelle operators and wavelet filters*, J. Fourier Anal. Appl. **21** (2015), no. 5, 1034–1052, DOI 10.1007/s00041-015-9396-z. MR3393694

[5] D. Alpay, P. Jorgensen, and I. Lewkowicz, *Characterizations of rectangular (para)-unitary rational functions*, Opuscula Math. **36** (2016), no. 6, 695–716, DOI 10.7494/OpMath.2016.36.6.695. MR3571417

[6] D. Alpay, P. Jorgensen, and I. Lewkowicz, *Characterizations of families of rectangular, finite impulse response, para-unitary systems*, J. Appl. Math. Comput. **54** (2017), no. 1-2, 395–423, DOI 10.1007/s12190-016-1015-x. MR3647183

[7] C. Anantharaman and J. Renault, *Amenable groupoids*, Groupoids in analysis, geometry, and physics (Boulder, CO, 1999), Contemp. Math., vol. 282, Amer. Math. Soc., Providence, RI, 2001, pp. 35–46, DOI 10.1090/conm/282/04677. MR1855241

[8] N. Aronszajn, *Theory of reproducing kernels*, Trans. Amer. Math. Soc. **68** (1950), 337–404, DOI 10.2307/1990404. MR0051437

[9] L. W. Baggett, *An abstract interpretation of the wavelet dimension function using group representations*, J. Funct. Anal. **173** (2000), no. 1, 1–20, DOI 10.1006/jfan.1999.3551. MR1760275

[10] L. W. Baggett, V. Furst, K. D. Merrill, and J. A. Packer, *Generalized filters, the low-pass condition, and connections to multiresolution analyses*, J. Funct. Anal. **257** (2009), no. 9, 2760–2779, DOI 10.1016/j.jfa.2009.05.004. MR2559716

[11] L. W. Baggett, V. Furst, K. D. Merrill, and J. A. Packer, *Classification of generalized multiresolution analyses*, J. Funct. Anal. **258** (2010), no. 12, 4210–4228, DOI 10.1016/j.jfa.2009.12.001. MR2609543

[12] L. W. Baggett, K. D. Merrill, J. A. Packer, and A. B. Ramsay, *Probability measures on solenoids corresponding to fractal wavelets*, Trans. Amer. Math. Soc. **364** (2012), no. 5, 2723–2748, DOI 10.1090/S0002-9947-2012-05584-X. MR2888226

[13] M. Bakonyi and T. Constantinescu, *Schur's algorithm and several applications*, Pitman Research Notes in Mathematics Series, vol. 261, Longman Scientific & Technical, Harlow; copublished in the United States with John Wiley & Sons, Inc., New York, 1992. MR1192547

[14] V. Baladi, *Positive transfer operators and decay of correlations*, Advanced Series in Nonlinear Dynamics, vol. 16, World Scientific Publishing Co., Inc., River Edge, NJ, 2000. MR1793194

[15] G. Baumslag and D. Solitar, *Some two-generator one-relator non-Hopfian groups*, Bull. Amer. Math. Soc. **68** (1962), 199–201, DOI 10.1090/S0002-9904-1962-10745-9. MR0142635

[16] S. Bezuglyi and P. E. T. Jorgensen, *Representations of Cuntz-Krieger relations, dynamics on Bratteli diagrams, and path-space measures*, Trends in harmonic analysis and its applications, Contemp. Math., vol. 650, Amer. Math. Soc., Providence, RI, 2015, pp. 57–88, DOI 10.1090/conm/650/13008. MR3441734

[17] N. M. Blachman. The information rate of a continued fraction (corresp.). *IEEE Trans. Inform. Theory*, 11:150–151, 1965.

[18] N. M. Blachman, *The continued fraction as an information source*, IEEE Trans. Inform. Theory **30** (1984), no. 4, 671–674, DOI 10.1109/TIT.1984.1056924. MR755797

[19] O. Bratteli and P. Jorgensen, *Wavelets through a looking glass*, Applied and Numerical Harmonic Analysis, Birkhäuser Boston, Inc., Boston, MA, 2002. The world of the spectrum. MR1913212

[20] P. Brémaud, *Markov chains*, Texts in Applied Mathematics, vol. 31, Springer-Verlag, New York, 1999. Gibbs fields, Monte Carlo simulation, and queues. MR1689633

[21] T. Constantinescu, *Schur parameters, factorization and dilation problems*, Operator Theory: Advances and Applications, vol. 82, Birkhäuser Verlag, Basel, 1996. MR1399080

[22] F. Cucker and S. Smale, *On the mathematical foundations of learning*, Bull. Amer. Math. Soc. (N.S.) **39** (2002), no. 1, 1–49, DOI 10.1090/S0273-0979-01-00923-5. MR1864085

[23] J. Cuntz, *Simple C^*-algebras generated by isometries*, Comm. Math. Phys. **57** (1977), no. 2, 173–185. MR0467330

[24] J. Cuntz and W. Krieger, *Topological Markov chains with dicyclic dimension groups*, J. Reine Angew. Math. **320** (1980), 44–51, DOI 10.1515/crll.1980.320.44. MR592141

[25] I. Daubechies, *Ten lectures on wavelets*, CBMS-NSF Regional Conference Series in Applied Mathematics, vol. 61, Society for Industrial and Applied Mathematics (SIAM), Philadelphia, PA, 1992. MR1162107

[26] P. Diaconis and D. Freedman, *Iterated random functions*, SIAM Rev. **41** (1999), no. 1, 45–76, DOI 10.1137/S0036144598338446. MR1669737

[27] D. E. Dutkay and P. E. T. Jorgensen, *Fourier duality for fractal measures with affine scales*, Math. Comp. **81** (2012), no. 280, 2253–2273, DOI 10.1090/S0025-5718-2012-02580-4. MR2945155

[28] D. E. Dutkay and P. E. T. Jorgensen, *Spectral measures and Cuntz algebras*, Math. Comp. **81** (2012), no. 280, 2275–2301, DOI 10.1090/S0025-5718-2012-02589-0. MR2945156

[29] D. E. Dutkay and P. E. T. Jorgensen, *The role of transfer operators and shifts in the study of fractals: encoding-models, analysis and geometry, commutative and non-commutative*, Geometry and analysis of fractals, Springer Proc. Math. Stat., vol. 88, Springer, Heidelberg, 2014, pp. 65–95, DOI 10.1007/978-3-662-43920-3_3. MR3275999

[30] D. E. Dutkay and P. E. T. Jorgensen, *Wavelets on fractals*, Rev. Mat. Iberoam. **22** (2006), no. 1, 131–180, DOI 10.4171/RMI/452. MR2268116

[31] H. Dym and V. Katsnelson, *Contributions of Issai Schur to analysis*, Studies in memory of Issai Schur (Chevaleret/Rehovot, 2000), Progr. Math., vol. 210, Birkhäuser Boston, Boston, MA, 2003, pp. xci–clxxxviii. MR1985190

[32] P. Erdös, *On a family of symmetric Bernoulli convolutions*, Amer. J. Math. **61** (1939), 974–976, DOI 10.2307/2371641. MR0000311

[33] P. Erdös, *On the smoothness properties of a family of Bernoulli convolutions*, Amer. J. Math. **62** (1940), 180–186, DOI 10.2307/2371446. MR0000858

[34] R. Exel and J. Renault, *Semigroups of local homeomorphisms and interaction groups*, Ergodic Theory Dynam. Systems **27** (2007), no. 6, 1737–1771, DOI 10.1017/S0143385707000193. MR2371594

[35] P. Faurre, M. Clerget, and F. Germain, *Opérateurs rationnels positifs* (French), Méthodes Mathématiques de l'Informatique [Mathematical Methods of Information Science], vol. 8, Dunod, Paris, 1979. Application à l'hyperstabilité et aux processus aléatoires. MR525380

[36] Ĭ. Ī. Gīhman and A. V. Skorohod, *Controlled stochastic processes*, Springer-Verlag, New York-Heidelberg, 1979. Translated from the Russian by Samuel Kotz. MR544839

[37] G. Hamel, *Eine charakteristische Eigenschaft beschränkter analytischer Funktionen* (German), Math. Ann. **78** (1917), no. 1, 257–269, DOI 10.1007/BF01457102. MR1511896

[38] O. Hernández-Lerma and J. B. Lasserre, *Markov chains and invariant probabilities*, Progress in Mathematics, vol. 211, Birkhäuser Verlag, Basel, 2003. MR1974383

[39] J. E. Hutchinson, *Fractals and self-similarity*, Indiana Univ. Math. J. **30** (1981), no. 5, 713–747, DOI 10.1512/iumj.1981.30.30055. MR625600

[40] M. Ionescu and P. S. Muhly, *Groupoid methods in wavelet analysis*, Group representations, ergodic theory, and mathematical physics: a tribute to George W. Mackey, Contemp. Math., vol. 449, Amer. Math. Soc., Providence, RI, 2008, pp. 193–208, DOI 10.1090/conm/449/08713. MR2391805

[41] M. Ionescu, P. S. Muhly, and V. Vega, *Markov operators and C^*-algebras*, Houston J. Math. **38** (2012), no. 3, 775–798. MR2970658

[42] Y. Jiang and D. Ruelle, *Analyticity of the susceptibility function for unimodal Markovian maps of the interval*, Nonlinearity **18** (2005), no. 6, 2447–2453, DOI 10.1088/0951-7715/18/6/002. MR2176941

[43] Y. Jiang and Y.-L. Ye, *Ruelle operator theorem for non-expansive systems*, Ergodic Theory Dynam. Systems **30** (2010), no. 2, 469–487, DOI 10.1017/S014338570900025X. MR2599889

[44] P. E. T. Jorgensen, *Ruelle operators: functions which are harmonic with respect to a transfer operator*, Mem. Amer. Math. Soc. **152** (2001), no. 720, viii+60, DOI 10.1090/memo/0720. MR1837681

[45] P. E. T. Jorgensen and A. M. Paolucci, *States on the Cuntz algebras and p-adic random walks*, J. Aust. Math. Soc. **90** (2011), no. 2, 197–211, DOI 10.1017/S1446788711001212. MR2821778

[46] P. E. T. Jorgensen and A. M. Paolucci, *q-frames and Bessel functions*, Numer. Funct. Anal. Optim. **33** (2012), no. 7-9, 1063–1069, DOI 10.1080/01630563.2012.682139. MR2966144
[47] S. Kakutani, *On equivalence of infinite product measures*, Ann. of Math. (2) **49** (1948), 214–224, DOI 10.2307/1969123. MR0023331
[48] T. Katsura, P. S. Muhly, A. Sims, and M. Tomforde, *Graph algebras, Exel-Laca algebras, and ultragraph algebras coincide up to Morita equivalence*, J. Reine Angew. Math. **640** (2010), 135–165, DOI 10.1515/CRELLE.2010.023. MR2629692
[49] F. Latrémolière and J. A. Packer, *Noncommutative solenoids and their projective modules*, Commutative and noncommutative harmonic analysis and applications, Contemp. Math., vol. 603, Amer. Math. Soc., Providence, RI, 2013, pp. 35–53, DOI 10.1090/conm/603/12039. MR3204025
[50] F. Latrémolière and J. A. Packer, *Explicit construction of equivalence bimodules between noncommutative solenoids*, Trends in harmonic analysis and its applications, Contemp. Math., vol. 650, Amer. Math. Soc., Providence, RI, 2015, pp. 111–140, DOI 10.1090/conm/650/13031. MR3441736
[51] P. D. Lax and R. S. Phillips, *Scattering theory for automorphic functions*, Bull. Amer. Math. Soc. (N.S.) **2** (1980), no. 2, 261–295, DOI 10.1090/S0273-0979-1980-14735-7. MR555264
[52] P. D. Lax and R. S. Phillips, *Scattering theory*, 2nd ed., Pure and Applied Mathematics, vol. 26, Academic Press, Inc., Boston, MA, 1989. With appendices by Cathleen S. Morawetz and Georg Schmidt. MR1037774
[53] M. Loève, *Probability theory*, Third edition, D. Van Nostrand Co., Inc., Princeton, N.J.-Toronto, Ont.-London, 1963. MR0203748
[54] G. W. Mackey, *The scope and history of commutative and noncommutative harmonic analysis*, History of Mathematics, vol. 5, American Mathematical Society, Providence, RI; London Mathematical Society, London, 1992. MR1171011
[55] A. E. Marrero and P. S. Muhly, *Groupoid and inverse semigroup presentations of ultragraph C^*-algebras*, Semigroup Forum **77** (2008), no. 3, 399–422, DOI 10.1007/s00233-008-9046-8. MR2457327
[56] J. Milnor, *Dynamics in one complex variable*, Friedr. Vieweg & Sohn, Braunschweig, 1999. Introductory lectures. MR1721240
[57] R. Nair, *On the metrical theory of continued fractions*, Proc. Amer. Math. Soc. **120** (1994), no. 4, 1041–1046, DOI 10.2307/2160215. MR1176073
[58] E. Nelson, *Topics in dynamics. I: Flows*, Mathematical Notes, Princeton University Press, Princeton, N.J.; University of Tokyo Press, Tokyo, 1969. MR0282379
[59] E. Nelson, *Quantum fields and Markoff fields*, Partial differential equations (Proc. Sympos. Pure Math., Vol. XXIII, Univ. California, Berkeley, Calif., 1971), Amer. Math. Soc., Providence, R.I., 1973, pp. 413–420. MR0337206
[60] J. Neveu, *Bases mathématiques du calcul des probabilités* (French), Masson et Cie, Éditeurs, Paris, 1964. MR0198504
[61] P. Niyogi, S. Smale, and S. Weinberger, *A topological view of unsupervised learning from noisy data*, SIAM J. Comput. **40** (2011), no. 3, 646–663, DOI 10.1137/090762932. MR2810909
[62] J. R. Norris, *Markov chains*, Cambridge Series in Statistical and Probabilistic Mathematics, vol. 2, Cambridge University Press, Cambridge, 1998. Reprint of 1997 original. MR1600720
[63] J. Renault, *AF equivalence relations and their cocycles*, Operator algebras and mathematical physics (Constanţa, 2001), Theta, Bucharest, 2003, pp. 365–377. MR2018241
[64] D. Ruelle, *Characteristic exponents and invariant manifolds in Hilbert space*, Ann. of Math. (2) **115** (1982), no. 2, 243–290, DOI 10.2307/1971392. MR647807
[65] D. Ruelle, *Thermodynamic formalism*, 2nd ed., Cambridge Mathematical Library, Cambridge University Press, Cambridge, 2004. The mathematical structures of equilibrium statistical mechanics. MR2129258
[66] C. Ryll-Nardzewski, *On the ergodic theorems. II. Ergodic theory of continued fractions*, Studia Math. **12** (1951), 74–79. MR0046583
[67] J. Schur, *Über Potenzreihen, die im Innern des Einheitskreises beschränkt sind* (German), J. Reine Angew. Math. **147** (1917), 205–232, DOI 10.1515/crll.1917.147.205. MR1580948
[68] S. Smale and D.-X. Zhou, *Online learning with Markov sampling*, Anal. Appl. (Singap.) **7** (2009), no. 1, 87–113, DOI 10.1142/S0219530509001293. MR2488871
[69] B. Solomyak, *On the random series $\sum \pm\lambda^n$ (an Erdős problem)*, Ann. of Math. (2) **142** (1995), no. 3, 611–625, DOI 10.2307/2118556. MR1356783

[70] H. S. Wall, *Analytic Theory of Continued Fractions*, D. Van Nostrand Company, Inc., New York, N. Y., 1948. MR0025596

Schmid College of Science and Technology, Chapman University, One University Drive, Orange, California 92866

Email address: alpay@chapman.edu

Department of Mathematics, University of Iowa, Iowa City, Iowa 52242

Email address: palle-jorgensen@uiowa.edu

Department of Electrical & Computer Engineering, Ben-Gurion University of the Negev, Beer-Sheva 84105, Israel

Email address: izchak@ee.bgu.ac.il

SELECTED PUBLISHED TITLES IN THIS SERIES

706 **Yeonhyang Kim, Sivaram K. Narayan, Gabriel Picioroaga, and Eric S. Weber, Editors,** Frames and Harmonic Analysis, 2018

702 **Mark Grant, Gregory Lupton, and Lucile Vandembroucq, Editors,** Topological Complexity and Related Topics, 2018

701 **Joan-Carles Lario and V. Kumar Murty, Editors,** Number Theory Related to Modular Curves, 2018

700 **Alexandre Girouard, Dmitry Jakobson, Michael Levitin, Nilima Nigam, Iosif Polterovich, and Frédéric Rochon, Editors,** Geometric and Computational Spectral Theory, 2017

699 **Mark L. Agranovsky, Matania Ben-Artzi, Catherine Bénéteau, Lavi Karp, Dmitry Khavinson, Simeon Reich, David Shoikhet, Gilbert Weinstein, and Lawrence Zalcman, Editors,** Complex Analysis and Dynamical Systems VII, 2017

698 **Alexander M. Blokh, Leonid A. Bunimovich, Paul H. Jung, Lex G. Oversteegen, and Yakov G. Sinai, Editors,** Dynamical Systems, Ergodic Theory, and Probability: in Memory of Kolya Chernov, 2017

697 **Fabrizio Broglia, Françoise Delon, Max Dickmann, Danielle Gondard-Cozette, and Victoria Ann Powers, Editors,** Ordered Algebraic Structures and Related Topics, 2017

696 **Ara S. Basmajian, Yair N. Minsky, and Alan W. Reid, Editors,** In the Tradition of Ahlfors–Bers, VII, 2017

695 **Katrina Barron, Elizabeth Jurisich, Antun Milas, and Kailash Misra, Editors,** Lie Algebras, Vertex Operator Algebras, and Related Topics, 2017

694 **Manjul Bhargava, Robert Guralnick, Gerhard Hiss, Klaus Lux, and Pham Huu Tiep, Editors,** Finite Simple Groups: Thirty Years of the Atlas and Beyond, 2017

693 **Michael Cwikel and Mario Milman, Editors,** Functional Analysis, Harmonic Analysis, and Image Processing, 2017

692 **Anatole Katok, Yakov Pesin, and Federico Rodriguez Hertz, Editors,** Modern Theory of Dynamical Systems, 2017

691 **Farrell Brumley, Maria Paula Gómez Aparicio, and Alberto Mínguez, Editors,** Around Langlands Correspondences, 2017

690 **Andrés Eduardo Caicedo, James Cummings, Peter Koellner, and Paul B. Larson, Editors,** Foundations of Mathematics, 2017

689 **Erica Flapan, Allison Henrich, Aaron Kaestner, and Sam Nelson, Editors,** Knots, Links, Spatial Graphs, and Algebraic Invariants, 2017

688 **Jeffrey Bergen, Stefan Catoiu, and William Chin, Editors,** Groups, Rings, Group Rings, and Hopf Algebras, 2017

687 **Fernanda Botelho, Raena King, and T. S. S. R. K. Rao, Editors,** Problems and Recent Methods in Operator Theory, 2017

686 **Alp Bassa, Alain Couvreur, and David Kohel, Editors,** Arithmetic, Geometry, Cryptography and Coding Theory, 2017

685 **Heather A. Harrington, Mohamed Omar, and Matthew Wright, Editors,** Algebraic and Geometric Methods in Discrete Mathematics, 2017

684 **Anna Beliakova and Aaron D. Lauda, Editors,** Categorification in Geometry, Topology, and Physics, 2017

683 **Anna Beliakova and Aaron D. Lauda, Editors,** Categorification and Higher Representation Theory, 2017

682 **Gregory Arone, Brenda Johnson, Pascal Lambrechts, Brian A. Munson, and Ismar Volić, Editors,** Manifolds and K-Theory, 2017